Practical Data
Communications

Practical Data Communications

Second Edition

Roger L. Freeman

A Wiley-Interscience Publication
JOHN WILEY & SONS, INC.
New York • Chichester • Weinheim • Brisbane • Singapore • Toronto

The texts extracted from the ITU material have been reproduced with the prior authorization of the Union as the copyright holder. The sole responsibility for selecting extracts for reproduction lies with the author alone and can in no way be attributed to the ITU. The complete volumes of the ITU material from which the texts reproduced are extracted, can be obtained from

International Telecommunication Union
General Secretariat--Sales and Marketing Service
Place de Nations
CH-1211 GENEVA 20 (Switzerland)
Telephone: +41 22 730 51 11   Telex: 421 000 ui ch
Telegram: ITU GENEVE   Fax: +41 22 730 51 94
X.400: S = Sales; P = itu; A = Arcom; C = ch Internet: Sales@itu.ch

This book is printed on acid-free paper. ∞

Published by John Wiley & Sons, Inc.

Published simultaneously in Canada.

For ordering and customer service, call 1-800-CALL-WILEY

Library of Congress Cataloging in Publication Data:

Freeman, Roger L.
    Practical data communications / Roger L. Freeman.--2nd ed.
       p. cm. -- (Wiley series in telecommunications and signal processing)
    Includes bibliographical references and index.
    ISBN 0471-39302-9 (cloth)
    ISBN 0471-39273-1 (paper)
    1. Data transmission systems. I. Title. II. Series.

  TK5105. F72 2001
  004.6--dc21   00-043692

Printed in the United States of America

10 9 8 7 6 5 4 3 2 1

For my son, Bob

CONTENTS

PREFACE TO
THE SECOND EDITION

This revised second edition of *Practical Data Communications* has been prepared with the IT technologist in mind, rather than the electrical engineer. I did this simply to reach a larger readership. The revision required a completely different perspective on my part. I found this experience refreshing.

But which came first—the chicken or the egg? In this case, where should the emphasis be placed? On the electrical data signal itself? Or the software that makes a data circuit work? I opted for the latter in this edition.

Still and yet, I have not lost sight of my principal purpose—to share with others my long learning experience in the field of data communications. It started in 1955 with data's predecessor, the automatic telegraph aboard a ship where I served as Radio Officer, using the old Teletype Corporation Model 15 clacking along at 60 words/minute (wpm). When ashore, I marveled at the Navy's 100 wpm circuits which operated at 75 bps. When I was station engineer at the USAF 2049th Communications Group, one of the first LogComNet circuits was established between McClellan AFB and Norton AFB. It was truly high speed at 150 bps. The circuit was asynchronous, better called a start–stop operation. That system became DoD's AutoDin and was indeed a data circuit, but based on telegraph technology.

My learning was based on setting up real circuits; the experience was hands-on. Thus my goal with this text is to facilitate an understanding of data communications that emphasizes practical application rather than theory. It is meant to serve as a basic reference for practitioners at all levels of this vital industry. It is tutorial as well, and will bring along the neophyte.

I have gone to great lengths to use careful language. The term *bandwidth* is one of which I am particularly conscious. It is measured in Hz, not in bits per second, which seems to be the prevailing belief. I will admit I have been forced to slip on the misuse of words when I quote somebody.

This text deals with generic systems. I have kept clear of any description of proprietary techniques. Ethernet and CSMA/CD are used synonymously. If some find fault with equating the two—so be it. When I prepared the first edition, there were three subsets of IEEE 802.3; today there are 22 quoted in the IEEE green tome, some 1200 pages thick. I have devoted one complete chapter to this subject because it pervades the marketplace. There are two reasons for this: cost and simplicity.

Some people refer to telecommunications as a technology dealing exclusively with voice communications. Data communication is a separate entity with its own rules and laws of nature; never the twain shall meet. From my perspective, they both fall under electrical communication, which is synonymous with telecommunication. In fact, later chapters of the text show the melding of voice, data and video into the larger field of telecommunication. For example, the underlying objective of the Integrated Services Digital Networks (ISDN) is the integration of all media into a single telecommunication system. ATM is based on the same premise, and in my opinion, truly accomplishes the goal. I did prepare one section on the philosophy of signaling showing where, in that one arena, data and traditional voice telecommunications differ.

I also have broken with tradition and included chapters on the description of the present underlying digital network, including SONET and SDH. Other texts on data do not do this. I firmly believe that to design a complete data network, more often than not, we have to take advantage of services available from the PSTN. This is common practice with WANs. In Chapter 17, "Last-Mile Data Distribution Systems," a customer can access any one or several of the media at his/her premises site, which are voice, data, and video. On the entry side, these circuits somewhere, somehow, will be transported on the PSTN. Knowledge of how the PSTN can either hurt or help, is vital for a complete understanding of data networks.

The first edition chapter sequence started with rudimentary electrical signals and later on introduced protocols. This edition does the reverse, starting with a description of a data-link layer protocol after introducing the layering concept and OSI. There are three chapters dealing with data transmission, which serve as a tutorial basis for the remainder of the book. LANs and WANs are treated in an area which I choose to call *Enterprise Networks*. I emphasize data networks and their protocols eliminating discussion of specialized networks, such as DQDB, SMDS, and HIPPI. IPv4 is stressed, but I do include a short overview of IPv6. An entire chapter is devoted to ATM, highlighting data application.

I am of the opinion that three generic system types, CATV, DSL and LMDS, will vie for last-mile distribution. One or two of these systems will continue the battle for leadership for broadband delivery during the lifetime of this edition. Internet connectivity is the driving force. A brief description of each system can be found in Chapter 17.

During the preparation phase of the second edition, the Internet was discovered to be an excellent source of information. This held particularly true for IP. There is a data bank available for all RFCs including the latest.

ACKNOWLEDGMENTS

Dr. Ernie Woodward gave me a new perspective on how to present practical data communications to a general readership. Data communications may be broken down into two basic elements: procedural and electrical. I was brought up more on the electrical engineering side of the fence. Ernie showed me that there was another way of doing it by following a procedural approach from the beginning, namely, the application of data protocols. The objective was to make the book more attractive to IT personnel. Ernie was previously a section manager at Motorola Chandler and now is with INTEL here in the Valley. While at Motorola, Ernie was my boss for some eight months. In that period, I prepared the Frame Relay Interface Control Document for the Celestri terminal segment. He was a great boss and is a fine human being.

Many other associates offered positive suggestions to improve this work. Dr. Ron Brown, an independent consultant, gave me a lot of encouragement for the new approach while the book was in the outline stage. I have gone out on a limb with several statements and Ron and John Lawlor supported me. John is another "associate" and independent consultant. Our son, Bob, who works in the field, gave me much encouragement, particularly in layer 3 switching, (or is it routing?). My wife, Paquita, kept the home fires burning. She had great patience and gave me warm love and support.

I also received support and encouragement from the crew I worked with at Motorola on the Celestri program (which eventually became Teledesic). In particular, the wife-and-husband engineering team, Jill and Dave Wheeler, gave me insight into network security. Doug White, "Mr. TCP/IP," willingly answered my many questions on such topics as "time to live." Thank you all. You are a great crew.

ROGER L. FREEMAN

Scottsdale, Arizona
December 2000

Practical Data
Communications

1

THE ENTERPRISE NETWORK
ENVIRONMENT

1.1 ENTERPRISE NETWORKS

An *enterprise network* is a highway that transports data from one user station
to another. An enterprise is an entity, which is either government or industry
in the broadest interpretation. Examples of government may be the USAF
Rome Laboratory or the Scottsdale, AZ, police department. Examples of
industry may be Motorola in Schaumberg, IL, the American Red Cross in
Washington, DC, or Len's auto repair shop in Warner, NH. Enterprise
networks pervade the world.

There are two types of enterprise networks: local area networks (LANs)
and wide area networks (WANs). The two types differ in coverage area. The
local area network, as its name implies, covers a small, restricted area.
Commonly, it covers a portion of one floor, the entire floor, many floors, or
the whole building; in some cases, there may be workstations in several
buildings. A WAN covers a much larger area, from across town to worldwide.

We can liken a LAN to a wide highway but with just one lane, in most
cases. It accommodates large vehicles that move very fast. The WAN, on the
other hand, is a comparatively narrow-lane highway, often with one lane
coming and another going. Its car-handling capacity is notably smaller than
that of a LAN.

Another major difference between a LAN and a WAN is that an entire
LAN has single ownership. It traverses an area that is under one owner, and
the LAN is operated for the exclusive benefit of that owner. A WAN
traverses comparatively long distances where right-of-way is required, or it
traverses over the network of another entity that is in the telecommunication
transport business that will carry the connectivity* to the desired distant end.
The connectivity could be carried out over a facility owned by the enterprise.

---

*For an introductory discussion of connectivities, consult Ref. 1.

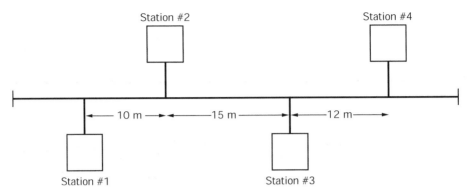

Figure 1.1. An example of a LAN. Single ownership is assumed over its entire route.

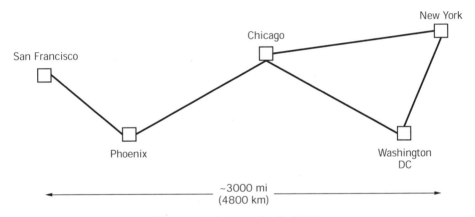

Figure 1.2. An example of a WAN.

This would be a private network. With the familiar arrangement, the connectivity could be facilitated by the owner's local telephone company or by a long-distance carrier such as Sprint, MCI, France Telecom, or AT & T.

An example of a LAN is illustrated in Figure 1.1, and an example of a WAN is shown in Figure 1.2.

The bit rate capacity of a LAN varies from 4 Mbps to 1000 Mbps; for a WAN, from 56/64 kbps to 1.5/2.0 Mbps up to 45 Mbps or greater. The LAN ordinarily is owned by the enterprise. Therefore there are no recurring costs besides maintenance. The WAN customarily involves leasing, commonly from the PSTN or other common carrier. Leasing is a recurring cost, and it increases exponentially with bit rate and more or less linearly with distance. Maintenance is customarily borne by the carrier leasing the circuits to the enterprise. The exception, of course, is when the enterprise can justify a private network. Power companies have private networks.

Often an enterprise will have many LANs. In some cases they may be interconnected by bridges that contain routing software. Some types of LANs have very severe distance limitations, typically Ethernet* LANs. Even with the 2.5-km maximum extension, such a LAN can accommodate hundreds of accesses. As the number of accesses increases, the efficiency of a LAN, especially an Ethernet LAN, decreases. Transactions slow down; and in the case of Ethernet, collisions and backoffs increase. Somewhere between an 18% to 35% activity factor, transactions will stop and the entire time is spent with collision resolution.

One method to alleviate the situation is to segment the LAN into areas of common community of interest. A bridge connects segments together. For example, the entire accounting department is on one segment; the operations department on another segment; and engineering on a third segment. Traffic intensity inside a segment is high; among segments it is low.

1.2 TYPES OF NETWORK TOPOLOGY

LAN topologies may differ in many respects with PSTN topologies. Two common topologies dominate in the LAN arena: the bus network and the ring network. We hasten to explain that a tree network is just an extension of a bus network.

The most elementary network topology is a point-to-point connectivity, which is illustrated in Figure 1.3. An extension of this scheme is the point-to-multipoint or just multipoint connectivity, illustrated in Figure 1.4. Point-to-point connectivity may be used either in the local area or in the

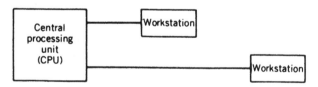

Figure 1.3. Point-to-point connectivity.

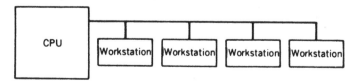

Figure 1.4. Multipoint connectivity (sometimes called multidrop).

*In this text we use Ethernet and CSMA/CD (carrier sense multiplex access with collision detection) interchangeably.

sense of a WAN. Figure 1.1 shows a simple bus network. A tree network is illustrated in Figure 1.5, and a ring network is shown in Figure 1.6. Figure 1.7 is a simple star network, and Figure 1.8 shows a higher-order star network. In some situations in the enterprise environment, star networks are used (e.g., where a PABX acts as a data switch). In the PSTN, the hierarchical network is nearly universal. A hierarchical network derives from the higher-order star network.

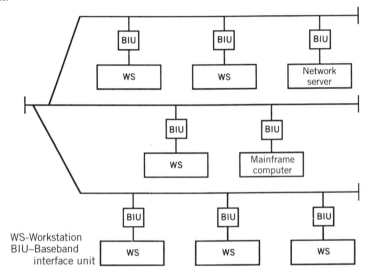

Figure 1.5. An example of a tree network. It is an extension of the bus network.

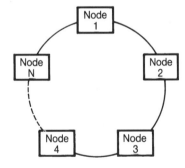

Figure 1.6. A ring network.

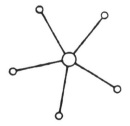

Figure 1.7. A simple star network.

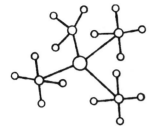

Figure 1.8. A higher-order star network. Such a network easily evolves into a hierarchical network, typical of the PSTN.

The two most common topologies found in the local area enterprise environment are the bus/tree configuration and the ring network. Ethernet utilizes a tree topology; token ring, of course, uses a ring; and FDDI is based on the ring configuration.

A LAN will have three or four types of terminals attached. These are workstations (PCs), one or more servers, and often a bridge and *hubs*. A bridge connects the LAN to other LAN(s)/WAN(s) or to another segment of the same LAN.

A *server* can be broadly defined as a device that performs a service for a user or for the network. For example, on a network running NT, a *network server* polices the network with logins and firewalls; it may also carry out complex processing functions for a user. A *print server* allows users to share a high-speed printer. *File servers* contain large core storage and high-speed processing capabilities.

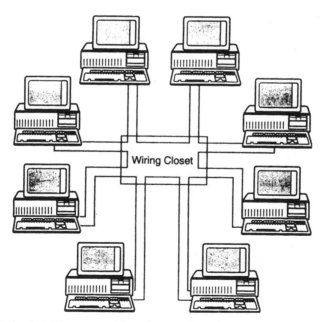

Figure 1.9. A simple LAN connected in a ring topology showing the use of a wiring closet.

A *hub* is a centralized connecting point for a LAN. It is a multiport device, where each LAN user is connected to a port. Physical rings or buses are formed by internally configuring the hub ports. A typical hub may have 8 or 16 ports. Hubs can be stacked (stackables) using one hub port for each interconnection. Hubs can also have a certain amount of intelligence such as the incorporation of a network management function. Repeaters may also be mounted in a hub. A repeater is a simple regenerator device.

The telecommunications wiring closet performs several important functions. Figure 1.9 graphically illustrates one of the functions. First, it is the centralized repository of common telecommunication (data communication) equipment. Among such common equipment are servers, hubs, bridges, and so on. Second, it is the centralized location on a floor for data and voice signal leads. These leads terminate on a distribution frame allowing easy access by technicians and other crafts persons.

1.3   NETWORK ACCESS: AN OVERVIEW

A LAN has one single pathway connecting all users. The pathway may be a wire pair, a length of coaxial cable, or a fiber-optic strand mounted in a cable. In the case of FDDI there will be two strands, one passing a light signal in one direction and the other passing light in the other direction. Either of these three is called a transmission medium.

The transmission medium is shared by all users. It should be shared equitably. There are two types of access to the medium: random and controlled. The random access may also be called contention because a station "contends" for use of the medium. Once a station achieves successful access with a random access technique, it may hold that access until all its transactions are complete. At first blush it would seem unfair and does not meet a criterion of "equitable," which we stated at the outset. Further examination will show that it is more equitable than it first appears. The secret lies in the average duration of a transaction. Ethernet is a typical example of random access. A typical bit rate is 10 Mbps. Simple arithmetic tells us that in 100 ms we have completed 1,000,000 bits of transaction or we have passed to the distant end 125,000 bytes* of data. The average duration of a transaction is generally not this long. Consider now that we have higher bit rate versions of Ethernet, namely, 100 Mbps and 1000 Mbps (gigabit Ethernet).

Token passing schemes offer controlled access. A token is a short data packet that is passed from user to user. A user holding the token, if there is traffic to transmit, has access to the medium for a fixed length of time. At the

---

*A *byte* in this text is assumed to have 8 bits. We prefer the term *octet* rather than byte because it does not need definition.

end of that time duration, the user must relinquish the token to the next user in line. If a user cannot finish transmitting the packets the station has in queue, the user must await the next turn on the token.

We discuss these access techniques in Chapters 9 and 10.

1.4   INITIAL NETWORK DESIGN CONSIDERATIONS

The network design selection will depend largely on cost, installation simplicity, traffic load, number of accesses, distances to be covered, and latency/turn-around time. Consider Table 1.1 in this regard.

Some 75% of installed LANs in the United States and Canada are of the Ethernet type. The driving factors influencing the decision to select an Ethernet-type LAN are cost and simplicity. Network topology is dictated by the LAN access technique. Of course, ring topologies are used with token ring and FDDI; and bus/tree topology is dictated for Ethernet. Once the access method has been selected, the topology type is dictated by default.

WAN network designs are usually based on cost and latency. X.25 and TCP/IP have high latency; frame relay and ATM have low latency. Latency will also depend on distance and number of switching nodes the circuit must pass through. The service level agreement (SLA) with the carrier should also include backup and survivability provisions. True packet networks should involve multiple routes. Leased circuits take on the aspect of point-to-point connectivity. See Figure 1.2. Frame relay circuits from the user's viewpoint when based on permanent virtual circuits (PVCs) can be seen as point-to-point circuits.

TABLE 1.1   Initial Selection Tradeoff Analysis: Random Access and Token Passing

Item / Parameter	Random Access (i.e., Ethernet)	Controlled Access (i.e., Token Passing)
Cost / user	Low	Medium–high
Installation simplicity	More simple	More complicated
Traffic load	Breaks down at 18% to 38% loading	Handles up to 100% loading
Number of accesses	Up to 1000 (in theory)	Up to 250 (token ring)
Maximum distance	Up to 2500 m	Up to 16-dB loss at 16 MHz or 640 ft (195 m) with CAT5 UTP. FDDI 100 km
Latency / turn-around time	Low when no collisions; collisions notably increase latency	Token passing time; function of number of users: 10 users, low; 200 users, medium–high

1.5   CONNECTION-ORIENTED AND CONNECTIONLESS SERVICE

There are two generic types of service used in data communications: connection-oriented service and connectionless service. In the OSI literature (Chapter 2) the terms are connection mode and connectionless mode of service. The IEEE (Ref. 1) defines the terms as follows:

> *Connection-Oriented Service.* A kind of delivery service where different virtual circuit configurations are used to transmit messages.
>
> *Connectionless Service.* A kind of delivery service that treats each packet as a separate entity. Each packet contains all protocol layers and destination address at each intermediate node in the network. *Note:* The order of arrival of packets is not necessarily the same as the order of transmission.

With connection-oriented service a real or virtual circuit can be traced from origin to destination throughout the duration of the call or interchange of packets. The analogy is the conventional telephone call.

Connectionless service can be likened to postal service. When a letter is placed in a mailbox for delivery, we have no idea of the letter's routing nor do we care. We believe there will be a true and honest effort by the postal service to deliver the letter to its destination. Most LANs use connectionless service; one variation of ATM* offers connectionless service.

There are two variants of these services. They deal with acknowledgment of receipt of packets or frames. There is connectionless service with or without acknowledgment, and there is connection mode service with or without acknowledgment. A common form of acknowledgment is by the use of sequence numbers for packets or frames. This subject is introduced in Chapter 3.

1.6   DATA PROTOCOLS: KEY TO NETWORK OPERATION

1.6.1   Introduction

To get the most out of a network, certain operational rules, procedures, and interfaces have to be established. A data network is a big investment, and our desire is to get the best return on that investment. One way to optimize return on a data network is by the selection of operational protocols. We must argue that there are multiple tradeoffs involved, all interacting one with another. Among these are data needs to be satisfied, network topology and architecture, selection of transmission media, switching and network management hardware and software, and operational protocols. In this section we focus on the protocols.

---

*ATM stands for asynchronous transfer mode. See Chapter 14.

In the IEEE dictionary [Ref. 2], one definition of a protocol is "a set of rules that govern functional units to achieve communication." We would add interfaces to the definition to make it more all-encompassing. In this section we trace some of the evolution of protocols up through the International Standards Organization (ISO) OSI and its seven layers. Emphasis will be placed on the first three layers because they are more directly involved in communication.

This section serves as an introduction to Chapter 2, the OSI Model, and Chapter 3, High-Level Data Link Control.

Before we delve into protocols, let us make some assumptions. First, a data message consists of a series of *bits*. A bit is the lowest denominator of information. It can be one of two possible states: a 1 or a 0. This is neat because we can apply binary arithmetic in such a situation. Bits can be expressed electrically fairly easily by many possible two-state arrangements. The current is on or it is off; the voltage is positive going or negative going just for starters.

In nearly all situations bits are taken eight at a time or in groups of eight sequential bits. Such a grouping is called an *octet* (preferred nomenclature) or *byte* (deprecated nomenclature). A *frame* consists of a number of octets (or bytes) in purposeful order. A *data message* may consist of one or more frames. With these assumptions, we should be able to make it through the first several chapters of the book without any problem.

Further discussion of elementary telecommunications may be found in Ref. 1.

1.6.2 Basic Protocol Functions

Typical protocol functions are:

- Segmentation and reassembly (SAR)
- Encapsulation
- Connection control
- Ordered delivery
- Flow control
- Error control

A short description of each of these functions is given below.

> *Segmentation and Reassembly. Segmentation* refers to breaking up the data message or file into packets, frames, or blocks with some bounded size. In asynchronous transfer mode (ATM—Chapter 15), a message is broken up into cells. *Reassembly* is the reverse of segmentation, because it involves putting the packets, frames, or blocks back into their original order. The device that carries out segmentation and reassembly in a packet network is called a *PAD* (packet assembler/disassembler).

Encapsulation. *Encapsulation* is the adding of header and control information to a frame in front of the text or info field and the adding of parity-check information. The parity-check field is generally carried after the text or info field of a frame. Use of the parity check is described in detail in Chapter 4 in the section on data integrity (Section 4.3.5).

Connection Control There are three stages of connection control:

1. Connection establishment
2. Data transfer
3. Connection termination

Of course we are talking about connection-oriented service. There is no connection control in connectionless service.

Some of the more sophisticated protocols also provide connection interrupt and recovery capabilities to cope with errors and other sorts of interruptions.

Ordered Delivery. Packets, frames, or blocks are often assigned sequence numbers to ensure ordered delivery of data at the destination. In a large network with many noses and possible routes to a destination, especially when operated in the packet mode, the packets can arrive at the destination out of order. With a unique packet or frame numbering plan using a simple numbered sequence, it is a rather straightforward task for a long file to be reassembled at the destination in its original order. Packet and frame numbering, as discussed in Chapter 4, is also a tool used to maintain data integrity.

Flow Control. *Flow control* refers to the management of the data flow from source to destination such that buffer memories do not overflow, but maintain full capacity of all facility components involved in the data transfer. Flow control must operate at several peer layers of protocols, as will be discussed later.

Error Control. *Error control* is a technique that permits recovery of lost or errored packets (frames or blocks). There are four possible functions involved in error control:

1. Numbering of packets (frames, blocks) to maintain a consistent accounting system. This will indicate at the receive end if there are packets missing. It is also a handy tool used in a data integrity acknowledgment scheme.
2. Incomplete octets. By definition an octet must have 8 bits.
3. Incomplete frames. When key elements of a frame are missing or are redundant. When the octet maximum in an information field is exceeded or when the octet minimum is not complied with.
4. Bit-error detection. Error detection and error correction are discussed in Sections 4.4 and 4.5.

REFERENCES

1. Roger L. Freeman, *Fundamentals of Telecommunications*, John Wiley & Sons, New York, 1999.
2. *The IEEE Standard Dictionary of Electrical and Electronic Terms*, IEEE Std. 100-1996, IEEE, New York, 1996.

2

THE OSI MODEL AND THE DATA-LINK LAYER

2.1 INTRODUCTION

The concept of Open System Interconnection (OSI) and its reference model were developed by the International Standards Organization (ISO; Geneva) in the early 1970s. The ISO impetus to develop OSI stemmed from the proliferation of proprietary systems (typically IBM 360 and UNIX, others) that could not interoperate. The concept was that if an operating system or protocol was developed around an open system rather than a proprietary system, interoperation of systems became more practicable.

The IEEE (Ref. 1) defines an open system as follows:

A system that implements sufficient specifications for interfaces, and supporting formats to enable properly engineered applications software to be ported across a wide range of systems with minimal changes, to interoperate with other applications on local and remote systems, and to interact with users in a style which facilitates user portability.

A key element of this definition is the use of the term *open specification*, which is defined (Ref. 1) as follows:

A public specification that is maintained by an open, public consensus process to accommodate new technology over time and that is consistent with standards.

CCITT Rec.(ommendation) X.200 (Ref. 2) states that OSI is a reference model used to ideally permit a full range of data communications among disparate data equipment and networks. In other words, OSI allows standardized procedures to be defined, enabling the interconnection and subsequent effective exchange of information between users. Such users are systems (i.e., a set of one or more computers, associated software, servers, peripherals, terminals, human operators, physical processes, information transfer means, etc.) that for an autonomous whole are capable of performing information processing and/or information transfer. The reference model, in particular, permits interworking between different networks, of the same or different types, to be defined such that communication may be achieved as easily over a combination of networks as over a single network.

2.2 LAYERING

2.2.1 Notation

In the layered OSI model the following notation is used:

(N)-layer: any specific layer
$(N + 1)$-layer: the next higher layer
$(N - 1)$-layer: the next lower layer

2.2.2 Basic Structuring Technique

The basic structuring technique in the reference model of OSI is layering. According to this technique, each open system is viewed as logically composed of an ordered set of subsystems, represented for convenience in the vertical sequence shown in Figure 2.1. Adjacent systems communicate through their common boundary. Subsystems of the same range (N) collectively form the (N)-layer of the reference model of OSI. An (N)-subsystem consists of

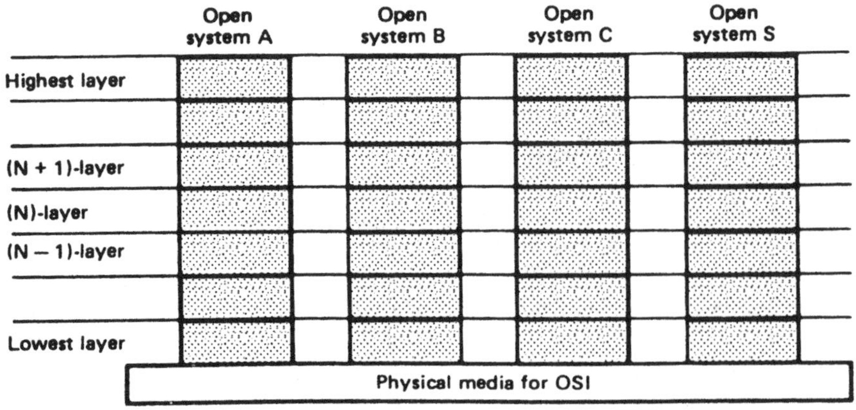

Figure 2.1. Layering in cooperating open systems.

one or several (N)-entities. Entities exist in each layer. Entities in the same layer are termed *peer entities*. Note that the highest layer does not have an ($N + 1$)-layer above it and the lowest layer does not have an ($N - 1$)-layer below it.

Not all peer (N)-entities need, nor have the ability, to communicate. There may be conditions that prevent this communication (e.g., they are not in interconnected open systems, or they do not support the same protocol subsets).

2.3   TYPE AND INSTANCE

The distinction between the *type* of some object and an *instance* is a distinction of significance for OSI. A type is a description of a class of objects. An instance of this type is any object that conforms to this description. The instances of the same type constitute a *class*. A type, and any instance of this type, can be referred to by an individual name. Each nameable instance and the type to which this instance belongs carry distinguishable names.

For example, given that a programmer has written a computer program, that programmer has generated a type of something, where instances of that type are created every time the particular program is invoked into execution by a computer. Thus a FORTRAN compiler is a type; and for each occasion where a copy of that program is invoked in a data processing machine, it displays an instance of that program.

Consider now an (N)-entity in the OSI context. It, too, has two aspects: a type and a collection of instances. The type of an (N)-entity is defined by the specific set of (N)-layer functions it is able to perform. An instance of that type of (N)-entity is a specific invocation of whatever it is when the relevant open system that provides the (N)-layer functions is called for by its type for a particular occasion of communication. It follows from these observations that (N)-entity types refer only to the properties of an association between peer (N)-entities, whereas an (N)-entity instance refers to the specific, dynamic occasions of actual information exchange.

It is important to note that actual communication occurs only between (N)-entity instances at all layers. It is only at connection establishment time (or its logical equivalent during a recovery process) that (N)-entity types are explicitly relevant. Actual connections are always made to specific (N)-entity instances, although a request for a connection may well be made for arbitrary (N)-entity instances of a specified type. Nothing in the OSI standard as reflected in CCITT Rec. X.200 (Ref. 2), however, precludes the request for a connection with a specific named instance of a peer (N)-entity. If an (N)-entity instance is aware of the name of its peer (N)-entity instances, it is able to request another connection to that (N)-entity instance.

2.4 POSSIBLE SUBLAYERS

It may be necessary to further divide a layer into smaller substructures called *sublayers** and to extend the technique of layering to cover other dimensions of OSI. A sublayer is described as a grouping of functions in a layer which may be bypassed. The bypassing of all sublayers of a layer is not allowed. A sublayer uses the entities and connections of its layer.

Except for the highest layer of OSI, each (N)-layer provides $(N + 1)$-entities in the $(N + 1)$-layer with (N)-services. The highest layer is assumed to represent all possible uses of the services which are provided by the lower layers.

Note 1. Note that all open systems provide the initial source or final destination of data; such open systems need not contain the higher layer of the architecture as shown in Figure 2.6.

Each service provided by an (N)-layer may be tailored by the selection of one or more (N)-facilities which determine the attributes of that service. When a single (N)-entity cannot by itself support a service requested by an $(N + 1)$-entity, it calls upon the cooperation of other (N)-entities to help complete the service requested. In order to cooperate, (N)-entities in any layer, other than those in the lowest layer, communicate by means of a set of services provided by the $(N - 1)$-layer (see Figure 2.2). The entities in the lowest layer are assumed to communicate directly via the physical media which connect them.

The services of an (N)-layer are provided to the $(N + 1)$-layer using the (N)-functions performed within the (N)-layer and, as necessary, the services available from the $(N - 1)$-layer.

An (N)-entity may provide services to one or more $(N + 1)$-entities and use the services of one or more $(N - 1)$-entities. An (N)-service access point (SAP) is the point at which a pair of entities in adjacent layers use or provide services. Cooperation between (N)-entities is governed by one or more (N)-protocols. The entities and protocols within a layer are shown in Figure 2.3.

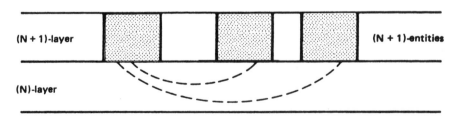

Figure 2.2. $(N + 1)$-entities in the $(N + 1)$-layer communicate through the (N)-layer.

---

*We will note that this is done with IEEE 802-type LANs where the data link layer is divided into two sublayers: logical link control sublayer and the medium access control sublayer. This is described in Chapter 9.

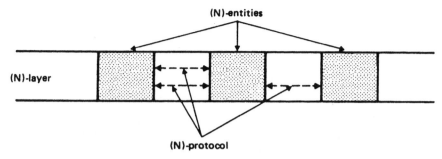

Figure 2.3. (N)-protocols between (N)-entities.

2.5 DATA UNITS

Information is transferred in various types of data units between peer entities and between entities attached to a specific SAP. The following are several definitions dealing with data units:

A *protocol data unit* (PDU) is a unit of data specified in an (N)-protocol and consisting of (N)-protocol control information and possibly (N)-user data.

(N)-*interface data* is information transferred from (N + 1)-entity to an (N)-entity for transmission to a correspondent (N + 1)-entity over an (N)-connection, or, conversely, information transferred from an (N)-entity to an (N + 1)-entity after being received over an (N)-connection from a correspondent (N + 1)-entity.

An (N)-*interface data unit* is a unit of information transferred across the (N)-service access point between an (N − 1)-entity and an (N)-entity in a single interaction. Each (N)-interface data unit contains (N)-interface control information and may also contain the whole or part of an (N)-service data unit.

The relationship among data units is shown in Figures 2.4 and 2.5.

	Control	**Data**	**Combined**
(N)-(N) peer entities	(N)-protocol control information	(N)-user data	(N)-protocol data units
(N + 1)-(N) adjacent	(N)-interface control information	(N)-interface data	(N)-interface data unit

Figure 2.4. Relationship among data units.

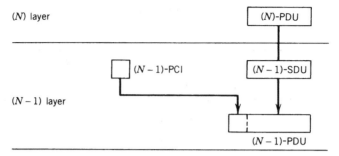

Figure 2.5. An illustration of mapping between data units in adjacent layers. PCI, protocol control information; PDU, protocol data unit; SDU, service data unit. *Note 1*. This figure assumes that neither segmenting nor blocking of (N)-service data units is performed. *Note 2*. This figure does not imply any positional relationship between protocol control information and user data in protocol data units. *Note 3*. An (N)-protocol data unit may be mapped one-to-one into an (N − 1)-service data unit, but other relationships are possible. (From Figure 9, page 16, ITU-T Rec. X.200, Ref. 2.)

Except for the relationships defined in Figures 2.4 and 2.5, there is no overall architectural limit to the size of data units. There may be other size limitations at specific layers. The size of (N)-interface data units is not necessarily the same at each end of the connection. Data may be held within a connection until a complete data unit is put into the connection.

2.6 SPECIFIC LAYERS OF THE OSI REFERENCE MODEL

The OSI reference model contains seven layers as shown in Figure 2.6. The highest layer, the layer at the top of the figure, is the application layer, which consists of application entities that cooperate in the OSI environment (OSIE). The lower layers provide services through which the application entities cooperate.

Layers 1–6 together with the physical media for OSI provide a step-by-step enhancement of communication services at which an OSI service recommendation is defined, while the functioning of the layers is governed by OSI protocol recommendations.

Not all open systems provide initial source or final destination of data. When the physical media for OSI do not link all open systems directly, some open systems act only as relay open systems,* passing data to other open systems. The functions and protocols that support the forwarding of data are then provided in the lower layers as shown in Figure 2.7. Note in the figure that the relay function only requires the first three OSI layers.

---

*A node that carries out a relay function.

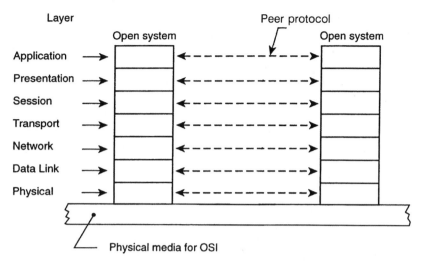

Figure 2.6. The seven-layer OSI reference model with peer protocols. (From Figure 11, page 28, ITU-T Rec. X.200, Ref. 2.)

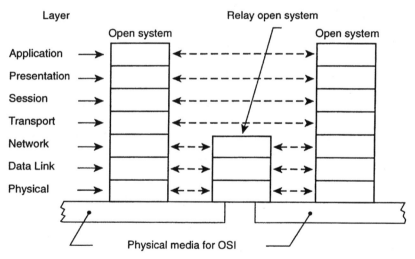

Figure 2.7. Communication involving relay open systems. (From Figure 12, page 29, ITU-T Rec. X.200, Ref. 2.)

2.7   LAYER DESCRIPTIONS

Application Layer. The purpose of the application layer is to serve as the window between correspondent application processes (actual user data) that are using the OSI to exchange meaningful information. The application layer necessarily provides all OSI services directly usable by application processes.

The application entity contains one user element and a set of application service elements. The user element represents that part of the application process which uses those application service elements needed to accomplish the communication objectives of that application process. Application service elements may call upon each other and/or upon presentation services (layer 6) to perform their function.

The only means by which user elements in different systems may communicate is through the exchange of application protocol data units. These application protocol data units are generated by application service elements.

Other services that may be included in the application layer are (a) identification of intended communication partners (e.g., by name and address, by definite description, and by generic description), (b) determination of the current availability of intended communication partners, (c) establishment of the authority to communicate, (d) authentication of intended communication partners, (e) determination of cost allocation methodology, (f) determination of adequacy of resources, (g) determination of acceptable quality of service (such as response time and error rate), (h) synchronization of cooperating applications, (i) selection of the dialogue discipline including the initiation and release procedures, (j) agreement on the responsibility for error recovery, (k) agreement on procedures for control of data integrity, and (l) identification of constraints on data syntax (character sets, data structure).

When a specific instance of an application process wishes to communicate with an instance of an application process in some other open system, it must invoke an instance of an application entity in the application layer of its own open system. It then becomes the responsibility of this instance of the application entity to establish an association with an instance of an appropriate application entity in the destination open system. This process occurs by invocation of instances of entities in the lower layers. When the association between the two application entities has been established, the application processes can communicate.

Presentation Layer. Two definitions need to be given here:

Concrete Syntax. Those aspects of the rules used in the formal specification of data which embody a specific representation.

Transfer Syntax. The concrete syntax used in the transfer of data between open systems.

The presentation layer provides for the representation of information that application entities either communicate or refer to in their communication. The presentation layer covers two complementary aspects of this representation of information: (i) the representation of data to be transferred between application entities and (ii) the representation of the data structure which application entities refer to in their communication, along with the representation of the set of actions which may be performed on this data structure.

The complementary aspects of the representation of information outlined above refer to the general concept of transfer syntax. The presentation layer is concerned with the syntax (meaning the representation of data) and not with its semantics (its meaning to the application layer), which is known only by the application entities.

The presentation layer provides for a common representation to be used between application entities. This relieves application entities of any concern with the problem of "common" representation of information, and thus it provides them with syntax independence. The syntax independence can be described in two ways:

1. The presentation layer provides common syntactical elements that are used by application entities.
2. The application entities can use any syntax, and the presentation layer provides the transformation between these syntaxes and the common syntax needed for communication between application entities. This transformation is performed inside the open systems. It is not used by other open systems and therefore has no impact on standardization of presentation protocols.

The approach in 2 above is the one commonly used.

The presentation layer provides session services (layer 5) and the following facilities: (a) transformation of syntax and (b) selection of syntax. Transformation of syntax is concerned with code and character set conversions, the modification of the layout of the data, and the adaptation of actions on the data structures. Selection of syntax provides the means of initially selecting a syntax and subsequently modifying the selection.

Session services are provided to application entities in the form of presentation services. The presentation layer performs the following functions to assist in accomplishing the presentation services: (a) session establishment request, (b) data transfer, (c) negotiation and renegotiation of syntax, (d) transformation of syntax including data transformation, formatting, and special-purpose transformations (e.g., compression), and (e) session termination request.

There are three syntactic versions of the data: the syntax used by the originating application entity, the syntax used by the receiving application entity, and the syntax used between presentation entities (the transfer syntax). Of course, any two or all three of these syntaxes may be identical. The presentation layer contains the functions necessary to transform between the transfer syntax and each of the other syntaxes as required.

To meet the service requirements specified by the application entities during the initiation phase, the presentation layer may use any transfer syntax available to it. To accomplish other service objectives (e.g., data volume reduction to reduce data transfer cost), syntax transformation may be performed either as a specific syntax-matching service provided to the application entities or as a function internal to the presentation layer.

Negotiation of syntax is carried out by communication between presentation entities on behalf of the application entities to determine the form that data will have while in the OSI environment. The negotiations will determine what transformations are needed (if any) and where they will be performed. Negotiations may be limited to the initiation phase, or they may occur any time during a session.

Session Layer. Two definitions need to be given here:

Interaction Management. A facility of the session service which allows correspondent presentation entities to control explicitly whose turn it is to exercise certain control functions.

Session-Connection Synchronization. A facility of the session service which allows presentation entities to define and identify synchronization points and to reset a session connection to a predefined state and to agree on a resynchronization point.

The purpose of the session layer is to provide the means necessary for cooperating presentation entities to organize and synchronize their dialogue and to manage their data interchange. To do this, the session layer provides services to (a) establish a session connection between two presentation entities and (b) support orderly data exchange interactions.

To implement the transfer of data between presentation entities, the session connection is mapped onto and uses a transport connection (transport layer, layer 4).

A session connection is created when requested by a presentation entity at a session-service access point. During the lifetime of the session connection, session services are used by the presentation entities to regulate their dialogue and to ensure orderly message exchange on the session connection. The session connection exists until it is released by either of the presentation entities or the session entities. While the session connection exists, session services maintain the state of the dialogue even over data loss by the transport layer.

A presentation entity can access another presentation entity only by initiating or accepting a session connection. A presentation entity may be associated with several session connections simultaneously. Both concurrent and consecutive session connections are possible between two presentation entities.

The initiating presentation entity designates the destination presentation entity by a session address. In many systems, a transport address may be used as the session address (i.e., there is a one-to-one correspondence between the session address and the transport address). In general, however, there is a many-to-one correspondence between the session address and the transport address. This does not imply multiplexing of session connections onto transport connections, but it does imply that at session-connection establishment time, more than one presentation entity is a potential target of a session-connection establishment request arriving on a given transport connection.

The following services are provided by the session layer to the presentation layer: (a) session-connection establishment, (b) session-connection release, (c) normal data exchange, (d) quarantine service, (e) expedited data exchange, (f) interaction management, (g) session-connection synchronization, and (h) exception reporting.

Transport Layer. The transport service provides transparent transfer of data between session entities and relieves them from any concern with the detailed way in which reliable and cost-effective transfer of data is achieved. The transport layer optimizes the use of the available network service to provide the performance required by each session entity at minimum cost. This optimization is achieved within the constraints imposed by the overall demands of all concurrent session entities and the overall quality and capacity of the network service available to the transport layer.

All protocols defined in the transport layer have end-to-end significance, where the ends are defined as correspondent transport entities. Therefore the transport layer is OSI end open system oriented, and transport protocols operate only between OSI end open systems.

The transport layer is relieved of any concern with routing and relaying because the network service provides network connections from any transport entity to any other including the case of tandem subnetworks (see section entitled "Network Layer" on page 24).

The transport functions invoked in the transport layer to provide a requested service quality depend on the quality of the network service. The quality of the network service depends on the way the network service is achieved (see section entitled "Network Layer").

The following are services provided to the session layer by the transport layer. The transport layer uniquely identifies each session entity by its transport address. The transport service provides the means to establish, maintain, and release transport connections. Transport connections provide duplex transmission between a pair of transport addresses.

More than one transport connection can be established between the same pair of transport addresses. A session entity uses transport-connection endpoint identifiers provided by the transport layer to distinguish between transport-connection endpoints. The operation of one transport connection is independent of the operation of all others except for the limitations imposed by the finite resources available to the transport layer.

The quality of service provided on a transport connection depends on the service class requested by the session entities when establishing the transport connection. The selected quality of service is maintained throughout the lifetime of the transport connection. The session entity is notified of any failure to maintain the selected quality of service on a given transport connection.

The following services are provided by the transport layer: (a) transport-connection establishment, (b) data transfer, and (c) transport-connection release.

The following are functions performed within the transport layer: (a) mapping transport address onto network address, (b) multiplexing (end-to-end) transport connections onto network connections, (c) establishment and release of transport connections, (d) end-to-end sequence control on individual connections, (e) end-to-end error detection and any necessary monitoring of the quality of service, (f) end-to-end error recovery, (g) end-to-end segmenting, blocking, and concatenation, (h) end-to-end flow control on individual connections, (i) supervisory functions, and (j) expedited transport-service data unit transfer.

Network Layer. The following definition needs to be given here:

Subnetwork. A set of one or more intermediate open systems which provide relaying and through which end systems may establish network connections.

The network layer provides (a) the means to establish, maintain, and terminate network connections between open systems containing communicating applications entities and (b) the functional and procedural means to exchange network-service data units between transport entities over network connections. It provides to the transport entities independence from routing and relay considerations associated with the establishment and operations of a given network connection. This includes the case where several subnetworks are used in tandem or in parallel. It makes invisible to transport entities how underlying resources such as data-link connections are used to provide network connections.

Any relay functions and hop-by-hop service enhancement protocols used to support the network service between the OSI end open systems are operating below the transport layer (i.e., within the network layer or below).

The basic service that the network layer provides to the transport layer is to provide the transparent transfer of data between transport entities. This service allows the structure and detailed content of submitted data to be determined exclusively by layers above the network layer.

The network layer contains functions necessary to provide the transport layer with a firm network–transport-layer boundary which is independent of the underlying communications media in all things other than quality of service. Thus the network layer contains functions necessary to mask the differences in the characteristics of different transmission and subnetwork technologies into a consistent network service. The service provided at each end of a network connection is the same even when a network connection spans several subnetworks, each offering dissimilar services.

The quality of service is negotiated between the transport entities and the network service at the time of establishment of a network connection. While this quality of service may vary from one network connection to another, it will be agreed upon for a given network connection and be the same at both network-connection endpoints.

The following are services provided by the network layer: (a) network addresses, (b) network connections, (c) network-connection endpoints identifiers, (d) network-service data unit transfer, (e) quality of service parameters, (f) error notification, (g) sequencing, (h) flow control, (i) expedited network-service data unit transfer, (j) reset, (k) release, and (l) receipt of confirmation.

Network Addresses. Transport entities are known to the network layer by means of network addresses. Network addresses are provided by the network layer and can be used by transport entities to identify uniquely other transport entities (i.e., network addresses are necessary for transport entities to communicate using the network service). The network layer uniquely identifies each of the end open systems, represented by transport entities and by their network addresses. This may be independent of the addressing needed by the underlying layers.

A network connection provides the means of transferring data between transport entities identified by network addresses. The network layer provides the means to establish, maintain, and release network connections. A network connection is point-to-point. More than one network connection may exist between the same pair of network addresses.

The network layer provides to the transport entity a network-connection endpoint identifier that identifies the network-connection endpoint uniquely with the associated network address. The network layer may provide sequenced delivery of network-service data units over a given network connection when requested by the transport entities.

The following are functions performed by the network layer: (a) routing and relaying, (b) network connections, (c) network-connection multiplexing, (d) segmenting and blocking, (e) error detection, (f) error recovery, (g) sequencing, (h) flow control, (i) expedited data transfer, (j) reset, (k) service selection, and (l) network layer management.

Routing and Relaying. Network connections are provided by network entities in end open systems but may involve intermediate open systems which provide relaying. These intermediate open systems may interconnect subnetwork connections, data-link connections, and data circuits. Routing functions determine an appropriate route between network addresses. In order to set up the resulting communication, it may be necessary for the network layer to use the services of the data-link layer to control the interconnection of data circuits.

The control of interconnection of data circuits (which are in the physical layer) from the network layer requires interaction between a network entity and a physical entity in the same open system. Because the OSI reference model permits direct interaction only between adjacent layers, the network entity cannot interact directly with the physical entity. This interaction is thus described through the data-link layer which intervenes "transparently" to convey the interaction between the network layer and the physical layer.

Data-Link Layer. The data-link layer provides functional and procedural means to establish, maintain, and release data-link connections among network entities and to transfer data-link service data units. A data-link connection is built upon one or several physical connections. The data-link layer detects and possibly corrects errors that may occur in the physical layer. In addition, the data-link layer enables the network layer to control the interconnection of data circuits within the physical layer.

SERVICES PROVIDED TO THE NETWORK LAYER. In the connection-mode, the facilities provided by the data-link layer are as follows:

(a) Data-link addresses
(b) Data-link connection
(c) Data-link service data units (SDUs)
(d) Data-link connection endpoint identifiers
(e) Error notification
(f) Quality of service (QoS) parameters
(g) Reset

In the connectionless mode, the facilities provided by the data-link layer are as follows:

(a) Data-link addresses
(b) Transmission of data-link service data units of a defined maximum size
(c) Quality of service (QoS) parameters

Data-Link Addresses. Network entities are known to the data-link layer by means of data-link addresses. Data-link addresses are provided by the data-link layer and can be used by network entities to identify other network entities which communicate using the data-link service. A data-link address is unique within the scope of the set of open systems attached to a common data-link layer.

Data-Link Connection. A data-link connection provides the means of transferring data between network entities identified by data-link addresses. A data-link connection is established and released dynamically.

Data-Link Service Data Units. The data-link layer allows for exchange of data-link service data units over a data-link connection or exchange of data-link-service data units (that bear no relation to any other data-link service data units) using the connectionless mode data-link service. The size of the data-link service data units may be limited by the relationship between the physical-connection error rate and the data-link layer error detection capability.

Data-Link Connection Endpoint Identifiers. If needed, the data-link layer provides data-link connection endpoint identifiers that can be used by a network entity to identify a correspondent network entity.

Error Notification. Notification is provided to the network-entity when any unrecoverable error is detected by the data-link layer.

Quality of Service Parameters. Quality of service parameters may be optionally selectable. The data-link layer establishes and maintains a selected quality of service for the duration of the data-link connection. The quality of service parameters include mean time between detected but unrecoverable errors, residual error rate (where errors may arise from alteration, loss, duplication, misordering, misdelivery of data-link service data units, and other causes), service availability, transit delay, and throughput.

Reset. The network-entity can force the data-link entity invocation into a known state by invoking the reset facility.

FUNCTIONS WITHIN THE DATA-LINK LAYER. In the connection mode and connectionless mode, the functions performed by the data-link layer are as follows:

(a) Data-link service data unit mapping
(b) Identification and parameter exchange
(c) Control of data-circuit interconnection
(d) Error detection
(e) Routing and relaying
(f) Data-link layer management

In the connection mode, the following functions are also performed by the data-link layer:

(a) Data-link connection establishment and release
(b) Connection mode data-link data transmission
(c) Data-link connection splitting

(d) Sequence control

(e) Delimiting and synchronization

(f) Flow control

(g) Error recovery

(h) Reset

Data-Link Connection Establishment and Release. These functions establish and release data-link connections on activated physical connection. When a physical connection has multiple endpoints (for example multipoint connection), a specific function is needed with the data-link layer to identify the data-link service access points without establishing a data-link connection.

Connectionless Mode Data-Link Data Transmission. The connectionless mode data-link transmission provides the means for transmission of data-link service data units between data-link service access points without establishing a data-link connection.

Data-Link Service Data Unit Mapping. This function maps data-link service data units into data-link protocol data units on a one-to-one basis.

Data-Link Connection Splitting. This function performs splitting of one data-link connection onto several physical connections.

Delimiting and Synchronization. These functions provide recognition of a sequence of physical service data units (i.e., bits—see Physical Layer) transmitted over the physical connection as a data-link protocol data unit. (*Note:* These functions are often called framing.)

Sequence Control. This function maintains the sequential order of data-link service data units across a data-link connection.

Error Detection and Recovery. Error detection detects transmission, format, and operational errors occurring either on the physical connection or as a result of a malfunction of the correspondent data-link entity. Error recovery attempts to recover from detection transmission, format, and operational errors and notifies the network entities of errors which are unrecoverable.

Flow Control. In the connection mode, each network entity can dynamically control (up to the agreed maximum) the rate at which it receives data-link service data units from a data-link connection. This control may be reflected in the rate at which the data-link layer accepts data-link service data units at the correspondent data-link connection endpoint. In the connectionless mode, there is service boundary flow control, but no peer flow control.

Identification and Parameter Exchange. This function performs data-link-entity identification and parameter exchange.

Reset. This function performs a data-link reset forcing the data-link-entity invocation to a known state.

Control of Data-Circuit Interconnection. This function conveys to network entities the capability of controlling the interconnection of data circuits within the physical layer.
Note: This function is used in particular when a physical connection is established/released across a circuit-switched subnetwork by relaying within an intermediate system between data circuits. These data circuits are elements of the end-to-end path. A network entity in the intermediate system makes the appropriate routing decisions as a function of the path requirements derived from the network signaling protocols.

Routing and Relaying. Some subnetworks, and particularly some configurations of LANs, require that routing and relaying between individual local networks be performed in the data-link layer.

Data-Link Layer Management. The data-link layer protocols deal with some management activities of the layer (such as activation and error control). See Section 8 of ITU-T Rec. X.200 (Ref. 2) and ITU-T Rec. X.700 (Ref. 3) for the relationship with other management aspects.

Physical Layer. The following definition needs to be given here:

Data Circuit. A communication path in the physical media for OSI between two physical entities together with the facilities necessary in the physical layer for the transmission of bits on it.

The physical layer provides mechanical, electrical, functional, and procedural means to activate, maintain, and deactivate physical connections for bit transmission between data-link entities. A physical connection may involve intermediate open systems, each relaying bit transmission within the physical layer. Physical layer entities are interconnected by means of a physical medium.

SERVICES PROVIDED TO THE DATA-LINK LAYER. The services or elements of services provided by the physical layer are as follows:

(a) Physical connection
(b) Physical-service data units
(c) Physical connection endpoints

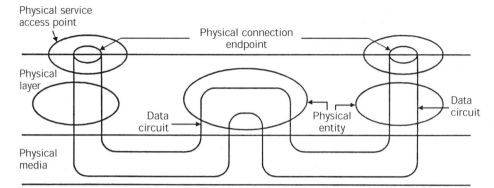

Figure 2.8. Interconnection of data circuits within the physical layer. (From Figure 17, page 50, ITU-T Rec. X.200, Ref. 2.)

(d) Data-circuit identification

(e) Sequencing

(f) Fault condition notification

(g) Quality of service parameters

Physical Connections. The physical layer provides for the transparent transmission of bit streams between data-link-entities across physical connections. A data circuit is a communication path in the physical media for OSI among two or more physical entities together with the facilities necessary in the physical layer for the transmission of bits on it.

A physical connection may be provided by the interconnection of data circuits using relay functions in the physical layer. The provision of a physical connection by such an assembly of data circuits is illustrated in Figure 2.8. The control of the interconnection of data circuit is offered as a service to data-link entities.

Physical Service Data Units. A physical service data unit consists of one bit or a string of bits.

Note: Serial or parallel transmission can be accommodated by the design of the protocol within the physical layer.

A physical connection may allow (full) duplex or half-duplex transmission of bit streams.

Physical Connection Endpoints. The physical layer provides physical connection endpoint identifiers that may be used by a data-link entity to identify physical connection endpoints.

Data Circuit Identification. The physical layer provides identifiers that uniquely specify the data circuits between two adjacent open systems.

Note: This identifier is used by network entities in adjacent open systems to refer to data circuits in their dialogue.

Sequencing. The physical layer delivers bits in the same order in which they were submitted.

Fault Condition Notification. Data-link entities are notified of fault conditions detected within the physical layer.

Quality of Service Parameters. The quality of service of a physical connection is derived from the data circuits forming it. The quality of service can be characterized by the following:

(a) Error rate, where errors may arise from alteration, loss, creation, and other causes
(b) Service availability
(c) Transmission rate
(d) Transit delay

FUNCTIONS WITHIN THE PHYSICAL LAYER. The functions of the physical layer are determined by the characteristics of the underlying medium and are too diverse to allow categorization into connection mode and connectionless mode.
The functions provided by the physical layer are as follows:

(a) Physical connection activation and deactivation
(b) Physical service data unit transmission
(c) Multiplexing
(d) Physical layer management

Physical Connection Activation and Deactivation. These functions provide for the activation and deactivation of physical connections between two data-link entities upon request from the data-link layer. These include a relay function that provides for interconnection of data circuits.

Physical Service Data Unit Transmission. The transmission of physical service data units (i.e., bits) may be synchronous or asynchronous. Optionally, the function of physical service data unit transmission provides recognition of the protocol data unit corresponding to a mutually agreed sequence of physical service data units that are being transmitted.

Multiplexing. This function provides for two or more physical connections to be carried on a single data circuit. This function provides the recognition of the framing required to enable identification of the physical layer PDUs

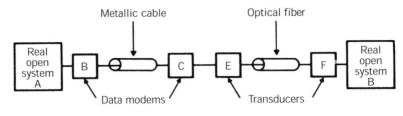

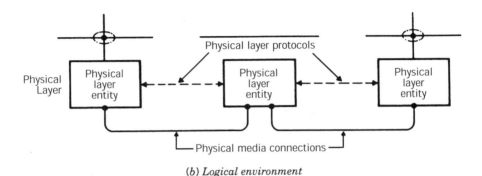

(a) *Real environment*

(b) *Logical environment*

Figure 2.9. Examples of interconnection. (From Figure 19, page 52, ITU-T Rec. X.200, Ref. 2.)

conveyed by the individual physical connections over the single data circuit. The multiplexing function is optional.

Note: A particular example of the use of multiplexing is offered when a transmission media is divided into data circuits in support of the different data link protocols used in the signaling phase and in the data transfer phase when using circuit-switched subnetworks. In such usage of multiplexing, flows of different nature are permanently assigned to different elements of the multiplex group.

Physical Layer Management. The physical layer protocols deal with some management activities of the layer (such as activation and error control). See Section 8 of Ref. 2 and ITU-T Rec. X.700 (Ref. 3) for the relationship with other management aspects.

Note: The above text deals with interconnection between open systems as shown in Figures 2.6 and 2.7. For open systems to communicate in the real environment, real physical connections should be made as, for example, in Figure 2.9a. Their logical representation is as shown in Figure 2.9b and is called the physical media connection. The mechanical, electromagnetic, and other media-dependent characteristics of physical media connections are defined at the boundary between the physical layer and the physical media. Definitions of such characteristics are specified in other standards which shall be discussed below.

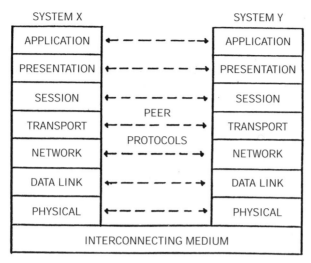

SYSTEM X   SYSTEM Y

Figure 2.10. The OSI reference model, conventional drawing.

2.8 SPECIFIC COMMENTS ON OSI

2.8.1 General

The following are comments regarding Section 2.7, which was extracted from
ITU-T Rec. X.200 (Ref. 2). This section consists of the author's interpretations of OSI layers 1 through 4, where one might consider layer 4 as the end
of the data communicator's responsibility.

Figure 2.10, the OSI reference model, repeats Figure 2.6, but is drawn in
the more conventional manner. The purpose of the model is to facilitate
communication data entities. It takes at least two to communicate. Thus we
consider the model in twos: one entity on the left side of the figure and one
on the right. The terms *peer* and *peers* are used. Peers are corresponding
entities on each side of Figure 2.10 (e.g., the left-side link layer is peer to the
right-side link layer), and they communicate by means of a common protocol.
We wish to add that there is no direct communication between peers except
at the physical layer (layer 1). That is, above the physical layer, each protocol
entity sends data *down* to the next lower layer to get data *across* to its peer
entity on the other side.

We can consider each layer of the OSI model as a functional module. It
depends on the module below it (layer $N - 1$) for service, and it provides
service to the module above it (layer $N + 1$).

As was pointed out in Section 2.7, each entity communicates above it and
below it across an interface. The interface is at a service access point (SAP),
which is shown in Figure 2.11. The (N)-connection endpoint identifier
distinguishes an (N)-connection when an $(N + 1)$-entity establishes an (N)-
connection with another $(N + 1)$-entity as illustrated in Figure 2.11.

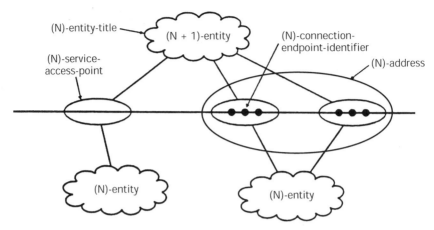

Figure 2.11. Entities, service access points (SAPs), and identifiers. (From Figure 7, page 14, ITU-T Rec. X.200, Ref. 2.)

The actual communication between entities is by means of *primitives.**
A primitive specifies a function to be performed and is used to pass data and control information. ITU-T Rec. X.210 (Ref. 4) describes four types of primitives used to define the interaction between adjacent layers of the OSI architecture. A brief description of each of these primitives follows:

Request. A primitive issued by a service user to invoke some procedure and to pass parameters needed to fully specify the service.

Indication. A primitive issued by a service provider either to invoke some procedure or to indicate that a procedure has been invoked by a service user at the peer service access point.

Response. A primitive issued by a service user to complete at a particular SAP some procedure invoked by an *indication* at the SAP.

Confirm. A primitive issued by a service provider to complete, at a particular SAP, some procedure previously invoked by a *request* at that SAP.

The ITU-T reference adds that *confirms* and *responses* can be positive or negative, depending on the circumstances.

2.9   DISCUSSION OF OSI LAYERS 1–4

2.9.1   The Physical Layer (Layer 1)

Layer 1 is the lowest layer of the OSI model. It provides the physical connectivity between two data end users who wish to communicate. The services it provides to the data-link layer (layer 2) are those required to

*We call these generic primitives. They are widely used in the world of protocols, which we consider to be the heart of data communications.

connect, maintain connected, and disconnect the physical circuits that form the physical connectivity. The physical layer describes the traditional interfaces between data terminal equipment (DTE) and data communication equipment (DCE). These interfaces are described in Section 6.4 (Chapter 6).

The physical layer has four important characteristics:

- Mechanical
- Electrical
- Functional
- Procedural

The mechanical aspects include the actual cabling and connectors necessary to connect the communications equipment to the transmission medium. Electrical characteristics include voltage, signal sense,* impedance, and balanced or unbalanced circuits for signal and control. Functional characteristics cover connector pin assignments at the interface and the precise meaning and interpretation of the various interface signals and data set controls. Procedures cover sequencing rules that govern the control functions necessary to provide higher-layer services such as establishing a connectivity across a switched network.

Some of the applicable standards for the physical layer are as follows:

- EIA/TIA†-232, EIA/TIA-422, EIA/TIA-423, and EIA/TIA-530
- ITU-T (CCITT) Recs. V.10, V.11, V.24, V.28, V.35, X.20, and X.21 bis
- ISO 2110, 2593, 4902, and 4903
- US Fed. Stds 1020, 1030, and 1031
- US Mil-STD-188-114B

The reader is reminded that often standards from one organization reflect the restatement of standards from another organization. For example, MIL-STD-114B restates much of EIA/TIA-232, EIA/TIA-422, and EIA/TIA-423.

2.9.2   The Data-Link Layer (Layer 2)

The data-link layer provides services for the reliable interchange of data across a data link established by the physical layer. Link layer protocols manage the establishment, maintenance, and release of data-link connections. Data-link protocols also control the flow of data and supervise data error recovery. A most important function of this layer is recovery from abnormal conditions. The data-link layer services the network layer (layer 3)

Signal sense: associating voltage polarity with a binary 1 or binary 0.
†EIA stands for Electronics Industries Alliance; TIA stands for Telecommunications Industry Association.

Flag	Address	Control	Information	FCS	Flag

Figure 2.12. Generalized data-link layer frame. The address and control fields taken together are often referred to as the *header*.

or logical link control (LLC in the case of LANs) and inserts a payload data unit into the INFO portion of the link layer. A generic data-link layer frame is illustrated in Figure 2.12.

There is a fair universality of this type of frame for the data-link layer across numerous data protocol families. Chapter 3 contains a detailed discussion of one application of a typical data-link layer frame.

Moving from left to right in Figure 2.12, the first field is the flag field. In nearly all data-link frames of this type, this field is the bit sequence 01111110. It tells the receiving processor of the frame to start counting bits/octets (bytes) at the end of the 8th bit of the flag sequence. The receiving processor knows a priori the length of the address and control fields so it can delineate these two. It also knows (by counting octets) that the end of the control field is the beginning of the information field.

The information field, depending on the protocol, may be of fixed length or variable length. If it is of fixed length, the processor will have this length information a priori. For the variable-length case, the control field will carry that length information which will tell the receiving processor what length to expect.

The processor continues its counting of bytes (octets) and will know the end boundary of the information field. It will now know that the next two octets (bytes) carry the frame check sequence (FCS) information for the case of a WAN, and it will know that the next four octets carry the FCS in the case of a LAN.

The functions of the fields are described as follows. The flag provides the marker for the beginning of operations. It delimits the frame at both ends. If only one frame is to be transmitted, the flag bit sequence not only starts the frame but it is also appended at the end of the frame to indicate termination. If another frame follows directly after the first frame, the ending flag sequence of the first frame is the beginning flag sequence of the second frame.

In some types of data-link layer protocols the address field only contains the destination address. However in most cases, it contains both the destination address(es) and the source address.* In other words, the address field tells the receiving processor who is the intended recipient(s) of the data message or frame. In some cases a second address field follows the destination address. This is the source address field that tells the receiving processor who originated the data message or frame.

*The *source* is the originator of the frame or packet.

The control field covers a number of functions such as sequence numbering, a command–response bit, and type of frame such as information frame, supervisory frame, and unnumbered frame. The control frame will often have a flow control responsibility, usually incorporated in the sequence numbering. In the cases of a variable-length frame, it will give some indication to the receiving processor the length of the information field.

The information field contains a payload service data unit inserted by the network layer. For fixed-length frames, the information field will contain padding if there are insufficient bytes (octets) inserted by the network layer to fulfill minimum byte (octet) requirements. There must be some sort of indicator to tell the processor that there is padding and how much to expect.

The frame check sequence (FCS) is the primary tool used to detect transmission errors. See Section 4.4 for a detailed discussion of the operation of an FCS.

2.9.3 The Network Layer (Layer 3)

The network layer moves data through the network. At relay or switching node along the traffic route, layering concatenates. In other words, the higher layers (above layer 3) are not required, and they are used only at user endpoints. The user endpoints and relay function are shown in Figure 2.7.

The network layer carries out the functions of switching and routing, sequencing, and logical channel control. We note the duplication of error recovery with the data-link layer. Link layer error recovery is only on the link it is responsible for. Network layer error recovery is network-wide.

The network layer also provides and manages logical channel connections between points in the network such as virtual circuits across the PSTN. On simpler data connectivities, such as point-to-point, point-to-multipoint, and LANs, the network layer is not required and can be eliminated. A packet-switched network, on the other hand, is a typical application where the network layer is most necessarily needed.

The best-known network layer standard is ITU-T Rec. X.25. Other protocols are ISO 8348 and the Internet Protocol (IP of the TCP/IP family of protocols). However, in this chapter we treat IP in its specialized function of interfacing between disparate networks in an artificial layer called layer 3.5.

2.9.4 Internet Protocol (IP) and Gateway Sublayer (Layer 3.5)

William Stallings in his *Handbook of Computer-Communication Standards* (Ref. 5), has introduced a sublayer called layer 3.5. This sublayer provides functions for interfacing different networks. When one of the networks is a LAN and the other a WAN, Stallings uses the term *gateway* as the device that performs the interface function. Commonly, this function is performed by a router or even a bridge.

A popular protocol for interfacing disparate networks is the Internet Protocol (IP), which is usually joined with a related protocol called *Transmis-*

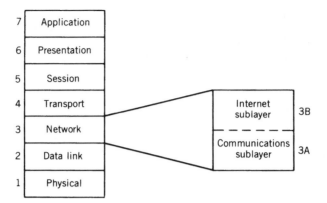

Figure 2.13. Sublayering of OSI layer 3 to achieve internetting.

sion Control Protocol (TCP).* Thus we have TCP/IP, which was developed
for the U.S. Department of Defense (DoD). It is now one of the most widely
used protocols in the commercial sector for OSI layers 3 and 4.

One applicable gateway protocol is ITU-T Rec. X.75 (Ref. 6), which is a
companion to X.25 (Ref. 7). Figure 2.13 shows the sublayering of OSI layer 3
for internetworking or internetting as suggested by Stallings (Ref. 5).

2.9.5 The Transport Layer (Layer 4)

The transport layer (layer 4) is the highest layer of the services associated
with the telecommunications provider. From our perspective, OSI layers 1–4
are the responsibility of the telecommunications system engineer, whereas
layers 5, 6, and 7 are the responsibility of the data end user.

The transport layer has the ultimate responsibility of providing a reliable
end-to-end data delivery service for higher-layer users. It is defined as an
end-system function, located in the equipment using network service or
services. In this way its operations are independent of the characteristics of
all the networks that are involved. Services that a transport layer provide are
as follows:

Connection Management. This includes establishing and terminating connections between transport users. It identifies each connection and
negotiates values of all needed parameters.

Data Transfer. This involves the reliable delivery of transparent data between users. All data are delivered in sequence with no duplication or
missing parts.

Flow Control. This is provided on a connection basis to ensure that data
are not delivered at a rate faster than the user's resources can accommodate.

---

*In fact TCP/IP is part of a rather large and popular family of protocols discussed in Section
11.4.

Transmission Control Protocol (TCP, Ref. 8) was the first working version of a transport protocol and was created, as was the rest of the TCP/IP family of protocols, by DARPA (Defense Advanced Research Projects Agency) for its DARPANET. All the features of TCP have been adopted in the ISO version. The applicable ISO references are ISO 8073, called *Transport Protocol Specification*, and ISO 8072, called *Transport Service Definition*.

Many readers will recognize the TCP/IP family of protocols for the part they play in the popular, and nearly ubiquitous, INTERNET.

2.10   PROCEDURAL VERSUS ELECTRICAL

2.10.1   Narrative

Up to this point we have discussed how to connect data facilities together forming a network based on a selected topology (Chapter 1). More emphasis has been placed on procedural issues in Chapters 2, 3, and 4. The objective is to transmit a serial bit stream from an origin to an intended destination. This involves both procedural issues and electrical circuit standards. The protocols which we will be dealing with mostly concentrate on standardized procedures. In the LAN arena, the Medium Access Control (MAC) is so closely allied to its electrical signal that the procedural and electrical cannot be separated and yet still present a reasonable argument.

To meet these requirements, Chapters 5 and 6 present information on basic data transmission, Chapter 7 introduces the reader to voice-channel modems, and Chapter 8 gives an overview of the public switched telecommunications network (PSTN). The PSTN is a major provider of long-distance data circuits. Chapters 9 and 10 pick up on the issues of LAN protocols; Chapters 10 through 13 deal with long-distance data communication including WANs.

REFERENCES

1. James Isaak, Kevin Lewis, Kate Thompson, and Richard Straub, *Open Systems Handbook*, IEEE Standards Press, IEEE, New York, 1994.
2. *Information Technology—Open Systems Interconnection—Basic Reference Model: The Basic Model*, ITU-T Rec. X.200, ITU Geneva, July 1994.
3. *Management Framework for Open Systems Interconnection (OSI) for CCITT Applications*, CCIT Rec. X.700, ITU Geneva, September 1992.
4. *Information Technology—Open Systems Interconnection—Basic Reference Model: Conventions for the Definition of OSI Services*, ITU-T Rec. X.210, ITU Geneva, November 1993.

5. William Stallings, *Handbook of Computer Communications*, Vol. 1, Macmillan, New York, 1987.

6. *Packet-Switched Signaling System Between Public Networks Providing Data Transmission Services*, ITU-T Rec. X.75, ITU Geneva, October 1996.

7. *Interface Between Data Terminal Equipment (DTE) and Data Circuit-Terminating Equipment (DCE) for Terminals Operating in the Packet Mode and Connected to Public Data Networks by Dedicated Circuit*, ITU-T Rec. X.75, ITU Geneva, October 1996.

8. *Transmission Control Protocol (TCO)*, RFC 793, DDN Network Information Center, SRI International, Menlo Park, CA, September 1981.

3

HIGH-LEVEL DATA-LINK CONTROL (HDLC) TYPICAL DATA-LINK LAYER PROTOCOL

3.1 INTRODUCTION

The majority of the data-link layer protocols in use today derive from HDLC (Ref. 1). It was developed by the International Standards Organization (ISO), and the relevant standard is ISO 3309. The following is a partial list of data-link layer standards spawned by HDLC:

1. Advanced Data Communications Control Procedures (ADCCP), developed by ANSI. It has been adopted by the U.S. government for official use. See FIPS Pub. 71-1 and FED-STD-1003A. It is identical to HDLC.
2. Link Access Procedure Balanced (LAPB) operates in the HDLC balanced mode only, otherwise it is almost identical to HDLC. It is supported by the ITU-T Organization and is incorporated as part of the ITU-T Rec. X.25 document (Ref. 2). See Section 11.3.2.
3. Link Access Procedure D-Channel (LAPD). Originally designed for the ISDN D-channel, Basic Rate Service. Variants of LAPD are used in the frame relay service (see Section 13.7).
4. Synchronous Data-Link Control (SDLC) was developed by IBM, particularly for SNA, and has wide use in that arena.
5. Logical Link Control (LLC) resides in the upper sublayer 2 of all IEEE 802 and FDDI LANs. See Section 9.4.

In the next section, we define some terms used in HDLC.

3.2   STATIONS AND CONFIGURATIONS

Primary Station. A logical primary station is an entity that has primary link control responsibility. It assumes responsibility for organization of data flow and for link level error recovery. Frames issued by a primary station are called *commands*.

Secondary Station. A logical secondary station operates under control of a primary station. It has no direct responsibility for control of the link but instead responds to primary station control. Frames issued by a secondary station are called *responses*.

Combined Station. A combined station combines the features of primary and secondary stations. It may issue both commands and responses.

Unbalanced Configuration. An unbalanced configuration consists of a primary station and one or more secondary stations. It supports both full-duplex and half-duplex operation and point-to-point and multipoint circuits.

Balanced Configuration. A balanced configuration consists of two combined stations in which each station has equal and complementary responsibility of the data link. A balanced configuration operates only in the point-to-point mode and supports full-duplex and half-duplex operation. Figure 3.1 shows HDLC stations and configurations.

There is also a *symmetric configuration* that connects two unbalanced stations multiplexed through a single point-to-point connection. Transmission may be in the half- or full-duplex mode over nonswitched or switched facilities. There are logically two separate streams of command and response frames. A station may be configurable in that it can operate as a different type in a different environment.

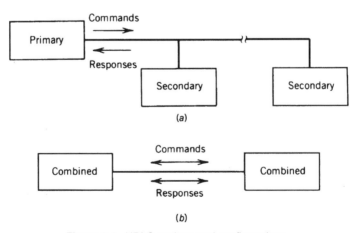

Figure 3.1. HDLC stations and configurations.

3.3 MODES OF OPERATION USED WITH HDLC

In order for a data-link mechanism to provide the needed services, different relationships between the involved stations must be established. HDLC provides for a variety of relationships as well as the means to change them in the most appropriate ways.

Response modes establish a relationship between stations that provides for information transfer. The response modes must be agreed upon before information can be transferred between stations.

Three data transfer modes are described:

Normal Response Mode (NRM). In this mode a primary station initiates data transfer to a secondary station. A secondary station transmits data only in response to a poll from a primary station. This mode of operation applies to an unbalanced configuration.

Asynchronous Response Mode (ARM). In this mode a secondary station may initiate transmission without receiving a poll from a primary station. It is useful on a circuit where there is only one active secondary station. The overhead of continuous polling is thus eliminated.

Asynchronous Balanced Mode (ABM). This is a balanced mode that provides symmetric data transfer capability between combined stations (Figure 3.1b). Each station operates as if it were a primary station, can initiate data transfer, and is responsible for error recovery. One application of this mode is hub polling, where a secondary station needs to initiate transmission.

3.4 HDLC FRAME STRUCTURE

Figure 3.2 shows the format of an HDLC frame. Moving from left to right in the figure in the order of transmission, we have the *flag field* (F), which delimits the frame at both ends with the unique bit pattern 01111110. If frames are sent sequentially, the closing flag of the first frame is the opening flag of the next frame (see Section 2.9.2).

The flag field is a special bit pattern, and receiving stations are constantly searching for the flag bit sequence to mark the beginning and end of a frame. There is no reason why this same sequence cannot appear somewhere in midframe. This, then, would incorrectly tell a receiving station that the frame has ended and that a new frame has begun. Of course, the misinterpretation corrupts the entire frame and probably some subsequent frames. To avoid this problem, no sequence of six consecutive 1s bracketed by 0s is permitted in other than the flag field locations.

Let's consider this a little more carefully. To ensure transparent transmission of everything between the delimiting flags, a *zero insertion* scheme is employed. This prevents a flag sequence from appearing inside a frame

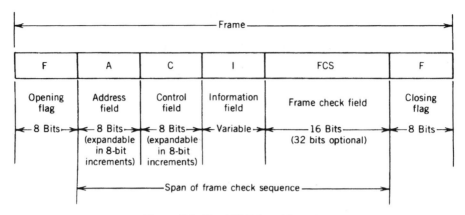

Figure 3.2. The HDLC frame format.

where it would be erroneously interpreted as an end of frame delimiter. There are three cases that must be dealt with: (a) a valid flag pattern, (b) a frame abort pattern (seven or more consecutive 1s), and (c) data that can look like either. The transmitter will insert a zero after 5 consecutive 1s in the data stream. The receiver will monitor the data stream and will look for a pattern of a binary 0 and 5 binary 1s. If the next bit is a binary 0, it is discarded and the binary 1s are passed as data. If the sixth bit was a binary 1, the next bit is tested. If it is a binary 0, a flag is received. If it was a binary 1, the frame reception is aborted. Transmission of a frame is aborted by sending between 7 and 14 consecutive binary 1s in the data stream. An abort sequence will cause the receiver to discard any accumulated data and ready itself for new frame reception.

By the use of zero insertion, HDLC is *transparent* to any 8-bit sequence.

The *address field* (A) immediately follows the opening flag of a frame and precedes the control field (C). Each station in a network normally has an individual address and a group (global) address. A group address identifies a family of stations. It is used when data messages must be accepted from or destined to more than one user. Normally, the address is 8 bits long, providing 256 bit combinations or addresses. In HDLC and ADCCP the address field can be extended in increments of 8 bits. When this is implemented, the least significant bit is used as an extension indicator. When that bit is 0, the following octet is an extension of the address field. The address is terminated when the least significant bit of an octet is 1. It thus can be seen that the address field can be extended indefinitely.

Commands always carry the address of the destination, and responses always carry the address of the source.

The *control field* (C) immediately follows the address field (A) and precedes the information field (I). The control field conveys commands, responses, and sequence numbers to control the data link. The basic control

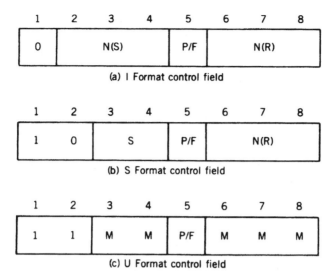

Figure 3.3. The three control field formats of HDLC.

field is 8 bits long and uses modulo-8 sequence numbering. There are three types of control field: (i) I frame (information frame), (ii) S frame (supervisory frame), and (iii) U frame (unnumbered frame). The three control field formats are shown in Figure 3.3.

The control field may be extended by the addition of a second contiguous octet immediately following the basic field. The extension increases the modulus count to 128. The extended control field is often used on long propagation delay circuits such as geostationary satellite circuits, basically to extend the sequence modulus number.

Consider the basic 8-bit form of the control field as shown in Figure 3.3. The information follows from left to right. If the first bit in the control field is a binary 0, the frame is an I frame (Figure 3.3a). If the bit is a 1, the frame is an S frame or a U frame (Figures 3.3b and 3.3c). If the first bit is a 1 followed by a 0, it is an S frame; and if the bit is a 1 followed by a 1, it is a U frame. These bits are called *format identifiers*.

3.4.1 Sequence Numbering in HDLC

Turning now to the I figure (Figure 3.3a), its purpose is to carry user data. Here bits 2, 3, and 4 of the control field in this case carry the *send* sequence of transmitted messages (I frames). Sequence numbers are used to detect lost or out-of-sequence frames. This provides one check on the integrity of the received data. Sequence numbers are controlled by send and receive state variables and are represented by send and receive sequence numbers that appear in the control field.

Each information frame is sequentially numbered and may have the value 0 through MODULUS minus 1, where MODULUS is the MODULUS of the sequence number. Modulus equals 8 for the basic control field and 128 for the extended control field. The sequence numbers cycle through the entire range.

The maximum number of sequentially numbered information format frames that a station may have outstanding (i.e., unacknowledged) at any given time may never exceed one less than the modulus of the sequence numbers. This restriction is to prevent any ambiguity in the association of transmission frames with sequence numbers during normal operation and/or error recovery actions.

Conceptually, sequence number control is often represented as a "window" that is *open* to the value equal to the number of unacknowledged frames allowed before transmission is halted (flow control). The two variables and two numbers involved are as follows:

Send State Variable $V(S)$. This is the sequence number of the next in-sequence information frame to be transmitted.

Send Sequence Number $N(S)$. The value of $V(S)$ is inserted in the control field of a transmitted information frame as $N(S)$.

Receive State Variable $V(R)$. This denotes the sequence number of the next expected in-sequence information frame to be received.

Receive Sequence Number $N(R)$. The value of $V(R)$ is inserted into the control field of a transmitted frame to indicate that frames numbered $N(R) - 1$ and lower have been received.

To describe *window*, consider data link X to Y. The receiver Y has seven buffers (the number was selected arbitrarily). Thus Y can accept seven frames (or messages) and X is allowed to send seven frames without acknowledgment. To keep track of which frames have been acknowledged, each is labeled with a sequence number 0–7 (modulo 8). Station Y acknowledges a frame by sending the next sequence number expected. For instance, if Y sends a sequence number 3, this acknowledges frame number 2 and is awaiting frame number 3. Such a scheme can be used to acknowledge multiple frames. As an example, Y could receive frames 2, 3, and 4 and withhold all acknowledgments until frame 4 arrives. By sending sequence number 5, it acknowledges the receipt of frames 2, 3, and 4 all at once. Station X maintains a list of sequence numbers that it is allowed to send, and Y maintains a list of sequence numbers it is prepared to receive. These lists are thought of as a *window of frames*.

HDLC allows a maximum window size of 7, or 127 frames. In other words, a maximum number of 7, or 127 unacknowledged frames, can be sent, or one less than the modulus 8 or 128. $N(S)$ is the sequence number of the next frame to be transmitted, and $N(R)$ is the sequence number of the frame to be received.

3.4.2 The Poll/Final Bit

Each frame carries a poll/final (P/F) bit. It is bit 5 in each of the three different types of (basic) control fields, and it is bit 9 in the extended control field. The P/F bit serves a function in both command and response frames. In a command frame it is referred to as a poll (P) bit, and in a response frame it is referred to as a final (F) bit. In both cases the bit is set to 1. The P bit is used to solicit a response or sequence of responses from a secondary or a balanced station. On a data link only one frame with a P bit set to 1 can be outstanding at any given time. Before a primary or a balanced station can issue another frame with a P bit set to 1, it must receive a response from a secondary or a balanced station with the F bit set to 1. In the NRM mode, the P bit is set to 1 in command frames to solicit response frames from the secondary station. In this mode of operation the secondary station may not transmit until it receives a command frame with the P bit set to 1.

The F bit is used to acknowledge an incoming P bit. A station may not send a final frame without prior receipt of a poll frame. As can be seen, P and F bits are exchanged on a one-for-one basis. Thus only one P bit can be outstanding at a time. As a result, the $N(R)$ count of a frame containing a P or F bit set to 1 can be used to detect sequence errors. This capability is called *check pointing*. It can be used not only to detect sequence errors but to indicate the frame sequence number to begin retransmission when required.

3.4.3 Supervisory Frames

Supervisory (S) frames, shown in Figure 3.3b, are used for flow and error control. Both go-back-n and selective ARQ can be accommodated.* There are four types of supervisory or (S) frames:

S Frame	Bits 1 to 8
1. Receive ready (RR)	1000 P/F $N(R)$
2. Receive not ready (RNR)	1001 P/F $N(R)$
3. Reject (REJ)	1010 P/F $N(R)$
4. Selective reject (SREJ)	1011 P/F $N(R)$

The RR frame is used by a station to indicate that it is ready to receive information and acknowledge frames up to and including $N(R) - 1$. Also a primary station may use the RR frame as a command with the poll (F) bit set to 1.

The RNR frame tells a transmitting station that it is not ready to receive additional incoming I frames. It does acknowledge receipt of frames up to

---

Go-back-n and *selective ARQ* are forms of error recovery. Errors and error recover are described in Section 4.4.

and including sequence number $N(R) - 1$. I frames with sequence number $N(R)$ and subsequent frames, if any, are not acknowledged. The REJ frame is used with go-back-n ARQ to request retransmission of I frames with frame sequence number $N(R)$, and $N(R) - 1$ frames and below are acknowledged.

3.4.4 Unnumbered (U) Frames

The unnumbered (U) frames are used for a variety of control functions. They do not carry sequence numbers, as the name indicates, and do not alter the flow of sequencing I frames. Unnumbered (U) frames can be grouped into the following four categories:

1. Mode-setting commands and responses
2. Information transfer commands and responses
3. Recovery commands and responses
4. Miscellaneous commands and responses

There are also *unnumbered information* (UI) frames that furnish a means of providing connectionless service at the data-link level. The UI frames are transmitted without disturbing the $N(S)$ and $N(R)$ values used in normal I frame transmissions. Because of this, the sequencing and verification facilities provided by I, RR (receive ready command), SREJ (selective reject command), and so on, are not applicable. The UI frame may be lost or duplicated without any notification to the sender. If a more reliable environment for UI is needed, similar mechanisms to the normal I frames must be constructed at an OSI layer higher than the data-link layer. The UI frames can be used for higher layer procedures or for interruption of the normal data transfer mechanisms without disturbing them. Prioritized or real-time transmissions can be accommodated by the UI frame.

3.4.5 Information Field

The *information field* (I) follows the control field, as illustrated in Figure 3.2. The information field is present only in information (I) frames and some unnumbered (U) frames. The I field may carry any number of bits in any code, related to character set structure or not. Its length is not specified in the ISO underlying standard [ISO 3309 (Ref. 1)]. Specific system implementations, however, usually place an upper limit on I field size. Some implementations of HDLC require that the I field contain an integral number of octets.

3.4.6 Frame Check Sequence (FCS) Field

The *frame check sequence field* (FCS) follows the information field (or the C field if there is no I field) and is carried in each frame. The FCS field detects errors due to transmission. The FCS field contains 16 bits, in most implementations, which are a result of a mathematical computation on the digital value of all bits excluding inserted 0s (zero insertion) in the frame and including the address, control, and information fields.

The CRC used in HDLC is based on the CRC-16 FCS developed by CCITT in Rec. V.41 (Ref. 7) using generating polynomial $X^{16} + X^{12} + X^5 + 1$. It should be noted, however, that in some situations that require more stringent undetected error conditions and/or because of frame length, a 32-bit CRC (FCS) may be used. This is similar to the 32-bit CRC used with LANs, described in Chapter 9.

Error recovery and the operation of the FCS/CRC are described in Section 4.4.

3.5 COMMANDS AND RESPONSES

Table 3.1 gives a listing of commands and responses used in the HDLC family of protocols.

3.5.1 Mode-Setting Commands

A set of eight mode-setting commands are described below. These are used to establish the mode of the connected receiving station. When an *E* is appended to the end of the abbreviation, it indicates an extended mode. These commands, of course, are used in the U format, and the mode setting information is carried in the M or modifier bits shown in Figure 3.3c, bits 3, 4, 6, 7, and 8.

Set normal response mode (*SNRM*) command is used to place the addressed secondary station in the normal response mode (NRM) where all control fields are one octet in length. This command places the secondary receiving station under control of the transmitting primary station. When in the normal response mode, no unsolicited transmissions are allowed from a secondary station. Bit sequence 1100P001.

Set asynchronous response mode (*SARM*) command places the addressed secondary station in asynchronous response mode (ARM). Bit sequence 1111P000.

Set asynchronous balanced mode (*SABM*) places the addressed balanced station in the ABM. Bit sequence 1111P100.

There are three extended modes. SNRME, SARME, and SABME are the previous three mode commands extending control fields to two octets in length. Bit sequences: SNRME 1111P011; SARME 1111P010; and SABME 1111P110.

SIM or set initialization mode command starts up system-specified link-level initialization procedures at a remote station.

TABLE 3.1 HDLC Commands and Responses

Format	Control Field Bit Encoding								Commands		Responses	
	1	2	3	4	5	6	7	8	(abbr.)		(abbr.)	
Information	0	—	N(S)	—	*	—	N(R)	—	I	Information	I	Information
Supervisory	1	0	0	0	*	—	N(R)	—	RR	Receive ready	RR	Receive ready
	1	0	0	1	*	—	N(R)	—	REJ	Reject	REJ	Reject
	1	0	1	0	*	—	N(R)	—	RNR	Receive not ready	RNR	Receive not ready
	1	0	1	1	*	—	N(R)	—	SREJ	Selective reject	SREJ	Selective reject
Unnumbered	1	1	0	0	*	0	0	0	UI	Unnumbered information	UI	Unnumbered information
	1	1	0	0	*	0	0	1	SNRM	Set normal response mode		
	1	1	0	0	*	0	1	0	DISC	Disconnect	RD	Request disconnect
	1	1	0	0	*	1	0	0	UP	Unnumbered poll		
	1	1	0	0	*	1	1	0	Nonreserved 0		UA	Unnumbered acknowledge
	1	1	0	1	*	0	0	1	Nonreserved 1		Nonreserved 0	
	1	1	0	1	*	0	1	0	Nonreserved 2		Nonreserved 1	
	1	1	0	1	*	1	0	0	Nonreserved 3		Nonreserved 2	
	1	1	1	0	*	0	0	0	SIM	Set initialization mode	RIM	Request initialization mode
	1	1	1	0	*	0	0	1	SARM	Set async response mode	FRMR	Frame reject
	1	1	1	1	*	0	0	1	RSET	Reset	DM	Disconnect mode
	1	1	1	1	*	0	1	0	SARME	Set arm extended mode		
	1	1	1	1	*	0	1	1	SNRME	Set NRM extended mode		
	1	1	1	1	*	1	0	0	SABM	Set async balanced mode		
	1	1	1	1	*	1	0	1	XID	Exchange identification	XID	Exchange identification
	1	1	1	1	*	1	1	0	SABME	Set ABM extended mode		

*P/F bit. P or F, "1" or "0" depending on whether command or response.

Disconnect (*DISC*) command performs a logical disconnect. It informs the receiving station that the transmitting station is suspending operation with the receiving secondary or balanced station. The DISC command may also be used to initiate a physical disconnect in a switched network. Secondary stations remain disconnected until receipt of an SNRM or SIM command.

3.5.2 Miscellaneous Commands

There are four miscellaneous commands, which are described below.

Unnumbered poll (UP) command solicits response frames from one or a group of secondary stations to establish a logical operation condition. Bit sequence: 1100P100.

TEST command is used to test the data-link layer. It causes the addressed station to reply with a TEST response. Bit sequence: 1100P111.

RSET or reset command is used by a combined station to reset the receive state variable and the FRMR condition. Bit sequence: 1111P001.

The exchange identification (XID) command causes the addressed station to report its station identification. It can be used optionally to provide the station identification of the transmitting station to the remote station. With this command an information (I) field is optional. If it is used, it will contain the identification (ID) of the transmitting station. If the unique address of the secondary station is unknown, the XID command may use the global (group) address. Bit sequence: 1111P101.

3.5.3 Responses to Unnumbered Commands

The request disconnect (RD) is used to indicate to the remote station that the transmitting station wishes to be placed in a disconnected mode, or it is used to request a physical disconnect operation on switched networks. Bit sequence: 1100F010.

The unnumbered acknowledge (UA) command acknowledges receipt and acceptance of numbered commands. Bit sequence: 1100F110.

The TEST response replies to a TEST command in any mode. Bit sequence: 1100F111.

The request for initialization (RIM) is used to inform another station of the need of a SIM command. No command transmissions are accepted until the RIM condition is reset by receipt of a RIM or DISC. The receipt of any command except a RIM or DISC will cause the station to repeat the RIM. Bit sequence: 1110F000.

When there is an errored condition that is not recoverable by retransmission of the errored frame, the frame reject response (FRMR) is used. These conditions are as follows: a command that is not implemented or is invalid, an information field which exceeds maximum length, and an invalid receive sequence number. Bit sequence: 1110F001.

The disconnected mode (DM) response reports a nonoperational status where the station is logically disconnected from the link. The DM response requests the remote station to issue a set mode command or, if sent in response to the reception of a set mode command, to inform the addressed primary station that the transmitting station still cannot respond to the set mode command. Bit sequence: 1111F000.

The exchange identification (XID) response replies to an XID command. An information field containing the ID of the transmitting secondary station is optional. Bit sequence: 1111F101.

Note: In Sections 3.5.1 through 3.5.3, where a control frame bit sequence was given, the bit position 5 was a letter, either a P or an F. The incisive reader would have deduced correctly that bit 5 was the P/F bit. If a P were given, then the P/F bit would be in the "poll" state; where it was F, of course it would be in the "final" state.

3.6   FRAME OPERATION

An HDLC link is in an *active state* when a station is transmitting frames, an abort sequence, or interframe time fill. When in an active state, the transmitting station reserves the right to continue transmission.

An HDLC link is in an *idle state* when it is not in an active state. An idle state is identified by the detection of a continuous transmission of 1s that persists for 15 or more bit times. This is called the *idle sequence*.

When a transmitting station wishes to maintain a link in the active state but without traffic, it uses the *interframe time fill* to maintain link synchronization. Time fill may also be used to avoid timeouts and to hold the authority to transmit. Interframe time fill consists of a series of contiguous flag sequences. They immediately follow the closing flag of the last frame transmitted and follow the opening flag when the next frame is to be transmitted.

Abort may be carried out during the transmission of a frame when the transmitting side decides, before the end of a frame, to terminate that frame and to force the receiver to discard the frame in question. The aborting is accomplished by transmitting at least 7 but less than 15 consecutive 1s with no zero insertion. The receipt of 7 contiguous 1s is interpreted as an abort.

An *invalid frame* is one that is not properly bounded by an opening and closing flag or one that is too short in length (i.e., less than 32 bits between flags). An aborted frame is an invalid frame. A receiving station ignores invalid frames.

In HDLC, bits are transmitted low-order bit first (a bit with a weighting of $2°$). This *order of bit transmission* is valid for addresses, commands, responses, and sequence numbers. The order of bit transmission for data contained within the information field is application-dependent and is not specified. The order of bit transmission for the FCS is most significant bit first.

3.7   ERROR RECOVERY

Checkpoint Recovery. This is the basic error recovery procedure in HDLC and in related bit-oriented protocols. Checkpoint error recovery uses poll and final frames to effect recovery. This recovery procedure is used when neither REJ nor SREJ reconvert is implemented. It also serves as a fall-back method for both REJ and SREJ because it detects unsuccessful attempts at REJ/SREJ recovery such as an errored REJ frame.

A checkpoint can be in information (I), supervisory (S), or unnumbered (U) format frame with the P or F bit set to 1. Each "poll" frame forces a "final" frame response. Another poll may be issued only when an error-free final frame is received in response. In the case of a lost poll or final frame, a poll may be issued after a system-defined timeout.

A checkpoint exchange is a pair of consecutive checkpoints—that is, a poll frame and its responding final frame.

When a receiving station receives a frame with the P/F bit set to 1, it starts retransmission of all unacknowledged I frames with sequence numbers less than the send variable (S) at the time the previous P/F frame was transmitted. Retransmission of such frames starts with the lowest numbered unacknowledged information (I) frame, and the frames are retransmitted sequentially. The term "checkpoint retransmission" is the retransmission of I frames.

Duplicate transmissions are avoided during a checkpoint cycle by any of the following conditions:

(a) If an REJ frame with POLL or FINAL set in the OFF state has already been received and acted upon

(b) If a combined station receives and acts on any REJ except one with the FINAL in the ON condition

(c) Recovery is also stopped by an SREJ or any unnumbered format with the POLL/FINAL in the ON condition

(d) An acknowledgment of any frame with POLL/FINAL in the ON condition before the next checkpoint occurs.

Retransmission of I frames can be specified by a station returning an RR which has an $N(R)$ less than the $N(S)$ number the sending station is using. Retransmission will begin with the indicated number.

Timeout Recovery. When the expected acknowledgment has not been received of a previously transmitted frame, timeout functions take over. Expiration of the timeout function causes initiation of timer recovery.

Suppose a receiving station, due to transmission error, does not receive (or receives and discards) a single I frame or the last I frame(s) in a sequence of I frames. It will not detect an out-of-sequence exception and therefore will not transmit SREJ/REJ. Following the completion of a system specified timeout period, the station that transmitted the unacknowledged I frame(s) takes the appropriate timeout recovery action to determine the sequence number at which retransmission must begin.

Generally, a recovery action results in the station detecting the timeout to send a status request. The receiving station responds with a frame indicating the value of its receive variable indicating the point at which retransmission begins.

Reject Recovery. Reject recovery is used to initiate faster retransmission response than is possible with checkpoint recovery following the detection of a sequence error. Its use is typical on full-duplex circuits with their fast turnaround.

Only one REJ exception condition may be established between two stations at any given time. The REJ command/response may only be transmitted once with a checkpoint cycle. The REJ exception condition is cleared (i.e., reset) upon receipt of an I frame with an $N(S)$ number equal to the $N(R)$ number of the REJ command response.

Selective Reject Recovery. Selective reject recovery provides more efficient error recovery by requesting the retransmission of a single information frame following the detection of a sequence error rather than requesting go-back-n frames and repeating all subsequent frames. It simply invokes selective ARQ and is used where only a single I frame is missing. This is determined on receipt of the out-of-sequence $N(S)$ number. When an I frame sequence error is detected, the SREJ is transmitted as soon as permissible. Only one SREJ exception condition may be outstanding at a time between two given stations. A "sent SREJ" exception condition is cleared when the requested I frame is received.

3.8 OTHER STATION MODES

The specifications of response modes were the means of activating the data-link relationship by proposing and agreeing to the rules to be used between stations. This relationship is deactivated by the use of the DISC command to disconnect a secondary/combined station. If the connection was over switched circuits, this can create a physical disconnection. If the station is still physically connected, it is either in normal disconnected mode (NDM) or asynchronous disconnected mode (ADM) based on agreed upon procedures for a network.

A station in disconnected mode has a limited set of possible actions. All commands received without POLL are simply discarded. Any mode setting command, such as SARM, is responded to in the normal manner. Any commands that are not mode setting and have a POLL will have either DM, RIM, or XID as the responses. These have been described in Section 3.5.3.

The final station, the initialization state, uses the set initialization mode (SIM) command, defined in Section 3.5.1, to provide a means for a secondary station to undergo a system-defined procedure. The procedure will vary according to the type of equipment and network parameters in a given implementation. It is intended to place a secondary station in a known state where it can proceed with data-link control procedures.

There is an unnumbered POLL (UP) that is an optional polling mechanism for a primary station. It can be used with a single secondary station or as a group poll. The group poll procedure is not specified. When the POLL bit is set, the secondary station must respond with at least one (usually RR if no I frames are ready) frame with FINAL bit set in the last frame. If the POLL bit is off, the secondary may respond if it has UI, I, unacknowledged frames, or a change in status to report. The final bit is off in all frames.

3.9 SDLC VARIATIONS WITH HDLC

IBM's SDLC (synchronous data link control) variations with HDLC are as follows:

1. It has no balanced mode of operation.
2. The SDLC loop operation is not an HDLC option.
3. It uses different names for its command and response modes.

Chapter 3 is based on Refs. 1–6.

REFERENCES

1. *High Level Data Link Control (HDLC) Procedures—Frame Structure*, ISO 3309, International Standards Organization, Geneva, 1976.
2. *High Level Data Link Control Procedures—Consolidation of Elements of Procedures*, ISO 4335, International Standards Organization, Geneva, 1977.
3. *High Level Data Link Control Procedures—Consolidation of Classes of Procedures*, ISO 7809, International Standards Organization, Geneva, 1978.
4. John McConnell, *Internetworking & Advanced Protocols*, Network Technologies Group, Inc., Boulder, CO, 1985.
5. Roger L. Freeman, *Reference Manual for Telecommunication Engineering*, 3rd edition, John Wiley & Sons, New York, 2001.
6. William Stallings, *Handbook of Computer Communication Standards*, Vol. 1, Macmillan, New York, 1987.
7. *Code Independent Error Control System*, CCITT Rec. V.41, Fascicle VIII.I, IXth Plenary Assembly, Melbourne, 1988.

4

DATA NETWORK OPERATIONS

4.1 CHAPTER OBJECTIVE

Data communications means to communicate data. The objective of this chapter is to discuss how we can carry out this function effectively. At the outset, we assume that there are two terminals in a data connectivity, a near end of the circuit and a far end. We could call this one-to-one pairing. There is also another possibility, a one-to-many connectivity, sometimes called *broadcast*. For this discussion, data connectivities are on a one-to-one basis.

The *data* to be interchanged consists of 8-bit sequences* whose meanings are understood at both ends of the connection.

This chapter, then, contains a general overview of data operations or the effective interchange of data between two stations. The word *effective* implies that data messages will be delivered to the destination user(s) error-free. The reader should appreciate that no data circuit is completely error-free. Thus the subject of data integrity is treated with considerable detail.

The final section gives an overview of data switching, along with the switching alternatives available to the network designer.

4.2 GENERAL REQUIREMENTS FOR THE INTERCHANGE OF DATA

To efficiently deliver a data message from source to destination, a number of conditions must be met or requirements must be satisfied:

1. A pathway needs to exist between the two ends. The pathway may or may not pass through one or more relay nodes. The pathway may or may not be shared by other users. We assume that only one pathway exists.

*Called octets or bytes.

2. A method to access the pathway must exist.
3. A method must be incorporated in the data message to direct the message to the intended far-end user. The format of the data message must be such that its content is useful to the intended recipient.
4. Data urgency is an issue.
5. There must be some means to assure data integrity at the distant end.

Data network operations involve, as a minimum, these five issues and requirements.

4.3   DISCUSSION OF ISSUES AND REQUIREMENTS

4.3.1   Pathway(s) Exist(s)

If a data connectivity is desired to far end-user Y, then a circuit must exist between near-end local user X and far-end user Y. The circuit may be real or virtual. If we are dealing with a LAN, there is a circuit available. However, it is shared by other users on the LAN. The assumption is, of course, that we are dealing with connection-oriented service.

When dealing with a WAN, the alternatives available are many. Some of these alternatives were covered in Chapter 1. We might lease a T1(DS1) or E1 from the local PSTN, or from some other entity providing equivalent service. Traffic volume may be so great that a private network is warranted. If traffic volume is not so great, a fractional T1(DS1) or E1 may be considered.

A T1(DS1) carries 24 time-division circuits. Each circuit can carry a full-duplex telephone call or 56/64 kbps of full-duplex data. This is called a DS0. If we view that circuit from a user's perspective, it only belongs to that user 1/24th of the time. The other 23/24ths of the time, other users occupy the circuit. Thus we call this a *virtual circuit*. A *real circuit* is one where the physical facilities or an equivalent frequency slot are assigned for use or to a call 100% of the time.

Another option is to lease bit rate capacity from other carriers or quasi-carriers. A quasi-carrier may be the local power company that has extensive telecommunication facilities or a cellular radio company. There are any number of other alternatives.

One such possibility is signing a service level agreement (SLA) for (a) frame relay service (Chapter 12) between the points of interest, (b) X.25 service (Chapter 11), or (c) ATM service with TCP/IP (Chapters 16 and 11). Providers of such services would include (a) the local telephone company (LEC, or local exchange carrier) or (b) interexchange carriers such as AT & T, BT, Telefonica (Spain), Cable and Wireless, MCI, Sprint, and so forth.

Another possibility is to set up a VSAT (very small aperture terminal) network. This is a private-line geosatellite network used for interchange of data traffic.

4.3.2 Access to the Pathway

In nearly all situations, the selected circuit (pathway) is not connected to our data set 100% of the time. Other data stations in the facility may want access to that same circuit. The media access control (MAC) carried out this function with LANs. In the case of a WAN, access will be determined by OSI layers 1 and 2. If the WAN is ATM-based (asynchronous transfer mode; see Chapter 16), other access methods will be employed. In either case (OSI or ATM), the physical layer interface must be available. Physical layer interface is discussed more in detail in Section 6.4. These access functions deal with protocols.

4.3.3 Directing the Data Message to Its Intended User(s)

This function is carried out by *addressing*. Addressing information is carried in the link layer and network layer address fields in message/frame/packet headers. In the case of LANs, there are MAC addresses and LLC (logical link control; see Chapter 9) addresses. The LLC address identifies the final recipients(s) of the message.

For a WAN, Section 3.4 (Chapter 3) discusses the HDLC address field. This, of course, is link layer addressing. The TCP/IP protocol family (Section 11.4) covers an entire realm of network layer addressing. Frame relay and ATM each has its own particular form of addressing. As described in Chapter 2, we can appreciate that addressing is yet another important function of layer 2 and layer 3 protocols.

4.3.4 Data Urgency

When we ship an item by some agency such as Federal Express, you would be asked when you want it to arrive at the destination: tomorrow morning, two-day service, five-working-days delivery, and so forth. There is, in essence, the same situation when dealing with the delivery of data.

The most economic means of shipping data is probably by the Postal Service or United Parcel Service. The data could be stored on a floppy disk or on a CD, which is very easy to ship. Delivery to the desired recipient is *almost* assured. It certainly is cost-effective, if there is not much urgency. Credit card verification is an example of high-urgency data. We may want to consider using Federal Express for long payroll files. The difference in urgency is dramatic.

Consider the situation from still another angle. There is a high investment in a data network. The more data we can push through that network, the better is the return on our investment. This brings up the term *latency*. This is the time it takes a data message to arrive at its destination from its origin. Based on averages, the time it takes a data message to cross the North American continent using packet transmission based on a popular protocol is about 200 ms; when employing frame relay for the same message, it takes about 20 ms.

There are several components in the latency equation:

- Propagation time (on WANs, this will be the largest contributor)
- Time in queue when switching is involved
- Processing time
- Circuit quality (noisy circuits will require message repeats on errored messages)
- Collision resolution on Ethernet LANs
- Circuit setup for virtual circuits or for physical circuits

The longer a message sits in queue or is being serviced by repeats or is delayed for other reasons, the less efficient the circuit becomes. This is because of the time lost when it could be used by other traffic. Both ATM and frame relay are low-latency systems.

4.3.5   Maintenance of Data Integrity

Ideally a data message/frame/packet should arrive at its destination(s) as an identical replica of the message transmitted. The following is a listing of features and parameters of a data message that may be checked to assure integrity:

- The message/frame does not have an integral number of octets (bytes)
- The message/frame has one or more bit errors
- Lost frames
- Out-of-sequence frames
- Incomplete frames (e.g., fields are missing or incomplete, noncompliance with minimum number of octets in information field, header errors, etc.)

In most situations where a frame or message is corrupted, the frame/message is discarded. In the some cases the originator is informed of the nondelivery. However, in most cases, where feasible, a repeat of the corrupted frame is requested.

The most common form of a corrupted frame is one in which one or more errors have occurred. Errors are detected by the receive-end through the processing the FCS field (see Figure 3.2). A certain known bit sequence is expected in the FCS if the frame is error-free. If some other sequence appears, the frame contains one or more bit errors.

Error performance of data circuits lies at the very heart of quality of service of a data network. The reader should consider the impact of just one bit in error in a data file. The goal is an error-free system. Remember this point: *No system is error-free*. Error performance can be improved such that we can approach an error-free condition. In the following sections, we cover some commonly used methods of error detection and error correction.

4.4 ERROR DETECTION SCHEMES

4.4.1 Parity Checks

One technique used in early data transmission systems was a simple parity check. A source code widely employed in data communications is the ASCII code* (see Section 5.3). It is a 7-bit code. There are two possible regimes for error detection: *odd parity* and *even parity*. For parity checking, we count only the "1s" in a bit sequence. In this case the sequence is 7 bits long. An eighth bit is added as a parity bit complying with the convention of odd parity or even parity.

Let's assume now that a particular data circuit is operating under an odd parity rule and that the information sequence is 7 bits long. Consider the following examples:

Example 1. 0101001. There are three 1s, and 3 is an odd number; thus we have complied with the convention. Add a 0 for the eighth bit.

Example 2. 1101100. There are four 1s, an even number. To comply with the odd-parity convention, the eighth bit will be a 1 to make the number of 1s an odd number.

This is called *vertical redundancy checking* (VRC). Its error detection capability is weak. In Example 1 a certain noise burst occurred where a 1 changed to a 0 and a 0 changed to a 1; the convention would be complied with, yet there were two errors that went undetected.

The other type of parity checking is called *longitudinal redundancy checking* (LRC). Generally, LRC was used in addition to VRC to provide a more robust error detection capability. We build a matrix as shown in Figure 4.1. The character bits go from left to right, where the 8th bit is the conventional parity bit derived from VRC. There are seven rows for the seven 7-bit characters, each with an 8th bit appended using odd parity for the row. Seven characters are represented, thus seven rows. An eighth row is derived on the

*ASCII stands for American Standard Code for Information Interchange.

Character #	1	2	3	4	5	6	7	8
1	1	0	1	0	0	1	0	0
2	0	0	0	1	1	1	1	1
3	0	1	1	1	0	1	0	1
4	1	1	0	0	1	1	0	1
5	1	0	0	0	0	1	1	0
6	0	0	0	1	0	0	1	1
7	1	1	1	0	1	1	0	0
Frame parity character	1	0	0	0	0	1	0	1

Figure 4.1. Sample matrix employing VRC and LRC together. The convention here is odd parity.

bottom from longitudinal redundancy checking (LRC), again using the odd-parity checking on each column. We could transmit the 8th row to the distant end where a similar matrix could be constructed to verify that the same 8-bit sequence is derived from LRC. Of course the VRC would also be checked. Using LRC and VRC together reduces the residual error rate from 1 in 100 to 1 in 10,000 (Ref. 1).

It can be demonstrated that VRC/LRC used in tandem leaves a fair amount of residual errors. The next section discusses *cyclic redundancy check*, which has excellent residual error performance.

4.4.2 Cyclic Redundancy Check (CRC)

A more powerful method of error detection involves the use of CRC. It is employed with HDLC (Chapter 3). It is used with messages that are data blocks, frames, or packets. Such a data message can be simplistically represented as shown in Figure 4.2.

The bit sequence in a data frame can be represented by a polynomial. Let's call this the message polynomial, $G(X)$. $G(X)$ is multiplied by X^{16}, which corresponds to shifting the message $G(X)$ 16 places to provide space for the 16 bits in the FCS. We divide the resulting product $[X^{16}(GX)]$ by a known *generating polynomial*. The result of this division is a quotient and a remainder. The remainder is 16 bits long, and it is placed in the FCS slot (Figure 4.2). At the receive side, essentially the same operation is repeated and a remainder is generated. If the remainder generated by the receiver is the same sequence as found in the FCS slot of the received frame, the frame is error-free. If it is not the same, there was at least one bit error in the received frame.

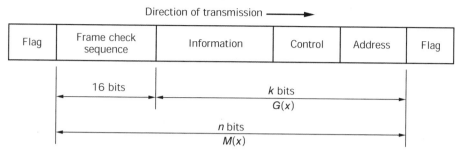

Figure 4.2. Coverage of CRC in a data-link frame (based on HDLC/LAPB). Note that there are 16 bits in the FCS field (Ref. 2).

Note: The description of CRC operation given above is valid for HDLC and its related link layer protocols such as ADCCP and LAPB. In fact, for error-free operation, the remainder is the sequence 0001110100001111. Traditional CRC systems that do not start with a preset value in the shift register (as HDLC does) have an all-zero remainder for error-free operation.

There are two generator polynomials accepted by the industry:

ANSI (CRC16): $X^{16} + X^{15} + X^2 + 1$
CCITT (ITU-T): $X^{16} + X^{12} + X^5 + 1$

ANSI (CRC-16) is applied to synchronous systems that use 8-bit characters. The FCS accumulation is 16 bits. It provides error detection of error bursts up to 16 bits in length. Additionally, more than 99% of error bursts greater than 16 bits can be detected.

The CCITT (ITU-T) standard generator polynomial also operates with 8-bit characters in synchronous bit streams. The FCS accumulation is 16 bits. It provides error detection of bursts up to 16 bits in length. In addition, more than 99% of error bursts greater than 16 bits can be detected. A typical shift register used with this CRC is shown in Figure 4.3.

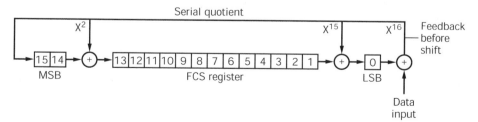

Figure 4.3. Shift register used with CRC-CCITT. MSB, most significant bit; LSB, least significant bit. ⊕, exclusive-OR gate modulo-2 addition.

To derive an algebraic polynomial from a string of bits, review the following: A polynomial is usually represented algebraically by a string of terms in powers of X such as $X^n + \cdots + X^3 + X^2 + X + X^0$ (or 1). In binary form, a 1 is placed in each position that contains a term; absence of a term is indicated by a 0. The convention used in the following presentation is to place the X^0 bit at the right. For example, if a polynomial is given as $X^4 + X + 1$, its binary representation is 10011 (3rd- and 2nd-degree terms are not present). The number 1 in the polynomial represents the X^0 power that was artificially added.

Suppose a message polynomial was 110011. This would be represented by $X^5 + X^4 + X + X^0$ (or $X^5 + X^4 + X + 1$).

4.5 ERROR CORRECTION SCHEMES

4.5.1 Automatic Repeat Request (ARQ)

The term "ARQ" (automatic repeat request) derives from the days of automatic telegraphy. ARQ is widely used to request the sending end to repeat a frame or several frames that have errors. It requires full-duplex or half-duplex operation. There are three variants of ARQ:

- Stop-and-wait ARQ
- Continuous, sometimes called selective, ARQ
- Go-back-n ARQ

Stop-and-Wait ARQ. Stop-and-wait ARQ is the most straightforward to implement and the least complex. However, it is inefficient. With stop-and-wait ARQ, a frame is transmitted to the distant end. At the completion of sending the frame, transmission ceases ("stop"). The receiving terminal processes the frame and then examines the FCS field. If it is the expected sequence for error-free operation, it sends an acknowledgment (ACK) to the sending end, which then promptly transmits the next frame. However, if the receiving end does not encounter the expected sequence in the FCS field as a result of its processing, it sends a negative acknowledgment (NACK) to the transmitting end. The transmitter then resends that frame.

Stop-and-wait ARQ is inefficient primarily due to the "stop" time. This is the wasted period of time waiting for the receive end to process the frame and to send the ACK or NACK. It can be particularly wasteful of circuit time on connectivities with long propagation delays such as geostationary satellite circuits, and there are long frames involved. It is, however, ideal for half-duplex operation.

Selective or Continuous ARQ. Selective or continuous ARQ eliminates the quiescent nonproductive time of waiting and then sending the appropriate signal back after each frame (or packet). With this type of ARQ, the transmit end sends a continuous string of frames (or packets), each with an identifying number in the header of each frame (packet). When the receive end detects a frame in error, it requests the transmit end to repeat that frame. The request that the receiver sends to the transmitter includes the identifying number (more commonly called *sequence number*) of the frame in error. The transmit end retrieves that frame from its buffer storage and retransmits it, with the appropriate identifying number, as another frame in the continuous string. The receive end receives the repeated frame (out of sequence order), reruns the processing procedure to ensure error-free reception, and then places the frame in proper numerical sequence for delivery to the end-user facility. Obviously, selective or continuous ARQ is more efficient regarding circuit usage. It requires full-duplex operation. Such operation uses bookkeeping at both ends of the link. It requires more buffer storage at both ends, but the cost of memory continues to erode.

Go-Back-n ARQ. Go-back-n ARQ is probably the most popular of the three varieties of ARQ. It, too, permits continuous frame transmission. As with continuous ARQ, when an errored frame is encountered, the receiving end informs the far-end transmitter by the return channel. The errored frame is identified by sequence number. The transmitter repeats the frame that was in error and all subsequent frames, even though they have been received before. This approach alleviates the problem at the receive end of inserting the errored block in its proper sequence. With this type of ARQ, buffers must be capable of storing two or three frames, depending on the propagation delay of the circuit and the frame length.

The reader should turn back to Section 3.4.1 for a description of how ARQ is implemented in HDLC. It will be noted that HDLC uses a go-back-n strategy with its window of frames and method of acknowledgment.

4.5.2 Forward Error Correction (FEC)

FEC is a method of error correction, or, if you will, a method of error reduction at the far-end receiver. No return channel is required. The concept is based on adding systematic redundancy at the transmit end of a link such that errors caused by the transmission medium can be corrected at the receiver by means of a decoding algorithm. The amount of redundancy is dependent on the type of code selected and the level of error correction capability desired. Unless an additional processing step is included, FEC will only correct random errors (i.e., not burst errors).

Forward error correction uses *channel coding* whereas the coding that many readers are familiar with (Section 5.3—ASCII code, EBCDIC, etc.) is *source coding*. The channel coding used in FEC can be broken down into two broad categories of codes: *block codes* and *convolutional codes*. Modulo-2 addition is widely used in both error detection and error correction. This is

Figure 4.4. Simplified diagram of FEC operation.

the same as binary addition but without carries. It is just an exclusive-OR operation (Figure 4.3).

Example

$$\begin{array}{r} 1111 \\ + 1010 \\ \hline 0101 \end{array}$$

A measure of error detection and correction capability of a specific code is given by the *Hamming distance*. The distance is the minimum number of digits in which two encoded words or bit sequences differ. Consider these binary sequences: 1010110 and 1001010. The Hamming distance is 3. If the Hamming distance between any two bit-sequences of a code is $e + 1$, then it is possible to *detect* the presence of e or fewer errors in a received sequence of binary words.

If the Hamming distance between any two words of a code is equal to $2e + 1$, then it is possible to *correct* any e or fewer errors occurring in a received binary sequence. A code with a minimum Hamming distance of 4 can correct a single error and detect two symbols in error (Ref. 3).

Figure 4.4 illustrates a digital communication system with FEC. The binary data source generates information bits at R_S bits per second. The information bits input to an FEC encoder with a code rate R. The output of the encoder is a serial bit stream with a symbol rate of R_C symbols per second. The code rate R is related to the encoder input and output by the following expression:

$$R = R_S/R_C$$

The code rate R is the ratio of the number of information bits to the number of encoded symbols for binary transmission.

Example 1. If the code rate were $\frac{1}{2}$ and the output of the data source were 14,000 bits per second, what would be the output of the encoder in symbols per second? (The symbol and bit are synonymous.) Answer = 28,000 symbols/second.

$$R = 14,000/28,000 = 1/2$$

Example 2. The output of the encoder is 36,000 symbols per second and the input to the encoder is 24,000 bits per second, what is the code rate (R)?

$$24,000/36,000 = 2/3$$

The encoder output sequence is then transmitted to the line via a modulator (e.g., modem) or modulator-transmitter. At the receive end, demodulation is performed to recover the coded symbols at R_C symbols per second. These symbols are then fed to a decoder with an output of R_S bits per second, which are then delivered to the data destination.

The major advantages of an FEC system are as follows:

- No feedback channel is required as with an ARQ system.
- There is constant information throughput (i.e., no stop-and-wait gaps).
- Decoding delay is generally small and constant.
- It is ideal for use on marginal transmission channels.

There are two basic disadvantages with an FEC system. To effect FEC for a fixed information bit rate, the bandwidth must be increased to accommodate the redundant symbols. For example, a rate $\frac{1}{2}$ FEC system would require twice the bandwidth of its noncoded counterpart. However, trellis coding, described later, ameliorates some of the bandwidth expansion requirement. There is also the cost and complexity of the added encoder and decoder. Nevertheless, with present VLSI/VHSIC technology, an encoder and decoder can be mounted on a chip, and they are readily available off-the-shelf for some of the more popular encoding/decoding configurations.

4.6 DATA SWITCHING

There are three approaches to data switching:

1. Circuit switching
2. Message switching (store-and-forward)
3. Packet switching

Plain old telephone service, affectionately called POTS, uses circuit switching. A data switch can operate in a similar manner. It will set up a circuit between origin and destination based on the message header. The circuit may be a physical circuit if the network is analog, or it may be a virtual circuit if the network is digital. A virtual circuit is a time slot or a series of time slots, possibly a different time slot on each link as a *virtual connection* is set up end-to-end. The virtual connection remains intact during the duration of a call or message, in the case of data circuit switching.

Message switching, often called store-and-forward switching, accepts data messages from originators, stores the messages, and then forwards each message to the next node or destination when circuits become available. We wish to make two points when comparing circuit and message switching.

Circuit switching provides end-to-end connectivity in near real time. Message switching does not. There is usually some delay as the message makes its way through the network to its destination. The second point deals with efficient use of expensive transmission links. A well-designed message-switching system keeps a fairly uniform traffic load throughout the working day and even into the night on each link. Circuit-switching systems, such as encountered with the public switched telecommunication network (PSTN), are designed for busy-hour loading, and the system tends to loaf when off the busy hour. Message switching, therefore, makes much more efficient use of expensive transmission links than does circuit switching. This improved efficiency is at the expense of some delay in service.

Telex is a good example of message switching. Another example is the older "torn-tape" systems. Such systems were labor-intensive and rather primitive by today's standards. With a torn-tape switching center, messages arriving from originators or relay nodes used paper tape to copy the messages, usually with the five-element ITA #2 code (Chapter 5) at somewhere between 50 and 75 bps. Headers were printed out laterally along the tape holes. Operators at the torn-tape center would read the header, tear off the tape, and then insert that same tape at the tape reader of the indicated outgoing circuit. Tape would often overflow onto the floor at either the receive terminal, when operators were very busy, or at the read head, when circuits were busy. Higher-priority messages moved up to the head of the queue. I've seen clothespins and paper clips used to piece tapes together in such queues.

Store-and-forward systems today use magnetic tape or other electronic storage rather than paper tape. A processor arranges messages by priority in the correct queue. Message processors "read" precedence and address information from the header and automatically pass the data traffic to the necessary outgoing circuits. In older systems, the floor was a storage bin of almost infinite size. In modern systems, a concern remains on memory overflow during peak traffic conditions.

Packet switching utilizes some of the advantages of message switching and some of circuit switching. It mitigates some of the disadvantages of both. A data "packet" is a comparatively short block of message data of fixed length. Each packet, of course, has its own header and FCS at the end of the packet. These packets may be sent over diverse routes to their eventual destination, and each packet is governed by an ARQ-type error-correction protocol. Because these packets are sent on diverse routes, they may not arrive at the far-end receiving node in proper sequential order. Thus the far-end node must have the capability of storing the packets and rearranging them in proper sequential order. The destination node then reformats the message as it was sent by the originator and forwards it to the final-destination end user.

A packet-switching network can show considerably greater efficiency when compared to circuit- and message-switching networks. There is a one caveat, however. For efficient packet-switching network operation, a multiplicity of paths (three or more) must exist from the originating local node to the destination local switching node. Well-designed packet-switching systems can reduce delivery delays when compared to conventional message switching. Expensive transmission facilities tend to be more uniformly loaded when compared with networks based on other data-switching methods. Adaptive routing becomes possible where a path between nodes has not been selected at the outset. Nodes are kept current on load conditions, and alternative-route traffic is directed to more lightly loaded trucks. Delivery delay is also reduced when compared to message switching because a packet switch starts forwarding packets before receipt of all constituent packets making up a message. Again, path selection is a dynamic function of real-time conditions of the network. Packets advancing through the network can bypass trunks and nodes that are congested or that have failed.

A packet switch is a message processor. Packets are forwarded over optimum routes based on route condition, delay, and congestion. It provides error control and notifies the originator of packet receipt at the destination. Packet switching approaches the intent of real-time switching. Of course, there is no *real* connection from originator to destination. The connection may be classified as one form of a *virtual* connection.

Table 4.1 summarizes the advantages and disadvantages of these three methods of data switching.

4.6.1 Philosophy of "Data Signaling" Versus Telephone Signaling

Data are delivered to their distant-end destination either by connection-oriented service or by connectionless service. Conventional telephony operates in a connection-oriented service by circuit switching described above. During call setup a circuit is established from the calling user to the called user. A signaling sequence is sent just once during the circuit setup period by means of address signaling. The circuit is held in its connected condition by means of supervisory signaling until one or the other party goes "on-hook" (i.e., hangs up).

In fact, frame relay and ATM have entire separate signaling regimes for setting up a permanent virtual circuit (PVC) and to maintain "health and welfare" of the circuit when it is quiescent.

When data are to be communicated, there are several ways to effect the connectivity for the communication:

1. A PVC is set up for the duration of the data call.
2. A PVC exists between each end of the call and there is no setup or clear-down of the call.
3. A switched virtual circuit (SVC) is used.

TABLE 4.1 Summary of Data Switching Methods

Switching Method	Advantages	Disadvantages

Circuit switching	Mature technology
	Near real-time
	  connectivity
		High cost of switch
		Lower system utilization,
		  particularly link
		  utilization

	Excellent for inquiry and response	

	Leased service
	  attractive
		Privately owned service
		  can only be justified
		  with high traffic
		  volume

Message switching	Efficient trunk
	  utilization
		Delivery delay may be a
		  problem
		Not viable for inquiry and
		  response

	Cost-effective for
	  low-volume leased
	  service
		Survivability
		  problematical
		
		Requires large storage
		  buffers

Packet switching	Efficiency	Multiple route and node network expensive
	Approaches near real- time connectivity	Processing intensive

	Highly reliable,
	  survivable
		Large traffic volume
		  justifies private
		  ownership

	Low traffic volume
	  attractive for
	  leased service	

4. Connectionless service is provided. Here data are communicated by packets, called datagrams in many standards. The circuits involved are usually SVCs.

5. A physical circuit is set up or in place.

A major difference between data communication and conventional telephony communication is that data communication has a header on each and every frame, packet, or ATM cell (see Chapter 16). In other words, address signaling is repeated over and over again, whereas in telephony it is only transmitted once to set up the call.

Frame relay (Chapter 7) has made an attempt to reduce the burden of the large header, yet its data-link connection identifier (DLCI) can be thought of in the light of a conventional header, directing the frame through the network to its proper destination. Comparatively, IP uses very long headers.

Data frame headers, along with the trailing FCS field, are overhead. The info field is the payload. It carries the revenue-bearing portion of a frame. Surely, some of us have gathered that even the info field carries overhead information. This is embedded information used to service the message or frame. The information derives from the upper OSI layers. A telephone call carries no such overhead except for the simple supervisory signaling. From this perspective, conventional telephony is more efficient than its data transmission equivalent.

4.6.2 Smart Bridges, Routers, and Switching Hubs

Each of these devices carries out some form of switching. The general application is with LANs. Some will say that each device carries out a routing function. The basis of routing is by means of a routing table which may or may not be dynamically updated. Bridges, routers, and hubs are discussed in Sections 10.2, 10.3, and 10.4, respectively.

REFERENCES

1. John E. McNamara, *Technical Aspects of Data Communication*, 2nd ed., Digital Equipment Corp., Bedford, MA, 1982.
2. *High Level Data Link Control (HDLC)—Frame Structure*, ISO 3309, International Standards Organization, Geneva, 1976.
3. *Reference Data for Radio Engineers*, 5th ed., ITT Howard W. Sams, Indianapolis, 1975.

5

DATA TRANSMISSION I

5.1 ELECTRICAL COMMUNICATION OF INFORMATION

Telecommunication transmission involves electrical communication of information. We communicate voice, image, and data. Our concern here is the electrical communication of data.

By *electrical communication* of data, we mean the formation of an electrical signal that connotes information which can be interpreted at some distant receiver with minimal or no ambiguity. One way of doing this is to impress a voltage on an electrical circuit. The "on" voltage can signal something, and the "off" voltage signals something else. Think of turning a light "on" and turning the light "off." The "on" condition could represent a binary 1 and the "off" condition can represent a binary 0. We now have related the very essence of expressing information electrically.

5.2 THE BIT AND BINARY TRANSMISSION OF INFORMATION

What was just described in the paragraph above was the bit. Some say the bit is an acronym for binary digit. One definition of the bit is "a contraction of the term binary digit; a unit of information represented by either a one or a zero" (Ref. 1).

The bit, therefore, has one of two possible states at any moment in time. In our everyday life we can think of many two-state situations. A light is on or it is off; a telephone line is idle or it is busy; a person is home or is out, a person is asleep or awake, just to name a very few.

On the one hand, we surmise that a two-state information system does not tell us very much; on the other we can find it very useful. For example, in telephone transmission and switching, it is imperative that we know whether a line is idle or busy. Nearly all telephone systems are built around this concept. It is called *supervisory signaling* or *line signaling*. Just one of two states can exist: idle or busy.

In another important system used in telecommunications, all we wish to know is whether a part, card, assembly, unit, or device is working or has failed. Again, just one of two states. Automatic trouble shooting of very sophisticated electronic equipment is based on this premise.

The earliest electrical representation of a bit followed the on/off concept. A "1" was represented by a positive voltage and a "0" was represented by no voltage (0 volts), as shown below:

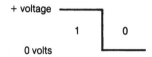

For the requirements of data communications, a single bit with its two possible states provides little information. Suppose now that two contiguous bits are transmitted rather than just one bit. We now have a four-state system or four possible bit combinations as follows:

00, 01, 10, and 11.

The electrical representation of these are shown below:

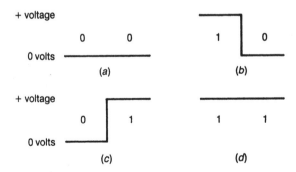

Let us carry this thinking forward one more step. This time each piece of information we wish to transmit contains three consecutive bits. Each of the bits, of course, can be a 1 or a 0. If we write down all the possibilities of 1s and 0s in a three-bit sequence, we find there are eight as follows: 000, 001, 010, 011, 100, 101, 110, and 111. Each bit sequence could represent a letter of the alphabet. However, we would only cover eight letters such as A, B, C, D,

E, F, G, and H. These bit sequences are shown with their electrical equivalents as follows:

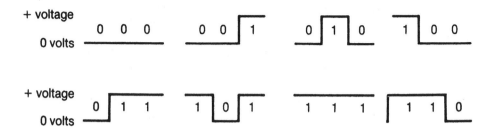

As we said, if we were to assign letters of the alphabet to these, we would only reach the letter H. Certainly, if we wish to develop a means of transmitting alphanumeric information using bits, we would want to cover the entire 26 letters of the alphabet. The digits 0 through 9 should also be included as well as punctuation. There are also nonprinting "characters" such as space (bar), carriage return, and line feed (word-wrap). There are many graphic symbols which are useful to express such as @, #, $, %, &, *, (and), = , +, and \.

It follows, therefore, that we must enrich the code.

Some readers may have caught the trend. Every time a binary code is extended 1 bit, the total number of different possibilities doubles. We then would expect that if we moved from a 3-bit code to a 4-bit code, we would have 16 possibilities rather than only 8. Now consider the following series:

One-bit code: $2^1 = 2$, or two distinct possibilities
Two-bit code: $2^2 = 4$, or four distinct possibilities
Three-bit code: $2^3 = 8$, or eight distinct possibilities
Four-bit code: $2^4 = 16$, or sixteen distinct possibilities
Five-bit code: $2^5 = 32$, or thirty-two distinct possibilities
Six-bit code: $2^6 = 64$, or sixty-four distinct possibilities
Seven-bit code: $2^7 = 128$, or one hundred twenty-eight possibilities

Then if n is the number of bits per character in a binary code set, the number of different characters, graphic symbols, and nonprinting characters (control functions) is simply 2^n.

Today we commonly use 5-, 6-, 7-, and 8-bit codes.

5.3 BINARY CODES FOR DATA COMMUNICATION

One of the earliest codes was the Baudot code, which is a 5-bit code. Thus there are 2^5, or 32, character possibilities. Now how can such a code accommodate the 26-letter alphabet, 10 digits (0 through 9), punctuation, and some nonprinting characters such as space (bar) with a total of only 32 characters?

The approach used was the same as that of the typewriter. The code is extended by using "uppercase." Two additional nonprinting characters were

Characters				Code elements[a]						
Letters Case	Communi-cations	Weather	CCITT #2[b]	Start	1	2	3	4	5	Stop
A	–	↑			▓	▓				▓
B	?	⊕			▓			▓	▓	▓
C	:	○				▓	▓	▓		▓
D	$	↗	WRU		▓			▓		▓
E	3	3			▓					▓
F	1	→	Unassigned		▓		▓	▓		▓
G	&	↘	Unassigned			▓		▓	▓	▓
H	STOP[c]	↓	Unassigned				▓		▓	▓
I	8	8				▓	▓			▓
J	'	↗	Audible signal		▓	▓		▓		▓
K	(	←			▓	▓	▓	▓		▓
L	)	↖				▓			▓	▓
M	.	.					▓	▓	▓	▓
N	,	⊕					▓	▓		▓
O	9	9						▓	▓	▓
P	θ	θ				▓	▓		▓	▓
Q	1	1			▓	▓	▓		▓	▓
R	4	4				▓		▓		▓
S	BELL	BELL	,		▓		▓			▓
T	5	5							▓	▓
U	7	7			▓	▓	▓			▓
V	;	⊕	=			▓	▓	▓	▓	▓
W	2	2			▓	▓			▓	▓
X	/	/			▓		▓	▓	▓	▓
Y	6	6			▓		▓		▓	▓
Z	"	+	+		▓				▓	▓
BLANK	–									▓
SPACE							▓			▓
CAR. RET.								▓		▓
LINE FEED						▓				▓
FIGURE					▓	▓		▓	▓	▓
LETTERS					▓	▓	▓	▓	▓	▓

[a] Blank, spacing element; crosshatched, marking element.

[b] This column shows only those characters which differ from the American "communications" version.

[c] Figures case H(COMM) may be stop or +.

Figure 5.1. The ITA#2 code.

required to do this: "uppercase" and "lowercase." Thus the net gain was to double the code possibilities minus the two nonprinting characters used to achieve the code extension. This left a code that could accommodate 62 characters, which met the requirements of the day.

A somewhat modified Baudot code is still in use today. It is the ITA#2 code. ITA stands for International Telegraph Alphabet. The code is employed chiefly on record traffic circuits, but has also been used for data communications. Figure 5.1 shows the ITA#2 code.

In the late 1950s, when data communications began to take hold, it was found that the ITA#2 code was not rich enough to support this emerging industry. Several ad hoc codes were being fielded, mostly 6-bit codes. By now we know that a 6-bit code provides $2^6 = 64$ possibilities, or a 64-character code. This was really not much better than the ITA#2 code with its upper- and lowercase characters.

Supported by some of the emerging giants in the data processing world, the American National Standards Institute (ANSI) developed a 7-bit code, which today has worldwide acceptance. This is the American Standard Code for Information Interchange, which goes by the acronym ASCII. The code provides 128 distinct code possibilities ($2^7 = 128$), or 128 characters. ASCII is illustrated in Figure 5.2. One should note that it provides a large number of nonprinting characters, many of which may be used for data-link control

Bits				b_7 b_6 b_5	0 0 0	0 0 1	0 1 0	0 1 1	1 0 0	1 0 1	1 1 0	1 1 1
b_4	b_3	b_2	b_1	Row ↓ / Column →	0	1	2	3	4	5	6	7
0	0	0	0	0	NUL	DLE	SP	0	@	P	`	p
0	0	0	1	1	SOH	DC1	!	1	A	Q	a	q
0	0	1	0	2	STX	DC2	"	2	B	R	b	r
0	0	1	1	3	ETX	DC3	#	3	C	S	c	s
0	1	0	0	4	EOT	DC4	$	4	D	T	d	t
0	1	0	1	5	ENQ	NAK	%	5	E	U	e	u
0	1	1	0	6	ACK	SYN	&	6	F	V	f	v
0	1	1	1	7	BEL	ETB	'	7	G	W	g	w
1	0	0	0	8	BS	CAN	(	8	H	X	h	x
1	0	0	1	9	HT	EM	)	9	I	Y	i	y
1	0	1	0	10	LF	SUB	*	:	J	Z	j	z
1	0	1	1	11	VT	ESC	+	;	K	[	k	{
1	1	0	0	12	FF	FS	,	<	L	\	l	¦
1	1	0	1	13	CR	GS	—	=	M	]	m	}
1	1	1	0	14	SO	RS	.	>	N	^	n	~
1	1	1	1	15	SI	US	/	?	O	_	o	DEL

Figure 5.2. American Standard Code for Information Interchange (ASCII). See Table 5.1 for definitions. (Basic source: Ref. 3; see also Ref. 4.)

purposes, the forerunner of present-day link-layer protocols. Among these we find EOT, end of text; SOH, start of heading; EOM, end of message; SOM, start of message; and NACK, negative acknowledgment. Table 5.1 gives the meanings of these control characters.

CCITT (now ITU-T) followed suit with a similar 7-bit code previously called International Alphabet No. 5 (IA5). It is now called International Reference Alphabet, which is shown in Figure 5.3, the basic code table.

TABLE 5.1   Definitions for Figure 5.2: Control Characters and Format Effectors

NUL: All zero characters. A control character used for fill.

SOH (Start of heading): A communication control character used at the beginning of a sequence of characters which constitute a heading. A heading contains address or routing information. An STX character has the effect of terminating a heading.

STX (Start of text): A communication control character that terminates a heading and indicates the start of text or the information field.

ETX (End of text): A communication control character used to terminate a sequence of characters started with STX and transmitted as an entity.

EOT (End of transmission): A communication control character used to indicate the conclusion of transmission which may have contained one or more texts and associated headings.

ENQ (Enquiry): A communication control character used as a request for response from a remote station. It may be used as a "Who are you" (WRU) to obtain identification, or may be used to obtain station status, or both.

ACK (Acknowledge): A communication control character transmitted by a receiver as an affirmative response to a sender.

BEL: A character for use when there is a need to call human attention. It may control alarm or attention devices.

BS (Backspace): A format effector which controls the movement of the active position one space backward.

HT (Horizontal tabulation): A format effector which controls the movement of the active position forward to the next character position.

LF (Line feed): A format effector which controls the movement of the active position advancing it to the corresponding position of the next line.

VT (Vertical tabulation): A format effector which controls the movement of the active position to advance to the corresponding character position on the next predetermined line.

FF (Form feed): A format effector which controls the movement of the active position to its corresponding character position on the next page or form.

CR (Carriage return): A format effector which controls the movement of the active position to the first character position on the same line.

SO (Shift out): A control character indicating the code combination which follows shall be interpreted as outside of the character set of the standard code table until a shift in (SI) character(s) is reached. It is also used in conjunction with the (ESC) escape character.

SI (Shift in): A control character indicating that the code combinations which follow shall be interpreted according to the standard code table (Figure 5.2).

DLE (Data-link escape): A communication control character which will change the meaning of a limited number of contiguously following bit combinations. It is used exclusively to provide supplementary control functions in data networks. DLE is usually terminated by a shift in (SI) character(s).

DC1, DC2, DC3, DC4 (Device controls): Characters for the control of ancillary devices associated with data or telecommunication networks. They switch these devices "on" or "off." DC4 is preferred for turning a device off.

NAK (Negative acknowledgment): A communication control character transmitted by a receiver as a negative response to a sender.

TABLE 5.1 (*Continued*)

SYN (Synchronous idle): A communication control character used by a synchronous transmission system in the absence of any other character to provide a signal from which synchronism may be achieved or retained.

ETB (End of transmission block): A communication control character used when block transmission is employed. It indicates the end of a block of data.

CAN (Cancel): A control character used to indicate that the data preceding it is in error or is to be disregarded.

EM (End of medium): A control character associated with the sent data which may be used to identify the physical end of the medium, or the end of the used or wanted portion of information recorded on a medium. It should be noted that the position of the character does not necessarily correspond to the physical end of the medium.

SUB (Substitute): A character that may be substituted for a character which is determined to be invalid or in error.

ESC (Escape): A control character intended to provide code extension (supplementary characters) in general information interchange. The ESC character itself is a prefix affecting the interpretation of a limited number of contiguously following characters. ESC is usually terminated by an SI (shift in) character(s).

FS (File separator), GS (Group separator), RS (Record separator), and US (Unit separator): These information separators may be used within data in optional fashion, except that their hierarchical relationship shall be FS is the most inclusive, then GS, then RS, and US is the least inclusive. (The content and length of a File, Group, Record, or Unit are not specified.)

DEL (Delete): This character is used primarily to "erase" or "obliterate" erroneous or unwanted characters in perforated tape. (In the strict sense DEL is not a control character.) DEL characters may also be used for fill without affecting the meaning or information content of the bit stream.

Note: SO, ESC, and DLE are all characters which can be used, at the discretion of the system designer, to indicate the beginning of a sequence of digits having special significance.

SP (Space): A nonprinting character used to separate words or sequences. It is a format effector that causes the active position to advance one character position.

Source: Ref. 3, also see Ref. 4.

CCITT Rec. T.50 (Ref. 2) points out that positions 0/0 to 1/15 (the first two columns, which are blank in the figure) are reserved for control characters. (Note the control characters in the first two columns of Figure 5.2).

IBM went one step further and defined an 8-bit code. It is called EBCDIC, which stands for Extended Binary Coded Decimal Interchange Code. This code is shown in Figure 5.4. It will be noted that EBCDIC is very rich in graphic characters. It should also be clear that some of the character possibilities are not used (i.e., left blank).

5.4 ELECTRICAL BIT DECISIONS

Electrical noise tends to corrupt a telecommunication system. Let us define noise as "unwanted disturbances superimposed upon a useful signal that tend to obscure its information content" (Ref. 1). Accept the fact that all electrical communication systems have some sort of noise component superimposed upon a wanted signal. All other signal impairments will be disregarded for now.

b7 → b6 → b5 →				0 0 0	0 0 1	0 1 0	0 1 1	1 0 0	1 0 1	1 1 0	1 1 1
b4	b3	b2	b1	0	1	2	3	4	5	6	7
0	0	0	0	0			SP	0	②	P	②	p
0	0	0	1	1			!	1	A	Q	a	q
0	0	1	0	2			"	2	B	R	b	r
0	0	1	1	3			#/②£	3	C	S	c	s
0	1	0	0	4			¤/①$	4	D	T	d	t
0	1	0	1	5			%	5	E	U	e	u
0	1	1	0	6			&	6	F	V	f	v
0	1	1	1	7			'	7	G	W	g	w
1	0	0	0	8			(	8	H	X	h	x
1	0	0	1	9			)	9	I	Y	i	y
1	0	1	0	10			*	:	J	Z	j	z
1	0	1	1	11			+	;	K	②	k	②
1	1	0	0	12			,	<	L	②	l	②
1	1	0	1	13			−	=	M	②	m	②
1	1	1	0	14			.	>	N	②	n	②
1	1	1	1	15			/	?	O	_	o	DEL

CO set

Figure 5.3. Basic code table for CCITT International Alphabet No. 5 (IA5), the International Reference Alphabet (IRA). (From Table 4, page 11, CCITT Rec. T.50, Ref. 2.)

We find that binary digital receivers are very forgiving of noise. The reason is that such a receiver is a device that must make a simple decision: whether a pulse is a "1" or a "0."

In the case of the scenario described in Section 5.2, a positive pulse is a "1" and no pulse is a "0." Consider Figure 5.5, which shows the ideal situation of a stream of alternating "1s" and "0s." The signals have not been corrupted by noise. Figure 5.6 shows the same signals corrupted by noise. What is meant by *corruption* is really the likelihood that an incorrect decision could be made. Conventionally, a binary digital receiver makes a decision at

B I T S			4	0	0	0	0	0	0	0	0	1	1	1	1	1	1	1	1
			3	0	0	0	0	1	1	1	1	0	0	0	0	1	1	1	1
			2	0	0	1	1	0	0	1	1	0	0	1	1	0	0	1	1
			1	0	1	0	1	0	1	0	1	0	1	0	1	0	1	0	1
8	7	6	5																
0	0	0	0	NUL				PF	HT	LC	DEL								
0	0	0	1					RES	NL	BS	IL								
0	0	1	0					BYP	LF	EOB	PRE			SM					
0	0	1	1					PN	RS	UC	EOT								
0	1	0	0	SP										¢	.	<	(	+	|
0	1	0	1	&										!	$	*	)	;	¬
0	1	1	0		/									∧	,	%	—	>	?
0	1	1	1											:	#	@	'	=	"
1	0	0	0		a	b	c	d	e	f	g	h	i						
1	0	0	1		j	k	l	m	n	o	p	q	r						
1	0	1	0			s	t	u	v	w	x	y	z						
1	0	1	1																
1	1	0	0		A	B	C	D	E	F	G	H	I						
1	1	0	1		J	K	L	M	N	O	P	Q	R						
1	1	1	0			S	T	U	V	W	X	Y	Z						
1	1	1	1	0	1	2	3	4	5	6	7	8	9						⌶

PF – Punch Off
HT – Horiz. Tab
LC – Lower Case
DEL – Delete
SP – Space
UC – Upper Case

RES – Restore
NL – New Line
BS – Backspace
IL – Idle
PN – Punch On
EOT – End of Transmission

BYP – Bypass
LF – Line Feed
EOB – End of Block
PRE – Prefix
RS – Reader Stop
SM – Start Message

Figure 5.4. The EBCDIC code.

Figure 5.5. A string of 0s and 1s uncorrupted—ideal signal.

mid-pulse. The decision will be made that it is a "1" if the voltage reaches at least half-amplitude in the negative direction (i.e., a negative polarity voltage); it is a "0" if the voltage has reached at least half-amplitude in the positive direction (i.e., a positive polarity voltage).

Suppose at the moment the decision is made, a noise spike is present that exceeds half-amplitude in the negative direction during the time a "0" was supposed to be received. Of course the decision circuit will confuse it for a binary "1." The bit will be in error.

Several measures could be taken to improve the chances of making the right decision. We could increase the negative and positive voltages of the

Figure 5.6. The same signal as in Figure 5.5, but corrupted with noise, cable pair capacitance, and loss. The capacitance causes "rounding," and the loss causes a reduction in amplitude.

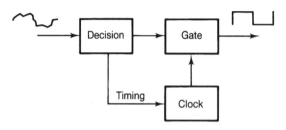

Figure 5.7. A simplified drawing of a binary digital regenerator.

pulses. We could set the decision level higher, say to $\frac{3}{4}$ amplitude, or use some other method of representing the 1 and the 0. There are many ways to improve performance of binary transmission of data in the presence of impairments such as noise. Often an apparent improvement will also have a drawback. The fact that binary data receivers have to make a decision of one of only two possibilities makes this type of transmission very forgiving in the presence of impairments. However, we must remember that the forgiveness is allowed only up to a certain point before performance suffers. That point will vary, of course, depending on the level of performance that is set for the system.

Another aspect of digital transmission is that the signal can be regenerated. A regenerator is simply a decision circuit and a clock. A regenerator accepts a corrupted pulse at its input and clocks out a "squared-up" "1" or "0." This concept is shown in Figure 5.7.

5.5   ELECTRICAL REPRESENTATION OF BINARY DATA

5.5.1   Neutral and Polar Waveforms

In Section 5.2 we used a "neutral" type of electrical representation of 1s and 0s. In this case, one state is active and the other state is passive. In our examples, a "1" was a positive voltage and a "0" was a no-voltage condition, a

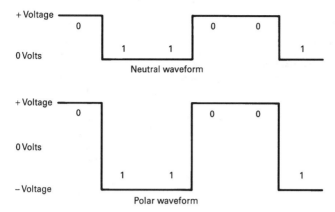

Figure 5.8. Neutral and polar waveforms compared.

passive state. A passive state can be easily corrupted by a noise spike such that a receiver would mistakenly consider it an active state if the noise level were of sufficient amplitude. To help avoid such situations, a *polar** representation of a binary 1 and 0 was devised, where both states are active. Neutral and polar representations of binary data are compared in Figure 5.8. In the future we will call these *neutral* and *polar* waveforms. These states are a negative voltage and a positive voltage. To be consistent with Section 5.2, we chose to call the positive pulse a 1 and a negative pulse a 0.

It should be noted that in modern Layer 1 standards, the states are reversed: a binary 1 is a negative voltage, and a 0 is a positive voltage typically as set out in EIA/TIA-232E (Ref. 5).

At times, while working with binary bit streams we can, either by accident or on purpose, reverse the *sense* of the bits. This simply means reversing the bits'* polarity, where the negative battery and positive battery are reversed. You might say where the 1s become 0s and the 0s become 1s. This would be highly undesirable for operational circuits, but we may want to reverse the sense to carry out some form of testing or troubleshooting.

5.5.2 Waveforms and Line Codes

There are yet other ways of representing 1s and 0s electrically, each waveform with its own advantages and disadvantages. Generally, binary data waveforms can be divided into two groups: non-return-to-zero (NRZ) and return-to-zero (RZ). These are shown and compared in Figure 5.9. Among the advantages and disadvantages are simplicity and the ability to extract timing, but at the expense of increasing the number of transitions per average unit time. This, in essence, increases the "frequency" of the waveform, which, under certain circumstances, may be highly undesirable (typically on a fiber-optics link).

*"Polar" is called *double-current* by CCITT/ITU-T.

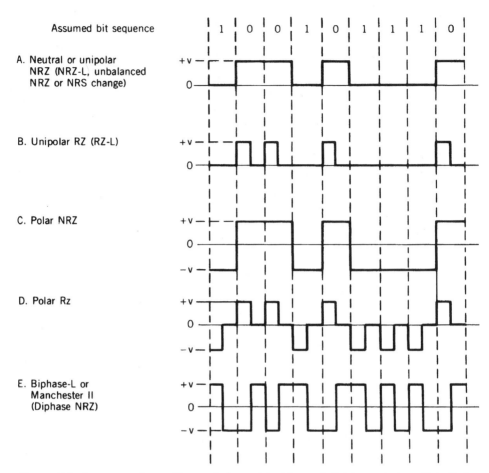

Figure 5.9. Some waveforms (line codes) used in binary digital transmission. Note that Manchester coding is widely used on baseband LANs.

5.6   BINARY CONVENTIONS

Another aspect of binary transmission is ambiguity. If we are to set up a data circuit, both ends of the circuit must agree on which of the two possible states is a "1" and which is a "0." For example, there is no reason why we cannot establish a "1" as a positive-going voltage and a "0" as a negative-going voltage (see reversing the sense in Section 5.4).

CCITT (ITU-T) has established some conventions to remove the ambiguity of which state is a 1 and which is a 0. This may be found in CCITT Rec. V.1. We have embellished upon the information in Rec. V.1 in Table 5.2. This table is called a *table of mark-space convention*. The terms *mark* and *space* derive from the days of telegraph to distinguish between the active and

TABLE 5.2   Table of Mark-Space Convention

Binary 1	Binary 0
Mark or marking	Space or spacing
Perforation (paper tape)	No perforation
Negative voltage[a]	Positive voltage
Condition Z	Condition A
Tone on (amplitude modulation)	Tone off
Low frequency (frequency shift keying [FSK])	High frequency
Opposite to the reference phase	Reference phase (phase modulation)
No phase inversion	Inversion of the phase
(Differential two-phase modulation)	

[a]CCITT Recs. V.10 and V.11 (Refs. 7 and 8, respectively) and EIA / TIA-232 (Ref. 5).
Source: CCITT Rec. V.1 (Ref. 6). Courtesy of ITU-CCITT.

passive state. To confuse matters, the mark is associated with the 1 and the space is associated with the 0; we call them synonymous, meaning that a binary 1 and a "mark" have the same meaning. Contrary to our discussion in Section 5.2, modern data systems use the convention that a negative-going voltage is a 1 and a positive-going voltage is a 0.

One key standard, probably familiar to many users of this text, namely EIA/TIA-232, states that a voltage of $+3$ volts or greater is a 0 and that a voltage of -3 volts or greater (e.g, -4 volts) is a 1. Voltages between -3 and $+3$ volts are indecisive. Many other standards generally follow suit such as EIA/TIA-422, EIA/TIA-423, and ITU-T Rec. V.10 and V.11, to name a few.

5.7   BIT-PARALLEL AND BIT-SERIAL

Data may be transferred in "serial" format over a single line one bit at a time contiguously or in "parallel" over several lines at once. In parallel transmission, bits are transmitted a character at a time, and each bit of the set of bits that represent a character has its own wire or line. An additional wire called the *strobe* or *clock* lead notifies the receiver unit that all the bits are present on their respective wires so that the voltages on the wires can be sampled. Parallel transmission is shown conceptually in Figure 5.10. Figure 5.11 illustrates a typical parallel-to-serial conversion circuit. The 1 and 0 bits at either extreme of the register are bits that are appended for control—typically start–stop operation, which is discussed in Chapter 6. A register is a digital storage device.

Computers and other digital data processors generally operate on parallel data, so data are transferred in parallel between these devices wherever they are in close proximity. However, as the distance between these devices increases, not only do the multiple wires become more costly, but the

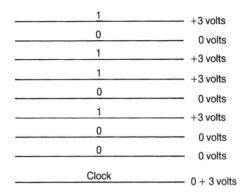

Figure 5.10. Bit-parallel transmission, 8-bit character.

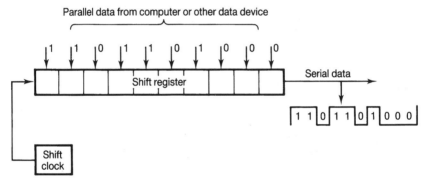

Figure 5.11. Serial data transmitter with parallel data input, configured for start–stop operation and an 8-bit character.

complexity of line drivers and receivers increases, due to the major difficulty of properly driving and receiving signals on long runs of wire pairs.

Bit-serial transmission is generally used where the cost of the communications medium is high enough to justify a relatively complex transmitter and receiver system that will serialize bits that represent a character, send over a single line, and reassemble them in parallel form at the receiving end. In most data transmission systems, serial transmission is preferable to parallel transmission.

5.8   BASEBAND

The term *baseband* is defined by the IEEE (Ref. 1) as "the band of frequencies occupied by the (data) signal before it modulates a carrier (or subcarrier) frequency to form the transmitted line or radio signal." Baseband, therefore, is the data signal as it leaves or enters a data processing or presentation device. The signal has frequency content extending into the direct current region. Baseband data can be transmitted hundreds and even thousands of feet. This is commonly done on wire pair. Figure 5.9 shows several common baseband data waveforms. Transmission distance is limited because of several factors. The data signal suffers loss due to the length and characteristics of the wire pair. It can be corrupted by noise, which often can be related to length of the wire pair. The signal itself will become distorted due to the electrical characteristics of the medium. Distances can be extended by the use of regenerators, which were described above.

Long distances can be achieved by having the baseband signal modulate a radio or light carrier signal. Some refer to this as *carrier transmission.*

5.9 DATA RATE

In this text, data rate and bit rate are synonymous. Both are defined as the transmission rate in bits per second (bps). Bit rate is constrained by the capability of the transmitting and receiving devices, the bandwidth of the intervening media, and a number of electrical impairments typical of each medium in question.* We might successfully transmit at a megabit per second, but the distant-end receiving device may be suffering an excessive error rate on receiving that signal. In other words, a high percentage of the characters received are in error, and the data circuit is useless to a user. Thus, when we use the terms *bit rate* or *data rate*, there is some implication of satisfactory operation end-to-end.

5.9.1 Error Rate

The most common measure of a data connectivity's performance is *error rate.* Often the unit of measurement is bit error rate (BER), which is the probability of bit error. BER is often expressed using the powers of 10. For example, a certain data circuit has a bit error rate of 1 in 10^5. This means that there is a probability that one error will occur for every 100,000 bits transmitted. More often such a BER is expressed as 1×10^{-5}.

When describing a bit rate on a data circuit, it often is most useful to give the bit rate with the accompanying BER. For example, the data circuit on the Boston to New York leg operates at 64 kbps with a BER of 3×10^{-7}. With a statement such as this, circuit performance is implicit.

Bit error rate performance in industrialized nations has shown dramatic improvements. CCITT specifies a bit error rate of 1×10^{-6} or better on ISDN circuits (ITU-T Rec. G.821). On WANs a user should specify 5×10^{-10} or better (see Bellcore TSGR, Ref. 9). Bit error rate requirements and other expected performance criteria are discussed at length in Section 6.3.3.

---

*These impairments are described in Chapters 6 and 8.

REFERENCES

1. *The IEEE Standard Dictionary of Electrical and Electronic Terms*, 6th ed., IEEE Standard 100-1996, IEEE, New York, 1996.

2. *International Reference Alphabet (IRA) Information Technology—7-bit Coded Character Set for Information Interchange*, CCITT Rec. T.50, ITU Geneva, September 1992.

3. *Common Long-Haul and Tactical Communication System Technical Standards*, MIL-STD-188-100, US Department of Defense, Washington DC, November 15, 1972.

4. American National Standard, *Coded Character Sets: 7-bit American National Standard Code for Information Interchange (7-bit ASCII)*, ANSI X3.4-1986, ANSI, New York, 1986.

5. *Interface Between Data Terminal Equipment and Data Circuit Terminating Equipment Employing Serial Binary Data Interchange*, EIA/TIA-232E, Telecommunications Industry Association, Washington DC, July 1991.

6. *Equivalence Between Binary Notation Symbols and the Significant Conditions of a Two-Condition Code*, CCITT Rec. V.1, Fascicle VIII.1, IXth Plenary Assembly, Melbourne, 1988.

7. *Electrical Characteristics for Unbalanced Double-Current Interchange Circuits Operating at Data Signaling Rates Nominally up to 100 kbps*, ITU-T Rec. V.10, ITU Geneva, March 1993.

8. *Electrical Characteristics for Balanced Double-Current Interchange Circuits Operating at Data Signaling Rates up to 10 Mbps*, ITU-T Rec. V.11, ITU Geneva, October 1996.

9. *Transport Systems Generic Requirements (TSGR): Common Requirements*, Telcordia (Bellcore) GR-499-CORE, Issue 1, Piscataway, NJ, December 1995.

6

DATA TRANSMISSION II

6.1 INTERPRETING A SERIAL STREAM OF BITS

6.1.1 The Problem

A serial bit stream consists of bits transmitted one at a time sequentially and contiguously. At the very instant one bit ends, the next bit begins. Consider the serial bit stream shown in Figure 6.1. It contains the ASCII character *A* preceded by an ASCII character and followed by another ASCII character. We presume that the receiver knows it is receiving ASCII characters at some nominal bit rate. Each character in the serial bit stream in the figure consists of 7 bits plus an added parity bit for a total of 8 bits per character. If we would allow that the receiver samples at the center of each bit, then the receiver would be in exact synchronism with the transmitter. Getting and keeping the receiver in exact synchronism with its companion far-end transmitter is yet another problem that must be solved. We address this issue in the next two subsections.

6.1.2 Start–Stop Transmission

One of the oldest methods of data circuit timing and synchronization is start–stop transmission. Many entities, including CCITT, call this type of transmission *asynchronous*. The IEEE calls this type of transmission *nonsynchronous*. Start–stop transmission dates back to the first automatic telegraphy circuits. In those days the "mark" or "1" was the active state and the "space" or "0" was the passive state. A typical ASCII start–stop character is shown in Figure 6.2. As illustrated in the figure, the beginning and end of each character is well-delineated.

Note: The 1s and 0s in Figure 6.2 are using today's convention: A positive-going voltage indicates a 0, and a negative-going voltage indicates a 1.

Figure 6.1. A serial ASCII bit stream.

Figure 6.2. Start–stop character format. SS, start-space; SM, stop-mark. ASCII letter *A* and the beginning of the ASCII letter *S*. *Note:* The duration of the stop-space is one unit interval (i.e., duration of 1 bit), and the stop-mark (in the figure) has a duration of two unit intervals (i.e., 2 bits).

In those old teleprinters, the convention was adopted that when the line is idle, current is flowing (e.g., voltage is present). The commencement of the transmission of a character or string of characters is indicated when this current is interrupted and the normal current (or voltage) state drops to 0, the space condition. To start the receiver teleprinter, the line condition changes from one of current to a no-current condition, and the "0" state holds for the duration of one bit (1-bit time). The "space" element is called the *start bit* or *start element*. The start element is followed by eight successive bit times or unit intervals (see Figure 6.2) which represent the character being sent. These bit times may be either a "1" (active state) or "0" (passive state) in any combination thereof for 8 consecutive bits. (Note that if 8 information bits are used, we could be dealing with the EBCDIC code or ASCII with the added parity bit.)

The last element transmitted is the "stop" element, which can have 1-, 1.5-, or 2-bit duration, depending on the convention accepted. The stop element allows the receiver to coast back to a known position in time until the beginning of the next "start space" element signifying the beginning of the next character.

With start–stop transmission it is true that precise synchronization between the far-end transmitter and the near-end receiver is not necessary. However, the receiver must sample the income bit stream at roughly the same rate as the transmitted bits are gated out on the transmit side so that the receiver can sample each incoming bit at the proper time.

When a line is idle in start–stop transmission, a continuous 1 or mark is transmitted and current is flowing in the data loop. Once the receiver detects a transition from current flow to no-current flow, it counts out 5, 6, 7, or 8 bits, depending on the code used, and then knows that the next element must be the stop-mark or current-flow state. If another character directly follows the first, there again is the mark-to-space transition right after the stop-mark and the "start-space" begins and lasts for the duration of 1 bit, and the receiver starts its count again. As one can see, the receiver must maintain some sort of synchronization with its companion far-end transmitter only during the 5-, 6-, 7-, or 8-unit intervals (or bits in this case) of one character. The receiver catches up with the transmitter or slows down to let the transmitter catch up with its clock during the time period of the stop-mark element.

The advantage of asynchronous (start–stop) transmission is that very loose requirements are placed upon the timing and synchronization of the data link. We can easily delineate the beginning of a character at the receiver by the mark-to-space transition that begins the start-space element. The end of a character is defined by the stop-mark. Timing starts at the start-space and ends at the stop-mark, not a very difficult feat to accomplish.

The disadvantage of this type of transmission is the overhead of the control elements: the start-space and stop-mark. Let us say that the code used is ASCII with 1 parity bit, for a total of 8 bits per character. Allow the start-space and stop-mark each to be of 1-bit duration. Thus it takes 10 bits per character to transmit 8 useful bits. There is a built-in 20% inefficiency.

6.1.3 Synchronous Transmission

Synchronous transmission avoids most of the inefficiency cited above. One method proposed to ensure that the receiver clock is kept in exact synchronism with the gating of transmitted bits was to extend a separate clocking lead from the transmitter to the receiver. In most situations this approach is untenable.

If we consider for just a moment the makeup of a serial bit stream, we can see that there is a built-in clocking system, namely the mark-to-space and space-to-mark transitions of the information bits themselves. If we slave the receiver clock to the incoming transitions, often accomplished by phase-locked loop techniques, the receiver will maintain exact synchronism with the transmitter, but will be delayed by the propagation time of the intervening circuit. Of course the bits we are receiving are delayed by the same amount as we would expect.

With synchronous transmission, characters are delineated by simple bit counting. Our most common data transmission codes are based on 8-bit characters, typically ASCII with a parity bit and EBCDIC. Some call an 8-bit character a byte, but the term *octet* is perhaps a better term. All bit-oriented protocols rigidly require that data frames be composed of an integer number of octets. One might imagine character markers at all integer multiples of eight bits. Eight-bit characters are also popular with processors because there are 8-bit, 16-bit, 32-bit, and now 64-bit processors. For example, a 64-bit processor can gobble up 8 characters at a time.

The incisive reader will immediately have picked up the next problem. How do we know when a message (a block, frame, packet, or cell) begins and ends when synchronous transmission is employed? A message always starts with a unique field or bit sequence called a *flag*. The majority of bit-oriented protocols use the sequence 01111110 as the starting and ending flag. At the end of the starting flag, we start counting increments of eight delineating bits. Of course we must start counting with the very first bit appearing after the flag sequence.

At this juncture we must define a *character*. The IEEE (Ref. 1) defines a character (data transmission) as "one of a set of elementary symbols which normally include both alpha and numeric codes plus punctuation marks, and any other symbol which may be read, stored or written, and are used for organization, control, or representation of data." With a few exceptions (such as the ITA #2 code), all of our characters are 8 bits or 1 octet long. A character can be an operational sequence performing the function of space bar, hard return, soft return, and so on.

Before proceeding further, let us consider what a generic data frame looks like. Depending on the type of system, some may call a data frame a *packet*, a *block*, or a *cell*. A generic data frame is shown in Figure 6.3. Note that the *flag* is shown as the first and last field of the frame. The address and control fields together are often called the *header*. Generally packets and cells have fixed length text or info field; frames and blocks may carry varying-length text or info fields. The term *varying* should not be confused. All text or info fields have length constraints—that is, maximum length and minimum length. Thus, it would appear we would like long frames (long info fields) to amortize overhead. However, on noisy circuits where a lot of frame repeats are required, short frames are more desirable.

Frame structure has had detailed explanation in Chapters 2 and 3. Our discussion here deals with synchronous transmission. With start–stop transmission, the beginning and end of a character were well-delineated by the start and stop elements. In the case of synchronous transmission, the flag carries out this function; the receiver-processor counts bits/octets after the initial flag. The processor has been programmed in advance for the link-layer protocol to be used. Thus, the processor knows a priori where each of the header fields begins and ends (and acts upon the contents). For fixed-length frames, the processor knows the length of the info field, where it ends and where the FCS begins, and where the frame ends, by means of the trailing flag.

To measure the efficiency of a certain data link, we sum all the bits in the header, FCS, and flag fields and divide this value by the total number of bits in a frame. We then subtract that value from 100 for a percent efficiency value.

Flag	Address	Control	Information	FCS	Flag

Figure 6.3. A generic frame or packet of data. See Chapters 2 and 3 for more detailed discussion of frame structure.

Example. A certain data link has 16 bits for the two flags, 16 bits of address information, 8 bits of control information, and 16 bits for the FCS field. The information field has 1600 octets (12,800 bits).

There are 56 bits of overhead, and the total number of bits in the frame is 56 + 12,800 or 12,856 bits, 56/12,856 = 0.004356.

$$100 - 0.004356 = 99.995\%$$

Without considering header overhead, start–stop transmission based on 11-bit characters—of which 3 bits are used for start and stop elements—will have the following efficiency:

$$3/11 = 0.2727$$

$$100 - 0.2727 = 99.727\%$$

On synchronous circuits that are idle (no traffic is being passed), *fill* or *idle patterns* are transmitted to maintain synchronization. If a circuit is dormant and at some moment in time traffic will be passed, a short training period should precede the traffic to allow synchronization and to initiate other features such as automatic equalizers.

6.2 TIMING DISTORTION IN A SERIAL BIT STREAM

We define distortion as the unwanted modification or change of signals from their form by some characteristic of the communication line or equipment. In data communications, distortion is usually manifested by improper timing.

One type of distortion is called *bias distortion*, where the duration of the mark pulse (the "1" pulse) is elongated a certain amount and the space pulse ("0" pulse) is shortened a corresponding amount. Of course we can have the reverse situation, where the space pulse is lengthened at the expense of the mark pulse.

One cause of bias distortion is wire-pair capacitance. Another cause is that it is likely that the resistance of the driver (transmitter end) and the receiver circuitry is different for the mark-to-space transition than for the space-to-mark transition. Thus, there will be a different amount of time required to charge the cable capacitance of the two transitions. One of the most well-known standards involved in this interface is EIA/TIA RS-232 (Ref. 2), which states that there should be no more than 2500 picofarads of cable capacitance on such a circuit. If the value of 2500 picofarads is exceeded, the increase in bias distortion can be dramatic. Figure 6.4 shows a binary bit stream with marking bias (bias distortion).

Figure 6.4. A binary data serial bit stream with marking bias. Ideally the 1 bits and the 0 bits should be of equal duration.

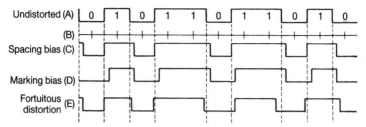

Figure 6.5. Types of timing distortion. *Note:* In bit stream drawings, such as this one, the current sense convention as established in EIA/TIA-232 and in similar standards may or may not be followed, depending on what is being expressed. This convention established the "1" as a negative-going voltage and the "0" as a positive-going voltage, as shown in Table 6.2. The sense is reversed in the drawing above.

The timing distortion of signal elements (bits) can be resolved into two components, commonly termed *systematic* and *fortuitous*. Systematic distortion is repetitive and is the average distortion as a data signal is transmitted through the system. The fortuitous component is the variation from the average. The sum of the systematic and fortuitous components makes up the *total distortion*.

Systematic distortion is further broken down into *bias* and *characteristic* distortion. Bias distortion is brought about by the lack of symmetry in the system, as was discussed above. Characteristic distortion is brought about by intersymbol interference (ISI) resulting from nonlinear phase versus frequency characteristic. Fortuitous distortion is caused by extraneous factors such as noise, crosstalk, and carrier phase effects, all of which may occur at random with respect to the signal bit sequence.

There are two ways of expressing the amount of distortion. It can be stated in time units, commonly milliseconds, or it can be expressed as a percent of a unit interval (a bit). The latter is more common in the industry. Of course when using percentage, we must state the speed of transmission (i.e., the data rate) at which the percent of distortion applies. The speed is almost universally given in bits per second. Figure 6.5 shows several types of distortion that have been discussed above.

6.3   THE TRANSMISSION OF DIGITAL DATA

6.3.1   Baseband Transmission

Data are transported electrically on a copper wire pair over comparatively short distances. For example, data may be transmitted across a room, across a floor, throughout a building, or, in some cases, throughout several contiguous buildings. This is commonly called *baseband* transmission. Distances are limited by the buildup of loss, distortion, and noise. Distance is also a function of bit rate.

As a wire pair is extended, resistance, capacitance, and inductance of the cable increase. As capacitance increases, it takes longer to charge the capacitance for each bit transmitted, thus limiting bit rate and/or distance. Keep in mind that we can view a wire pair as a very long capacitor. It meets the basic requirements because it consists of two conductors separated by an insulator.

Hysteresis due to inductance is another limiting factor. All of these factors limit bit rate and distance for baseband transmission.

One way to increase distance for baseband applications is to use a balanced pair rather than an unbalanced pair. Remember that with an unbalanced pair, one lead is grounded. In the case of balanced transmission, neither lead is grounded, and one lead is used for voltage reference, eliminating problems with ground potential differences from one end of the circuit to the other. Baseband distance can also be extended by the use of regenerative repeaters. The use of larger-diameter wire can also offer some advantages because it reduces loss and resistance per unit length. The use of low-capacitance wire pairs is yet another approach to achieve higher bit rates over longer distances.

6.3.2 Transmission of Data Over Longer Distances

6.3.2.1 *The Voice Channel.* We look to our ubiquitous telephone facilities to transport data electrically over longer distances. It was cost-effective because the telephone was everywhere—that is, worldwide. To do this we had to accept certain shortcomings of telephone facilities, particularly the intervening network. The telephone network consists of voice channels.

ITU-T defines a voice channel as occupying the voice-frequency band from 300 to 3400 Hz. Two important shortcomings can immediately be seen. First, the voice channel cannot accommodate DC (direct current) transmission, which we have called *baseband transmission*. Of course, the 0-Hz DC component of baseband transmission cannot be coupled to the voice channel passband. Second, its 3100-Hz bandwidth (i.e., 300 Hz–3400 Hz) places a severe limitation on data rate. If we were to allow 1 bit per hertz of bandwidth, then the maximum bit rate over a voice channel is 3100 bps. In practice, it is really less than this value. Certainly most readers know that considerably greater bit rates are handled by the voice channel by using sophisticated coding and modulation schemes. We discuss these important techniques at length in Chapter 8. Of major importance at this juncture is that we must convert the DC baseband serial bit stream to a signal that is compatible with the voice channel. The device that carries out this function is the popular *modem*, which is an acronym for modulator–demodulator.

6.3.2.2 *Modems.* A modem is any device that carries out (i) a modulation function on the transmit side of a circuit and (ii) a demodulation function on the receive side. However, for our discussion we define a modem as a device

TABLE 6.1   Data Modulation Terminology

Conventional Terminology	Data Modulation Terminology	Abbreviation Commonly Used
Amplitude modulation	Amplitude shift keying	ASK
Frequency modulation	Frequency shift keying	FSK
Phase modulation	Phase shift keying	PSK

that generates an audio tone whose frequency is compatible with the voice channel. This tone is modulated by a baseband digital data signal in either the frequency, amplitude, or phase domain (or combinations of two of these), and the resultant signal is coupled to the voice channel. On the receive side, a modem converts the modulated audio tone back to a digital data serial bit stream.

There are three generic types of modulation: amplitude modulation (AM), frequency modulation (FM), and phase modulation (PM). As we remember, a wave (e.g., a sine wave) is defined in terms of amplitude, phase, and frequency. Information can be transported on such a wave by varying the value of any one of these properties in accordance with an electrical baseband signal which contains that information. This, in essence, is modulation.

Many data transmission terms derive from its predecessor, the telegraph. When dealing with data transmission, consider Table 6.1. The term *keying* derives from Morse code operation of a transmitter using the Morse key. The key was held down (i.e., contact was made) just for a moment for a "dot," and it was held down for a longer period for a "dash." When the key was pressed down, an electrical contact was made and the transmitter was turned on, thereby emitting a carrier frequency (CW) wave. When the key was released, the transmitter was turned off and the carrier wave ceased. The terminology further evolved through automatic telegraph, a true forerunner of data communications.

When these automatic telegraph systems were introduced, the radio-frequency carrier remained on all the time, and the information was transmitted by means of "shifting" (changing) the carrier frequency upward in frequency some hundreds of hertz for a "mark" or "1" and downward in frequency an equal amount for a "space" or "0." Thus the term *shift keying* came about; that is, the carrier was *shifted*.

For ASK modulation, a carrier is turned on for a mark or 1 and is turned off for a space or 0, PSK systems came later, and in this case a binary 1 is transmitted at reference phase, and a binary 0 is transmitted with the phase retarded, usually 180 degrees.

Data modems employ FSK modulation almost universally for all systems operating at 1200 bps or less. Above 1200 bps, some form of PSK is used. For higher data rate modems, above 4800 bps (typically CCITT Rec. V.29; see Ref. 3), a combination of ASK and PSK is used.

6.3.2.3 *Voice Channel Impairments to Data Transmission.* The voice
channel remains the basic building block of the public switched telecommunication network (PSTN). Figure 6.6 illustrates the concept of a voice channel
traversing the telecommunications network.

The voice channel is constrained in frequency because it will be multiplexed with other voice channels to share the expensive transmission media.
As a minimum, there is a low-pass filter at every voice channel input of the
multiplexing channel bank and another low-pass filter at the companion
demultiplexing bank at the other end of the circuit. There may be other
filters in the intervening circuit.

The transmission media can be wire pair, coaxial cable, fiber-optic cable,
line-of-sight microwave radio, and satellite links. On long circuits we would
expect a mix of media.

There will be one or more modulation stages which the voice channel will
be subjected to as it traverses the circuit end-to-end over the transmission
media. Each modulation and amplification stage generates noise, and each
will cause errors in the case of digital transmission.

The intervening equipment and media through which the voice channel
passes cause transmission impairments to the voice channel. There are three
basic impairments:

- Amplitude distortion (frequency response)
- Phase distortion (delay distortion)
- Noise

In fact, these three impairments will affect any transmission medium, especially metallic media (e.g., wire pair and coaxial cable). There are two other
impairments that affect the voice user much more than the data user. These
are echo and singing.

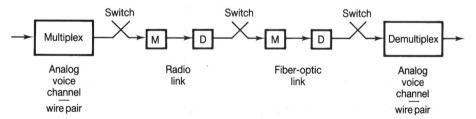

Figure 6.6. A simplified conceptual illustration of a voice channel traversing the telecommunications network. Note that the multiplexing may be in the time domain or in the frequency
domain. If the multiplexing is in the time domain, the switching is digital using time division
techniques. M, modulator; D, demodulator.

Amplitude Distortion. Amplitude distortion is called *frequency response* by audio enthusiasts. It is the distortion caused by the variation of transmission loss with frequency. The frequency band of interest here is the voice band or voice channel, that band between 300 and 3400 Hz. Suppose we were to place a 1000-Hz signal at the near-end input of a voice channel at a −10-dBm level. The level measured at the far-end output of the voice channel might be −16 dBm. If the channel were free of amplitude distortion, or we had what is called a *flat channel*, we would expect to get −16 dBm for any frequency measured at the far-end output between 300 and 3400 Hz with a near-end input of −10 dBm. A typical amplitude–frequency distortion characteristic for a voice channel is shown in Figure 6.7. Figure 6.8 shows amplitude distortion for a voice channel as specified (crosshatch) and as measured with equipment back-to-back. Low-pass and band-pass filters are a leading cause of amplitude distortion, particularly when they are in tandem, as one might expect in the PSTN.

The effect of nonuniform attenuation across the voice channel on data communications is to distort the received spectrum and, in turn, the data signal waveform. A reasonable amount of amplitude distortion can be tolerated on a voice channel transporting data. Provided that the loss is not actually infinite at any important signal frequency, it is theoretically possible to construct linear compensating networks that equalize amplitude characteristics over the frequency band of interest. However, if the variation of loss is very great, there are practical difficulties in supplying sufficient signal power to override noise in the high-loss portion of the voice band. The resultant

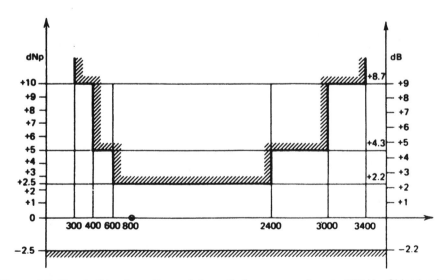

Figure 6.7. Permissible attenuation variation with frequency, reference 800 Hz. Objective for worldwide four-wire chain of 12 circuits in tandem, terminal service. (From Ref. 4. Courtesy of ITU-CCITT.)

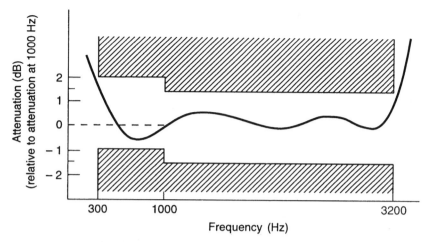

Figure 6.8. Typical attenuation distortion curve for a voice channel. This is a North American-defined, 200- to 3200-Hz voice channel with reference frequency at 1000 Hz. (From Ref. 5. Reprinted with permission.)

effect is a fairly definite limit on useful bandwidth and, consequently, data transmission rate.

Phase Distortion. Phase distortion constitutes the most limiting impairment to data transmission, particularly over the telephone voice channel. What we mean here is that it places an upper limit on data rate. Speech communications over a voice channel can tolerate phase distortion far in excess of that which is severe for high data rate transmission over that same channel.

Phase distortion results from different velocities of propagation at different frequencies across the voice channel. A signal takes a finite time to traverse the telecommunications network end-to-end. If two tones are keyed exactly at the same time at the near end, one at 400 Hz and one at 1800 Hz, the 1800-Hz tone will arrive at the far end before the 400-Hz tone. We can say that the 400-Hz tone was delayed relative to the 1800-Hz tone. Thus we have come to use the term *delay distortion*. Figure 6.9 illustrates typical differential delay across a voice channel.

The measurement of *envelope delay distortion* (EDD) is a common method used to determine the deleterious effects of phase distortion on data transmission. One technique (Ref. 6) injects a test tone at several test frequencies in the voice channel. The test tone is amplitude-modulated with a low-frequency signal (83.33 Hz). At the far end of the circuits, three signals appear: the carrier, an upper sideband, and a lower sideband. The three signals arrive at different times—that is, an envelope of arrival times.

Our interest here is relative envelope delay—that is, the delay at one frequency compared to another where one of these frequencies is a reference

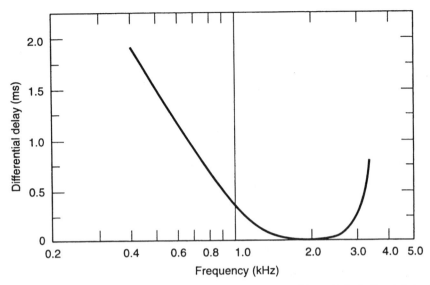

Figure 6.9. Typical differential delay across a voice channel. (From Ref. 5. Reprinted with permission.)

frequency. It is usually selected to be at a point of minimum delay. This occurs on voice channels between 1500 and 2000 Hz. The reference frequency is assigned a delay of 0 μs, and signal energy at all other discrete frequencies above and below the reference arrive later.

Excessive phase distortion gives rise to ISI (intersymbol interference). That is, signal energy components of the first symbol remain, impairing the capability of a decision circuit to determine whether a 1 or a 0 has been received for the second contiguous symbol. ISI can be a major contributor of errors in a received data bit stream.

Noise. There are four types of noise one must consider as impairments to data transmission. These are:

- Thermal noise
- Crosstalk
- Intermodulation noise
- Impulse noise

Thermal noise, as defined by the IEEE (Ref. 1), is the noise occurring in electric conductors and resistors and resulting from random movement of free electrons contained in the conducting material. The noise derives from the fact that such random motion depends on the temperature of the material. Thermal noise has a flat power spectrum out to extremely high frequencies.

Thermal noise is always present in an electrical circuit. It can be characterized by an equivalent noise temperature measured in kelvins (temperature scale based on absolute zero) or by noise figure. Noise figure (in decibels) of a device can be defined as the difference of the signal-to-noise ratio between the output of a device and its input.

In this text, unless otherwise specified, we will be dealing with thermal noise. Thermal noise is random in nature and is treated statistically (a Gaussian distribution). Thermal noise thresholds (noise floors) can be calculated based on a thermal noise floor of an ideally perfect receiver at absolute zero where the noise level is -228.6 dBW in 1 Hz of bandwidth. For the same receiver at room temperature (290 K), this value is -204 dBW/Hz.

Crosstalk is defined by the IEEE (Ref. 1) as the undesired energy appearing in one signal path as a result of coupling from another signal path. Path in this context implies wires, waveguides, or other localized or constrained transmission systems.

For example, on telephone circuits we may hear extraneous voice transmissions. Data circuits can be even more susceptible to extraneous signal energy. Excessive crosstalk, in this case, manifests itself usually as error bursts or, if crosstalk coupling loss is very low, in an unacceptable error rate.

Intermodulation noise is extraneous signal energy derived from some portion of the wanted energy when the signal energy is passed through a nonlinear device. It is characterized by the appearance at the output of a device of frequencies equal to the sums and differences of integral multiples of two or more component frequencies present in the output signal. In Figure 6.10 the input of the device shown has frequencies A and B. (For the voice channel, often 1000 and 1200 Hz are used as examples of frequencies A and B.) The terms *intermodulation noise* and *intermodulation products* are synonymous, sometimes referred to as *IM noise*.

Intermodulation products (IM noise) are generated in nonlinear devices such as amplifiers, transformers, and coils which use ferrous materials, even waveguides. Although many amplifiers operate in a linear region, they display what are called *weak nonlinear characteristics*. And as such we have to contend with their intermodulation products.

Of principal concern are second- and third-order products. For the case of a two-frequency input to a nonlinear device as shown in Figure 6.10, there are the second-order products. The first of these is simply the second harmonic ($2F_1$ and/or $2F_2$), which is the same as that obtained if each input

Figure 6.10. Examples of Intermodulation (IM) products. F_1, frequency 1; F_2, frequency 2.

were applied separately. The other type of second-order product consists of the sum and difference frequencies of each pair of applied frequencies. It can be shown (Ref. 7) that if each of the input frequencies has the same amplitude, the power of the sum and difference products will be 6 dB higher than the power of the second harmonic products.

There are three different types of third-order products. There are the third harmonics $3F_1$ and $3F_2$. There are the $2F_1 \pm F_2$ products, which consist of the sum and difference of one frequency and the second harmonic of another. Compared with the third harmonic amplitude, these products are larger by a factor of $3F_2/F_1$. Thus, if each of the input frequencies has the same amplitude, the power of the $2F_1 \pm F_2$ products will be 9.6 dB higher than that of the third harmonic power. The third type of third-order product is the three-frequency situation considering their sums and differences such as $F_1 \pm F_2 \pm F_3$.

Impulse noise is defined as any burst of noise which exceeds the rms noise level by a given magnitude. The magnitude is nominally 12 dB for a 3-kHz bandwidth (Ref. 7). Impulse noise is the result of electrical sparks or flash-overs of a switch being turned on and off, car ignition noise, lightning, operation of mechanical telephone switches (e.g., step-by-step, crossbar, and rotary), and electrical motor brushes. Impulse noise spikes occur randomly and have a broad spectral content.

In most cases, impulse noise can be neglected for telephone speech transmission. However, for data transmission it can be a major source of errors. Impulse noise objectives are based primarily on the error susceptibility of data signals. The susceptibility depends on the type of modulation used on a data circuit, the bit rate, and the characteristics of the transmission medium employed. The important characteristics of the transmission medium are (a) the amount of phase and attenuation distortion and (b) the rms signal-to-peak impulse noise amplitude ratio.

Impulse noise parameters are based on the fact that digital error rates in the absence of other impairments are approximately proportional to the number of impulses which exceed the rms data signal by a certain number of decibels. Reference 7 defines impulse noise as a voltage increase of 12 dB or more above rms noise for 10 milliseconds or less. Impulse noise is measured by a counter which responds to waveform excursions above a settable threshold. One objective for the PSTN is no more than 15 counts (hits) in 15 minutes for 80% of all connections at a threshold of 5 dB below the received-signal level. The received-signal level is established by transmitting a 1004-Hz tone on the voice channel (Ref. 7, page 157).

6.3.3 Data Circuit Performance

6.3.3.1 Introduction to Error Performance. Certainly the most common performance measure of a data circuit is its error performance. This concept was introduced in Section 5.9.1. There is bit error performance expressed as bit error rate, and there is block error performance. However, with the latter, one must express how long a block is either in bits or in octets. Block error rate is coming more into favor with ATM and its fixed-length cells. A data block is a set of contiguous bits or octets which make up a definable quantity of information. The term *frame* can be used interchangeably with a block.

Bit error rate (ratio) (BER) is defined as the ratio of the total bits received at a data-link destination to the bits received in error; or the contrary, the bits received in error to the total number of bits received. It is expressed as 1 in 10^n or 1×10^{-n}. For example, if 100 bits are received and one of those bits is in error, the bit error rate (BER) is 1×10^{-2}, or 1 error in 10^2.

Let it be understood that there is no data circuit completely free of errors. However, we may have to wait awhile for an error to occur, but eventually an error will occur. There may be periods of time when a circuit is free of errors. The common unit of performance in this case is the error-free second (EFS).

Error rate should be handled statistically, and thus we should talk about probability of error. This is because one of the most common causes of bit errors is random noise peaks, which are treated statistically. So when we express BER, we really are describing the probability of error.

Another measure of data circuit performance is *throughput*. To many of us, throughput defines the number of correct bits, blocks, or frames received which are error-free. To others, it means the number of *useful* bits received that are not in error. Unless we concretely define "useful bits," it is a comparatively meaningless measure of performance. For example, should the start and stop bits be included as useful bits? As we delve further into our discussion of data communications, we will encounter many other bit group-ings that are necessary to telecommunication engineers but have no use for the end user, who may be a programmer. Header information is an example in point. Headers direct traffic to a particular destination and carry out other control functions, but carry no useful information for the data end user.

There are two generic categories of errors. These are random errors and burst errors. If the cause of error is insufficient signal-to-noise ratio, such errors will be random if the noise in question is random in nature such as thermal noise.

An error burst is a string of contiguous bits with a very high probability of being in error. There are two general causes of these error bursts: impulse noise spikes and signal fading. If some portion of a data circuit traverses a line-of-sight microwave link, fading may be expected. Fading may also occur during heavy rainfall events on certain satellite circuits as well as on certain line-of-sight microwave links. There may be other causes of error bursts such as crosstalk.

6.3.3.2 *Error Performance from an ITU-T Perspective.* CCITT in the 1970s recommended a bit error rate no worse than 1×10^{-5}. In the 1980s with the advent of ISDN, CCITT tightened this value to 1×10^{-6} end-to-end for the ISDN 64-kbps channel. In this case the 64-kbps digital channel is a voice channel used as a bearer of data. CCITT Rec. G.821 (Ref. 8) states that fewer than 10% of 1-minute intervals should have a bit error rate worse than

TABLE 6.2 CCITT Error Performance Objectives for International ISDN Connections

Performance Classification	Objective (Note 3)
(a) (Degraded minutes) (Notes 1, 2)	Fewer than 10% of 1-min intervals to have a bit error ratio worse than 1×10^{-6} (Note 4)
(b) (Severely errored seconds) (Note 1)	Fewer than 0.2% of 1-s intervals to have a bit error ratio worse than 1×10^{-3}
(c) (Errored seconds) (Note 1)	Fewer than 8% of 1-s intervals to have any errors (equivalent to 92% error-free seconds)

Note 1. The terms *degraded minutes, severely errored seconds,* and *errored seconds* are used as a convenient and concise performance objective "identifier." Their usage is not intended to imply the acceptability, or otherwise, of this level of performance.

Note 2. The 1-min intervals mentioned in Table 1/G.821 and in the notes (i.e., the periods for $M > 4$ in Annex B) are derived by removing unavailable time and severely errored seconds from the total time and then consecutively grouping the remaining seconds into blocks of 60. The basic 1-s intervals are derived from a fixed period.

Note 3. The time interval T_L, over which the percentages are to be assessed, has not been specified because the period may depend upon the application. A period of the order of any one moment is suggested as a reference.

Note 4. For practical reasons, at 64 kbit/s, a minute containing four errors (equivalent to an error ratio of 1.04×10^{-6}) is not considered degraded. However, this does not imply relaxation of the error ratio objective of 1×10^{-6}.

Source: CCITT Rec. G.821, Fascicle III.5, Table 1/G.821, page 29, CCITT Blue Books, IXth Plenary Assembly, Melbourne, November 1988.

1×10^{-6}. This 10% of 1-minute intervals is classified as "degraded minutes." Table 6.2 provides an overview of error performance recommended by CCITT in Rec. G.821.

ITU-T Rec. G.826 makes use of block-based measurements to make in-service measurements (ISMs) easier to carry out. We review some terminology of ITU-T Rec. G.826 (Ref. 9). Consider Figure 6.11, which shows some of the relationships to be discussed. The following definitions apply:

$$M = \text{number of blocks per second}$$

$$R = \text{number of bits per second}$$

$$\Delta t = 1/M$$

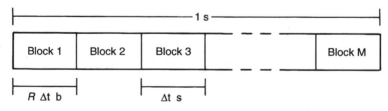

Figure 6.11. Block representation of signals. (From Figure 1, page 57, Ref. 10. Reprinted with permission.)

We introduce the following definitions:

Errored Block (EB). A block in which one or more bits are in error.

Errored Second (ES). A 1-s period with one or more errored blocks.

Severely Errored Second (SES). A 1-s period that contains $\geq 30\%$ EBs or at least one severely disturbed period (SDP). For out-of-service measurements, an SDP occurs when, over a minimum period of time equivalent to four contiguous blocks, either (a) all the contiguous blocks are affected by a high binary error density of 10^{-2} or (b) a loss of signal information is observed. For in-service monitoring purposes, an SDP is estimated by the occurrence of a network defect. CCITT defines *network defect* as the annexes to Rec. G.826 for different network formats such as SDH, PDH, or cell-based (see Chapter 16, ATM).

Background Block Error (BBE). An EB not occurring as part of an SES.

ES Ratio (ESR). The ratio of ESs to total seconds available time during a fixed measurement interval.

SES Ratio (SESR). The ratio of SESs to total seconds in available time during a fixed measurement interval.

BBE Ratio (BBER). The ratio of EBs to total blocks during a fixed measurement interval, excluding all blocks during SESs and unavailable time.

Error Performance Objectives (EPOs). The EPOs are summarized in Table 6.3. These objectives are measured over *available* time in a fixed measurement interval (e.g., 1 month recommended). All three objectives (i.e., ESR, SESR, and BBER) must hold concurrently to satisfy G.826, and they apply end-to-end for a 27,500-km hypothetical reference path (HRP), which is shown in Figure 6.12.

Availability. The concept of available time is the same as was defined in ITU-T Rec. G.821, except that the BER $> 1 \times 10^{-3}$ criterion is replaced by an SES criterion. Thus, unavailable time begins at the start of a block of 10

TABLE 6.3 Error Performance Objectives (EPOs) for G.826

Rate (Mb/s)	Bits / block	ESR	SESR	BBER
1.5–5	2000–8000	0.04	0.002	3×10^{-4}
> 5–15	2000–8000	0.05	0.002	2×10^{-4}
> 15–55	4000–20,000	0.075	0.002	2×10^{-4}
> 55–160	6000–20,000	0.16	0.002	2×10^{-4}
> 160–3500	15,000–30,000	—[a]	0.002	10^{-4}
> 3500	FFS[b]	FFS	FFS	FFS

[a]No objective given due to the lack of available information.
[b]FFS, for further study.
Source: Reference 10, Table 2, page 58.

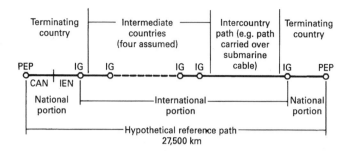

PEP: Path end point
IG: International gateway

Figure 6.12. Hypothetical reference path (HRP). (From Figure 2, page 57, Ref. 10. Reprinted with permission.)

TABLE 6.4   Apportionment Rules

Percent of EPO			
National Portion[a]		International Portion[a]	
Block Allowance	Distance Allowance	Transit Allowance	Distance Allowance
17.5% to both terminating countries	1% per 500 km	2% per intermediate country 1% per terminating country	1% per 500 km

[a]Satellite hops each receive 35%, but the distance of the hop is removed from the distance allowance.
[b]Four intermediate countries are assumed.
Source: Reference 10, Table 3, page 59.

consecutive SESs. The unavailable time stops (and of course available time begins) at the start of 10 consecutive ESs (or better), each of which is not severely errored.

Apportionment. The apportionment rules contained in G.826 are different from those used in G.821. These rules are given in Table 6.4, and one should refer to Figure 6.12 for the HRP.

The IEEE reference article (Ref. 10) notes that it is not immediately obvious that the EPO percentage figures of Table 6.4 yield 100%. However, under the assumption of four intermediate countries and no satellite hops, the following breakdown can be obtained:

Terminating countries:
$$2 \times 17.5\% + 2 \times 1\% = 37\%$$

Intermediate counties:
$$4 \times 2\% = 8\%$$

Distance allowance:

$$27{,}500 \text{ km} = 55 \times 500 \text{ km} (55 \times 1\%) = \underline{55\%}$$
$$\text{Total} \quad \overline{100\%}$$

If one satellite hop is employed, it uses 35%, corresponding to a nominal hop distance of 17,500 km.

The identification of an SES event during ISM is not so straightforward. This is because the definition of an SES involves SDPs. SDP events can only be measured during an out-of-service condition. Thus, "equivalent" ISM events need to be defined if SDPs are to be detected while in-service. It is recognized that there is not an exact $1:1$ correspondence between SDP events measured in-service and measured out-of-service. The objective of annexes 2, 3, and 4 to G.826 is to provide ISM events that are reasonably close to the G.826 error events.

It should be noted that G.826 generally applies to underlying transport systems such as E1/T1 (PDH—Chapter 7), SDH/SONET (Chapter 15), and cell-based transport (Chapter 16). Of course the impact of G.826 will be felt on data systems being carried by that digital transport network.

The material in Section 6.3.3.2 was derived from Refs. 8–10.

6.3.3.3 Error Performance from a Bellcore Perspective. Reference 11 states that the bit error rate at digital cross-connects up through DS3 shall be less than 2×10^{-10}, excluding burst errored seconds.* This value, of course, is that of the underlying digital network, which is described in Chapter 7. This value will be the best-case value where data bits are being mapped in PCM time slots. Specific data offerings on the underlying digital network are described in Section 7.6.

6.4 INTERFACE AT THE PHYSICAL LAYER

6.4.1 Introduction

A data link consists of equipment with resident software and related transmission facilities that allow two physically separated terminals data communication capability. Care must be taken when using this definition in the multilayered setting of open systems interconnection (OSI) or similar environments. What we describe below is the actual implementation of OSI layer 1.

Let us consider a simple point-to-point data connectivity. It will consist of (a) data equipment on one side of the link and (b) similar equipment on the other side of the link. The two sides are connected by some transmission facility. This facility may be made up of wire-pair, coaxial cable, fiber-optic

---

*Burst errored second is any errored second that contains at least 100 errors.

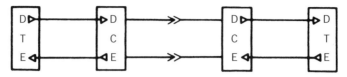

▷ Interface generator
▷ Interface load
≫ Telecommunication channel

Figure 6.13. Functional diagram of data terminal equipment (DTE) and data communication equipment (DCE) in a point-to-point configuration. The electrical and functional interface between DTE and DCE is discussed in this section.

cable, or a radio system or combinations thereof. The interconnection may or may not involve one or more switching nodes.

Our concern here is not the transmission media itself. That portion of the link is adequately covered in Chapters 7, 8, 14, and 15. In this section we discuss two key functional entities found in every data link. These are the *data terminal equipment* (DTE) and the *data communication equipment* (DCE). A simplified diagram of the DTE and DCE in a point-to-point configuration is illustrated in Figure 6.13.

Data terminal equipment includes the familiar PCs/workstations, computers/processors, mass storage devices, printers, and servers. We sometimes refer to these items of equipment as *input–output* (I/O) devices.

The DCE is a functional entity such as a modem that makes the electric signal from the DTE compatible with the transmission medium. It also provides a number of control and interface functions. When we think of DCE, most of us relate it to a modem, a device that generates a digitally modulated audio tone compatible with the analog network. However, those devices that condition a data signal for the digital network are also in the category of DCE. Among these are the North American (ATT) Digital Data System (DDS) DSU/CSU and the CCITT data conversion/conditioning equipment typically meeting ITU-T Rec. V.110 requirements (Ref. 12). These latter items of equipment map a data bit stream into a 64-kbps digital voice channel.

6.4.2 The DTE–DCE Interface

This all-important interface is broken down into a functional interface and an electrical interface. It is defined by many standards. The most familiar of these standards with worldwide acceptance is the Electronic Industries Alliance/Telecommunication Industry Association (U.S.) EIA/TIA-232 (Ref. 2). In this case the functional and electrical interfaces are found under the same cover. They are detailed separately in many other standards. For example, EIA/TIA-530 defines a 25-pin functional interface between DTE

TABLE 6.5   EIA/TIA-232 Pin Assignments

Pin Number	CCITT Number	Circuit	Description
1	—	—	Shield
2	103	BA	Transmitted data
3	104	BB	Received data
4	105/133	CA/CJ (Note 1)	Request to send/Ready for receiving
5	106	CB	Clear to send
6	107	CC	DCE ready
7	102	AB	Signal common
8	109	CF	Received line signal detector
9	—	—	(Reserved for testing)
10	—	—	(Reserved for testing)
11	126	(Note 4)	Unassigned
12	122/112	SCF/CI (Note 2)	Secondary received line signal Detector/Data signal rate selector (DCE source)
13	121	SCB	Secondary clear to send
14	118	SBA	Secondary transmitted data
15	114	DB	Transmitter signal element timing (DCE source)
16	119	SBB	Secondary received data
17	115	DD	Receiver signal element timing (DCE source)
18	141	LL	Local loopback
19	120	SCA	Secondary request to send
20	108/1,/2	CD	DTE ready
21	140/110	RL/CG	Remote loopback/Signal quality detector
22	125	CE	Ring indicator
23	111/112	CH/CI (Note 2)	Data signal rate selector (DTE/DCE source)
24	113	DA	Transmit signal element Timing (DTE source)
25	142	TM	Test mode
26		(Note 3)	No connection

Note 1: When hardware flow control is required Circuit CA may take on the functionality of Circuit CJ.

Note 2: For designs using interchange circuit SCF, interchange circuits CH and CI are assigned to pin 23. If SCF is not used, CI is assigned to pin 12.

Note 3: Pin 26 is contained on the Alt A connector only. No connection is to be made to this pin.

Note 4: Pin 11 is unassigned. It will not be assigned in future versions of EIA/TIA-232. However, in international standard ISO 2110, this pin is assigned to CCITT Circuit 126, Select Transmit Frequency.

Source: Reference 2. Copyright Electronic Industries Association/Telecommunications Industries Association. Reprinted with permission.

TABLE 6.6   EIA/TIA-232 Circuits by Category

Circuit Mnemonics	CCITT Number	Circuit Name	Circuit Direction	Circuit Type
AB	102	Signal Common	—	Common
BA	103	Transmitted data	To DCE	Data
BB	104	Received data	From DCE	Data
CA	105	Request to send	To DCE	Control
CB	106	Clear to send	From DCE	Control
CC	107	DCE ready	From DCE	Control
CD	108/1,/2	DTE ready	To DCE	Control
CE	125	Ring indicator	From DCE	Control
CF	109	Received line signal detector	From DCE	Control
CG	110	Signal quality detector	From DCE	Control
CH	111	Data signal rate selector (DTE)	To DCE	Control
CI	112	Data signal rate selector (DCE)	From DCE	Control
CJ	133	Ready for receiving	To DCE	Control
RL	140	Remote loopback	To DCE	Control
LL	141	Local loopback	To DCE	Control
TM	142	Test mode	From DCE	Control
DA	113	Transmitter signal element timing (DTE)	To DCE	Timing
DB	114	Transmitter signal element timing (DCE)	From DCE	Timing
DD	115	Receiver signal element timing (DCE)	From DCE	Timing
SBA	118	Secondary transmitted data	To DCE	Data
SBB	119	Secondary received data	From DCE	Data
SCA	120	Secondary request to send	To DCE	Control
SCB	121	Secondary clear to send	From DCE	Control
SCF	122	Secondary received line signal detector	From DCE	Control

Source: Reference 2. Copyright Electronic Industries Association / Telecommunications Industries Association. Reprinted with permission.

and DCE, and EIA/TIA-422 and EIA/TIA-423 uniquely define the electrical interface.

Table 6.5 gives the functional interface of EIA/TIA-232. The table shows the assignment and function of each pin in a 25-pin plug for a cable connecting the DTE to the DCE. The related CCITT standard is CCITT Rec. V.24 (Ref. 13). Table 6.6 gives the EIA/TIA-232 interchange circuits by category and provides equivalent CCITT Rec. V.24 circuit nomenclature.

6.4.2.1   *EIA/TIA-232 Electrical Interface.* For data interchange circuits, the signal is considered in the marking condition (i.e., binary 1) when the voltage on the interchange circuit, measured at the interface point, is more than -3 volts with respect to signal ground (circuit AB). The signal is considered in the spacing condition (i.e., binary 0) when the voltage is more

TABLE 6.7   Standard Signal Sense and Terminology

Notation	Interchange Voltage	
	Negative	Positive
Binary state	1	0
Signal condition	Marking	Spacing
Function	**OFF**	**ON**

positive than $+3$ volts with respect to signal ground (circuit AB). The region between $+3$ volts and -3 volts is defined as the *transition region*. The signal state is not uniquely defined when the voltage is in this transition region.

During the transmission of data, the marking condition is used to denote the binary state *one*, and the spacing condition is used to denote the binary state *zero*.

For the timing and electrical control interchange circuits, the function is considered *on* when the voltage on the interchange circuit is more positive than $+3$ volts with respect to circuit AB, and it is considered *off* when the voltage is more negative than -3 volts with respect to circuit AB. The function is not uniquely defined in the transition region between $+3$ volts and -3 volts. Table 6.7 reflects standard signal sense and terminology.

It should be noted that there are three types of interchange signals described in the standard: data, control, and timing.

EIA/TIA-232 places the following limitations on the interchange signals transmitted across the interface point, exclusive of external interference:

1. All interchange signals entering into the transition region shall proceed through the transition region to the opposite signal state and shall not reenter the transition region until the next significant change of signal condition.
2. There shall be no reversal of direction change while the signal is in the transition region.
3. For control interchange circuits, the time required for the signal to pass through the transition region during a change in state shall not exceed 1 ms.
4. For data and timing interchange circuits, the time required for the signal to pass through the transition region shall be in accordance with the table below:
5. The maximum instantaneous rate of voltage change should not exceed 30 volts per microsecond.

Duration of Unit Interval	Maximum Allowable Transition Time
UI $\geq$ 25 ms	1 ms
25 ms $>$ UI $\geq$ 125 μs	4% of a unit interval
125 μs $\geq$ UI	5 μs

Definition of Unit Interval (UI) (Also called "Signal Element"). The part of a signal that occupies the shortest interval of a signaling code. It is considered to be of unit duration in building up signal combinations (1). "Unit duration" in this context is the duration of 1 bit, and thus the UI is a bit.

Note: Good engineering practice requires that the rise and fall times of data and timing signals should be approximately equal (within a range of $2:1$ or $3:1$).

6.4.2.2 *Functional Description of Selected Interchange Circuits.* The circuit identifiers are shown in Table 6.6.

Circuit BA—Transmitted Data. Direction: To DCE. Signals on this circuit are generated by the DTE and are transferred to the local DCE for transmission of data to remote DCE(s) or for maintenance or control of the local DCE.

The DTE shall hold circuit BA (transmitted data) in the marking condition at all times when no data are being transmitted. In all systems, the DTE shall not transmit data unless *on* condition is present on all of the following circuits, where implemented:

1. Circuit CA (request to send)
2. Circuit CB (clear to send)
3. Circuit CC (DCE ready)
4. Circuit CD (DTE ready)

Circuit BB—Received Data. Direction: From DCE. Signals on this circuit are generated by the DCE in response to data signals received from remote DCE(s), or by the DCE in response to maintenance or control data signals. Circuit BB (received data) shall be held in the binary *one* (marking) condition at all times when circuit CF (received line signal detector) is in the *off* condition.

On a half-duplex channel, circuit BB shall be held in the binary *one* (marking) condition when circuit CA (request to send) is in the *on* condition and for a brief interval following the *on* to *off* transition of circuit CA to allow for the completion of transmission and decay of line reflections.

Circuit CA—Request to Send. Direction: To DCE. This circuit is used to condition the local DCE for data transmission and, on a half-duplex channel, to control the direction of data transmission of the local DCE.

For simplex or duplex operation, the *on* condition maintains the DCE in the transmit mode. The *off* condition maintains the DCE in a nontransmit mode.

For half-duplex operation, the *on* condition maintains the DCE in the transmit mode and inhibits the receive mode. The *off* condition maintains the DCE in a receive mode.

A transition from *off* to *on* instructs the DCE to enter the transmit mode. The DCE responds by taking such action as may be necessary and indicates completion of such actions by turning *on* circuit CB (clear to send), thereby

TABLE 6.8   EIA/TIA-530 Connector Contact Assignments

Contact Number	Circuit	CCITT Number	Interchange Points	Circuit Category	Direction
1	Shield	—	—		
2	BA	103	A-A'	I	To DCE
3	BB	104	A-A'	I	From DCE
4	CA/CJ (Note 1)	105/133	A-A'	I	To DCE
5	CB	106	A-A'	I	From DCE
6	CC (Note 3)	107	A-A'	II	From DCE
7	AB	102A	C-C'	—	
8	CF	109	A-A'	I	From DCE
9	DD	115	B-B'	I	From DEC
10	CF	109	B-B'	I	From DCE
11	DA	113	B-B'	I	To DCE
12	DB	114	B-B'	I	From DCE
13	CB	106	B-B'	I	From DCE
14	BA	103	B-B'	I	To DCE
15	DB	114	A-A'	I	From DCE
16	BB	104	B-B'	I	From DCE
17	DD	115	A-A'	I	From DCE
18	LL	141	A-A'	II	To DCE
19	CA/CJ (Note 1)	105/133	B-B'	I	To DCE
20	CD (Note 3)	108/1, /2	A-A'	II	To DCE
21	RL	140	A-A'	II	To DCE
22	CE	125	A-A'	II	From DCE
23	AC	102B	C-C'		
24	DA	113	A-A'	I	To DCE
25	TM	142	A-A'	II	From DCE
26	(Note 2)	—	—	—	—

Note 1: When hardware flow control is required, circuit CA may take on the functionality of circuit CJ.

Note 2: Contact 26 is contained on the Alt A connector only. No connection is to be made to this contact.

Note 3: In ANSI/EIA-530-87, circuits CC and CD were category I circuits. Interoperation between category I and II circuits is not possible without circuitry similar to that in Figure B.1 in Ref. 14.

Source: Reference 16, Figure 3.9, page 16. Courtesy of Electronic Industries Alliance/Telecommunications Industry Association. Reprinted with permission.

indicating to the DTE that data may be transferred across the interface point on interchange circuit BA (transmitted data).

A transition from *on* to *off* instructs the DCE to complete the transmission of all data which were previously transferred across the interface point on interchange circuit BA and then assume a nontransmit mode or a receive mode as appropriate. The DCE responds to this instruction by turning *off* circuit CB (clear to send) when it is prepared to again respond to a subsequent *on* condition of circuit CA (request to send).

Note: A nontransmit mode does not imply that all line signals have been removed from the communications channel.

When circuit CA (request to send) is turned *off*, it shall not be turned *on* again until circuit CB (clear to send) has been turned *off* by the DCE.

6.4.2.3 EIA/TIA Interface Standards: EIA/TIA-530, EIA/TIA-422, and EIA/TIA-423. EIA/TIA-530 (Ref. 14) defines a 25-pin, and alternatively a 26-pin, interface between DTE and DCE. It is designed for data rates up to 2.1 Mbps with serial bit streams. The electrical interfaces are described in EIA/TIA-422 (Ref. 15) and EIA/TIA-423 (Ref. 16). Connector contact assignments are shown in Table 6.8, and Table 6.9 gives the interchange circuits.

TABLE 6.9   EIA/TIA-530 Interchange Circuits

Circuit Mnemonics	CCITT Number	Circuit Name	Circuit Direction	Circuit Type
AB	102	Signal common		COMMON
AC	102B	Signal common		
BA	103	Transmitted data	To DCE	DATA
BB	104	Received data	From DCE	
CA	105	Requested to send	To DCE	
CB	106	Clear to send	From DCE	
CF	109	Received line signal detector	From DCE	
CJ	133	Ready for receiving	To DCE	
CE	125	Ring indicator	From DCE	CONTROL
CC	107	DCE ready	From DCE	
CD	108/1, /2	DTE ready	To DCE	
DA	113	Transmit signal element timing (DTE source)	To DCE	
DB	114	Transmit signal element timing (DCE source)	From DCE	TIMING
DD	115	Receiver signal element timing (DCE source)	From DCE	
LL	141	Local loopback	To DCE	
RL	140	Remote loopback	To DCE	
TM	142	Test mode	From DCE	

Source: Reference 14, Figure 4.1, page 17. Courtesy of Electronic Industries Alliance/Telecommunication Industry Association. Reprinted with permission.

EIA/TIA-422 (Ref. 15) and EIA/TIA-423 (Ref. 16) give the electrical characteristics for interchange circuits and are companion standards for EIA/TIA-530 (Ref. 14). EIA/TIA-422 provides for balanced voltage digital interface circuits, and EIA/TIA-423 provides for unbalanced voltage digital interface circuits. Remember that an unbalanced connection means that one signal lead is grounded.

The transition region at the receiver for EIA/TIA-422 is between $+200$ mV and -200 mV. Thus any voltage more positive than $+200$ mV represents a spacing or 0 condition ("on" condition) and any voltage more negative

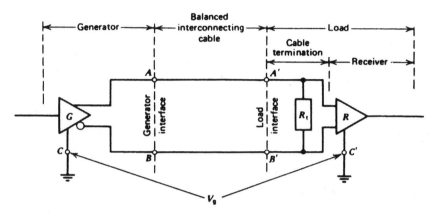

Figure 6.14. Balanced digital interface circuit, EIA/TIA-422. A, B = generator interface A', B' = load interface; C = generator circuit ground; C' = load circuit ground; R = optional cable termination resistance; V_g = ground potential difference. (From Ref. 17. Copyright Electronic Industries Alliance.)

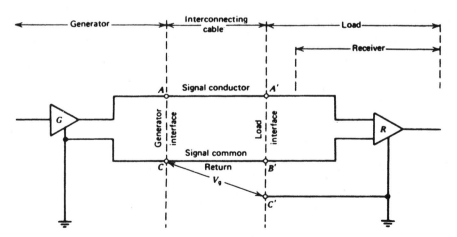

Figure 6.15. Unbalanced digital interface circuit, EIA/TIA-423. A, C = generator interface; A', B' = load interface; C' = load circuit ground; C = generator circuit ground; V_g = ground potential difference. (From Ref. 16. Copyright Telecommunication Industry Association.)

than -200 mV represents a marking or 1 condition. The nominal signal voltages at the receiver are -6 and $+6$ volts DC.

EIA/TIA-423 is a similar standard but is used for unbalanced circuits. The receiver transition range is the same as EIA/TIA-422, and the binary 1 and binary 0 significant states are the same. The nominal data rate maximum is 100 kbps.

Figure 6.14 shows the EIA/TIA-422 balanced digital interface circuit, and Figure 6.15 illustrates the EIA/TIA-423 unbalanced digital interface circuit.

6.5   THE QUESTION OF BANDWIDTH

6.5.1   Bandwidth Versus Bit Rate

When I started in the field of data communications, I always wondered *how many bits per second of data could fit into a unit of bandwidth*. The question still seems to bother a lot of people as my seminar experience bears out. We first will define bandwidth, at least from one perspective. Then there will be a very brief review by two well-known theoreticians. We then will delve into some practical matters and attempt to answer the question.

6.5.2   Bandwidth Defined

Bandwidth provides a vital role in data communications. It can be defined as the range between the lowest and highest frequencies used for a particular purpose. For our purposes we should consider bandwidth as the range of frequencies within which the performance characteristics of a device are above certain specified limits. For filters, attenuators, amplifiers, transmission lines, and other electromagnetic active and passive equipment, these limits are generally taken where a signal will fall 3 dB below the average (power) level in the passband, or below the level at a reference frequency. The voice channel is one notable exception. In North America it is specifically defined at the 10-dB points about the reference frequency of 1000 Hz, approximately in the band 200–3300 Hz. The CCITT voice channel occupies the band 300–3400 Hz.

The 3-dB bandwidth concept illustrated in Figure 6.16.

6.5.3   First and Second Bandwidth Approximations

If we remain in the binary domain (i.e., we deal only with two states called 1s and 0s), we often will approximate bandwidth as 1 bit per hertz of bandwidth. This first approximation roughly allows for nominal impairments such as amplitude–frequency response, phase distortion, and noise. It is a *rough* approximation.

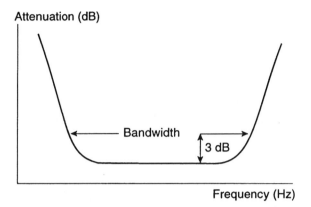

Figure 6.16. The concept of the 3-dB power bandwidth.

H. Nyquist of Bell Telephone Laboratories (Ref. 17) was a pioneer in information theory. He essentially stated that theoretical error-free transmission of two bits of information can be related to the 1 Hz of bandwidth. This is often called the *Nyquist rate* or *Nyquist bandwidth*. In other words, if we have 1000 Hz of bandwidth, theoretically it can transport 2000 bps "error-free."

6.5.4 Shannon

Claude Shannon, also of Bell Telephone Laboratories, came up with a theory contradicting* Nyquist's 2-bit rule (Ref. 18). Shannon's traditional formula relates channel capacity C (binary digits or bits/second) to bandwidth W with the following relationship:

$$C = W \log_2(1 + S/N)$$

where S/N is the signal-to-noise ratio, which, in this case, must be expressed as a numeric. If S/N were 30 dB, its numeric equivalent would be 1000. Let the bandwidth be 3000 Hz, then

$$C = 3000 \times 10 \text{ (approximately)}$$

$$= 30,000 \text{ bps}$$

Often we will hear of the *Shannon limit* in reference to the above relationship. As we mentioned previously, one of the principal impairments in the transmission channel is phase distortion. It is mainly this impairment that

---

*"Contradicting" may be a little abrupt. Where Nyquist basically considered intersymbol interference, Shannon's only parameter was signal-to-noise ratio.

restricts us from reaching the Shannon limit. In a bandwidth-limited channel, as the bit rate increases beyond a certain point, intersymbol interference begins to impact bit error rate.

6.5.5 Summary

As we describe in the next chapter, we can pack more bits per hertz of bandwidth by turning to multilevel systems or *M*-ary techniques, rather than binary techniques.

Thus, we can turn to the original idea of 1 bit per hertz for binary transmission, admitting it is a very rough first approximation. We warn that the conventional habit of equating bandwidth to bit rate can be fraught with danger.

REFERENCES

1. *The IEEE Standard Dictionary of Electrical and Electronic Terms*, 6th ed., IEEE Std. 100-1996, IEEE, New York, 1996.
2. *Interface Between Data Terminal Equipment and Data Circuit Terminating Equipment Employing Serial Binary Data Interchange*, EIA/TIA-232F, Telecommunication Industry Association, Washington DC, 1997.
3. *9600 Bits per Second Modem Standardized for Use of Point-to-Point 4-Wire Telephone-Type Circuits*, CCITT Rec. V.29, Vol. VIII.1, IXth Plenary Assembly, Melbourne, 1988.
4. *Attenuation Distortion*, CCITT Rec. G.132, IXth Plenary Assembly, Melbourne, 1988.
5. Roger L. Freeman, *Telecommunication Transmission Handbook*, 4th ed., John Wiley & Sons, New York, 1998.
6. *Data Communications Using Voiceband Private Line Channels*, Bell System Technical Reference Publication 41004, AT & T, New York, October 1973.
7. *Transmission Systems for Communications*, 4th ed., Bell Telephone Laboratories, Holmdel, NJ, 1970.
8. *Error Performance of an International Connection Forming Part of an ISDN*, CCITT Rec. G.821, Vol. III, Fascicle III.3, IXth Plenary Assembly, Melbourne, 1988.
9. *Error Performance Parameters and Objectives for International Constant Bit Rate Digital Paths at or above the Primary Rate*, ITU-T Rec. G. 826, ITU Geneva, 1993.
10. Monsoor Shafi and Peter J. Smith, The Impact of G.826. *IEEE Communications Magazine*, September 1993.
11. *Transport Systems Generic Requirements: Common Requirements*, Bellcore/Telcordia GR-499-CORE Issue 1, Bellcore Piscataway, NJ, December 1995.
12. *Support of Data Terminal Equipment (DTEs) with V-Series Type Interfaces by an Integrated Services Digital Network (ISDN)*, CCITT Rec. V.110, Vol. VIII.1, IXth Plenary Assembly, Melbourne, 1988.

13. *List of Definitions for Interchange Circuits Between Data Terminal Equipment (DTE) and Data Circuit-Terminating Equipment (DCE)*, CCITT Rec. V.24, IXth Plenary Assembly, Melbourne, 1988.

14. *High Speed 25-Position Interface for Data Terminal Equipment and Data Circuit-Terminating Equipment, Including Alternative 26-Position Connector*, EIA/TIA-530, Telecommunication Industry Association, Washington DC, May 1992.

15. *Electrical Characteristics of Balanced Voltage Digital Interface Circuits*, EIA/TIA-422, Telecommunication Industry Association, Washington DC, December 1978.

16. *Electrical Characteristics of Unbalanced Voltage Digital Interface Circuits*, EIA/TIA-423, Telecommunication Industry Association, Washington DC, December 1978.

17. H. Nyquist, Certain Topics in Telegraph Transmission Theory, *AIEE Transactions*, Vol. 47, April 1928.

18. C. E. Shannon and W. Weaver, *The Mathematical Theory of Communications*, University of Illinois Press, Urbana, 1963.

THE TELECOMMUNICATIONS NETWORK AS A VEHICLE FOR DATA TRANSPORT

7.1 THE PUBLIC SWITCHED TELECOMMUNICATION NETWORK

7.1.1 Introduction

A major means of longer-distance data transport is the public switched telecommunication network (PSTN). It is essentially ubiquitous and is often the most cost-effective alternative for such transport. To intelligently decide when and how to use the PSTN to serve data communication needs, we must know how the network operates, as well as its advantages and limitations. The intent of this section is to provide the reader with a review of the operation of the PSTN from the point of view of a potential data user. Emphasis is placed on the digital network, both transmission and switching.

The PSTN consists of a user access configuration, local switching, tandem switching in selected situations with interconnecting trunks, and a long-distance network with its transit exchanges (switches). The PSTN access configuration primarily consists of wire-pair subscriber loops that enter a local serving telephone exchange via a main distribution frame. The main distribution frame facilitates cross-connects. The loops then may be time-division-multiplexed into groups of 24 or 30 channels in pulse code modulation (PCM) configurations if the local exchange is digital. For most digital exchanges, each switch port accepts such a channel group.

If the local serving exchange is an analog switch, each switch port accepts one single voice channel. For the analog situation, at some point there will be conversion to a PCM format to interface with higher levels of the local network (local exchange carrier in the United States) and the toll (long distance) network (IXC or interexchange carrier network).

Given a particular desired connectivity, several switches in tandem may be required to intervene to set up a call. In network design, the trunks that connect switches are called *links*. CCITT Rec. E.171 (Ref. 1) states that there shall not be more than 12 links in tandem for an international connection. The value of 12 is broken down into four links for the international portion of the connection, four for the national side on one end, and four for the national side on the other end of a call. Each time a link is added to a connection, the quality of service (QoS) tends to deteriorate. For digital connections, the deterioration is a degradation in bit error rate (BER). For this reason, the number of links in tandem which make up a telecommunication connectivity must be limited.

It should be noted that for voice operation over the digital PSTN, the requirements for BER are far less stringent than those for data operation. Voice operation requires a BER of $< 1 \times 10^{-3}$, whereas for data operation the value is better than 1×10^{-6}.

Our concerns here are as follows:

1. What services are available on the PSTN for the data end user?
2. How, and to what extent, will the PSTN degrade our data signal? This includes other "degradations" such as loss of traffic, out-of-sequence delivery, and so on.
3. What limitations are there in data rate and speed of service?

7.1.2   Access to the PSTN: The Subscriber Network

The subscriber plant presents a major constraint to effective data rate. The subscriber plant consists of wire-pair terminations on the user premises extending to the local serving exchange. We can expect the characteristic impedance (Z_0) of the typical wire pair as connected to be 600 or 900 ohms.

Ordinarily, wire pair is an excellent transmission medium. It can support more than 100 Mbps over limited distances. However, our in-place subscriber plant probably cannot support such data rates. We must consider the following:

1. Whether a particular pair has inductive loading and bridged tap(s). Both limit useful bandwidth, add to group delay, and notably constrain data rate.
2. The age of the portion of the plant in question and the condition of the loop under consideration. Some plants have been in operation well over 50 years. Cables get wet; they dry and get wet again. Insulation deteriorates. It has been said that only about 50% of the subscriber plants in the United States can support the ISDN basic rate (160 kbps in the United States).
3. Wire pairs can also be noisy. One cause is the accidental grounding of one side of the pair. Another cause is the deterioration of balance to ground.

If the wire pair extends to an analog switch, we can expect an increase in impulse noise as well as degradation of both amplitude and phase characteristics of a particular connectivity. Poor grounding in the total wire plant can be another noise source.

The voice channel deriving from the wire pair at some point will enter multiplex equipment. As local serving switches are modernized to digital exchanges, that point will be on PCM channel banks between the switch subscriber line and main distribution frame or be incorporated at the input of the switch itself. Once the voice channel enters the channel bank, it is restricted in frequency by a low-pass filter to the band 300–3400 Hz.

Delay is another consideration for the PSTN. If a (geostationary) satellite link is involved in a connectivity, the one-way propagation time to the satellite is about 125 μs and another 125 μs for the link to the distant earth station. This totals 250 μs for one-way propagation delay alone. Delay is troublesome for voice and some image transmission systems. It notably adds to the latency on data circuits, especially on those that require many hand-shakes* and turn-arounds. Special problems may arise if a distant data terminal is emulating a local area network (LAN) user. Depending on the MAC protocol involved, delay may be an overriding consideration.

7.2 INTRODUCTION TO DIGITAL NETWORKS

7.2.1 Rationale

The PSTN on a worldwide basis will shortly be all-digital. How will this digital network impact the data user? Equally important is the private network that will most certainly be digital. Many medium- and large-size enterprises are turning to the development and implementation of their own private networks. So our question is equally pertinent. It is therefore incumbent upon the data user to be fully knowledgeable of digital networks. That is what we set out to do in this subsection.

The digital network is based on PCM. We discuss the development of a PCM signal from the analog voice channel, its coding, the frame, and higher-level multiplexing, among other factors. We then cover digital network topologies and impairments. European and North American systems are reviewed. This is followed by interfacing a data bit stream to the digital 64-kbps channel.

7.2.2 Development of a PCM Signal

When PCM was originally invented (Reeves, ITT STL 1937), the only requirement was to transmit voice (speech). Thus it was built around the nominal 4-kHz voice channel. With the invention of the transistor, PCM

---

*Typically on X.25 and IP connectivities unless compensated for. Frame relay is probably the most forgiving.

became a feasible product (Shockley, Bell Laboratories 1948) and voice transmission was still the only requirement. Today's digital network is optimized for speech transmission, and data transmission is a secondary issue.

There are three steps in the development of a PCM signal:

1. Sampling
2. Quantization
3. Coding

Remember that we are going to digitize an analog voice channel. We are going from a continuously varying amplitude signal (analog) to a signal made up of discrete elements that are digital.

Sampling. The first step is sampling the voice channel. The sampling is based on Nyquist's sampling theorem. It states: "If a signal $f(t)$ is sampled at regular intervals and at a rate higher than twice the highest significant signal frequency, then the samples contain all the information of the original signal. The function $f(t)$ may be reconstructed from these samples by the use of a low pass filter."

If we were to sample the nominal 4-kHz voice channel, the sampling rate would be 8000 samples per second because the highest significant frequency is 4000 Hz (i.e., 4000 Hz × 2). For a 15-kHz program channel, the sampling rate would be 30,000 times a second (i.e., 15,000 × 2). An NTSC video channel, if sampled at the Nyquist rate, would be sampled 8.4×10^6 times a second. The NTSC TV video channel has a nominal bandwidth of 4.2 MHz.

The output of the sampler is a pulse amplitude modulation (PAM) wave. Such a wave, which is the result of sampling a simple sinusoid, is shown in Figure 7.1.

With some notable exceptions, practical PCM systems involve time division multiplexing (TDM) carried out in the sampling process. Sampling in these cases does not involve just one voice channel, but many. In practice, one system to be discussed samples 24 voice channels in sequence (T1), and another samples 30 voice channels (E1). The result is a more complex PAM wave than is shown in Figure 7.1.

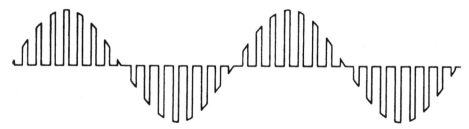

Figure 7.1. A PAM wave as a result of sampling a simple sinusoid.

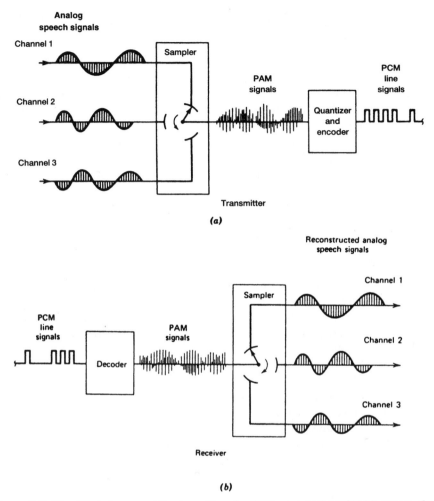

Figure 7.2. Simplified diagram of the formation of a PAM wave and a PCM bit stream. Only three channels are sampled sequentially in this example. (a) Transmitter, (b) receiver. (Courtesy of GTE Lenkurt Inc., San Carlos, California.)

A simplified diagram of the processing involved is shown in Figure 7.2. Here just three channels are sampled sequentially.

A sample is simply a measurement of the voltage level of a channel at a moment in time. This is done by gating. For the case of a 24-channel system, samples are taken sequentially 8000 times per second. The time it takes to sample 24 channels just once is 125 μs. If we are to sample each channel 8000 times per second, the time to sequence one sampling of 24 channels is 1/8000 s, or 125 μs. If we are to sample 24 channels, the gate could be open as long as ~ 5.2 μs per channel (125/24 = ~ 5.2). Sequencing through each channel just once is called a *frame*. And, of course, a frame is 125 microseconds in duration.

Quantization and Coding. The next step in the formation of a PCM signal is quantization. In modern PCM systems, coding is carried out simultaneously with the quantization process. Realize that quantization is simply setting discrete voltage level possibilities. Each voltage level is assigned a binary code, a sequence of "1s" and "0s."

The number of bits required to represent a character is called a *code length*, or, more properly, a *coding level*. For instance, as we have learned in Chapter 5, a binary code with four discrete binary elements (a four-level code) can code 2^4 separate and distinct meanings, or 16 characters ($2^4 = 16$); a five-level code would provide 2^5, or 32, distinct characters or meanings. In modern PCM, we use an eight-level code deriving 256 different code possibilities. Thus we must quantize the sample voltage levels into just 256 possibilities.

The PAM wave has an infinite number of code level (voltage level) possibilities. For all intents and purposes it is still an analog signal. If the positive excursion of a PAM wave is between 0 and +1 volt, the reader should ask: How many discrete values are there between 0 and 1? All values must be considered, including 0.4578046 volts and an infinite number of other values.

The intensity range of voice signals over an analog telephone channel is of the order of 50 dB. The −1 to 0 to +1 volt of the PAM wave at the excursion coder input should represent that 50-dB range. Furthermore, we defeat the purpose of digitizing if we assign an infinite number of values to satisfy every level encountered in the 50-dB range (or a range from −1 to +1 volt). The key, of course, is to assign *discrete* levels from −1 volt to +1 volt (the 50-dB range).

Actually, the assignment of these discrete values is what we call *quantization*. To cite an example, consider Figure 7.3. Between −1 and +1 volt, 16 quantum steps exist and are coded as shown in Table 7.1.

Figure 7.3 shows that step 12 is used twice, but in neither instance is it the true value of the impinging sinusoid. It is a rounded-off value. These rounded-off values are shown in the figure with a dashed line that follows the general outline of the sinusoid. The horizontal dashed lines show the points where the quantum step changes to the next higher or lower level if the sinusoid curve is above or below that value. Take step 14, for example. The curve, dropping from its maximum, is given two values of 14 consecutively. For the first, the curve is above 14, and for the second, below. That error, in the case of 14 for instance, from the quantized value to the true value is called *quantizing distortion* or *quantization noise*. This distortion or "noise" is a major form of imperfection found in PCM systems.

In Figure 7.3, maintaining the −1 to 0 to +1 volt relationship, let us double the number of quantum steps from 16 to 32. What improvement

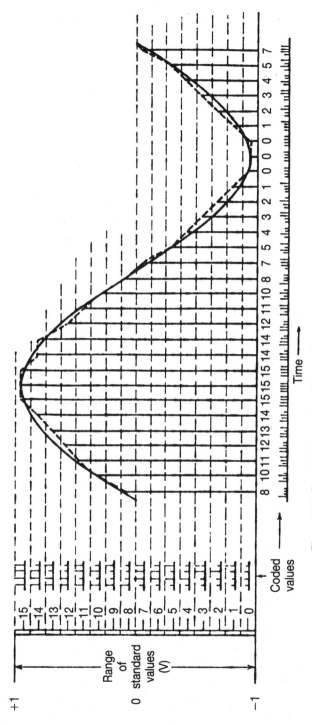

Figure 7.3. Quantization and resulting coding using 16 uniform quantizing steps.

TABLE 7.1   Sixteen Coding Steps and Values

Step	Code	Step	Code
0	0000	8	1000
1	0001	9	1001
2	0010	10	1010
3	0011	11	1011
4	0100	12	1100
5	0101	13	1101
6	0110	14	1110
7	0111	15	1111

would we achieve in quantization distortion? First determine the step increment in millivolts in each case. The total excursion is 2000 millivolts (i.e., -1000 to 0 to $+1000$ mV). This excursion is divided into 16 steps, or 125 mV per step ($2000/16 = 125$). In the second case we have 32 steps, thus $2000/32$ or 62.5 mV per step. Or we have 62.5 mV above or below the nearest quantizing step. For the 32-step case, the worst quantizing distortion case would be at the half-step level or 31.25 mV, whereas for the first case it was 62.5 volts at mid-step level.

Thus the improvement in decibels for doubling the number of quantizing steps is

$$20 \log(62.5/31.25) = 20 \log 2 = 6 \text{ dB}$$

(Remember it is $20 \log$ because we are in the voltage domain.)

This expression is valid for linear quantization only. Thus, increasing the number of quantizing steps for a fixed range of input values reduces quantization distortion accordingly. Experiments with linear quantization have shown that if 2048 uniform steps are provided, sufficient voice signal quality is achieved.

For the 2048-quantum-step (samples) case, an 11-element code is required ($2^{11} = 2048$). With a sampling rate of 8000 samples per second per voice channel, binary information rate per voice channel will be 88,000 bps.

Consider that equivalent bandwidth is a function of the information rate (data rate); the desirability of reducing this figure (i.e., 11 bits per sample) is obvious when bandwidth is at a such a premium.

There are two different types of PCM systems in the world: the European system with its E1 family of waveforms, and the North American system with its T1 (DS1) family of waveforms.

The North American PCM system uses 256 quantum steps for PCM voltage samples—except every sixth frame, where only 128 quantum steps are utilized. Each sample consists of an 8-bit PCM "word." So a quantum step is represented by an 8-bit word except at every sixth frame where only a 7-bit word is used. The 8th bit in that frame is reserved for signaling.

European systems use 8-bit words for all samples, allowing a full 256 quantum steps.

How do we reduce 2048 quantum steps to only 256? It was mentioned above that 2048 quantum steps would be required for sufficient voice fidelity, assuming linear quantization. The reduction is accomplished by *companding*. Companding is a contraction for "compression" and "expansion." In conventional telephone practice, a compressor compresses the intensity of speech signals at the input circuit of a communication channel by imparting more gain to weak signals than to strong signals. At the far-end output of the communication circuit, the expander performs the reverse function. It restores the intensity of the signal to its original dynamic range.

The standard approach, both European and North American, is to impart larger quantizing steps to high-level signals and smaller steps for lower-level signals. The system takes advantage of the fact that for speech transmissions there is a much greater likelihood of encountering signals with small amplitudes than those of large amplitudes.

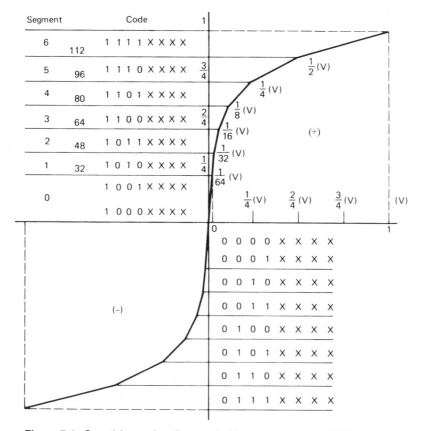

Figure 7.4. Quantizing and coding used with the European E1 PCM system.

If we allow a range of 0 to $+1$ volt to cover all input signal amplitude ranges in the positive direction (see Figure 7.4), then from 0 to 1/64th-volt the signal amplitude will have 32 step gradations available for assignment and the range from $\frac{1}{2}$ volt to 1.0 volt will have only 16 step gradations.

It can be appreciated that this nonuniform companding is carried out during the coding process. The compression and expansion functions are logarithmic and follow one of two companding laws: the A-law and the μ-law. The μ-law is used for the North American PCM system, and the A-law is used for the European system. As we can see in Figures 7.4 and 7.6, the companding circuitry does not provide an exact replica of a logarithmic curve. The circuitry produces approximate equivalents using a segmented curve, with each segment being linear.

To help understand how a PCM system works, consider that the near-end transmitter simply measures the voltage level of the signal sample at a known moment in time. It transmits this value as an 8-bit message to the distant receiver which reconstructs the signal with this information. Of course the signal is reconstructed over a series of many samples—in fact, 8000 samples in one second. The 8-bit message is the "PCM word." A PCM word occupies a time slot. The concept of the time slot is important to understand for data communications. In one method of data transmission, the 8-bit sample is replaced by an 8-bit data byte or octet.

Let's return to Figure 7.4 to learn how the distant end receiver interprets the 8-bit message it receives. By inspection of the figure, one will note that all messages (8-bit sequences) starting with a binary 1 represent positive voltages; all negative voltages start with a binary "0" and are below the ordinate of the curve.

Figure 7.4 has 13 linear segments closely approximating the A-law logarithmic curve, where $A = 87.6$. However, it should be noted that the segment passing through the origin has four steps: two above the ordinate and two below. Thus, really, a 16-segment representation of the curve is used.

As discussed above, the first bit of an 8-bit sequence message tells us if the sample is a positive or negative voltage. The next three bits of the sequence indicate the segment where the sample lies. The last four bits of the sequence tell us where in the segment the voltage point is located.

Suppose the receiver reads an 8-bit sequence as 1101XXXX, where XXXX are the last 4 bits. Reading from left to right, the first bit (a "1") tells us the sample is a positive voltage; it lies above the ordinate. The next three code elements (bits) indicate that it is the fourth step, or:

	0	1000 and 1001
	1	1010
	2	1011
	3	2200
→	4	1101
	5	1110
	6	1111

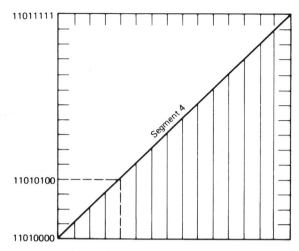

Figure 7.5. Coding of segment 4 (positive) of the E1 system. Dashed line shows position of sample 8-bit sequence.

Figure 7.5 shows a "blowup" of the uniform quantizing and subsequent straightforward coding of step 4; it illustrates the final segment coding. Note that the quantum steps are of uniform size inside that segment, providing 16 ($2^4 = 16$) coded quantum steps.

The North American DS1 (T1) system uses a 15-segment approximation of the logarithmic μ-law. Again, there are 16 segments. The segments cutting the origin are collinear and are counted as 1, but are actually two: one above and one below the ordinate. This leads to the 16 value (i.e., there are 16 segments). See Figure 7.6.

The quantization of the DS1 system is shown in Figure 7.6 for the positive portion of the curve. Segment 5 representing quantum steps 64 to 80 is shown blown up in the figure. Table 7.2 shows the DS1 (T1) coding. As can be seen again, the first code element, whether a 1 or a 0, indicates whether the quantum step is above or below the horizontal axis (the ordinate). As with E1, the next three code elements identify the segment, and the last four code elements (bits) identify the actual quantum level (voltage point) inside the segment.

7.2.3   The Concept of Frame

As illustrated in Figure 7.2, PCM multiplexing is carried out in the sampling process, sampling several sources sequentially. These sources may be nominal 4-kHz voice channels or other analog information sources such as a data tone or video. The final result of the sampling and subsequent quantization and coding is a series of pulses, a serial bit stream that requires some indication or identification of the beginning of a scanning sequence. This identification tells the far-end receiver when each full sampling sequence starts and ends; it

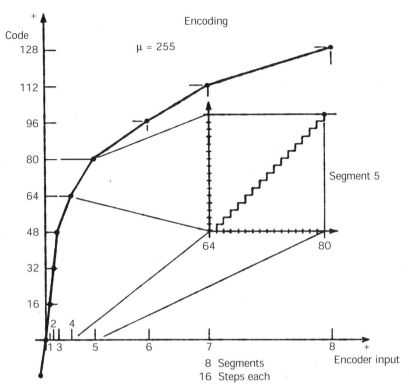

Figure 7.6. Positive portion of the segmented approximation of the μ-law ($\mu = 255$) quantizing curve used in the North American DS1 PCM system.

times and aligns the receiver. Such identification is called *framing*. A full sequence or cycle of samples is called a *frame* in PCM terminology.

Note that there are only 255 quantizing steps because steps 0 and 1 use the same bit sequence, thus avoiding a sequence of all zeros (no transitions).

CCITT Rec. G.701 (Ref. 2) defines a frame as "a set of consecutive digit time slots in which the position of each digit time slot can be identified by reference to a frame alignment signal. The frame alignment signal does not necessarily occur, in whole or in part, in each frame."

The North American PCM system carries out frame alignment in a different way from the E1 system. As its European counterpart, the North American system uses 8-bit time slots for each sample and samples 24 channels sequentially. Sampling each of the 24 channels just once sequentially is a frame. To each frame one bit is added, called the *framing bit*. One frame therefore has 193 bits, or

$$8 \times 24 + 1 = 193$$

TABLE 7.2   Eight-Level Coding of the North American DS1 (T1) System

Code Level		1	2	3	4	5	6	7	8
255	(Peak positive level)	1	0	0	0	0	0	0	0
239		1	0	0	1	0	0	0	0
223		1	0	1	0	0	0	0	0
207		1	0	1	1	0	0	0	0
191		1	1	0	0	0	0	0	0
175		1	1	0	1	0	0	0	0
159		1	1	1	0	0	0	0	0
143		1	1	1	1	0	0	0	0
127	(Center levels)	1	1	1	1	1	1	1	1
126	(Nominal zero)	0	1	1	1	1	1	1	1
111		0	1	1	1	0	0	0	0
95		0	1	1	0	0	0	0	0
79		0	1	0	1	0	0	0	0
63		0	1	0	0	0	0	0	0
47		0	0	1	1	0	0	0	0
31		0	0	1	0	0	0	0	0
15		0	0	0	1	0	0	0	0
2		0	0	0	0	0	0	1	1
1		0	0	0	0	0	0	1	0
0	(Peak negative level)	0	0	0	0	0	0	1[a]	0

[a]One digit is added to ensure that the timing content of the transmitted pattern is maintained.

The framing rate (i.e., sampling rate) is 8000 times per second. This is the so-called "Nyquist rate." To compute the bit rate for the North American DS1 (T1) system, multiply 8000 by 193:

$$193 \times 8000 = 1,544,000 \text{ bps, or } 1.544 \text{ Mbps}$$

The E1 system is a 32-channel system where 30 channels carry "speech" on voice channels and the remaining two channels carry supervisory signaling and framing/synchronization information. Each channel is an 8-bit time slot, and we speak of time slots 0–31 as follows:

Time Slot Number	Type of Information or Function
0	Synchronizing (framing)
1–15	Speech
16	Signaling
17–31	Speech

In time slot 0 a synchronizing code or word is transmitted every second frame, occupying digits 2–8 as follows:

0011011

In those frames without the synchronizing word, the second bit of time slot 0 is frozen at a 1 so that in these frames the synchronizing word cannot be

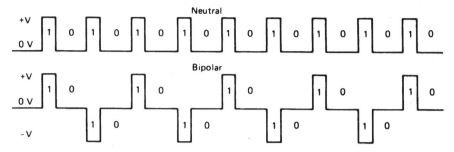

Figure 7.7. Neutral versus bipolar bit streams. *Top:* Alternating 1s and 0s transmitted in a neutral mode. *Bottom:* Alternating 1s and 0s transmitted with alternate mark inversion (AMI), also called bipolar.

imitated. The remaining bits of time slot 0 can be used for the transmission of supervisory information signals.

Framing and timing should be distinguished. Framing ensures that the PCM receiver is aligned regarding the beginning (and end) of a sequence or frame; timing refers to the synchronization of the receiver clock—specifically, that it is in step with its far-end (companion) transmit clock. Timing at the receiver is corrected via the incoming mark–space (and space–mark) transitions. It is important, then, that long periods without transitions do not occur. This was equally important in our discussion of data transmission in Chapters 5 and 6.

7.2.4 The Line Code

PCM signals transmitted on metallic-pair cable are in the bipolar mode (biternary) as shown in Figure 7.7. The marks or 1s have only a 50% duty cycle.* There are several advantages to this mode of line transmission:

1. No direct-current return is required; hence transformer coupling can be used on the line.
2. The power spectrum of the transmitted signal is centered at a frequency equivalent to half of the bit rate.

It will be noted in bipolar transmission that the 0s are coded as absence of pulses, and the 1s are alternately coded as positive or negative pulses, with the alternation taking place at every occurrence of a 1. This mode of transmission is more often called *alternate mark inversion* (AMI).

---

*With NRZ bit streams, which we have assumed up to now, each bit has 100% duty cycle. Here we mean that the active part of a bit occupies the entire time slot assigned. In the case of AMI (bipolar) transmission, the 1s occupy only half of the slot assigned.

One drawback to straightforward AMI transmission is that when a long string of 0s is transmitted (i.e., there are no transitions), a timing problem may come about because repeaters and decoders have no way of extracting timing without transitions. The problem can be alleviated by forbidding long strings of 0s. Codes have been developed that are bipolar, but with *N* zeros substitution; they are called *BNZS* codes. For instance, a B6ZS code substitutes a particular signal for a string of six 0s.

Another such code is the HDB3 (high-density binary-3), where the 3 indicates that it substitutes for binary formations with more than three consecutive 0s. With HDB3 the second and third zeros of the string are left unchanged. The fourth 0 is transmitted to the line with the same polarity of the previous mark sent, which is a "violation" of the AMI concept. The first 0 may or may not be modified to a 1 to ensure that the successive violations are of opposite polarity. HDB3 is used with the European E-1 system.

7.2.5 Regenerative Repeaters

Pulses passing down a digital transmission line or over a radio system suffer attenuation and are badly distorted by the frequency characteristic of the line and, in the case of radio, by the transmission medium and filters in the radio equipment. A *regenerative repeater* amplifies and reconstructs such badly distorted digital signals and develops a nearly perfect replica of the original signal at its output. Regenerative repeaters are the essential key to digital transmission in that we can say that "the noise stops at the repeater." The principal advantage of digital systems over their analog counterparts is that noise does not accumulate over a digital system.

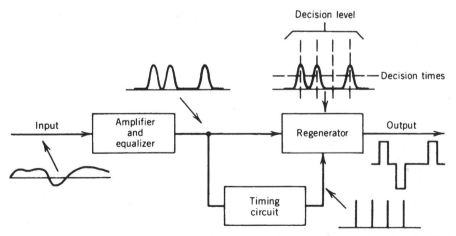

Figure 7.8. Simplified functional block diagram of a regenerative repeater for use with PCM metallic cable systems.

Figure 7.8 is a simplified block diagram of a regenerative repeater and shows typical waveforms corresponding to each functional stage of signal processing. As shown in the figure, the first stage of signal processing is amplification and equalization. Equalization is often a two-step process.

The first step is a fixed equalizer that compensates for the attenuation-frequency characteristic of the nominal section, which is the standard length of transmission line between repeaters (often 6000 ft or 1850 m). The second equalizer is variable and compensates for departures between nominal repeater section length and the actual length and loss variations due to temperature. The adjustable equalizer uses automatic line build-out (ALBO) networks that are automatically adjusted according to characteristics of the received signal.

The signal output of the regenerative repeater must be accurately timed to maintain specified pulse width and space between pulses. The timing is derived from the incoming bit stream. The incoming signal is rectified and clipped, producing square waves that are applied to the timing extractor, which is a circuit tuned to the timing frequency. The output of the circuit controls a clock-pulse generator that produces an output of narrow pulses that are alternatively positive and negative at the zero crossings of the square-wave input.

The narrow positive clock pulses gate the incoming pulses of the regenerator, and the negative pulses are used to run off the regenerator. Thus the combination is used to control the width of the regenerated pulses.

Regenerative repeaters are the major source of timing jitter in a digital transmission system. Jitter is one of the principal impairments in a digital network, giving rise to pulse distortion and intersymbol interference.

Most regenerative repeaters transmit a bipolar (AMI) waveform as shown in Figure 7.7. Such signals can have one of three possible states in any instant of time—positive, zero, or negative—and are often designated $+$, 0, and $-$. The threshold circuits are gated to admit the signal at the middle of the pulse interval. For example, if the signal is positive and exceeds a positive threshold, it is recognized as a positive pulse. If it is negative and exceeds a negative threshold, it is regenerated as a negative pulse. If it has a value between the positive and negative thresholds, it is recognized as a 0 (i.e., no pulse).

7.2.6   Higher-Order PCM Multiplex Systems

The PCM multiplex hierarchy is built up based on a 24-channel (DS1) or 30-channel (E1) group. By channel we mean the digital voice channel. These basic groups are called level 1 in the hierarchy. Subsequent multiplex levels are then developed (i.e., levels 2, 3, and 4; and in two cases, level 5). These are not to be confused with the synchronous optical network (SONET) and the synchronous digital hierarchy (SDH) levels covered in Chapter 15. It should be noted that the DS1 (T1) and E1 families of digital multiplex are often called the *plesiochronous digital hierarchy* (PDH), whereas SONET and its European counterpart, SDH, are called the *synchronous digital hierarchy*.

Table 7.3 summarizes and compares these multiplex levels for the North American, Japanese, and European systems. The development of the North American hierarchy is shown in Figure 7.9, which gives respective DS line rates and multiplex nomenclature. Regarding nomenclature, we see from the figure that M12 accepts level 1 inputs, delivering level 2 outputs to the line. It

TABLE 7.3   PCM Multiplex Hierarchy Comparison[a]

System Type	Level				
	1	2	3	4	5
North America	1	2	3	4	
Number of voice channels	24	96	672	4032	
Line bit rate (Mbps)	1.544	6.312	44.736	274.176	
Japan					
Number of voice channels	24	96	480	1440	5760
Line bit rate (Mbps)	1.544	6.312	32.064	97.728	400.352
Europe					
Number of voice channels	30	120	480	1920	7680
Line bit rate (Mbps)	2.048	8.448	34.368	139.264	564.992

[a]See Ref. 3.

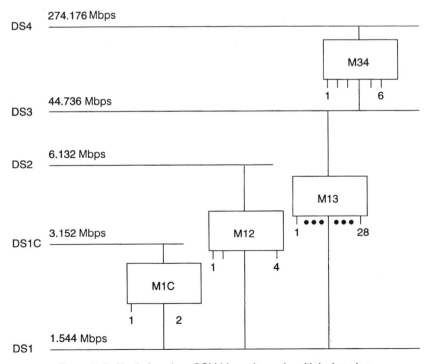

Figure 7.9. North American PCM hierarchy and multiplexing plan.

actually accepts 4 DS1 inputs deriving a DS2 output (6.312 Mbps). M13 accepts level 1 inputs, delivering a level 3 output to the line. In this case, 28 DS1 inputs form one DS3 output (44,736 Mbps). The M34 takes six DS3 inputs (level 3) to form one DS4 line rate (274.176 Mbps). DS1C is a special case where two 1.544 Mbps DS1 rates are multiplexed to form a 48-channel group with a line rate of 3.152 Mbps.

By simple multiplication we can see that the higher-order line rate is a multiple of the lower inputs rate plus some number of bits. The DS1C is an example. Here the line rate is 3152 kbps, which is 2×1544 kbps + 64 kbps. The additional 64 kbps are used for multiplex synchronization and framing.

Multiplex and demultiplex timing is very important, as one might imagine. The two DS1 signal inputs are each 1.544 Mbps plus or minus some speed tolerance (specified as ±130 ppm). The two input signals must be made alike in repetition rate and in a rate suitable for multiplexing. This is done by *bit stuffing*. In this process, bits (pulses) are added to each signal in sufficient quantity to make the signal operate at a precise rate controlled by a common clock circuit in the multiplex. Pulses are inserted (stuffed) into multiplex slots but carry no information. Thus it is necessary to code the signal in such a manner that these noninformation bits can be recognized and removed at the receiving demultiplex terminal.

Consider the more general case, using CCITT terminology. If we wish to multiplex several lower-level PCM bit streams deriving from separate tributaries into a single PCM bit stream at a higher level, a process of *justification* is required (called *bit stuffing* above). In CCITT Rec. G.701 (Ref. 2) we find the following definition of justification: "a process of changing the rate of a digital signal in a controlled manner so that it can accord with a rate different from its own inherent rate usually without loss of information."

Positive justification (as above) adds or stuffs digits; negative justification deletes or slips digits. "Stuffing" is a North American term; "justification" is European. If we think of computer displays and print copy, a page may be right-hand justified, left-hand justified, or right- and left-justified as this page

TABLE 7.4   North American DS Series of Line Rates, Tolerances, and Line Codes (Format)

Signal	Repetition Rate (Mbps)	Tolerance (ppm)[a]	Format	Duty Cycle (%)
DS0	0.064	[b]	Bipolar	100
DS1	1.544	±130	Bipolar	50
DS1C	3.152	±30	Bipolar	50
DS2	6.312	±30	B6ZS	50
DS3	44.736	±20	B3ZS	50
DS4	274.176	±10	Polar	100

[a]Parts per million.
[b]Expressed in terms of slip rate.

TABLE 7.5   Summary of CEPT30 + 2-Related Line Rates and Codes

Level	Line Data Rate (Mbps)	Tolerance (ppm)	Code	Mark Peak Voltage (V)
1	2.048	±50	HDB3	2.37 or 3[a]
2	8.448	±30	HDB3	2.37
3	34.368	±20	HDB3	1.0
4	139.264	±15	CMI[b]	1 ± 0.1[c]

[a]2.37 V on coaxial pair; 3 V on symmetric wire pair.
[b]Coded mark inversion.
[c]Peak-to-peak voltage.

is. Spacing is added or deleted for such justification. With this in mind, justification is a better term.

7.2.7   Line Rates and Codes

Table 7.4 summarizes North American DS series of PCM multiplex line rates, tolerances of these rates, and the types of line code used. Table 7.5 gives similar information for the E1 family of multiplex line rates.

7.3   A BRIEF OVERVIEW OF DIGITAL SWITCHING

7.3.1   Advantages and Issues of PCM Switching

There are both economic and technical advantages to digital switching, and it is in this context that we refer to PCM switching. The economic advantages of time-division PCM switching include the following:

1. There are notably fewer equivalent cross-points for a given number of lines and trunks than in an analog space-division switch.
2. A PCM switch is of considerably smaller size, and it requires less prime power than its analog counterpart.
3. It has much more common circuitry (i.e., common modules).
4. It is easier to achieve full availability* within economic constraints.

The technical advantages include the following:

1. It is regenerative (i.e., the switch does not distort the signal; in fact, the output signal is "cleaner" than the input).
2. It is noise-resistant.

Full availability means that each and every switch inlet can connect to each and every switch outlet.

3. It is computer based and thus incorporates all the advantages of SPC (stored program control).
4. The binary message format (in principle) is compatible with digital computers and related data communications. It is also compatible with signaling.
5. A digital exchange is lossless. There is no insertion loss as a result of a switch being inserted in the network.
6. IT exploits the continuing cost erosion of digital logic and memory: LSI, VLSI, and VHSIC insertion.

Two technical issues may be listed as disadvantages:

1. A digital switch degrades error performance of the system. A well-designed switch may only impact network error performance minimally, but it still does it.
2. Switch and network synchronization and the reduction of jitter and wander can be gating issues in system design.

7.3.2 Approaches to PCM Switching

A digital switch's architecture is made up of two elements called T and S for time-division switching (T) and space-division switching (S). We can describe a digital switch architecture with a sequence of Ts and Ss. For example, the AT & T No. 4ESS is a TSSSST switch, GTE No. 3 EAX is an SSTSS switch, and the Northern Telecom DMS 100/200 is TSTS-folded. Some switches designed to serve local networks are simply TST switches (e.g., AT & T's 5ESS). Let us now examine these basic elements, the time (T) and space (S) switch networks.

7.3.3 Time Switch

In a most simplified way, Figure 7.10 is a time switch, or what many call a *time slot interchanger*. From Section 7.2 we know that a time slot represents 8 bits, the 8-bit sample of voice level for speech operation. The duration of the PCM frame is 125 μs. For the DS1 format the basic frame contains 24 time

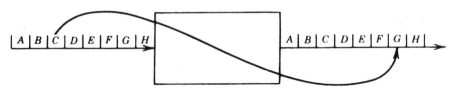

Figure 7.10. A time-division switching showing time slot interchange. Connectivity is from incoming time slot *C* to user *G* (outgoing time slot *G*).

slots, and the E1 format contains 32 time slots. The time duration of a time slot for DS1 (T1) is $125/24 = 5.2803$ μs, and for E1 it is $125/32 = 3.906$ μs. Time slot interchanging involves moving the information contained in each time slot from the incoming bit stream location to an outgoing bit stream, but in a different location. The outgoing location of the time slot is in accordance with the destination of the time slot, where the information contained in the time slot may be voice, data, video, or facsimile. Of course, at the output port of the switch a whole new frame is generated for transmission. To accomplish this, at least one time slot must be stored in memory (write) and then called out of memory in a changed sequence position (read). The operations must be controlled in some manner; and some of these control actions must also be kept in memory, such as the "idle" and "busy" condition of time slots. Now we can identify three basic blocks of a time switch:

1. Memory for speech (by time slots)
2. Memory for control
3. Time slot counter or processor

These three blocks are shown in Figures 7.11 and 7.12. The incoming time slots can be written into the speech memory either sequentially (i.e., as they appear in the incoming bit stream) (Figure 7.11) or randomly (Figure 7.12) where the order of appearance in memory is the same as the order of

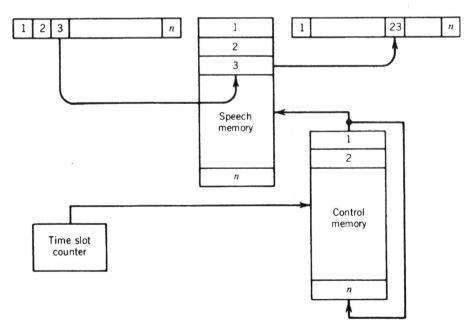

Figure 7.11. Time slot interchange time switch; sequential write, random read. (From Ref. 4.)

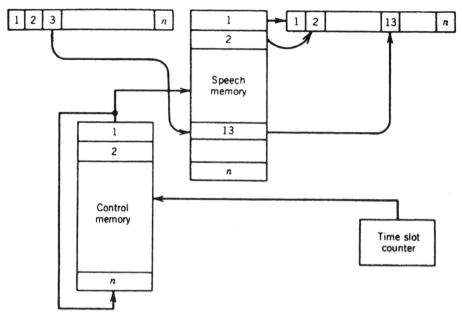

Figure 7.12. Time switch, time slot interchanger, random write, sequential read. (From Ref. 4.)

appearance in the outgoing bit stream. Now with the second version the outgoing time slots are read sequentially because they are in the proper order of the outgoing bit stream (i.e., random write, sequential read). This means that the incoming time slots are written into memory in the desired *output order*. The writing of incoming time slots into speech memory can be controlled by a simple time slot counter and can be sequential (i.e., in the order in which they appear in the incoming bit stream (Figure 7.11). The readout of the speech memory is controlled by the control memory. In this case the readout is random where the time slots are read out in the desired output order. The memory has as many cells as there are time slots. For the DS1 example, there would be 24 cells. This time switch works well for a single inlet–outlet switch.

Consider a multiple port switch, which may be a tandem switch, for example, handling multiple trunks. Here time slots from an inlet port are destined for multiple outlet ports. Enter the space switch (S). Figure 7.13 is a simple illustration of this concept. For example, time slot B_1 is moved to the Z trunk into time slot Z_1, and time slot C_n is moved to trunk W into time slot W_n. However, we see that in each case of a space switch, there is no change in time slot position.

7.3.4 Space Switch

A typical time-division space switch is shown in Figure 7.14. It consists of a cross-point matrix made up of logic gates that allow the switching of time

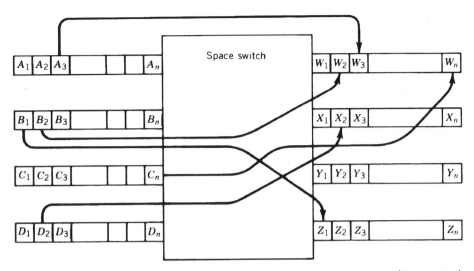

Figure 7.13. Space switch allows time slot interchange among multiple trunks. (From Ref. 4.)

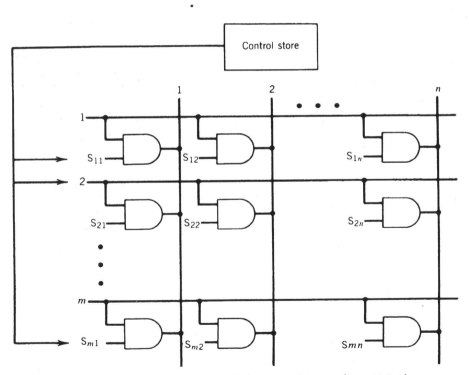

Figure 7.14. Time-division space switch cross-point array. (From Ref. 4.)

slots in the PCM time slot bit streams in a pattern required by the network connectivity for that moment in time. The matrix consists of a number of input horizontals and a number of output verticals with a logic gate at each cross-point. The array, as shown in the figure, has m horizontals and n verticals, and we call it an $m \times n$ array. If $m = n$, the switch is nonblocking. If $m > n$, the switch concentrates; and if $n > m$, the switch expands.* Sometimes other notation is used in the literature, such as n and k.

In Figure 7.14, the array consists of a number of (m) input horizontals and (n) output verticals. For a given time slot, the appropriate logic gate is enabled and the time slot passes from its input horizontal to the desired output vertical. The other horizontals, each serving a different serial stream of time slots, can have the same time slot (e.g., a time slot from time slots 1–24, 1–30, or 1–n; for instance, time slot 7 on each stream) switched into other verticals by enabling their gates. In the next time slot position (i.e., time slot 8), a completely different path configuration could occur, again allowing time slots from horizontals to be switched to selected verticals. The selection, of course, is a function of how the traffic is to be routed at that moment for calls in progress or calls being set up. The space array (cross-point matrix) does not switch time slots in the time domain as does a time switch (time slot interchanger). This is because the occurrences of time slots are identical on the horizontal and on the vertical. The control store (memory) in Figure 7.14 enables the cross-point gates in accordance with its stored information.

In Figure 7.14, it is desired to transmit a signal from input 1 (horizontal) to output 2 (vertical), and the gate at that interface would be activated by placing an enable signal on S_{12} during the desired time slot. Then the information bits of that time slot would pass through the logic gate onto the vertical. In the same time slot, an enable signal S_{m1} on the mth horizontal would permit that particular time slot to pass to vertical. From this we can see that the maximum capacity of the array during one time slot interval measured in simultaneous call connections is the smaller value of m or n. For example, if the array is 20×20 and a time slot interchanger is placed on each input (horizontal line) and the interchanger handles 32 time slots, the array can serve $20 \times 32 = 640$ different time slots. The reader should note how the time slot interchanger (TSI) multiplies the call-handling capability of the array when compared to its analog counterpart.

7.3.5 Time–Space–Time Switch

Digital switches are composed of time- and space-switching stages. The letter S designates a space-switching stage, and the letter T designates a time-switching stage. For instance, a switch composed of a time-switching stage, a space-switching stage, and a time-switching stage is called a TST switch.

---

*Concentration and expansion are the terms used, particularly for local switches, where there are notably fewer trunks than lines.

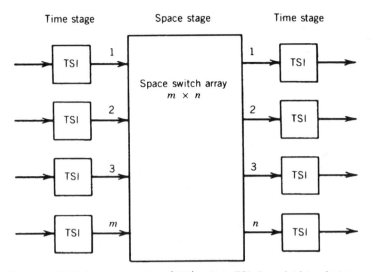

Figure 7.15. A time–space–time (TST) switch. TSI, time slot interchanger.

A switch consisting of a space-switching stage, a time-switching stage, and a space-switching stage is called an STS switch. As we discussed previously, there are other examples such as the AT & T No. 4ESS switch, which is a TSSSST switch.

Figure 7.15 illustrates the time–space–time (TST) concept. The first stage of the switch is a time slot interchanger (TSI) or time stage which interchanges information between external incoming channels and the subsequent space stage. Better said, there is a TSI on every input port, and it reassembles time slots on each bit stream into a desired order prior to placing them on the horizontals of the space array (the space stage). The space stage provides connectivity between time stages at the input and output ports. We saw earlier that space-stage time slots need not have any relation to either external incoming or outgoing time slots regarding number, numbering, or position. For instance, an incoming time slot 4 can be connected to outgoing time slot 19 via space network time slot 8.

If the space stage of a TST switch is nonblocking, blocking in such a switch can still occur if there is no internal space-stage time slot during which the link from the inlet time stage and the link to the outlet time stage are both idle. The blocking probability can be minimized if the number of space-stage time slots is large. A TST switch is strictly nonblocking if

$$1 = 2c - 1$$

where 1 is the number of space-stage time slots and c is the number of external TDM time slots (8).

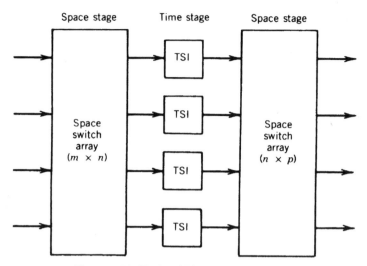

Figure 7.16. Typical STS switch architecture.

7.3.6 Space–Time–Space Switch

A space–time–space (STS) switch reverses the architecture of a TST switch. The STS switch consists of a space cross-point matrix at the input followed by an array of time slot interchangers whose outputs feed another cross-point matrix. Such a switch is illustrated in Figure 7.16. Consider this operational example with an STS switch. Suppose that incoming time slot 5 must be connected to output time slot 12. This requires a time slot interchanger that is available to interchange time slots 5 and 12. This can be accomplished by n time slot interchangers in the time stage.

7.4 DIGITAL NETWORK STRUCTURE

In the United States and Canada, the long-distance PSTN network is now all-digital. The connecting network that provides subscriber access (LECs* and independent telephone companies in the United States) will become 99.9% digital by 2002.

The network topology is fairly much the same as its prior analog counterpart. The trend is now away from a multilayer hierarchical structure to one that may be considered either two- or three-layer hierarchy in the United States and moving toward dynamic nonhierarchical routing (DNHR). All reference to hierarchical structures was removed from CCITT in the 1988 Plenary Assembly. There is little to lead one to believe that the reason for

---

*LEC stands for local exchange carrier (the local telephone company).

this is because of the change to digital. It is probably because of the improvements in transmission which have taken place in the past 30 years, including vast implementations of both satellite communications and fiber-optic links. The structure of the U.S. network probably has more to do with divestiture of the Bell System/AT & T and the advent of competing long-distance carriers. Blocking probability objectives remain essentially the same—that is, a probability of 1 in 100 of encountering blockage during the busy hour (BH). The busy hour is the hour of highest traffic density during a working day.

7.5 DIGITAL NETWORK IMPAIRMENTS AND PERFORMANCE REQUIREMENTS

7.5.1 Error Performance

Error performance requirements of the digital network are driven by the needs of data users. The trend is toward bit-by-bit mapping of data into digital bit streams. When this is the case, there is also a bit-by-bit impact on data error rate. In other words, the data channel error rate will be the same as the digital transport channel. We must also keep in mind that errors accumulate across a digital built-up connection. Thus error rates must be specified on an end-to-end basis to satisfy the data end user.

There are many definitions of *links* in telecommunications, but if we define a link as that portion of the network between digital regenerators, we will specify the BER for such a link as 1×10^{-9} or now even 1×10^{-10}. We must do this to meet the error performance criteria of Section 6.3.3.

Suppose we had 10 such links in tandem which make up a built-up connection and we specified the error performance of each link as 1×10^{-9}. What would be the error performance at the end of the 10-link connectivity (random errors assumed)? It would be 1×10^{-8} when measured over some averaging period, say 1 week or even 24 hours. It assumes that each link is of uniform structure such as a fairly long fiber-optic line and that errors are random.

7.5.2 Slips

7.5.2.1 Definition. A *slip* is a synchronization impairment of a digital network. Slips occur because no two free-running digital network clocks are exactly synchronized. Consider the simple case where two switches are connected by digital links. An incoming link to a switch terminates in an elastic store (memory) to remove transmission timing jitter. The elastic store at the incoming interface is *written into* at the recovered clock rate $R(2)$ but *read out* by the local clock rate $R(1)$. If the average clock rate at the recovered line clock $R(2)$ is different from $R(1)$ (i.e., network synchronization has been lost), the elastic store will eventually overflow or underflow.

When $R(2)$ is greater than $R(1)$, an overflow will occur, causing a loss of data. When $R(1)$ is greater than $R(2)$, an underflow occurs causing extraneous data being inserted into the bit stream. These disruptions of data are called *slips*.

Slips must be controlled to avoid loss of frame synchronization. The most common approach is to purposely add an artificial frame or delete a traffic frame that does not affect framing synchronization and alignment.

Slips do not impair voice transmission per se. A slip to a listener sounds like a click. The impairments to signaling and framing due to slips depend on the PCM system, whether it is DS1 or E1. Consider the latter (E1). It uses channel-associated signaling in channel 16 and may suffer multiframe misalignment due to a slip. Alignment may take up to 5 ms to be reacquired, and calls in the process of being set up may be lost. Common channel signaling such as CCITT Signaling System No. 7 is equipped with error detection and retransmission features so that a slip will cause increased retransmit activity, but the signaling function will be otherwise unaffected.

One can imagine the effect of slips on data transmission that is riding on one or more digital 64-kbps channels. At least one octet (8 bits) of data is lost per channel. If frame alignment is lost, all data are lost, including the slip, for the period required to reacquire alignment. As an example, consider just one 64-kbps channel. If 5 ms is required (including one time slot slip) and it takes 125 μs to transmit a frame, the number of lost bits is

$$8(5 \times 10^{-3})/125 \times 10^{-6} = 320 \text{ bits}$$

If the slip and realignment period occur right in the midst of where one data frame ends and the next begins, then two frames are lost and must be recovered. Besides, many data link protocols depend on "bit counting" to determine field and frame boundaries. Imagine what would happen to the counting if we drop 8 bits or add 8 extraneous bits!

7.5.2.2 *North American Slip Objectives.* The North American end-to-end slip objective is 0 slips. However, when a higher-level network synchronization clock suffers a fault, a slip rate of 255 slips per day is possible (Ref. 5). This could happen if the master network synchronizing clock is lost and all subsidiary switch clocks are left free-running.

7.5.2.3 *CCITT Slip Objectives.* The CCITT slip objective is 1 slip per 70 days per plesiochronous link. Note that plesiochronous operation implies free-running, high-stability clocks. By high stability, we mean a stability on the order of $\pm 1 \times 10^{-11}$ per month or better (Ref. 6).

The slip objective per link is 1 slip in 5.8 days. This is based on the CCITT hypothetical digital connection (HRX) which consists of 13 nodes interconnected by 12 links. Thus we have $70/12$ or 5.8 days. CCITT Rec. G.822 (Ref. 7) states that acceptable slip rate performance on a 64-kbps international connection is less than 5 slips in 24 hours for more than 98.9% of the time. Forty percent of the slip rate is allocated to the local network on each end of the connection, 6% is allocated for each national transit (toll) exchange, and 8% is allocated for the international transit portion.

7.6   DATA TRANSMISSION ON THE DIGITAL NETWORK

7.6.1   The Problem

To transmit binary data on the digital network, two problems arise. One is the waveform. It is quite simple to fix. Computer data are usually in a serial format with an NRZ waveform. Pulse code modulation uses AMI (alternative mark inversion, also called bipolar).

The second problem is data rate. Computer data sources have data rates related to 75×2^n, such as 75, 150, 300, 600, 1200, 2400, 4800, 9600, and 14,400 bps, and these rates have no relationship whatsoever to common digital network transmission rates of 64 kbps (or the North American 56 kbps).

7.6.2   Some Solutions

One of the most commonly used and least elegant approaches is to use analog modems such as those found in the CCITT V series recommendations discussed in Chapter 8. In essence here, we are impressing, say, 1200 bps on a 64-kbps DS0 or E0 channel. It seems a waste of valuable bit rate capacity.

Present PCM channel banks can come equipped with a 14.4- or 28.8-kbps data port. In other words, 14.4 or 28.8 kbps are mapped into a 64/56-kbps digital channel.* The result again is a waste of bits (i.e., $64,000 - 14,400 = 49,600$ wasted bits). Wasted bits in our context are bits not used to transport user data.

The propensity of data users only have analog line terminations on premises. They have no other alternative than to use standard analog data modems such as we describe in Chapter 8 (e.g., modems based on ITU-T Recs. V.34 and V.90).

There are users that have ISDN which provides 64-kbps digital channels at the desktop. Some PABXs such as Nortel SL-1 permit digital operation directly to the desktop.

Unfortunately, our standardized data rates (by ITU-T and TIA) are based on the relationship 75×2^n, resulting in incompatibility with the digital 64,000-bps data rate. Rate adaptation is required. One approach is to base the adaptation on CCITT Rec.V. 110 (Ref. 8). A two-step process is used.

---

*In certain circumstances the North American network can only provide a 56-kbps "clear" channel.

First the user data rate is converted to an appropriate intermediate rate expressed by the relation $2k \times 8$ kbps, where $k = 0$, 1, or 2. The next step performs the second conversion from the intermediate rate(s) to 64 kbps.

7.6.2.1 The North American Digital Data System (DDS).

The North American DDS provides duplex point-to-point and multipoint private-line digital data transmission at a number of synchronous data rates. It is based on the standard 1.544-Mbps DS1 PCM line rate, where individual bit streams have data rates that are submultiples of that line rate (i.e., based on 64 kbps). However, pulse slots (bit slots) are reserved for identification in the demultiplexing of individual user bit streams as well as for certain status and control signals and to ensure that sufficient line pulses (bits) are transmitted for clock recovery and pulse regeneration. The maximum data rate to the subscriber for this system is 56 kbps, some 87.5% of the theoretical 64-kbps maximum.

As we remember, the 1.544-Mbps line signal as applied to the DDS service consists of 24 sequential 8-bit words (i.e., channel time slots) plus one additional framing bit. The entire sequence is repeated 8,000 times per second. Note that again we have $8000(192 + 1) = 1.544$ Mbps, where the value 192 is 8×24 (see Section 7.2.2). Thus the line rate of a DDS facility is compatible with the DS1 (T1) PCM line rate and offers the advantage of allowing a mix of voice (PCM) and data when the full dedication of a DS1 facility to data would be inefficient.

In the descriptive material (Refs. 9, 10, and 11) of DDS, the 8-bit word is called a *byte*. One bit of each 8-bit byte is reserved for network control and for stuffing to meet the nominal line rate requirements. This control bit is called the *C-bit*. With the C-bit removed, we see where the standard channel bit rate is derived, namely 56 kbps or 8000×7. Four subrates or submultiples data rates are available to the user besides the basic 56 kbps. These are 2.4, 4.8, 9.6, and 19.2 kbps. The actual conversion is carried out in a CSU/DSU. DDS today is being outpaced by more modern conversion techniques such as the use of the CCITT Rec. V.110 techniques. Others returning to the direct use of a 64/56 kbps channel or groups of fractional DS1 channels all the way to leasing a direct DS1 connection. Usually, on such a connection only 1.536 Mbps are usable.

7.6.2.2 Bit Integrity in a Time Slot.

If we are to map data bits (usually an octet at a time) into PCM time slots and those time slot with data bits are intermixed in a DS1 with time slots carrying voice samples, a word of caution is advised. Those time slots carrying voice samples may require A-law to μ-law conversion or vice versa, bit integrity will be lost. In other words, in all probability bit values will be changed, completely destroying the data being carried. Likewise, if digital padding (adding loss) is employed, data integrity will also be lost. If the data time slots can be segregated from those carrying voice samples, then integrity can be maintained in either case.

7.7   INTERCONNECTS AND BYPASS

Bypass has the connotation of "bypassing the local telephone company." This can prove very attractive, particularly to organizations that interexchange large quantities of data on a daily basis. In North America the driving factor is economic. Can some form of bypass save money? Another factor is whether there is some service not available through the local telephone company or IXC (interexchange carrier) which would be available through bypass.

There are two subsets of bypass: (i) bypass by means of a private network and (ii) bypass by means of a specialty carrier, or a specialized common carrier. Also, alternative access is becoming more and more prevalent in North America. Here we mean the existence of two or more local telephone companies, or organizations that offer competing services for a particular local area.

The most straightforward way to effect bypass is via satellite, followed by establishing private local network. Most common of these is a VSAT network. (VSAT stands for very-small-aperture terminal, a type of satellite communications network.) VSAT networks are discussed in Chapter 11.

The building of a private local network requires good planning by telecommunication professionals. Switching is assumed. Building an analog network would be a step backwards; thus the network should be digital. It would be wise to have full compatibility with the serving PSTN for backup or extensions (see "interconnect" below). This move returns us to a network optimized for voice operation. Would we have the courage to use an ATM-based regime and design the transport and switching fabric optimized for that operation? (See Chapter 16 for an extensive discussion of ATM.)

We can argue the means of transmission transport: LOS microwave or fiber optics. Line-of-sight microwave in many instances may be more economic, at least in the short- or mid-term. Fiber has almost unlimited digital capacity; LOS microwave does not. Local "town fathers" often object to LOS microwave, with the mistaken notion that such microwave systems impair the health of residents. Just this fact alone has brought many corporations to the decision to opt for buried or aerial fiber links.

Interconnect is the ability of a private network to "interconnect" with the public switched telecommunications network (PSTN). It is highly recommended that the design of any private network be made compatible with the PSTN. If nothing else, we may want to use the PSTN for backup during failure, or use it for overflow. There is one dichotomy: the LAN. LANs operate, most generally, at much higher speeds than can be accommodated on the PSTN, namely, 10 Mbps or better. However, the PSTN in North America is even getting prepared to handle data transmission rates in this range or greater with switched multimegabit data service (SMDS). Frame relay is yet another attempt to meet this need.

7.8 BYPASS IN ECONOMICALLY EVOLVING NATIONS

In many countries, bypass, if permitted, is simply a way of getting telecommunication services. There are many locations in the world where there is no telecommunication service whatsoever, or there is a 4-year wait for a simple telephone. In many instances, the QoS of a PSTN is so poor that a private network (where permitted) can provide acceptable service.

Bypass seems very desirable. Two points should be made, however. If the initial, or even upgraded, design of a private network does not meet expectations, the network owner remains responsible. The owner cannot turn to the PSTN for damages.

The second point is maintenance. This will be the owner's responsibility unless the maintenance is outsourced. Outsourcing is becoming more and more attractive.

When sizing a private network, overbuild should be considered. How much overbuild? An economic study should answer that question. This overbuild is not only for growth of traffic over the planning period, but also the possibility of leasing network capacity. One of the best customers for spare capacity may well be the PSTN. This leasing of spare capacity has become a major source of income for electric power companies in Canada and the United States.

REFERENCES

1. *International Telephone Routing Plan*, CCITT Rec. E.171, Fascicle II.2, IXth Plenary Assembly, Melbourne, 1988.
2. *Vocabulary of Digital Transmission and Multiplexing, and Pulse Code Modulation (PCM) Terms*, CCITT Rec. G.701, ITU Geneva, March 1993.
3. *Digital Hierarchy Bit Rates*, CCITT Rec. G.702, Fascicle III.4, IXth Plenary Assembly, Melbourne, 1988.
4. Roger L. Freeman, *Telecommunication System Engineering*, 3rd ed., John Wiley & Sons, New York, 1996.
5. *Bellcore Notes on the Networks*, Issue 3, SR-2275, Bellcore, Morristown, NJ, December 1997.
6. *Timing Characteristics of Primary Reference Clocks*, ITU-T Rec. G.811, ITU Geneva, September 1997.

7. *Controlled Slip Rate Objectives on an International Digital Connection*, CCITT Rec. G.822, Fascicle III.3, IXth Plenary Assembly, Melbourne, 1988.

8. *Support by an ISDN of Data Terminal Equipments with V-Series Types Interfaces*, ITU-T Rec. V. 110, ITU Geneva, October 1996.

9. *Digital Data System—Channel Interface Specification*, PUB 62310, Bellcore, Morristown, NJ, September 1983.

10. *Digital Data Special Access Service*, Bellcore Technical Reference TR-NPL-000341, Issue 1, Morristown, NJ, March 1989.

11. *Digital Data System, Data Service Unit Interface Specification*, PUB 41450, AT & T, New York, 1981.

8

THE TRANSMISSION OF DATA OVER THE ANALOG VOICE CHANNEL

8.1 BACKGROUND

Access to the PSTN is commonly by means of the standard analog telephone channel. That access is usually a wire pair connecting the telephone to the local serving exchange (central office). In the business environment, the telephone wire pair is usually terminated in a PABX where there is a conversion to a digital regime internal to the PABX. When a call is routed to a distant end-user, the PABX will place that call on a DS1 time slot, which is then passed to the local serving exchange in a DS1 configuration (Chapter 7). Some PABXs extend digital service directly to an end-user's telephone.

In most cases, whether residential or commercial, end-users are constrained by the typical analog voice channel. Baseband data transmission, which we can typically expect as input/output of the DTE (see Section 6.3), is incompatible with the analog voice channel. The device that makes the baseband data signal compatible with the analog voice channel, is the DCE, which we have called a *modem*.

In this chapter we first discuss two-wire and four-wire transmission, because many data sets are connected to the network on a four-wire basis. This is followed by a review of the basic impairments we can expect as our data bit stream traverses the PSTN. The principal theme of the chapter is the description of how modern data modems work.

Before tackling that question, we have to step back a bit to discuss the evolution of modem technology. The principal constraint of the analog voice channel is its limited bandwidth. Remember that it occupies the band from 300 to 3400 Hz, a 3100-Hz bandwidth. If we allow 1 bit per hertz of bandwidth for data occupancy, the maximum data rate it can support is only 3100 bps, far less than the 34- and 56-kbps service that current modems provide. This, really, is the basic thrust of the chapter.

8.2 TWO-WIRE VERSUS FOUR-WIRE OPERATION

8.2.1 What is Two-Wire and Four-Wire Operation?

Our telephone at home or in the office involves two-wire operation. It connects to the serving central office (exchange) as a subscriber loop or as an equivalent subscriber loop to the serving PABX, or key system in many office situations. A wire pair means two wires. We talk on it, and we listen on it—both talk and listen on the same pair of wires.

Four-wire operation means we talk on one pair and listen on the other pair. There are few situations where this actually happens. The U.S. Department of Defense AutoVon network carries four-wire operation to the subscriber (in theory), as one example.

Four-wire operation is a concept internal to a telephone network; it is used very often with higher-speed data modems, such as the CCITT V.29 modem discussed in Section 8.5.3. Private telecommunication networks will commonly resort to four-wire operation for all but local service—that is, on the user side of a PABX.

In the PSTN, typical systems using four-wire operation are PCM (TDM), FDM, radio, coaxial cable, some wire-pair, and fiber-optic basic transport systems.

To further our data skills, we must have a firm understanding of two-wire and four-wire operation and of the impairments involved.

8.2.2 Two-Wire Transmission

By its very nature a telephone conversation requires transmission in both directions. When both directions are carried on the same wire pair, we call it two-wire transmission. A more proper definition of two-wire transmission is that when oppositely directed portions of a single telephone conversation occur over the same electrical transmission channel or path, we call this two-wire operation.

8.2.3 Four-Wire Transmission

Carrier, fiber optics, and radio systems require that oppositely directed portions of a single conversation occur over separate transmission channels or paths (or using mutually exclusive time periods). Thus we have two wires for the transmit path and two wires for the receive path, or a total of four wires for a full-duplex (two-way) telephone conversation.

Nearly all long-distance (toll) telecommunication connections traverse four-wire links. In many, if not most, cases, from the near end-user the connection to the long-distance network is partially, or in some cases totally, two-wire. Likewise, the distant end-user is also connected to the long-

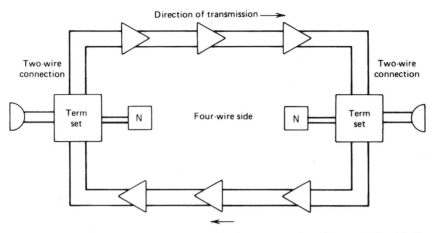

Figure 8.1. Typical long-distance telecommunications connection. Conceptually, this figure has been drawn to show operation of a hybrid or term set. N is the matching network.

distance network via a similar two-wire link. Such a long-distance connection is shown conceptually in Figure 8.1. Schematically, the four-wire connection shown in the figure is as though it were a wire-line, single channel with amplifiers. More than likely it would be one channel of a multichannel carrier* on cable and/or multiplex on radio. However, the amplifiers and their implied direction of transmission serve to convey the ideas of this section.

As shown in Figure 8.1, conversion from two-wire to four-wire operation is carried out by a terminating set, more commonly referred to in the industry as a *term set*. This set contains a four-winding balanced transformer or a resistive network, the latter being less common. The balanced transformer or resistive network is called a *hybrid*. The hybrid is the key element in a term set. A term set may also include some signaling devices, such as a break-out of E & M signaling. This is a carrier-derived supervisory signaling system. Our concern here is the operation of the hybrid.

Operation of a Hybrid. A hybrid, for telecommunications work, is a transformer. For a simplified description, a hybrid may be viewed as a power splitter with four sets of wire-pair connections (i.e., a four-port device). Figure 8.2 shows a functional block diagram of a hybrid. Two of the wire-pair connections belong to the four-wire path, one pair connecting to the transmit port and one pair connecting to the receive port. The third pair is the

---

*In all probability, one channel input on a PCM channel bank.

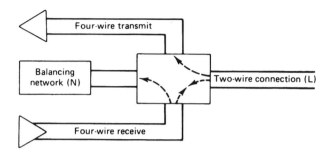

Figure 8.2. Operation of a hybrid transformer.

connection to the two-wire link (L), which we can imagine connecting, probably through a local switch, to a subscriber. The last wire pair connects the hybrid to a resistance–capacitance balancing network (N) which electrically balances the hybrid with the two-wire connection to the subscriber's subset over the frequency range of the balancing network. An artificial line may also be used for this purpose.

The hybrid function permits signals to pass from any pair through the transformer to both adjacent pairs but blocks signals to opposite pairs as shown in Figure 8.2. Signal energy entering from the four-wire side divides equally, half dissipating in the balancing network and half going to the desired two-wire connection. Ideally, no signal energy in this path crosses over to the four-wire transmit side. This is an important point, which is discussed in the following text.

Signal energy entering from the two-wire side, connecting (eventually) to a subset, divides equally, half dissipating in the impedance of the four-wire side receive path and half going to the four-wire transmit path. Here the ideal situation is that no energy is to be dissipated in the balancing network (i.e., there is a perfect balance). The balancing network is supported to display the characteristic impedance of the two-wire line (subscriber connection) to the hybrid.

The reader notes that in the description of the hybrid, in every case, ideally half the signal energy entering the hybrid is used to advantage and half is dissipated, wasted. Also keep in mind that any passive device inserted in a circuit such as a hybrid has an insertion loss. As a rule of thumb we say that the insertion loss of a hybrid is 0.5 dB. Hence there are two losses here, of which the reader must not lose sight (Ref. 1):

ybrid insertion loss: 0.5 dB
Hybrid dissipation loss: 3.0 dB (half power)
                          _____
                           3.5 dB (total)

8.3   ECHO AND SINGING: TELECOMMUNICATION NETWORK IMPAIRMENTS

The operation of the hybrid with its two-wire connection on one end and four-wire connections on the other leads us to the discussion of two phenomena that, if not properly designed for, may lead to major impairments to the data bit stream being transported. These impairments are echo and singing.

Echo. As the name implies, echo in telecommunication systems is the return of a talker's voice for speech operation. When the operation is data transmission, it is the return of data energy in the form of a modulated tone on the four-wire receive side entering the data modem. If of sufficient level, the modulated tone will corrupt the desired receive signal. We can expect the echo to have the same effect if that modem is operating in a two-wire mode. In this mode, the local echo cancellers will not be able to control the interference.

Singing. Singing is the result of sustained oscillations due to positive feedback in amplifying circuits (i.e., the amplifiers in Figure 8.1). Singing can be thought of as echo which is completely out of control. Singing will have the same effect as echo on the return path to the modem but can be much more disruptive.

The primary cause of echo and singing generally can be attributed to the mismatch in a hybrid between the balancing network (see Figure 8.2) and its two-wire connection associated with the subscriber loop. It is at this point that the major mismatch usually occurs. The problem is particularly acute when there is two-wire switching between the two-wire side of the hybrid and subscriber loops.

When subscriber loops terminate in a channel bank, which is more or less the rule on modern networks, echo and singing are much less of a problem for the data user. If both sides of a data circuit are operating in a four-wire mode, echo and singing are negligible.

On long-distance (toll) circuits, echo and singing are controlled by adding loss to the circuit. Up to 6 dB may be added with modern all-digital networks up to a certain round-trip delay, where the echo is controlled by echo cancellers.

8.4   AMPLITUDE DISTORTION AND PHASE DISTORTION

8.4.1   Introduction

As mentioned above, the analog voice channel, as defined by CCITT, occupies the band from 300 to 3400 Hz, providing a bandwidth of 3100 Hz.

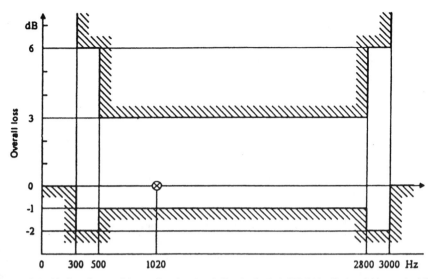

Figure 8.3. Limits of overall loss of a circuit relative to that at 1020 Hz. It should be noted that the figure expresses an amplitude distortion specification. (From Figure 1/M.1020, page 56, CCITT Rec. M.1020, Ref. 4.)

Not all of the hertz are useful for data transmission. As we approach either band edge, amplitude distortion and phase distortion become intolerable.

In this section we review amplitude and phase distortion as an introduction and background for our subsequent discussion of conditioning and equalization. Section 6.3.2.3 gave a more in-depth review of these impairments.

8.4.2   Amplitude Distortion

Amplitude distortion, attenuation distortion, and amplitude response are different names for the same phenomenon. Let us define this impairment as the relative attenuation at any frequency with respect to that at reference frequency. Common reference frequencies for the voice channel are 1000 Hz in North America and 800 Hz in Europe.

CCITT Rec. G.132 (Ref. 2) recommends no more than 9-dB attenuation distortion relative to 800 Hz between 400 and 3000 Hz on an international connection. Such distortion would be untenable for a data circuit. Figure 8.3 shows the typical limits required of attenuation distortion on a data circuit.

8.4.3   Phase Distortion

Many sources in the literature tell us that phase distortion is the principal bottleneck to data rate. Phase distortion is the result of delay. Delay varies from band center to band edge. In other words, our data waveform travels the fastest at band center, and as we approach band edge, it starts to slow down. What we are saying is that the velocity of propagation is not constant across the band of interest.

Of course delay is minimal, and delay distortion (i.e., phase distortion) is the flattest around band center, which is in the vicinity of 1700 and 1800 Hz. By *flattest* we mean where there is a comparatively minimum variation in delay. Delay distortion is often expressed by a related parameter called *envelope delay distortion* (EDD). For example, a certain specification may read that the EDD between 1000 and 2600 Hz will not exceed 1000 μs. Note that 1800 Hz is right in the center of this band.

One rule of thumb often used is that envelope delay distortion in the band of interest should not exceed the period of one bit. With a non-return-to-zero (NRZ) waveform, the period of a bit is simply calculated by

$$\text{Bit period (seconds)} = 1/(\text{bit rate})$$

Some Examples: The period of a bit (i.e., the time duration of one bit) for 2400 bps bit stream is $1/2400$ s or 416.66 μs; for 56,000 bps, it is $1/(56,000)$ = 17.857 μs; for 1 Mbps, it is $1/(1 \times 10^6)$ s or 1 μs; for 1.544 Mbps, it is $1/(1.544 \times 10^6)$ or 0.6476 μs. In a similar manner, given the period of a bit in seconds, we can calculate the bit rate in bps, assuming an NRZ waveform. For this discussion, we will assume that envelope delay distortion values are the same as *group delay distortion* values.

If the envelope delay distortion is 1 ms from 1000 to 2600 Hz, the maximum bit rate (assumes binary transmission) is 1 kbps. We will find that there are two ways to get around this dilemma of a limited bit rate. The first is to use equalization (discussed below), and the second is to reduce the modulation rate. One way to accomplish the latter is to reduce the bit rate. Another approach would be to use an M-ary wave form which reduces the modulation rate without sacrificing the bit rate. M-ary waveforms are discussed in Section 8.5. For further reading see CCITT Rec. G.133 (Ref. 3).

8.4.4   Conditioning and Equalization

Line conditioning in the United States is a tariffed offering by the serving common carrier to the data end user. In other words, the telephone company or administration guarantees that a certain leased line or lines will meet some minimum standards regarding amplitude distortion and envelope delay distortion. Because these lines are special, the telephone company charges the customer accordingly. Table 8.1 shows bandwidth parameter limits for typical conditioned line offerings for the United States.

CCITT Rec. M.1020 provides requirements for special conditioning of leased lines for data transmission. Figure 8.3 shows requirements for amplitude distortion (i.e., overall loss relative to 1020 Hz), and Figure 8.4 shows limits of group delay relative to the minimum measured group delay in the

TABLE 8.1   Bandwidth Parameter Limits[a, b]

Channel Conditioning	Attenuation Distortion (Frequency Response) Relative to 1004 Hz		Envelope Delay Distortion	
	Frequency Range (Hz)	Variation (dB)[c]	Frequency Range (Hz)	Variation (microseconds)
Basic	500–2500	−2 to +8	800–2600	1750
	300–3000	−3 to +12		
C1	1000–2400[d]	−1 to +3	1000–2400[d]	1000
	300–2700[d]	−2 to +6	800–2600	1750
	300–3000	−3 to +12		
C2	500–2800[d]	−1 to +3	1000–2600[d]	500
	300–3000[d]	−2 to +6	600–2600[d]	1500
			500–2800[d]	3000
C3 (access line)	500–2800[d]	−0.5 to +1.5	1000–2600[d]	110
	300–3000[d]	−0.8 to +3	600–2600[d]	300
			500–2800[d]	650
C3 (Trunk)	500–2800[d]	−0.5 to +1	1000–2600[d]	80
	300–3000[d]	−0.8 to +2	600–2600[d]	260
			500–2800[d]	500
C4	500–3000[d]	−2 to +3	1000–2600[d]	300
	300–3200[d]	−2 to +6	800–2800[d]	500
			600–3000[d]	1500
			500–3000[d]	3000
C5	500–2800[d]	−0.5 to +1.5	1000–2600[d]	100
	300–3000[d]	−1 to +3	600–2600[d]	300
			500–2800[d]	600

[a]C-conditioning applies only to the attenuation and envelope delay characteristics.
[b]Measurement frequencies will be 4 Hz above those shown. For example, the basic channel will have −2- to +8-dB loss, with respect to the 1004-Hz loss, between 504 and 2504 Hz.
[c](+) means loss with respect to 1004 Hz. (−) means gain with respect to 1004 Hz.
[d]These specifications are tariffed items.

Source: Reference 5.

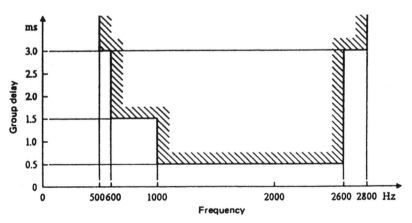

Figure 8.4. Limits of group delay relative to the minimum measured group delay in the 500- to 2800-Hz band. (From Figure 2/1020, page 56, CCITT Rec. M.1020, Ref. 4.)

band from 500 to 2800 Hz. For our discussion, group delay (distortion) values are very close to envelope delay values, and thus we may use them interchangeably. For example, in Figure 8.4, between 1000 and 2600 Hz, the envelope delay distortion is 500 μs.

Conditioning, therefore, is what the telephone company (administration) can do for the data end user. Equalization is what the data end-user can do to improve either raw (if nonconditioned lines are used) or residual amplitude distortion and envelope delay distortion in the case of conditioned lines.

There are several methods of performing equalization. One of the most common is to use several networks in tandem. Such networks tend to flatten response. In the case of amplitude distortion, they add attenuation increasingly toward band center where end-to-end loss is the least, and less and less is added as band edges are approached. Essentially the amplitude equalizer provides a mirror response opposite to that of the channel response, making the net amplitude response comparatively flat.

The delay equalizer operates in a fairly similar fashion. Voice channel delay increases toward channel edges parabolically from channel center. The delay equalizer adds delay in the center, with less and less delay added as band edges are approached, much like an inverted parabola. Thus the delay response is flattened at some small cost in absolute delay, which, in most data systems, has no effect on performance. However, care must be taken with the effect of a delay equalizer on an amplitude equalizer, and, conversely, the effect of an amplitude equalizer on a delay equalizer. Their design and adjustment must be such that the flattening of the channel for one parameter does not entirely distort the channel for the other parameter. Figure 8.5 shows typical envelope delay response in a voice channel along with the opposite response of an equalizer to flatten the envelope delay characteristics of the voice channel.

Another type of equalizer is the transversal type of filter, which is shown in Figure 8.6. This equalizer is useful where it is necessary to select among, or to adjust, several attenuation (amplitude) and phase characteristics. As shown in Figure 8.6, the basis of the filter is a tapped delay line to which the input is presented. The output is taken from a summing network which adds or sums the outputs of the taps. Such a filter is adjusted to the desired response (equalization of both phase and amplitude) by adjusting the tap contributions.

If the characteristics of a line are known, which is probably the case for a leased facility, another method of equalization is predistortion of the output signal of the data set (e.g., modem). Some devices designed for this requirement consist of a shift register and a summing network. If the equalization needs to be varied, then a feedback circuit from the receiver to the transmitter would be required to control the shift register. Such a type of active predistortion is valid for binary transmission only.

A major drawback to the equalizers discussed (with the exception of the last with a feedback circuit) is that they are useful only on dedicated or

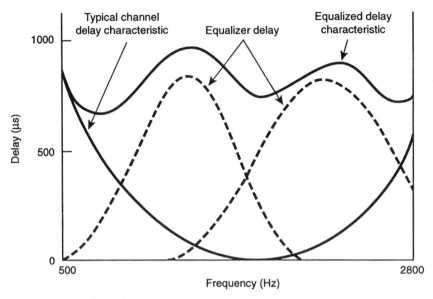

Figure 8.5. A delay (phase) equalizer tends to flatten delay characteristic of a voice channel. (From Ref. 6. Copyright BTL. Reprinted with permission.)

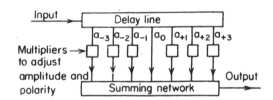

Figure 8.6. A transversal filter.

leased circuits where the circuit characteristics are known and remain fixed. Obviously, if a switched circuit is used, such as on a dial-up connection, a variable automatic equalizer is required, or manual conditioning would be necessary on every circuit in the switched systems that would possibly be transmitting data.

Circuits are conventionally equalized on the receiving end. This is called post-equalization. Equalizers must be balanced and present the proper impedance to the line. Administrations (telephone companies) may choose to condition trunks and attempt to eliminate the need to equalize station lines. The economy of considerably fewer equalizers is obvious. In addition, each circuit that would possibly carry comparatively high-speed data in the system would have to be equalized, and the equalization must be good enough that any possible circuit combination would meet the overall requirements. If equalization requirements become greater (i.e., parameters are more stringent), then consideration may have to be given to the restriction of the maximum number of circuits (trunks) in tandem. CCITT (Rec. E.171) recommends no more than 12 circuits in tandem on an international connection.

Equalization to meet amplitude–frequency response (attenuation distortion) requirements is less exacting on the overall system than is EDD. With modern digital networks, considering the network in isolation, the principal contributors to amplitude and phase distortion are the input and output 3.4-kHz filters. The measured performance for amplitude and phase distortion should prove out considerably better than the requirements of CCITT Rec. M.1020. Now add a subscriber loop to each end of the connection to permit a DCE/DTE access, and we may have some fairly different results.

Automatic equalization for both amplitude and delay is widely used. All CCITT-recommended modems operating at 2400 bps and higher have automatic equalizers built in. These equalizers are self-adaptive and require a short adaptation or training period after a circuit is switched, on the order of 1 s or less. This can be carried out during the transmission of synchronization sequences. Not only is the modem clock being "averaged" for the new circuit on the transmission of a synchronization idle signal, but the self-adaptive equalizer adjusts for optimum equalization at the same time.

8.5 DATA MODEMS

8.5.1 Where We Are and Where We Are Going

In Section 6.3.2.2 modems were introduced. In Section 8.1 we showed that a data rate up to 3100 bps could optimistically be achieved using some form of binary transmission. This was a hypothetical case where we had the 3100-Hz bandwidth of the standard voice channel. In this assumption, we did not cover the fact that those frequencies near band edge had excessive amplitude and phase distortion. For all intents and purposes, the really useful part of the band is 600–2800 Hz, which could derive perhaps some 2200 bps.

2200 bps is quite far from today's speeds of 28 and 56 kbps. How do we get from here to there? Where we fall down is that we were considering just one bit per hertz of occupancy. How can we improve the *bit packing* to 2, 3, 4, or 5 or more bits per hertz?

8.5.2 Getting More Bits per Hertz

In Section 6.3.2 modulation was introduced where the three generic modulation types available to the modem designer were discussed. These are amplitude modulation (AM), frequency modulation (FM), and phase modulation (PM).

Because of the influence of automatic landline and radio telegraphy, the nomenclature has been changed to the following:

- Amplitude shift keying (ASK) for AM
- Frequency shift keying (FSK) for FM
- Phase shift keying (PSK) for PM

In each case an audio tone is used to carry the information. It is called a tone because if we connect a loudspeaker to the tone generator, the tone can be heard having a musical quality. This happens because the tone, when detected, is in the audio-frequency range of the human ear. It is also in the range of the standard voice channel, between 300 and 3400 Hz as one would expect.

In the case of ASK, for binary transmission, a binary 1 is "tone on" and a binary 0 is "tone off." A binary sequence, then, is transmitted as a series of tone-on and tone-off sequences.

The two binary states for FSK represent a higher-frequency tone for the mark or binary 1 and a lower-frequency tone for the space or binary 0. At the FSK receiver, two filters are used, one centered on each tone frequency. If there is tone energy in the higher-frequency filter, then a mark or binary 1 has been received; if there is tone energy in the lower-frequency filter slot, then a space or binary 0 has been received.

Phase-shift keying modems use only one tone, usually centered around 1700 or 1800 Hz, where phase distortion is flattest. The phase of the tone can be retarded. Suppose a binary 0 is assigned to the condition of no phase retardation and a binary 1 is assigned to 180° phase retardation. This now defines BPSK or binary phase shift keying. For example, if we receive a signal with 180° of phase retardation, we receive a binary 1; if we receive a signal (an 1800-Hz tone) with no phase retardation, we receive a binary 0. Phase modulation is usually illustrated diagrammatically by a circle. Figure 8.7 shows our BPSK example.

Carry this concept one step further. Retard the phase in 90° increments, or 0°, 90°, 180°, and 270°. Figure 8.8 shows where we divide the circle into

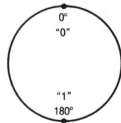

Figure 8.7. A spatial representation of a data tone with BPSK modulation. Commonly a circle is used when we describe phase modulation. The frequency of the data tone is usually 1800 Hz.

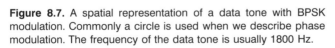

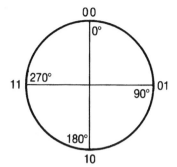

Figure 8.8. A spatial representation of a data tone with QPSK modulation. Such a spatial representation is often called *signal constellation*.

quadrants, representing the following:

Phase Change (Degrees)	Equivalent Binary Number
0	00
90	01
180	10
270	11

We have increased the *bit packing* capacity to 2 bits per hertz of bandwidth. This is so because for every transition, 2 bits are transmitted. For instance, if there were a transition to 270°, the bit sequence 11 would be transmitted.

This can be carried still one step further, where the circle has eight positions at 45° increments. This is illustrated in Figure 8.9. Now each transition represents 3 bits. The bit packing has been increased to 3 bits per hertz of bandwidth. If there were 3000 Hz of bandwidth available, its theoretical capacity is 9000 bps. We call this waveform *8-ary PSK*.

The bit packing values discussed here (e.g., 1 bit/Hz, 2 bits/Hz, 3 bits/Hz) are theoretical values. Practical values are less optimistic. For example, QPSK may achieve no more than 1.2 bits per hertz in the real world.

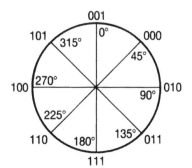

Figure 8.9. A spatial representation of 8-ary PSK. Note that the binary sequences have been selected arbitrarily. For an interface with the distant end, a certain specification or ITU recommendation is dictated, such as ITU-T Rec. V.34. Thus there is agreement by all parties concerned regarding the values of the signal points in the constellation.

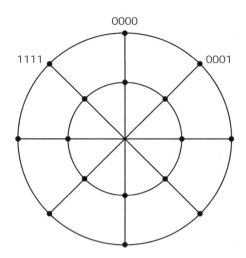

Figure 8.10. An example of a 16-QAM signal constellation (i.e., 8 phases and 2 levels.)

A very important concept here is that bandwidth is dictated by transitions per second, which we call *bauds*. Only in the binary regime (e.g., BPSK) are bauds and bits per second the same.

Up to this point we have described how to achieve 2 and 3 bits per hertz of bit packing. Suppose now we not only vary the phase as we did above, but we also vary to amplitude or level of the signal. Of course we will vary it in fixed increments. Figure 8.10 shows a simple example of this idea. We have eight phases (as we did with 8-ary PSK) *and* two amplitudes or levels. This illustrates a method of achieving 4 bits per hertz of bit packing. The CCITT V.29 modem uses this technique. It is a 9600-bps modem operating at 2400 baud. For every transition in this case, 4 bits are transmitted. The waveform is called 16-QAM, where QAM stands for *quadrature amplitude modulation.*

8.5.3 Specific High-Speed Modems

A well-known data modem used for 9600-bps operation is described in CCITT Rec. V.29 (Ref. 7). It is standardized for use in point-to-point four-wire telephone-type circuits. The main characteristics of this modem are as follows:

1. Fall back data rates of 7200 and 4800 bps
2. Full-duplex and half-duplex operation
3. Combined amplitude and phase modulation with synchronous mode operation
4. Automatic equalizer
5. Optional inclusion of a multiplexer for combining data rates of 7200, 4800, and 2400 bps
6. Line signal at 1700 Hz ± 1 Hz

7. Signal space coding. At 9600 bps the scrambled data stream to be transmitted is divided into groups of four consecutive data bits (called quad bits). The first bit, Q_1, of each quadbit is used to determine the signal element amplitude to be transmitted. The second quadbit element Q_2, third Q_3, and fourth Q_4 are encoded as a phase change relative to the phase of the immediate preceding element.

It should be noted that this modem achieves a theoretical bit packing of 4 bits per hertz. Three of these bits derive from using 8-ary PSK, and the fourth derives from using the equivalent of two amplitude levels. Thus it is a 2400-baud modem capable of transmitting 9600 bps.

Table 8.2 shows the CCITT Rec. V.29 phase encoding, Table 8.3 gives the amplitude–phase relationships, and Figure 8.11 is the signal space diagram for 9600 bps operation. The figure has 16 points in space representing the 16 quadbit possibilities, $2^4 = 16$. Thus the modem achieves 4 bits per hertz of theoretical bit packing.

The CCITT Rec. V.32 (Ref. 8) modem is designed for full-duplex operation at 9600 bps on two-wire circuits for use on the PSTN and on leased telephone-type circuits. Some of the highlights of the V.32 modem are as follows:

1. Full-duplex operation over the PSTN and on two-wire point-to-point leased circuits
2. Channel separation by echo cancellation techniques (i.e., separation of transmit and receive channels on same wire pair)

TABLE 8.2 Phase Encoding for the CCITT Rec. V.29 Modem

Q_2	Q_3	Q_4	Phase Change (deg)
0	0	1	0
0	0	0	45
0	1	0	90
0	1	1	135
1	1	1	180
1	1	0	225
1	0	0	270
1	0	1	315

TABLE 8.3 Amplitude–Phase Relationships with the CCITT Rec. V.29 Modem

Absolute Phase (deg)	Q_1	Relative Signal Element Amplitude
0, 90, 180, 270	0	3
	1	5
45, 135, 225, 315	0	$\sqrt{2}$
	1	$3\sqrt{2}$

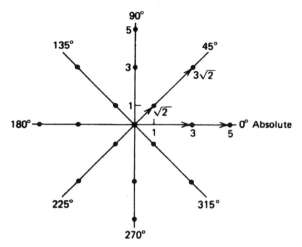

Figure 8.11. Signal space diagram for the CCITT V.29 modem when operating at 9600 bps.

3. Quadrature amplitude modulation (QAM) for each channel with synchronous line transmission at 2400 bauds.

4. Any combination of the following data signaling rates may be implemented: 9600-, 4800-, and 2400-bps synchronous.

5. At 9600-bps operation, two alternative modulation schemes are provided. One uses 16 carrier states, and the other uses trellis coding with 32 carrier states. However, modems providing the 9600-bps data signaling rate shall be capable of interworking using the 16-state alternative.

6. There is an exchange of rate sequences during start-up to establish the data rate, coding, and any other special facilities requiring coordination.

7. There is an optional provision for an asynchronous (start–stop) mode of operation in accordance with CCITT Rec. V.14.

8. The carrier (tone) frequency is 1800 Hz $\pm$ 1 Hz. The receiver can operate with frequency offsets up to ± 7 Hz.

CCITT Rec. V.32 offers two alternatives for signal element coding of the 9600-bps data rate: (i) nonredundant and (ii) trellis coding.

With nonredundant coding, the scrambled data bit stream to be transmitted is divided into groups of four consecutive bits. The first 2 bits in time, $Q1_n$ and $Q2_n$, where the subscript n designates the sequence number of the group, are differentially encoded into $Y1_n$ and $Y2_n$, in accordance with Table 8.4. Bits $Y1_n$, $Y2_n$, $Q3_n$, and $Q4_n$ are then mapped into coordinates of the signal state to be transmitted according to the signal space diagram shown in Figure 8.12 and as listed in Table 8.5.

When using the second alternative with trellis coding, the scrambled data stream to be transmitted is divided into two groups of four consecutive data

TABLE 8.4   Differential Quadrant Coding for 4800-bps Operation and for Nonredundant Coding at 9600 bps

Inputs		Previous Outputs		Phase Quadrant Change (deg)	Outputs		Signal State for 4800 bps
$Q1_n$	$Q2_n$	$Y1_{n-1}$	$Y2_{n-1}$		$Y1_n$	$Y2_n$	
0	0	0	0	+90	0	1	B
0	0	0	1		1	1	C
0	0	1	0		0	0	A
0	0	1	1		1	0	D
0	1	0	0	0	0	0	A
0	1	0	1		0	1	B
0	1	1	0		1	0	D
0	1	1	1		1	1	C
1	0	0	0	+180	1	1	C
1	0	0	1		1	0	D
1	0	1	0		0	1	B
1	0	1	1		0	0	A
1	1	0	0	+270	1	0	D
1	1	0	1		0	0	A
1	1	1	0		1	1	C
1	1	1	1		0	1	B

Source: Reference 8, Table 1/V.32, page 236.

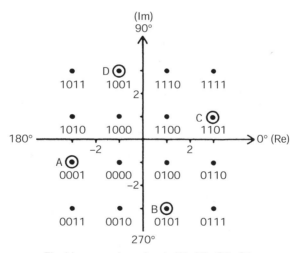

The binary numbers denote $Y1_n$ $Y2_n$ $Q3_n$ $Q4_n$

Figure 8.12. Sixteen-point signal constellation with nonredundant coding for 9600-bps operation with subset A, B, C, and D states used for 4800-bps operation and for training. (From Figure 1/V.32, page 236, CCITT Rec. V.32, Ref. 8.)

TABLE 8.5 Two Alternative Signal State Mappings for 9600-bps Operation

Coded Inputs[a]					Nonredundant Coding		Trellis Coding	
(Y0)	Y1	Y2	Q3	Q4	Re	Im	Re	Im
0	0	0	0	0	−1	−1	−4	1
	0	0	0	1	−3	−1	0	−3
	0	10	1	0	−1	−3	0	1
	0	0	1	1	−3	−3	4	1
	0	1	0	0	1	−1	4	−1
	0	1	0	1	1	−3	0	3
	0	1	1	0	3	−1	0	−1
	0	1	1	1	3	−3	−4	−1
	1	0	0	0	−1	1	−2	3
	1	0	0	1	−1	3	−2	−1
	1	0	1	0	−3	1	2	3
	1	0	1	1	−3	3	2	−1
	1	1	0	0	1	1	2	−3
	1	1	0	1	3	1	2	1
	1	1	1	0	1	3	−2	−3
	1	1	1	0	3	3	−2	1
1	0	0	0	0			−3	−2
	0	0	0	1			1	−2
	0	0	1	0			−3	2
	0	0	1	1			1	2
	0	1	0	0			3	2
	0	1	0	1			−1	2
	0	1	1	0		3		−2
	0	1	1	1		−1		−2
	1	0	0	0			1	4
	1	0	0	1			−3	0
	1	0	1	0			1	0
	1	0	1	1			1	−4
	1	1	0	0			−1	−4
	1	1	0	1			3	0
	1	1	1	0			−1	0
	1	1	1	1			−1	4

[a]See Tables 8.4 and 8.6.

Source: Reference 8, Table 3/V.32, page 238.

bits. As shown in Figure 8.13, the first 2 bits in time $Q1_n$ and $Q2_n$ in each group, where the subscript n designates the sequence number of the group, are first differentially encoded into $Y1_n$ and $Y2_n$ in accordance with Table 8.6. The two differentially encoded bits $Y1_n$ and $Y2_n$ are used to input a systematic convolutional coder which generates redundant bit $Y0_n$. The redundant bit and the four information carrying bits $Y1_n$, $Y2_n$, $Q3_n$, and $Q4_n$ are then mapped into coordinates of the signal element to be transmitted according to the signal space diagram shown in Figure 8.13 and as listed in Table 8.5.

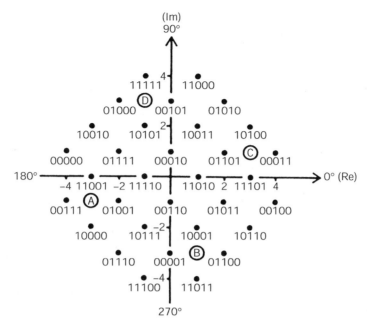

Figure 8.13. A 32-point signal constellation with trellis coding for 9600-bps operation and states A, B, C, and D used for 4800-bps operation and for training. The binary numbers denote $Y0_n$, $Y1_n$, $Y2_n$, $Q3_n$, and $Q4_n$. (From Figure 3/V.32, page 239, CCITT Rec. V.32, Ref. 8.)

TABLE 8.6   Differential Encoding for Use with Trellis-Coded Alternative at 9600 bps

Inputs		Previous Outputs		Outputs	
$Q1_n$	$Q2_n$	$Y1_{n-1}$	$Y2_{n-1}$	$Y1_n$	$Y2_n$
0	0	0	0	0	0
0	0	0	1	0	1
0	0	1	0	1	0
0	0	1	1	1	1
0	1	0	0	0	1
0	1	0	1	0	0
0	1	1	0	1	1
0	1	1	1	1	0
1	0	0	0	1	0
1	0	0	1	1	1
1	0	1	0	0	1
1	0	1	1	0	0
1	1	0	0	1	1
1	1	0	1	1	0
1	1	1	0	0	0
1	1	1	1	0	1

Source: Reference 8, Table 2/V.32, page 237.

For 4800-bps operation the data bit stream to be transmitted is divided into groups of two consecutive data bits $Q1_n$ and $Q2_n$, where $Q1_n$ is first in time and the subscript n designates the sequence number of the group, differentially encoded into $Y1_n$ and $Y2_n$ according to Table 8.4. Figure 8.13 shows subsets A, B, C, and D of the signal states used for 4800-bps operation.

CCITT Rec. V.33 (Ref. 9) defines a 14,400-bps modem standardized for use on point-to-point four-wire leased telephone-type circuits. The modem is intended to be used primarily on special-quality leased circuits that are typically defined in CCITT Rec. M.1020 (see Section 8.4.4). The modem's principal characteristics are:

- A fallback data rate of 12,000 bps
- A capability of operating in a duplex mode with continuous carrier
- Combined amplitude and phase modulation (i.e., QAM) with synchronous mode of operation
- Inclusion of eight-state trellis-coded modulation
- Optional inclusion of a multiplexer for combining data rates of 12,000, 9600, 7200, 4800, and 2400 bps
- A carrier frequency of 1800 Hz $\pm$ 1 Hz

At the 14,400-bps data rate, the scrambled data stream to be transmitted is divided into groups of six consecutive data bits. As illustrated in Figure 8.14, the first 2 bits in time $Q1_n$ and $Q2_n$ in each group are first differentially coded into $Y1$ and $Y2$ in accordance with Table 8.7. The two differentially encoded bits $Y1_n$ and $Y2_n$ are used as input to a systematic convolutional encoder which generates redundant bit $Y0_n$. This redundant bit and the six information-carrying bits $Y1_n$, $Y2_n$, $Q3_n$, $Q4_n$, $Q5_n$, and $Q6_n$ are then mapped into coordinates of the signal element to be transmitted in accordance with the signal space diagram shown in Figure 8.14. Figure 8.15 shows the signal constellation and mapping for trellis-coded modulation at 12,000 bps. The self-synchronizing generating polynomial is $1 + X^{-18} + X^{-23}$.

8.5.4   Scrambling and Its Rationale

Many of the more sophisticated higher-speed modems include a scrambler. This is a device that pseudorandomly shuffles bits in the serial bit stream before modulation. There is a companion descrambler at the other end of the circuit that shuffles these bits back to their proper order after demodulation in the receive modem. The generating polynomial is given for each modem. Thus the receive modem knows a priori the scrambling. Before it can properly descramble the receive bit stream, it must be time-slaved to the transmit modem to effect proper start of descrambling and to maintain synchronization throughout the scrambling period.

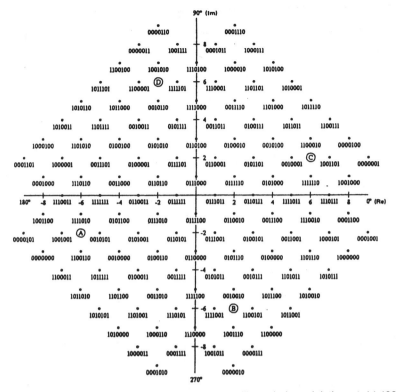

Figure 8.14. Signal constellation and mapping for trellis-coded modulation at 14,400 bps. Binary numbers refer to $Q6_n$, $Q5_n$, $Q4_n$, $Q3_n$, $Y2_n$, $Y2_n$, $Y1_n$, and $Y0_n$. A, B, C, and D refer to synchronizing signal elements. (From Figure 2/V.33, page 255, CCITT Rec. V. 33, Ref. 9.)

The rationale for scrambling is to optimize transmission performance by maintaining a minimum number of signal element transitions. Remember that we derive timing from transitions. Scrambling also minimizes the probability of repetitive data patterns. We also will find scrambling used on higher-level PCM digital systems to maintain the "1s" density. The "1s" density in this case just ensures that there will be a high probability of a minimum number of transitions. In PCM transmission the "1" is the active state; the "0" is the passive, or no-voltage state.

The scrambler here is not to be confused with a link encryption system (or scrambler) that is used to maintain data security (i.e., to prevent unauthorized "listening in"). They operate in a similar manner. However, in the case of the link encryption system, such parameters as the generating polynomial are confidential.

TABLE 8.7 Differential Encoding for Use with Trellis Coding

Inputs		Previous Outputs		Outputs	
$Q1_n$	$Q2_n$	$Y1_{n-1}$	$Y2_{n-2}$	$Y1_n$	$Y2_n$
0	0	0	0	0	0
0	0	0	1	0	1
0	0	1	0	1	0
0	0	1	1	1	1
0	1	0	0	0	1
0	1	0	1	0	0
0	1	1	0	1	1
0	1	1	1	1	0
1	0	0	0	1	0
1	0	0	1	1	1
1	0	1	0	0	1
1	0	1	1	0	0
1	1	0	0	1	1
1	1	0	1	1	0
1	1	1	0	0	0
1	1	1	1	0	1

Source: Reference 9, Table 1A/V.33, page 253.

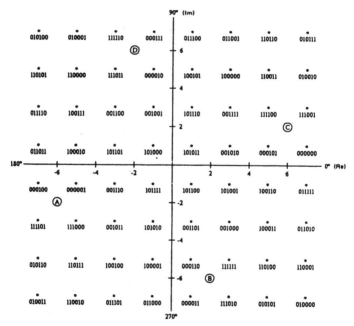

Figure 8.15. Signal space diagram and mapping for trellis-coded modulation at 12,000 bps. Binary numbers refer to $Q5_n$, $Q4_n$, $Q3_n$, $Y2_n$, $Y1_n$, and $Y0_n$. A, B, C, and D refer to synchronizing signal elements. (From Figure 3/V.33, page 256, CCITT Rec. V.33, Ref. 9.)

8.5.5   Introduction to Trellis-Coded Modulation (TCM)

In Section 4.5.2, forward error correction (FEC) was introduced. FEC allows the system designer to reduce signal-to-noise ratio a certain amount depending on the BER requirement. That amount of reduction (in dB) is called *coding gain*. For example, for a certain modulation type, an S/N of 9.6 dB is required for a BER of 1×10^{-5}. Using FEC with given parameters, the same BER can be achieved with an S/N of 5 dB. The results is a coding gain of $9.6 - 5$, or 4.6 dB. We pay for this by an increase in symbol rate (ergo bandwidth) due to the redundancy added.

When working with the analog voice channel, we have severe bandwidth limitations, nominally 3 kHz. When implementing FEC and maintaining a "high" data rate, our natural tendency is to use bit packing schemes, turning from binary modulation to *M*-ary modulation, typically QAM schemes. A good example is the V.29 modem described above, which has a two-dimensional signal constellation: one dimension in phase and the other in amplitude as illustrated in Figures 8.10 and 8.11. These figures have 16 points. As the number of points increases, the receive modem becomes more prone to errors introduced by noise hits and intersymbol interference. The problem can be mitigated by the design of the *M*-ary signal set.

In the introductory discussion of Section 8.5.2, coding and modulation were carried out separately. With trellis-coded modulation (TCM), there is a unified concept. Here coding gain is achieved without bandwidth expansion by expanding the signal set.

In conventional, multilevel (amplitude and/or phase) modulation systems, during each modulation interval, the modulator maps m binary symbols (bits) into one of $M = 2^m$ possible transmit signals, and the demodulator recovers m bits by making an independent *M*-ary nearest-neighbor decision on each signal received. Previous codes were designed for maximum *Hamming free distance*. Hamming distance refers to the number of symbols in which two codes' symbols or blocks differ regardless of how these symbols differ (Ref. 10).

TCM broke with this tradition and maximized free Euclidean distance (distance between maximum likely neighbors in the signal constellation) rather than Hamming distance. The redundancy necessary for TCM would have to come from expanding the signal set to avoid bandwidth expansion. Ungerboeck (Ref. 10) states that, in principle, TCM can achieve coding gains of about 7–8 dB over conventional uncoded multilevel modulation schemes. Most of the achievable coding gain would be obtained by expanding the signal sets only by a factor of 2. This means using signal set sizes 2^{m+1} for transmission of m bits per modulation interval.

All TCM systems (Ref. 10) achieve significant distance gains with as few as 4, 8, and 16 states. Roughly speaking, it is possible to gain 3 dB with 4 states, 4 dB with 8 states, nearly 5 dB with 16 states, and up to 6 dB with 128 or more states. Doubling the number of states does not always yield a code with larger free distance. Generally, limited distance growth and increasing numbers of nearest neighbors and neighbors with next-largest distances are the two mechanisms that prevent realizing coding gains from exceeding the ultimate limit set by channel capacity. This limit can be characterized by the signal-to-noise ratio at which the channel capacity of a modulation system with $(2m + 1)$-ary signal set equals m bps/Hz (Ref. 10; also see Ref. 11). Practical TCM systems can achieve coding gains of 3–6 dB at spectral efficiencies equal to or larger than 2 bits/Hz.

8.5.6   The V.34 Modem: 28,800 bps and 33,600 bps

8.5.6.1   General. The V.34 modem (Ref. 12) is designed to operate in the PSTN and on point-to-point leased telephone-type circuits. The following are some of the salient characteristics of the V.34 modem:

1. Duplex and half-duplex modes
2. Channel separation by echo cancellation techniques
3. Quadrature amplitude modulation (QAM) for each channel with synchronous line transmission at selectable symbol rates including the mandatory rates of 2400, 3000, and 3200 symbols per second and optional rates of 2743, 2800, and 3429 symbols per second
4. Synchronous primary channel data rates of:

33,600 bps	31,200 bps
28,800 bps	26,400 bps
24,000 bps	21,600 bps
19,200 bps	16,800 bps
14,400 bps	12,000 bps
9600 bps	7200 bps
4800 bps	2400 bps

5. Trellis coding for all data rates
6. An optional auxiliary channel with a synchronous data rate of 200 bps, a portion of which may be provided to the user as an asynchronous secondary channel
7. Adaptive techniques that enable the modem to achieve close to the maximum data rate the channel can support in each direction
8. Exchange of rate sequence during start-up to establish the data rate
9. Automoding to V-Series modems supported by ITU-T Rec. V.32 bis automode procedures and Group 3 facsimile machines

8.5.6.2   Selected Definitions

Constellation Shaping. A method for improving noise immunity by introducing a nonuniform two-dimensional probability distribution for transmitted signal points. The degree of constellation shaping is a function of the amount of constellation expansion.

Data Mode Modulation Parameters. Parameters determined during start-up and used during data mode transmission.

Frame Switching. A method for sending a fractional number of bits per mapping frame, on average, by alternating between sending an integer $b - 1$ bits per mapping frame and b bits per mapping frame according to a periodic switching pattern.

Line Probing. A method for determining channel characteristics by sending periodic signals, which are analyzed by the modem and used to determine data mode modulation parameters.

Nonlinear Encoding. A method of improving distortion immunity near the perimeter of a signal constellation by introducing a nonuniform two-dimensional (2D) signal point spacing.

Precoding. A nonlinear equalization method for reducing equalizer noise enhancement caused by amplitude distortion. Equalization is performed at the transmitter using precoding coefficients provided by the remote modem.

Preemphasis. A linear equalization method where the transmit signal spectrum is shaped to compensate for amplitude distortion. The preemphasis filter is selected using a filter index provided by the remote modem.

Shell Mapping. A method of mapping data bits to signal points in a multidimensional signal constellation, which involves partitioning a two-dimensional signal constellation into rings containing an equal number of signal points.

Trellis Encoding. A method of improving noise immunity using a convolutional coder to select a sequence of subsets in a partitioned signal constellation. The trellis encoders employed in the V.34 modem are all four-dimensional (4D) and they are used in a feedback structure where the inputs to the trellis encoder are derived from the signal points.

8.5.6.3   An Overview of Selected Key Areas of Modem Operation

Line Signals. The primary channel supports synchronous data rates of 2400–33,600 bps in multiples of 2400 bps. An auxiliary channel with a synchronous data rate of 200 bps may also be optionally supported. The primary and auxiliary data rates are determined during phase 4 of modem start-up according to the procedures described in paragraphs 11.4 and 12.4 of the

TABLE 8.8   Symbol Rates

Symbol Rate, S	a	c
2400	1	1
2743	8	7
2800	7	6
3000	5	4
3200	4	3
3429	10	7

Source: Reference 12, Table 1/V.34, ITU-T Rec. V.34.

reference document (Ref. 12). The auxiliary channel is used only when the call and answer modems have both declared this capability. The primary channel data rates may be asymmetric.

Symbol Rates. The symbol rate is $S = (a/c) \times 2400 \pm 0.01\%$ 2D symbols per second, where a and c are integers from the set given in Table 8.8 (in which symbol rates are shown rounded to the nearest integer). The symbol rates 2400, 3000, and 3200 are mandatory; 2743, 2800, and 3429 are optional. The symbol rate is selected during phase 2 of modem start-up according to procedures described in paragraph 11.2 or 12.2 of the reference publication. Asymmetric symbol rates are optionally supported and are used only when the call and answer modems have both declared this capability.

Carrier Frequencies. The carrier frequency is $(d/e) \times S$ Hz, where d and e are integers. One of two carrier frequencies can be selected at each symbol rate, as given in Table 8.9, which provides the values of d and e and the corresponding frequencies rounded to the nearest integer. The carrier frequency is determined during phase 2 of modem start-up according to the procedures given in paragraph 11.2 or 12.2 of the reference publication. Asymmetric carrier frequencies are supported.

TABLE 8.9   Carrier Frequencies Versus Symbol Rate

Symbol Rate, S	Low Carrier Frequency	d	e	High Carrier Frequency	d	e
2400	1600	2	3	1800	3	4
2743	1646	3	5	1829	2	3
2800	1680	3	5	1867	2	3
3000	1800	3	5	2000	2	3
3200	1829	4	7	1920	3	5
3429	1959	4	7	1959	4	7

Source: Reference 12, Table 2/V.34, ITU-T Rec. V.34.

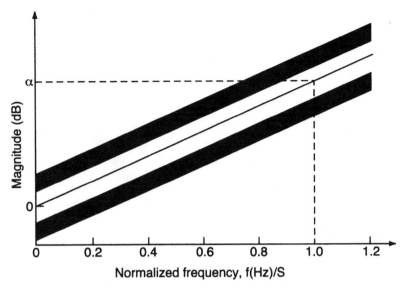

Figure 8.16. Transmit spectra templates for indices 0 to 5. *Note:* Tolerance for transmit spectrum is ±1 dB. (From Figure 1/V.34, ITU-T Rec. V.34, Ref. 12.)

Preemphasis

TRANSMIT SPECTRUM SPECIFICATIONS. The transmit spectrum specifications use a normalized frequency, which is defined as the ratio f/S, where f is the frequency in hertz and S is the symbol rate. The magnitude of the transmitted spectrum conforms to the templates shown in Figures 8.16 and 8.17 for normalized frequencies in the range from $(d/e = 0.45)$ to $(d/e + 0.45)$. The transmitted spectrum is measured using a 600-Ω pure resistive load.

Figure 8.16 requires parameter α, which is given in Table 8.10; and Figure 8.17 requires parameters β and τ, which are given in Table 8.11.

SELECTION METHOD. The transmitted spectrum is specified by a numerical index. The index is provided by the remote modem during phase 2 of start-up procedures as defined in paragraph 11.2 or 12.2 of the reference publication.

Electrical Characteristics of Interchange Circuits. For the primary channel, where an external physical interface is provided, the electrical characteristics will conform to ITU-T Rec. V.10 or V.11. The connector and pole assignments specified in ISO 2110 Amd. 1.0 or ISO/IEC 11569, column "V-Series > 20,000 bps," is used. Alternatively, when the DTE–DCE interface speed is not designed to exceed 116 kbps, these same connectors may be used with characteristics conforming to ITU-T Rec. V.10 only.

Where an external physical interface is provided for the secondary channel, electrical characteristics are in accordance with ITU-T Rec. V.10.

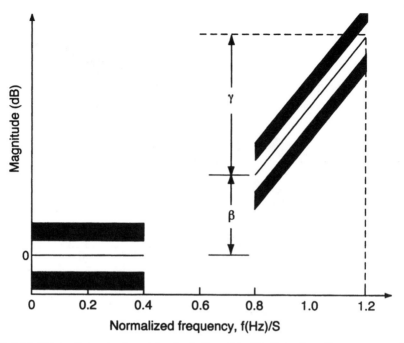

Figure 8.17. Transmit spectra templates for indices 6 to 10. *Note:* Over the range specified, the tolerance for the transmit spectrum magnitude is ±1 dB. (From Figure 2/V.34, ITU-T Rec. V.34, Ref. 12.)

TABLE 8.10   Parameter α for Indices 0 to 5

Index	α
0	0 dB
1	2 dB
2	4 dB
3	6 dB
4	8 dB
5	10 dB

Source: Reference 12, Table 3/V.34, ITU-T Rec. V.34.

TABLE 8.11   Parameters β and τ for Indices 6 to 10

Index	β	τ
6	0.5 dB	1.0 dB
7	1.0 dB	2.0 dB
8	1.5 dB	3.0 dB
9	2.0 dB	4.0 dB
10	2.5 dB	5.0 dB

Source: Reference 12, Table 4/V.34, ITU-T Rec. V.34.

Scrambler. A self-synchronizing scrambler is included in the modem for the primary channel. Auxiliary channel data are not scrambled. Each direction of transmission uses a different scrambler. According to the direction of transmission, the generating polynomial is

Call mode: $1 + X^{-18} + X^{-23}$
Answer mode: $1 + X^{-5} + X^{-23}$

8.5.6.4 *Framing.* Figure 8.18 gives an overview of the frame structure. The duration of a superframe is 280 ms and consists of J data frames, where $J = 7$ for symbol rates 2400, 2800, 3000, and 3200, and $J = 8$ for symbol rates 2743 and 3429. A data frame consists of P mapping frames, where P is given in Table 8.12. A mapping frame consists of four four-dimensional (4D) symbol intervals. A 4D symbol interval consists of two 2D symbol intervals. A bit inversion method is used for superframe synchronization.

Mapping frames are indicated by the time index i, where $i = 0$ for the first mapping frame of signal B1 (as defined below) and is incremented by 1 for each mapping frame thereafter.

Sequence B1 consists of one data frame of scrambled 1's at the end of start-up using selected data mode modulation parameters. Bit inversions for superframe synchronization are inserted as if the data frame were the last

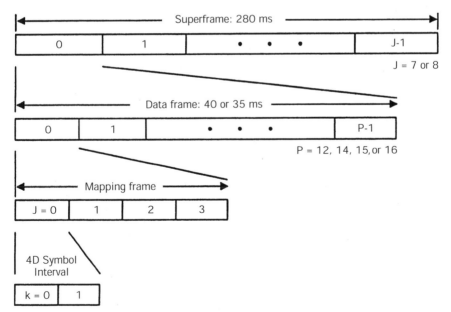

Figure 8.18. Overview of framing and indexing. (From Figure 3/V.34, ITU-T Rec. V.34, Ref. 12.)

TABLE 8.12   Framing Parameters

Symbol Rate, S	J	P
2400	7	12
2743	8	12
2800	7	14
3000	7	15
3200	7	16
3429	8	15

Source: Reference 12, Table 7/V.34, ITU-T Rec. V.34.

data frame in a superframe. Prior to transmission of B1, the scrambler, trellis encoder, differential encoder, and precoding filter tap delay line are initialized to zeros.

The 4D symbol intervals are indicated by the time index $m = 4i$, where j ($= 0, 1, 2, 3$) is a cyclic time index that indicates the position of the 4D symbol interval in a mapping frame. The 2D symbol intervals are indicated by the time index $n = 2m + k$, where k ($= 0, 1$) is a cyclic time index that indicates the position of the 2D symbol interval in a 4D symbol interval.

Mapping Frame Switching. An integer number of data bits are transmitted in every data frame. The total number of primary and auxiliary channel data bits transmitted in a data frame is denoted by

$$N = R \times 0.28/J$$

where R is the sum of the primary channel and auxiliary channel data rates.

The total number of (primary and auxiliary) data bits transmitted in a mapping frame varies between $b - 1$ ("low frame") and b ("high frame") bits according to a periodic switching pattern (SWP), of period P, such that the average number of data bits per mapping frame is N/P. the value of b is defined as the smallest integer not less than N/P. The number of high frames in a period is the remainder:

$$r = N - (b - 1)P \qquad \text{for } 1 \leq r \leq P$$

SWP is represented by 12–16-bit binary numbers where 0 and 1 represent low and high frames, respectively. The leftmost bit corresponds to the first mapping frame in a data frame. The rightmost bit is always 1.

SWP may be derived using an algorithm that employs a counter as follows: Prior to each data frame the counter is set to zero. The counter is incremented by r at the beginning of each mapping frame. If the counter is less than P, send a low frame; otherwise, send a high frame and decrement the counter by P.

TABLE 8.13 *b* and Switching Pattern (SWP) as a Function of Data Rate and Symbol Rate

Data Rate, *R*	2400 sym/s P = 12		2743 sym/s P = 12		2800 sym/s P = 14		3000 sym/s P = 15		3200 sym/s P = 16		3429 sym/ P = 15	
	b	SWP	*b*	SWP	*b*	SWP	*b*	SWP	*b*	SWP	*b*	SWP
2400	8	FFF	—	—	—	—	—	—	—	—	—	—
2600	9	6DB	—	—	—	—	—	—	—	—	—	—
4800	16	FFF	14	FFF	14	1BB7	13	3DEF	12	FFFF	12	0421
5000	17	6DB	15	56B	15	0489	14	1249	13	5555	12	36DB
7200	24	FFF	21	FFF	21	15AB	20	0421	18	FFFF	17	3DEF
7400	25	6DB	22	56B	22	0081	20	3777	19	5555	18	0889
9600	32	FFF	28	FFF	28	0A95	26	2D6B	24	FFFF	23	14A5
9800	33	6DB	29	56B	28	3FFF	27	0081	25	5555	23	3F7F
12,000	40	FFF	35	FFF	35	0489	32	7FFF	30	FFFF	28	7FFF
12,200	41	6DB	36	56B	35	1FBF	33	2AAB	31	5555	29	1555
14,400	48	FFF	42	FFF	42	0081	39	14A5	36	FFFF	34	2D6B
14,600	49	6DB	43	56B	42	1BB7	39	3FFF	37	5555	35	0001
16,800	56	FFF	49	FFF	48	3FFF	45	3DEF	42	FFFF	40	0421
17,000	57	6DB	50	56B	49	15AB	46	1249	43	5555	40	36DB
19,200	64	FFF	56	FFF	55	1FBF	52	0421	48	FFFF	45	3DEF
19,400	65	6DB	57	56B	56	0A95	52	3777	49	5555	46	0889
21,600	72	FFF	63	FFF	62	1BB7	58	2D6B	54	FFFF	51	14A5
21,800	73	6DB	64	56B	63	0489	59	0081	55	5555	51	3F7F
24,000	—	—	70	FFF	69	15AB	64	7FFF	60	FFFF	56	7FFF
24,200	—	—	71	56B	70	0081	65	2AAB	61	5555	57	1555
26,400	—	—	—	—	—	—	71	14A5	66	FFFF	62	2D6B
26,600	—	—	—	—	—	—	71	3FFF	67	5555	63	0001
28,800	—	—	—	—	—	—	—	—	72	FFFF	68	0421
29,900	—	—	—	—	—	—	—	—	73	5555	68	36DB

Source: Reference 12, Table 8/V.34, ITU-T Rec. V.34.

Table 8.13 gives the values for *b* and SWP for all combinations of data rate and symbol rate. In the table, SWP is represented by a hexadecimal number. For example, at 19,200 bps and symbol rate 3000, SWP is 0421 (hex) or 0000 0100 0010 0001 (binary).

Multiplexing of Primary and Auxiliary Channel Bits. The auxiliary channel bits are time division multiplexed with the scrambled primary channel bits.

The number of auxiliary channel bits transmitted per data frame is $W = 8$ at symbol rates 2400, 2800, 3000, and 3200, and $W = 7$ at symbol rates 2743 and 3429. In each mapping frame, the bit $I1_{i,0}$ is used to send either an auxiliary channel bit or a primary channel bit according to the auxiliary channel multiplexing pattern, AMP, of period P (see Figure 8.19). AMP can be represented as a P-bit binary number where a 1 indicates that an auxiliary channel bit is sent and a 0 indicates that a primary channel bit is sent. AMP depends only on the symbol rate and is given in Table 8.14 as a hexadecimal number. The leftmost bit corresponds to the first mapping frame in a data frame.

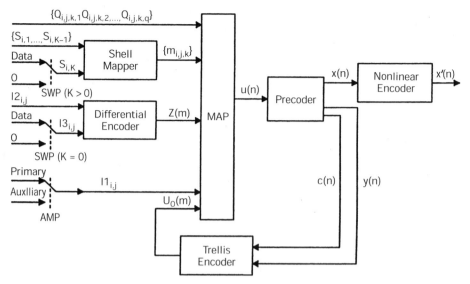

Figure 8.19. Encoder block diagram. (From Figure 4/V.34, ITU-T Rec. V.34, Ref. 12.)

TABLE 8.14   Auxiliary Channel Multiplexing Parameters

Symbol Rate, S	W	P	AMP
2400	8	12	6DB
2743	7	12	56B
2800	8	14	15AB
3000	8	15	2AAB
3200	8	16	5555
3429	7	15	1555

Source: Reference 12, Table 9/V.34, ITU-T Rec. V.34.

The auxiliary channel multiplexing pattern may be derived using an algorithm similar to the algorithm for SWP, the frame switching pattern. Prior to each data frame, a counter is set to zero. The counter is incremented by W at the beginning of each mapping frame. If the counter is less than P, a primary channel bit is sent; otherwise, an auxiliary channel bit is sent, and the counter is decremented by P.

8.5.6.5   *Encoder.* Figure 8.19 is a functional block diagram of the encoder.

Signal Constellation. Signal constellations consist of complex-valued signal points that lie on a two-dimensional rectangular grid.

All signal constellations used with the V.34 modem are subsets of a 960-point superconstellation. Figure 8.20 shows one-quarter of the points in

	-43	-39	-35	-31	-27	-23	-19	-15	-11	-7	-3	1	5	9	13	17	21	25	29	33	37	41	45
45									408	396	394	400	414										
41								398	375	349	339	329	326	335	347	359	386						
37							412	371	340	314	290	279	269	265	273	281	302	322	353	390			
33						401	357	318	282	257	236	224	216	212	218	228	247	270	298	337	378		
29					406	350	306	266	234	206	185	173	164	162	170	181	197	220	253	288	327	379	
25				360	310	263	226	193	165	146	133	123	121	125	137	154	179	207	242	289	338	391	
21			384	324	277	229	189	156	131	110	96	87	83	92	100	117	140	172	208	254	299	354	
17		355	294	243	201	160	126	98	79	64	58	54	62	71	90	112	141	180	221	271	323	387	
13	392	330	274	222	177	135	102	77	55	41	35	31	37	48	65	91	118	155	198	248	303	361	
9	380	316	255	203	158	119	84	60	39	24	17	15	20	30	49	72	101	138	182	230	283	348	415
5	367	304	244	194	148	108	75	50	28	13	6	4	8	21	38	63	93	127	171	219	275	336	402
1	362	296	238	186	142	103	69	43	22	9	1	0	5	16	32	56	85	122	163	213	267	328	395
-3	365	300	240	190	144	106	73	45	25	11	3	2	7	18	36	59	88	124	166	217	272	331	397
-7	372	307	251	199	152	113	80	52	33	19	12	10	14	26	42	66	97	134	174	225	280	341	409
-11	388	320	261	210	167	128	94	67	47	34	27	23	29	40	57	81	111	147	187	237	291	351	
-15	410	343	284	232	183	149	115	89	68	53	46	44	51	61	78	99	132	168	209	258	315	376	
-19		369	311	259	214	175	139	116	95	82	74	70	76	86	104	129	157	195	235	285	342	399	
-23			403	345	292	249	205	176	150	130	114	107	105	109	120	136	161	191	227	268	319	373	
-27				382	332	287	250	215	184	169	153	145	143	151	159	178	202	231	264	308	358	413	
-31					377	333	293	260	233	211	200	192	188	196	204	223	245	278	312	352	404		
-35						383	346	313	286	262	252	241	239	246	256	276	295	325	363	407			
-39							405	370	344	321	309	301	297	305	317	334	356	385					
-43								411	389	374	366	364	368	381	393								

Figure 8.20. One-quarter of the points in the superconstellation. (From Figure 5 / V.34, page 14, ITU-T Rec. V.34, Ref. 12.)

the superconstellation. These points are labeled with decimal integers between 0 and 239. The point with the smallest magnitude is labeled 0, the point with the next larger magnitude is labeled 1, and so on. When two or more points have the same magnitude, the point with the greatest imaginary component is taken first. The full superconstellation is the union of the four quarter-constellations obtained by rotating the constellation in Figure 8.20 by 0°, 90°, 180°, and 270°.

A signal constellation with L points consists of $L/4$ points from the quarter-constellation in Figure 8.20 with labels 0 through $L/4 - 1$, along with the $3L/4$ points that are obtained by 90°, 180°, and 270° rotations of these signal points.

8.5.7   The V.90 Modem—Maximum Data Rate: 56 kbps

8.5.7.1.   *Principal Characteristics.*   The V.90 modem (Ref. 12) is designed for full duplex operation over the PSTN. It normally operates on a two-wire basis where interference between the send and receive channels is minimized

using local echo cancellation techniques. The modulation downstream (toward the end-user) is PCM at 8000 symbols per second. It provides synchronous data signaling rates in the downstream direction from 28,000 to 56,000 bps in increments of 8000/6 bps.

In the upstream direction (toward the local serving exchange or other, similar node) V.34 (Section 8.5.6) modulation is employed. The bit stream is synchronous with upstream data rate from 4800 bps to 28,800 bps in increments of 2400 bps, with optional support of 31,200 and 33,600 bps.

Adaptive techniques are used which enable the modem to achieve close to the maximum data signaling rates the channel can support on each connection. The modem will also negotiate V.34 operation if a connection will not support V.90 operation. When the modem is started up, it exchanges rate sequences to establish the operational data signaling rate.

8.5.7.2 Definitions

Analog Modem. The analog modem is the modem of the pair that, when in data mode, generates ITU-T Rec. V.34 signals and receives ITU-T Rec. G.711 PCM signals that have been passed through a G.711 decoder. The modem is typically connected to the PSTN.

Note: ITU-T G.711, "Pulse Code Modulation of Voice Frequencies," deals with the coding of PCM signals derived from audio signals, as well as with the decoding of those signals back to their audio equivalents. See Chapter 7 for details.

Digital Modem. The digital modem is the modem of the pair, that, when in the data mode, generates ITU-T Rec. G.711 signals and receives ITU-T Rec. V.34 signals that have been passed through a Rec. G711 encoder. The modem is connected to a digitally switched network through a digital interface [e.g., a basic rate interface (BRI) or a primary rate interface (PRI)]. (For a discussion of BRI and PRI, see Chapter 13.)

Downstream. Transmission in the direction from the digital modem toward the analog modem.

Uchord. Ucodes are grouped into eight Uchords. $Uchord_1$ contains Ucodes 0 to 15; $Uchord_2$ contains Ucodes 16 to 31; . . . ; and $Uchord_8$ contains Ucodes 112 to 127.

Ucode. The universal code used to describe both a μ-law and A-law PCM codeword. All universal codes are given in decimal notation in Table 8.15. The μ-law and A-law codewords are the octets to be passed to the digital interface by the digital modem and are given in hexadecimal notation. All modifications defined in ITU-T Rec. G.711 have already been made. The MSB in the μ-law PCM and A-law PCM columns in Table 8.15 correspond to the polarity bit of the Rec. G.711 character signals. A linear representation of each PCM code word is also given.

TABLE 8.15   The Universal Set of PCM Codewords

Ucode	μ-Law PCM	μ-Law Linear	A-Law PCM	A-Law Linear	Ucode	μ-Law PCM	μ-Law Linear	A-Law PCM	A-Law Linear
0	FF	0	D5	8	33	DE	428	F4	560
1	FE	8	D4	24	34	DD	460	F7	592
2	FD	16	D7	40	35	DC	492	F6	624
3	FC	24	D6	56	36	DB	524	F1	656
4	FB	32	D1	72	37	DA	556	F0	688
5	FA	40	D0	88	38	D9	588	F3	720
6	F9	48	D3	104	39	D8	620	F2	752
7	F8	56	D2	120	40	D7	652	FD	784
8	F7	64	DD	136	41	D6	684	FC	816
9	F6	72	DC	152	42	D5	716	FF	848
10	F5	80	DF	168	43	D4	748	FE	880
11	F4	88	DE	184	44	D3	780	F9	912
12	F3	96	D9	200	45	D2	812	F8	944
13	F2	104	D8	216	46	D1	844	FB	976
14	F1	112	DB	232	47	D0	876	FA	1008
15	F0	120	DA	248	48	CF	924	E5	1056
16	EF	132	C5	264	49	CE	988	E4	1120
17	EE	148	C4	280	50	CD	1052	E7	1184
18	ED	164	C7	296	51	CC	1116	E6	1248
19	EC	180	C6	312	52	CB	1180	E1	1312
20	EB	196	C1	328	53	CA	1244	E0	1376
21	EA	212	C0	344	54	C9	1308	E3	1440
22	E9	228	C3	360	55	C8	1372	E2	1504
23	E8	244	C2	376	56	C7	1436	ED	1568
24	E7	260	CD	392	57	C6	1500	EC	1632
25	E6	276	CC	408	58	C5	1564	EF	1696
26	E5	292	CF	424	59	C4	1628	EE	1760
27	E4	308	CE	440	60	C3	1692	E9	1824
28	E3	324	C9	456	61	C2	1756	E8	1888
29	E2	340	C8	472	62	C1	1820	EB	1952
30	E1	356	CB	488	63	C0	1884	EA	2016
31	E0	372	CA	504	64	BF	1980	95	2112
32	DF	396	F5	528	65	BE	2108	94	2240

TABLE 8.15 (*Continued*)

Ucode	μ-Law PCM	μ-Law Linear	A-Law PCM	A-Law Linear	Ucode	μ-Law PCM	μ-Law Linear	A-Law PCM	A-Law Linear
66	BD	2236	97	2368	97	9E	8828	B4	8960
67	BC	2364	96	2496	98	9D	9340	B7	9472
68	BB	2492	91	2624	99	9C	9852	B6	9984
69	BA	2620	90	2752	100	9B	10364	B1	10496
70	B9	2748	93	2880	101	9A	10876	B0	11008
71	B8	2876	92	3008	102	99	11388	B3	11520
72	B7	3004	9D	3136	103	98	11900	B2	12032
73	B6	3132	9C	3264	104	97	12412	BD	12544
74	B5	3260	9F	3392	105	96	12924	BC	13056
75	B4	3388	9E	3520	106	95	13436	BF	13568
76	B3	3516	99	3648	107	94	13948	BE	14080
77	B2	3644	98	3776	108	93	14460	B9	14592
78	B1	3772	9B	3904	109	92	14972	B8	15104
79	B0	3900	9A	4032	110	91	15484	BB	15616
80	AF	4092	85	4224	111	90	15996	BA	16128
81	AE	4348	84	4480	112	8F	16764	A5	16896
82	AD	4604	87	4736	113	8E	17788	A4	17920
83	AC	4860	86	4992	114	8D	18812	A7	18944
84	AB	5116	81	5248	115	8C	19836	A6	19968
85	AA	5372	80	5504	116	8B	20860	A1	20992
86	A9	5628	83	5760	117	8A	21884	A0	22016
87	A8	5884	82	6016	118	89	22908	A3	23040
88	A7	6140	8D	6272	119	88	23932	A2	24064
89	A6	6396	8C	6528	120	87	24956	AD	25088
90	A5	6652	8F	6784	121	86	25980	AC	26112
91	A4	6908	8E	7040	122	85	27004	AF	27136
92	A3	7164	89	7296	123	84	28028	AE	28160
93	A2	7420	88	7552	124	83	29052	A9	29184
94	A1	7676	8B	7808	125	82	30076	A8	30208
95	A0	7932	8A	8064	126	81	31100	AB	31232
96	9F	8316	B5	8448	127	80	32124	AA	32256

Source: Table 1 / V.90, ITU-T Rec. V. 90, ITU Geneva, September 1998, Ref. 13.

8.5.7.3 *Overview of V.90 Operation.* As we know, the analog voice channel is limited to the frequency band of 300–3400 Hz. This severely limits the maximum data rate to be carried on that channel. In more advanced countries the analog voice channel terminates the PCM channel banks at the local serving switch (central office). Advantage is taken of this fact by the V.90 modem, which emulates the PCM analog-to-digital conversion and coding in the downstream direction. Depending on the quality of the subscriber loop, nearly 56 kbps can be achieved.

The V.90 modem operates asymmetrically providing the higher bit rate downstream. In the upstream direction, the V.90 modem operates just like the V.34 modem (see Section 8.5.6). In this case, the theoretical maximum bit rate achievable is 33,600 bps.

The V.90 modem operates in a digital mode in the upstream direction. It is very sensitive to quantization noise of the connecting circuits to the distant end-user. If it detects additional digital-to-analog/analog-to-digital signal processing (i.e., a hybrid network) in those intervening circuits, it will back off and operate in the V.34 mode in the upstream direction.

As we will remember from Chapter 7, PCM transmits samples of voltage levels with a discrete coding process. There are a theoretical 256 coding steps to express a voltage (255 steps in the real world). Companding is utilized, either A-law or μ-law, where, as a voltage level measurement approaches zero, more and more companding steps are available. The PCM gradations (i.e., distance between discrete steps) are so small that they can be obliterated by noise. Thus, with the V.90 modem, only the higher levels (steps 127–255), are utilized. There are 128 such levels. The V.90 ships data in chunks of 8 bits (which some may wish to call a *byte*). The transport rate is 8000 bytes or octets per second (that's the Nyquist sampling rate). Of course this is the same rate used by the PSTN. The 8000 bytes per second of the V.90 must be synchronized with the 8000 time slots a second of the digital PSTN.

The V.90 constantly probes the downstream bit stream operation. When it detects degradation, it drops back 92 levels for 52-kbps operation. Using fewer levels provides more robust operation, but at a lower rate.

The V.90 consists of two modems: the digital modem and the analog modem.

8.5.7.4 *V.90 Encoder.* Figure 8.21 is an overview of the encoder and represents one data frame. Data frames in the digital modem have a six-symbol structure. Each symbol position within the data frame is called a *data frame interval* and is indicated by a time index, $I = 0, \ldots, 5$, where $I = 0$ is the first in time. Frame synchronization between the digital modem transmitter and analog modem receiver is established during training procedures.

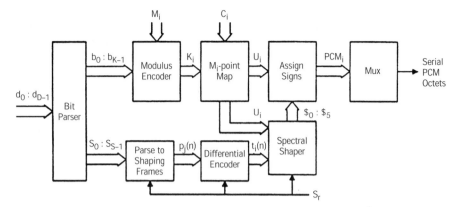

Figure 8.21. Digital modem encoder block diagram. (From Figure 1/V.90, ITU-T Rec. V.90, Ref. 13.)

8.5.7.5 *Mapping Parameters.* Mapping parameters established during training or rate renegotiation procedures are as follows:

- Six PCM code sets, one for each data frame interval 0–5, where data frame interval I has M_i members
- K, the number of modulus encoder input data bits per data frame
- S_r, the number of PCM code sign bits per data frame used as redundancy for spectral shaping and
- S, the number of spectral shaper input data bits per data frame, where $S + S_r = 6$. Table 8.16 shows the data signaling rates achieved by the valid combinations of K and S during data mode. Table 17/V.90 (in reference publication) shows the valid combinations of K and S used during phase 4 and rate renegotiation procedures

8.5.7.6 *Input Bit Parsing.* D (equal to $S + K$) serial data bits, d_0 to d_{D-1} where d_0 is first in time, are parsed into S sign input bits and K modulus encoder bits. D_0 to d_{S-1} form S_0 to s_{S-1} and d_S to d_{D-1} form b_0 to b_{K-1}. The K modulus encoder bits and the S sign bits are used as specified in the paragraphs below.

8.5.7.7 *Modulus Encoder.* K bits enter the modulus encoder. The data signaling rates associated with each value of K are tabulated in Table 8.16. There are six independent mapping moduli, M_0 to M_5, which are the number of members in the PCM code sets defined for data frame interval 0 to data frame interval 5, respectively. M_i is equal to the number of positive levels in the constellation to be used in data frame interval I as signaled by the analog

TABLE 8.16   Data Signaling Rates for Different K and S

K, Bits Entering Modulus Encoder	S, Sign Bits Used for User Data		Data Signaling Rate, kbit/s	
	From	To	From	To
15	6	6	28	28
16	5	6	28	29 1/3
17	4	6	28	30 2/3
18	3	6	28	32
19	3	6	29 1/3	33 1/3
20	3	6	30 2/3	34 2/3
21	3	6	32	36
22	3	6	33 1/3	37 1/3
23	3	6	34 2/3	38 2/3
24	3	6	36	40
25	3	6	37 1/3	41 1/3
26	3	6	38 2/3	42 2/3
27	3	6	40	44
28	3	6	41 1/3	45 1/3
29	3	6	42 2/3	46 2/3
30	3	6	44	48
31	3	6	45 1/3	49 1/3
32	3	6	46 2/3	50 2/3
33	3	6	48	52
34	3	6	49 1/3	53 1/3
35	3	6	50 2/3	54 2/3
36	3	6	52	56
37	3	5	53 1/3	56
38	3	4	54 2/3	56
39	3	3	56	56

Source: Table 2/V.90, ITU-T Rec. V.90, ITU-Geneva, September 1998, Ref. 13.

modem using the CP sequences defined in paragraph 8.5.2 of the reference document (Ref. 13).

The values of M_i and K shall satisfy the inequality $2^K \leq \prod_{i=0}^{5} M_i$.

The modulus encoder converts K bits into six numbers, K_0 to K_5, using the following algorithm.

Note: Other implementations are possible, but the mapping function must be identical to that given in the algorithm described below.

1. Represent the incoming K bits as an integer, R_0:

$$R_0 = b_0 + b_1*2^1 + b_2*2^2 + \cdots + b_{K-1}*2^{K-1}$$

2. Divide R_0 by M_0. The remainder of this division gives K_0, and the quotient becomes R_1 for use in the calculation for the next data frame

interval. Continue for the remaining five data frame intervals. This gives K_0 to K_5 as

$$K_i = R_i \text{ modulo } M_i, \qquad \text{where } 0 \leq K_i < M_i; \; R_{i+1} = (R_i - K_i)/M_i$$

3. The numbers $K_0, \ldots, K_5$ are the output of the modulus encoder, where K_0 corresponds to data frame interval 0 and K_5 corresponds to data frame interval 5.

8.5.7.8 *Mapper.* There are six independent mappers associated with the six data frame intervals. Each mapper uses a tabulation of M_i PCM codes that make up the positive constellation points of data frame interval i denoted C_i. The PCM codes to be used in each data frame interval are specified by the analog modem during training procedures. The PCM code that is denoted by the largest (smallest) Ucode is herein called the largest (smallest) PCM code. The members of C_i shall be labeled in descending order so that label 0 corresponds to the largest PCM code in C_i, while label $M_i - 1$ corresponds to the smallest PCM code in C_i. Each mapper takes K_i and forms U_i by choosing the constellation point in C_i labeled by K_i.

8.5.7.9 *Spectral Shaping.* The digital modem output line signal spectrum shall be shaped, if spectral shaping is enabled. Spectral shaping only affects the sign bits of transmitted PCM symbols. In every data frame of six symbol intervals, S_r sign bits are used as redundancy for spectral shaping while the remaining S sign bits carry user information. The redundancy, S_r, is specified by the analog modem during training procedures and can be 0, 1, 2, or 3. When $S_r = 0$, spectral shaping is disabled.

Note: The initial state of the spectral shaper does not affect the performance of the analog modem and is therefore left to the implementor.

$S_r = 0, S = 6$. The PCM code sign bits, $\$_0$ to $\$_5$, shall be assigned using input sign bits s_0 to s_5 and a differential coding rule:

$$\$_0 = s_0 \oplus (\$_5 \text{ of the previous data frame}); \text{ and}$$

$$\$_i = s_i \oplus \$_{i-1} \qquad \text{for } i = 1, \ldots, 5$$

where "$\oplus$" stands for modulus-2 addition.

$S_r = 1, S = 5$. Sign bits S_0 to S_4 are parsed to one 6-bit shaping frame per data frame in accordance with Table 8.16.

The odd bits are differentially encoded to produce the output p_j' in accordance with Table 8.17.

Finally, a second differential encoding shall be performed to produce the initial shaping sign bit assignment, $t_j(0)$ to $t_j(5)$, using the rule

$$t_j(k) = p_j'(k) \oplus t_{j-1}(k)$$

TABLE 8.17 Parsing Input Sign Bits to Shaping Frames

Data Frame Interval	$S_r = 1, S = 5$	$S_r = 2, S = 4$	$S_r = 3, S = 3$
0	$p_j(0) = 0$	$p_j(0) = 0$	$p_j(0) = 0$
1	$p_j(1) = s_0$	$p_j(1) = s_0$	$p_j(1) = s_0$
2	$p_j(2) = s_1$	$p_j(2) = s_1$	$p_{j+1}(0) = 0$
3	$p_j(3) = s_2$	$p_{j+1}(0) = 0$	$p_{j+1}(1) = s_1$
4	$p_j(4) = s_3$	$p_{j+1}(1) = s_2$	$p_{j+2}(0) = 0$
5	$p_j(5) = s_4$	$p_{j+1}(2) = s_3$	$p_{j+2}(1) = s_2$

Source: Table 3 / V.90, ITU-T Rec. V.90, ITU Geneva, September 1998, Ref. 13.

TABLE 8.18 Odd Bit Differential Coding

Data Frame Interval	$S_r = 1, S = 5$	$S_r = 2, S = 4$	$S_r = 3, S = 3$
0	$p'_j(0) = 0$	$p'_j(0) = 0$	$p'_j(0) = 0$
1	$p'_j(1) = p_j(1) \oplus p'_{j-1}(5)$	$p'_j(1) = p_j(1) \oplus p'_{j-1}(1)$	$p'_j(1) = p_j(1) \oplus p'_{j-1}(1)$
2	$p'_j(2) = p_j(2)$	$p'_j(2) = p_j(2)$	$p'_{j+1}(0) = 0$
3	$p'_j(3) = p_j(3) \oplus p'_j(1)$	$p'_{j+1}(0) = 0$	$p'_{j+1}(1) = p_{j+1}(1) \oplus p'_j(1)$
4	$p'_j(4) = p_j(4)$	$p'_{j+1}(1) = p_{j+1}(1) \oplus p'_j(1)$	$p'_{j+2}(0) = 0$
5	$p'_j(5) = p_j(5) \oplus p'_j(3)$	$p'_{j+1}(2) = p_{j+1}(2)$	$p'_{j+2}(1) = p_{j+2}(1) \oplus p'_{j+1}(1)$

Source: Table 4 / V.90, ITU-T Rec. V.90, ITU Geneva, September 1998, Ref. 13.

The spectral shaper converts each bit $t_j(k)$, to a PCM code sign bit $\$_k$ as described in Section 8.5.6.9.

$S_r = 2, S = 4$. Sign bits S_0 to S_3 are parsed to 3-bit shaping frames per data frame as shown in Table 8.17. The odd bit in each shaping frame is differentially encoded to produce differentially coded outputs p'_j and p'_{j+1} in accordance with Table 8.18.

Finally, a second differential encoding shall be performed on each shaping frame to produce the initial shaping sign bit assignments $t_j(0)$ to $t_j(2)$ and $t_{j+1}(0)$ to $t_{j+1}(2)$ using the differential encoding rule:

$$t_j(k) = p'_j(k) \oplus t_{j-1}(k)$$

$$t_{j+1}(k) = p'_{j+1}(k) \oplus t_j(k)$$

The spectral shaper converts each bit $t_j(k)$ to PCM code sign bit $\$_k$ and converts each bit $t_{j+1}(k)$ to PCM code sign bit $\$_{k+3}$ as described in Section 8.5.6.9.

$S_r = 3, S = 3$. The sign bits S_0 to S_2 are parsed to three 2-bit shaping frames per data frame as shown in Table 8.17. The odd bits in each shaping frame are differentially encoded to produce differentially encoded outputs p'_j, p'_{j-1} and p'_{j-2} as shown in Table 8.18.

Finally, a second differential encoding shall be performed on each shaping frame to produce the initial shaping sign bit assignments $t_j(0)$ to $t_j(1)$, $t_{j+1}(0)$ to $t_{j+1}(1)$, and $t_{j+2}(0)$ to $t_{j+2}(1)$ using the differential encoding rule:

$$t_j(k) = p'_j(k) \oplus t_{j-1}(k)$$

$$t_{j+1}(k) = p'_{j+1}(k) \oplus t_j(k)$$

$$t_{j+2}(k) = p'_{j+2}(k) \oplus t_{j+1}(k)$$

The spectral shaper converts each bit $t_j(k)$ to PCM code sign bit $\$_k$, each bit $t_{j+1}(k)$ to PCM code sign bit $\$_{k+2}$, and each bit $t_{j+2}(k)$ to PCM code sign bit $\$_{k+4}$ as described in Section 8.5.6.9.

8.5.7.10 *Spectral Shaper.* The spectral shaper operates on a spectral shaper frame basis. For the cases $S_r = 2$ and $S_r = 3$, there are multiple shaper frames per six-symbol data frame. Spectral shaper operation for each shaper frame within a data frame (called shaping frame j in this section) is identical except that they affect different data frame PCM sign bits as shown in Table 8.19.

The spectral shaper modifies the initial sign bits $(t_j(0), t_j(1), \ldots)$ to corresponding PCM code sign bits $(\$_0, \$_1 \ldots)$ without violating the constraint described below, so as to optimize a spectral metric. The constraint of the spectral is described using the 2-state trellis diagram shown in Figure 8.22.

TABLE 8.19 Shaping Frame to Data Frame Signal Relationship

Data Frame Interval	$S_r = 1, S = 5$	$S_r = 2, S = 4$	$S_r = 3, S = 3$	Data Frame PCM Sign Bit
0	$t_j(0)$	$t_j(0)$	$t_j(0)$	$\$_0$
1	$t_j(1)$	$t_j(1)$	$t_j(1)$	$\$_1$
2	$t_j(2)$	$t_j(2)$	$t_{j+1}(0)$	$\$_2$
3	$t_j(3)$	$t_{j+1}(0)$	$t_{j+1}(1)$	$\$_3$
4	$t_j(4)$	$t_{j+1}(1)$	$t_{j+2}(0)$	$\$_4$
5	$t_j(5)$	$t_{j+1}(2)$	$t_{j+2}(1)$	$\$_5$

Source: Table 5/V.90, ITU-T Rec. V.90, ITU Geneva, September 1998, Ref. 13.

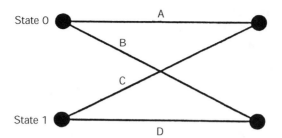

Figure 8.22. Trellis diagram used to constrain the spectral shaper.

In a given spectral shaping frame j, the spectral shaper shall modify the initial sign sequence, $t_j(k)$, according to one of the following four sign inversion rules:

- Rule A: Do nothing.
- Rule B: Invert all sign bits in spectral shaping frame j.
- Rule C: Invert even-numbered $(t_j(0), t_j(2)$, etc.) sign bits in spectral shaping frame j.
- Rule D: Invert odd-numbered $(t_j(1), t_j(3)$, etc.) sign bits in spectral shaping frame j.

The trellis diagram describes the sequence of sign inversion rules that are allowable. For example, when the spectral shaper is in state $Q_j = 0$ at the beginning of frame j, only rules A and B are allowable in frame j. The current state Q_j together with the sign inversion rule selected for frame j determine the next state Q_{j+1} according to the trellis diagram.

The look-ahead depth parameter, $1d$, is an integer between 0 and 3 selected by the analog modem during training procedures. $1d$ of 0 and 1 are mandatory in the digital modem. $1d$ of 2 and 3 are optional.

To select the sign inversion rule for the jth spectral shaping frame, the spectral shaper shall use the PCM symbol magnitudes produced by the mapper for spectral shaping frames $j, j + 1, \ldots, j + 1d$. It shall compute the spectral metric that would result if each of the allowable sequences of sign inversion rules for frames j through $j + 1d$, starting from the current state Q_j in frame j, were to be used. The shaper shall select the sign inversion rule for frame j that minimizes the spectral metric, $w[n]$, defined in Section 8.5.6.9 up to and including the final symbol of spectral shaping frame $j + 1d$. The selection determines the next state Q_{j+1}.

The shaper shall then set PCM code signs $\$_i$ for shaping frame j according to the selected sign inversion rule for shaper frame j.

8.5.7.11 *Spectral Shape Filter.* The analog modem determines the spectral shape filter function used in the digital modem by selecting parameters of the following transfer function:

$$T(z) = \frac{(1 - a_1 z^{-1})(1 - a_2 z^{-1})}{(1 - b_1 z^{-1})(1 - b_2 z^{-1})}$$

where a_1, a_2, b_1, and b_2 are parameters having absolute values less than or equal to 1. The parameters, a_1, a_2, b_1, and b_2, are specified by the analog modem during training procedures and are represented in the 8-bit two's-complement format with 6 bits after the binary point. The digital modem shall perform the spectral shaping according to the spectral shaping metric, $w[n]$ characterized by the filter:

$$F(z) = \frac{1}{T(z)} = \frac{(1 - b_1 z^{-1})(1 - b_2 z^{-1})}{(1 - a_1 z^{-1})(1 - a_2 z^{-1})}$$

The filter input, $x[n]$, shall be a signed signal proportional to the linear value corresponding to PCM codes being transmitted. The relationship between PCM codes and corresponding linear values is given in Table 1/V.90. $w[n]$ shall be computed as follows:

1. $y[n] = x[n] - b_1 x[n-1] + a_1 y[n-1]$
2. $v[n] = y[n] - b_2 y[n-1] + a_2 v[n-1]$
3. $w[n] = v^2[n] + w[n-1]$

8.5.7.12 *Sign Assignment.* Six sign bits generated by the spectral shaper described above are attached to six unsigned mapper outputs U_0 to U_5 to complete the mapping of the data frame intervals. A sign bit 0 means that the transmitted PCM codeword represents a negative voltage and a sign bit 1 means it will represent a positive voltage.

8.5.7.13 *MUX.* The signed PCM codewords, PCM_i, are transmitted from the digital modem sequentially with PCM_0 being first in time.

8.5.7.14 *Analog Modem.* The characteristics of the analog modem described in this section apply when in the V.90 mode. After fallback to the V.34 mode, the analog modem has the characteristics as defined in ITU-T Rec. V.34 (Section 24-7.9).

Data Signaling Rates. The modem supports data rates of 4800 to 28,800 bps in increments of 2400 bps, with optional support for 31,200 and 33,600 bps.

The 200-bps V.34 auxiliary channel is not supported. The data rate is determined during phase 4 of the modem start-up in accordance with the procedures described in paragraph 9.4 of the reference document.

Symbol Rates. The analog modem supports the symbol rate 3200. It may also support 3000 and the optional symbol rate 3429 as defined in ITU-T Rec. V.34 (Section 8.5.5). The other V.34 symbol rates—2400, 2743, and 2800—are not supported. The symbol rate is selected by the analog modem during phase 2 of modem start-up in accordance with paragraph 9.2 of the reference publication.

Carrier Frequencies. The analog modem supports the carrier frequencies specified in ITU-T Rec. V.34, paragraph 5.3 (Section 8.5.5) for the appropriate symbol rate. The carrier frequency is determined during phase 2 of modem start-up in accordance with the procedures specified in paragraph 9.2 of the reference document.

Preemphasis. The analog modem supports the preemphasis filter characteristics specified in ITU-T Rec. V.34, paragraph 5.4. The filter selection is provided by the digital modem during phase 2 of modem start-up in accordance with the procedures specified in paragraph 9.2 of the reference document.

Scrambler. The analog modem includes a self-synchronizing scrambler as specified in ITU-T Rec. V.34, Section 7, using the generating polynomial, GPA, given in Equation 7.2/V.34 (Section 8.5.5).

Framing. The analog modem uses the framing method specified for the V.34 primary channel described in Section 8 of ITU-T Rec. V.34.

Encoder. The analog modem uses the encoder specified for the V.34 primary channel in Section 9, ITU-T Rec. V.34.

Interchange Circuits. The requirements of this section apply to both modems.

List of Interchange Circuits (V.34/V.90). Reference in this section to ITU-T Rec. V.24 interchange circuit numbers are intended to refer to the functional equivalent of such circuits and are not intended to imply the physical implementation of such circuits. For example, references to circuit 103 should not be understood to refer to the functional equivalent of circuit 103 (see Table 8.20).

TABLE 8.20 Interchange Circuits

No.	Description	Notes
\multicolumn{2}{l}{Interchange Circuit}		
102	Signal ground or common return	
103	Transmitted data	
104	Received data	
105	Request to send	
106	Ready for sending	
107	Data set ready	
108/1 or	Connect data set to line	
108/2	Data terminal ready	
109	Data channel received line signal detector	1
125	Calling indicator	
133	Ready for receiving	2

Note 1: Thresholds and response times are not applicable because a line signal detector cannot be expected to distinguish received signals from talker echoes.
Note 2: Operation of circuit 133 shall be in accordance with § 4.2.1.1 of Recommendation V.43.
Source: Table 6/V.90, ITU-T Rec. V.90, ITU Geneva, September 1998, Ref. 13.

Asynchronous Character-Mode Interfacing. The modem may include an asynchronous-to-synchronous converter interfacing to the DTE in an asynchronous (or start-stop character) mode. The protocol for the conversion is in accordance with ITU-T Recs. V.14, V.42, and V.80. Data compression may also be employed.

Portions of Section 8.5.7 have been extracted from ITU-T Rec. V.90 of 9/98, Ref. 13.

REFERENCES

1. Roger L. Freeman, *Telecommunication Transmission Handbook*, 4th ed., John Wiley & Sons, New York, 1998.
2. *Attenuation Distortion*, CCITT Rec. G.132, Fascicle III.1, IXth Plenary Assembly, Melbourne, 1988.
3. *Group Delay Distortion*, CCITT Rec. G.133, Fascicle III.1, IXth Plenary Assembly, Melbourne, 1988.
4. *Characteristics of Special Quality International Leased Circuits with Special Bandwidth Conditioning*, CCITT Rec. M.1020, ITU Geneva, March 1993.
5. *Data Communications Using Voiceband Private Line Channels*, Bell System Technical Reference, PUB 41004, AT & T, New York, 1973.
6. *Transmission Systems for Communications*, 5th ed., Bell Telephone Laboratories, Holmdel, NJ, 1982.
7. *9600 Bits per Second Modem Standardized for Use on Point-to-Point 4-Wire Leased Telephone Circuits*, CCITT Rec. V.29, Fascicle, VIII.1, IXth Plenary Assembly, Melbourne, 1988.

8. *A Family of 2-Wire, Duplex Modems Operating at Data Signaling Rates up to 9600 Bits/s for Use on the General Switched Telephone Network and on Leased Telephone-Type Circuits*, CCITT Rec. V.32, ITU Geneva, March 1993.

9. *14,400 Bits per Second Modem Standardized for Use on Point-to-Point 4-Wire Leased Telephone-Type Circuits*, CCITT Rec. V.33, Fascicle VIII.1, IXth Plenary Assembly, Melbourne, November 1988.

10. Gottfried Ungerböck, Trellis-Coded Modulation with Redundant Sets—Parts I and II, *IEEE Communications Magazine*, Vol. 27, No. 2, February 1987.

11. Andrew J. Viterbi et al., A Pragmatic Approach to Trellis-Coded Modulation, *IEEE Communication Magazine*, Vol. 27, No. 7, July 1987.

12. *A Modem Operating at Data Signaling Rates of up to 33,600 Bits/s for Use on the General Switched Telephone Network and on Leased Point-to-Point 2-Wire Telephone-Type Circuits*, ITU-T Rec. V.34, ITU Geneva, February 1998.

13. *A Digital Modem and Analog Modem Pair for Use on the Public Switched Telephone Network (PSTN) at Data Signaling Rates of Up to 56,000 Bits/s Downstream and Up to 33,600 Bits/s Upstream*, ITU-T Rec. V.90, ITU Geneva, September 1998.

9

DATA COMMUNICATIONS IN THE OFFICE ENVIRONMENT, PART 1

9.1 INTRODUCTION

In the present office environment the personal computer (PC) is nearly ubiquitous. Some may be stand-alone devices, but in the majority of cases they are interconnected by one means or another. They may also connect, via a server, to a mainframe computer, high-speed printers, and mass storage devices.

A *workstation* consists of a PC, and an interconnection capability is implied. It is defined by the IEEE (Ref. 1) as a device used to perform tasks such as data processing and word processing. Often a workstation is just a PC with keyboard and display. An *office* may be viewed as a conglomeration of management information systems (MIS). MIS represents the entire electronic data processing capabilities of an organization.

The interconnection of MIS assets on a local level is the subject of this chapter. This interconnection is almost universally carried out by means of a *local area network* (LAN). The IEEE (Ref. 1) defines a LAN as a nonpublic data network in which serial transmission is used without store and forward techniques for direct data communication among data stations located on the user's premises.

A LAN is made cost-effective because a single transmission medium interconnects all devices on the network. There are several simple assumptions on why we can do this with just one transmission medium such as a wire pair. First, transactions are usually of very short duration. Consider a LAN transmitting at a 10-Mbps data rate. It transmits 10,000 bits in 1 ms, or one page of text of this book. Now if each user only transmits for a period of 1 ms, and we allow 1 ms between transactions for circuit setup, then in 1 second, 500 users could be accommodated. It is also apparent that a user would not have to wait very long to access the medium.

Another reason that this can be accomplished with one medium such as a wire pair is that distances are very short, probably never more than 200 m (500 ft). Furthermore, a wire pair over comparatively short distances can handle considerably more than a 10-Mbps data rate. 100 Mbps and even 1 Gbps are not uncommon rates with modern systems such as Ethernet (CSMA/CD).

In the following sections we discuss several of the more popular (nonproprietary) LANs which will likely be encountered in an office, factory, or campus environment.

9.2   DISTINGUISHING CHARACTERISTICS OF LANS

A LAN covers a limited geographical area. In fact the maximum extension of a LAN is a major factor in its design and selection. A LAN may have an extension of less than 100 ft covering a portion of a floor in a building. It may cover not only an entire floor but several floors. Some LANs can extend over a large university campus. A LAN has a single owner; it is a private network. Its principal purpose is cost sharing of high-value computer assets.

LANs have much higher data rates when compared to wide-area networks (WANs). Many operational WANs use transport systems optimized for voice; data transmission is a compromise. In other words, WANs are constrained by the intervening media; LANs are only constrained by the laws of physics.

A common bit rate for LANs still is 10 Mbps. 100 Mbps and 1 Gbps are becoming equally prevalent. These systems are optimized for data, and excellent BER performance values can be achieved without error correction, from 1×10^{-8} to 1×10^{-12} and better. There are three different transmission media used with LANs: wire pair, coaxial cable, and fiber optics.* The popularity of coaxial cable is diminishing. Wire pair is the transmission medium of choice, although comparatively distance-limited. Fiber optics, however, is rapidly taking its rightful place in the enterprise environment. It is the least constrained for data rate among the three transmission media.

Another advantage of LANs is that they permit coupling of multivendor equipments into an integral system, particularly when based on open system interconnection (OSI) (Chapter 2). However, it should be pointed out that connectivity via OSI layer 1 does not guarantee compatibility of layer 2 and above with equipment of different vendors. Such compatibility can often be achieved by conversion software.

LANs can be extended using repeaters; LANs can also be interconnected among themselves using devices called *bridges*. Smart bridges and routers can direct traffic to other disparate LANs or to a WAN (see Section 10.3).

---

*There are radio-based LANs, now popularly called wireless LANs (Section 10.1.5).

There are two generic transmission techniques utilized by LANs: baseband and broadband. Baseband transmission in this context can be loosely defined as the direct application of the baseband signal to the transmission medium.[†] Broadband transmission (regarding LANs) is where the baseband signal from the data device is translated in frequency to a particular band of frequencies in radio-frequency spectrum and applied to a coaxial cable medium. Broadband transmission requires a modem to carry out the translation. Baseband transmission may require some sort of signal conditioning device. With broadband transmission we usually think of simultaneous multiple carriers that are separated in the frequency domain. Broadband LANs derive from cable TV technology.

Baseband LANs have completely dominated the marketplace. For this reason, considerably more space in this chapter is dedicated to baseband LANs than to broadband LANs. Different LANs are defined by the transmission medium, the bit rate(s), and the protocol.

Many of the widely used LAN protocols have been developed or standardized in North America through the offices of the Institute of Electrical and Electronic Engineers (IEEE). The American National Standards Institute (ANSI) has subsequently accepted and incorporated these standards, and they now bear the ANSI imprimatur. A notable exception is FDDI, which is completely ANSI-sponsored. The IEEE develops LAN standards in the IEEE 802 committee, which is currently organized into these subcommittees:

802.1*	High-Level Interface
802.2*	Logical Link Control (LLC)
802.3*	CSMA/CD Networks
802.4*	Token Bus Networks
802.5*	Token Ring Networks
802.6*	Metropolitan Area Networks
802.7*	Broadband Technical Advisory Group
802.8	Fiber-Optic Technical Advisory Group
802.9*	Integrated Data and Voice Networks
802.10*	Standard for Interoperable LAN Security
802.11*	Wireless LANs
802.12*	Demand Priority Working Group
802.13	—
802.14*	Cable TV Working Group
802.15	Wireless Personal Area Working Group
802.16	Broadband Wireless Access (LMDS) Working Group

(Asterisk (*) indicates published standards)

[†]In some applications, the "baseband" modulates a radio frequency and then the resulting signal is applied to the cable medium.

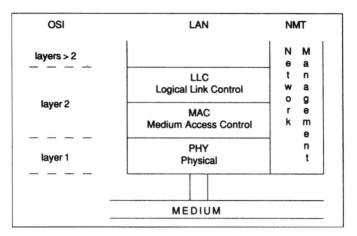

Figure 9.1. LAN architecture related to OSI.

9.3   HOW LAN PROTOCOLS RELATE TO OSI

LAN protocols utilize only OSI layers 1 and 2, the physical and data-link layers, respectively. The data-link layer is split into two sublayers: medium access control (MAC) and logical link control (LLC). These relationships to OSI are shown in Figure 9.1.

The sublayers, logical link control and medium access control, carry out four basic functions:

1. They provide one or more service access points (SAPs). An SAP is a logical interface between two adjacent layers.
2. Before transmission, they assemble data into a frame with address and error-detection fields.
3. On reception, they disassemble the frame and perform address recognition and error detection.
4. They manage communications over the link.

The first function and those related to it are performed by the LLC sublayer. The last three functions listed are handled by the MAC sublayer. Section 9.4 provides a brief description of LLC operation. Section 9.5 gives overviews of several MACs and their related protocols.

9.4   LOGICAL LINK CONTROL (LLC)

The LLC sublayer constitutes the top sublayer in the data-link layer as shown in Figure 9.2. The LLC is common to various MAC methods that are defined and supported by IEEE 802 committees [ISO 8802 activity (Ref. 2)].

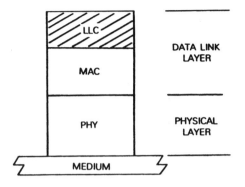

Figure 9.2. LLC relationship in the LAN reference model.

The LLC is a service specification to the network layer (OSI layer 3), to the MAC sublayer, and to the LLC sublayer management function. The service specification to the network layer provides a description of the various services that the LLC sublayer, plus underlying layers and sublayers, offers to the network layer. The service specification to the MAC sublayer provides a description of the services that the LLC sublayer requires of the MAC sublayer. These services are defined so as to be independent of both (a) the form of the medium access methodology and (b) the nature of the medium itself. The service specification to the LLC sublayer management function provides a description of the management services that are provided to the LLC sublayer. All of these service specifications are given in the form of primitives that represent in an abstract way the logical exchange of information and control between the LLC sublayer and the identified service function (i.e., network layer, MAC sublayer, or LLC sublayer management function).

The LLC provides two types of data-link operation. The first type of operation is a data-link connectionless mode of service across a data link with minimum protocol complexity. This type of operation is attractive when higher layers provide any essential recovery and sequencing services so that these do not need replicating in the data-link layer. In addition, this type of operation may prove useful in applications where it is not essential to guarantee the delivery of every data-link layer data unit. The connectionless mode is described in terms of "logical data links."

The second type of operation provides a data-link connection-mode service across a data link which is comparable to such protocol procedures as high-level data-link control (HDLC; see Chapter 3). This service includes support of sequenced delivery of data units and a comprehensive set of data-link layer error recovery techniques. This second type of service is described in terms of "data-link connections."

The LLC standard (Ref. 2) identifies two distinct "classes" of LLC operation. Class I provides data-link connectionless-mode service only. Class II provides data-link connection-mode service plus data-link connectionless-mode service. Either class of operation may be supported.

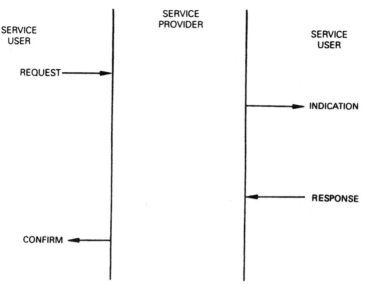

Figure 9.3. Service primitives.

The basic LLC protocols are peer protocols for use in multistation, multiaccess environments. Because of the multistation, multiaccess environment, it is possible for a station to be involved in a multiplicity of peer protocol data exchanges with a multiplicity of different stations over a multiplicity of different *logical* data links and/or data-link connections that are carried by a single physical layer (PHY) over a single physical medium. Each unique to–from pairing at the data-link layer defines a separate logical data link or data-link connection with separate logical parameters and variables.

9.4.1 LLC Sublayer Services and Primitives

The services of a layer (or sublayer) are the capabilities which it offers to a user in the next higher layer (or sublayer). In order to provide its service, a layer (or sublayer) builds its functions on the services which it requires from the next lower layer (or sublayer). Figure 9.3 illustrates the four generic primitives in relation to service providers and service users.

Services are specified by describing the information flow between the N-user and the N-layer (or sublayer). The information flow is modeled by discrete, instantaneous events, which characterize the provision of service. Each event consists of passing a service primitive from one layer (or sublayer) to the other through an N-layer (or sublayer) SAP associated with the N-user. Service primitives convey the information required in providing a particular service. These service primitives are an abstraction in that they specify only the service provided rather than the means by which the service is provided. This definition of service is independent of any particular interface implementation.

Services are specified by describing the service primitives and parameters that characterize each service. A service may have one or more related primitives that constitute the activity that is related to the particular service. Each service primitive may have zero or more parameters that convey the information required to convey the service.

There are four generic primitives described as follows:

Request. The request primitive is passed from the N-user to the N-layer (or sublayer) to request that a service be initiated.

Indication. The indication primitive is passed from the N-layer (or sublayer) to the N-user to indicate an internal N-layer (or sublayer) event which is significant to the N-user. This event may be logically related to a remote service request, or may be caused by an event internal to the N-layer (or sublayer).

Response. The response primitive is passed from the N-user to the N-layer (or sublayer) to complete a procedure previously invoked by an indication primitive.

Confirm. The confirm primitive is passed from the N-layer (or sublayer) to the N-user to convey the results of one or more associated previous service request(s).

(Note the similarity with HDLC described in Chapter 3.)

Possible relationships among primitive types are shown by means of time-sequence diagrams in Figure 9.4. The figure also shows the logical relationship of primitive types. Primitive types that occur earlier in time (connected by dotted lines in the figure) are the logical antecedents of subsequent primitive types.

9.4.1.1 *Unacknowledged Connectionless-Mode Service.* This is defined as the data transfer service that provides the means by which network entities (LAN accesses) can exchange link service data units (LSDUs) without the establishment of a data-link connection. The data transfer can be point-to-point, multicast, or broadcast.

9.4.1.2 *Connection-Mode Services.* This set of services provides the means for establishing, using, resetting, and terminating data-link layers connections. These connections are point-to-point connections between link-layer service access points (LSAPs).

The connection establishment service provides a means by which a network entity can request, or be notified of, the establishment of data-link layer connections.

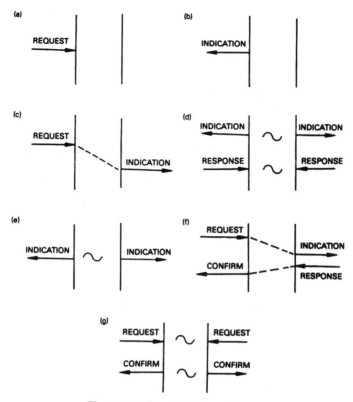

Figure 9.4. Time-sequence diagram.

The connection-oriented data transfer service provides the means by which a network entity can send or receive LSDUs, over a data-link layer connection. This service also provides data-link layer sequencing, flow control, and error recovery.

The connection reset service provides the means by which established connections can be returned to their initial state.

The connection termination service provides the means by which a network entity can request, or be notified of, the termination of data-link layer connections.

The connection flow control service provides the means to control the flow of data associated with a specified connection, across the network layer/data-link layer interface.

9.4.1.3   Type 1 and Type 2 Operation

Type 1 Operation. With Type 1 operation, protocol data units (PDUs) are exchanged between LLCs without the need for establishment of a data-link connection. In the LLC sublayer these PDUs are not acknowledged, nor is there any flow control or error recovery in the Type 1 procedures.

DSAP Address	SSAP Address	Control	Information
8 bits	8 bits	8 or 16 bits	M * 8 bits

Figure 9.5. LLC PDU format. DSAP Address = destination service access point address field; SSAP Address = source service access point address field; Control = control field (16 bits for formats that include sequence numbering, and 8 bits for formats that do not; Information = information field; * = multiplication; M = an integer value equal to or greater than 0. (Upper bound of M is a function of the medium access control methodology used.)

Type 2 Operation. With Type 2 operation, a data-link connection is established between two LLCs prior to any exchange of information-bearing PDUs. The normal cycle of communication between two Type 2 LLCs on a data-link connection consists of the transfer of PDUs containing information from the source LLC to the destination LLC, acknowledged by PDUs in the opposite direction.

9.4.2 LLC PDU Structure

9.4.2.1 Overview. Figure 9.5 shows the LLC PDU format. As the figure indicates, each LLC PDU contains two address fields: the destination service access point (DSAP) address field and the source service access point (SSAP) address field. Each of these fields contains only a single address. The DSAP address field identifies one or more SAPs for which the LLC information field is intended. The SSAP address field identifies the specific SAP from which the LLC information field was initiated. Figures 9.6 and 9.7 show the address field formats.

An individual address is usable as both an SSAP and a DSAP address; a null address is usable as both an SSAP and DSAP address; and a group address is usable only as a DSAP address.

All 1s in the DSAP address field (i.e., the address type designation bit set to 1 and the seven address bits set to 1) is predefined to be the "global" DSAP address. The DSAP address designates a group consisting of all DSAPs actively being serviced by the underlying MAC SAP address(es).

All 0s in the DSAP or SSAP address field (i.e., the address type designation bit set to 0 and seven address bits set to 0) is predefined to be the "null" address. The null SAP address designates the LLC that is associated with the underlying MAC SAP address and is *not* used to identify any SAP to the network layer or any SAP to an associated layer management function. Addresses 01000000 and 11000000 are designated as the individual and group addresses, respectively, for an LLC sublayer management function at the station. Other addresses with the next to low-order bit set to 1 are reserved for International Standards Organization (ISO) definition.

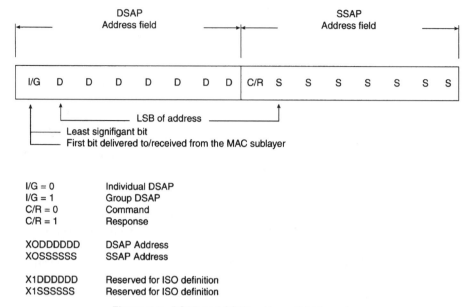

Figure 9.6. DSAP and SSAP address field formats.

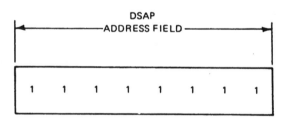

Figure 9.7. Global DSAP address field format.

The control field consists of one or two octets that are used to designate command and response functions and that contain sequence numbers when required. The content of this field is described below.

The information field consists of any integral number (including zero) of octets.

The *bit order* is as follows. Addresses, commands and responses, and sequence numbers are delivered to/received from the MAC sublayer least significant bit first (i.e., the first bit of a sequence number that is delivered/received has the weight 2**0). The information field is delivered to the MAC sublayer in the same bit order as received from the network layer. The information field is delivered to the network layer in the same bit order as received from the MAC sublayer.

Definition of an Invalid LLC PDU. An invalid LLC PDU is defined as one which meets at least one of the following conditions:

1. It is identified as such by the PHY or MAC sublayer.
2. It is not an integral number of octets in length.
3. It does not contain two properly formatted address fields, one control field, and, optionally, an information field in their proper order.
4. Its length is less than 3 octets (one-octet control field) or 4 octets (two-octet control field).

Invalid LLC PDUs are ignored.

9.4.2.2 *Control Field Formats.* The three formats defined for the control field, shown in Figure 9.8, are used to perform numbered information transfer, numbered supervisory transfer, unnumbered control, and unnumbered transfer functions. The numbered information transfer and supervisory transfer functions apply only to Type 2 operation. The unnumbered control and unnumbered information transfer functions apply to either Type 1 or Type 2 operation (but not both) depending on the specific function selected.

Information Transfer Format I. The I-format PDU is used to perform a numbered information transfer in Type 2 operation. Except where otherwise

LLC PDU Control field bits

	1	2	3	4	5	6 7 8	9	10-16
Information transfer command/response (I-format PDU)	0			N(S)			P/F	N(R)
Supervisory commands/responses (S-format PDUs)	1	0	S	S	X	X X X	P/F	N(R)
Unnumbered commands/responses (U-format PDUs)	1	1	M	M	P/F	M M M		

N(S)	=	Transmitter send sequence number (Bit 2 = low-order bit)
N(R)	=	Transmitter receive sequence number (Bit 10 = low-order bit)
S	=	Supervisory function bit
M	=	Modifier function bit
X	=	Reserved and set to zero
P/F	=	Poll bit—command LLC PDU transmissions
		Final bit—response LLC PDU transmissions
		(1 = Poll/final)

Figure 9.8. LLC PDU control field formats.

specified (e.g., UI, TEST, FRMR, and XID), it is the only LLC PDU that may contain an information field. The functions $N(S)$, $N(R)$, and poll/final (P/F) are independent [i.e., each I-format PDU has an $N(S)$ sequence number, an $N(R)$ sequence number that acknowledges or does not acknowledge additional I-format PDUs at the receiving LLC, and a P/F bit that is set to "1" or "0"]. One notes the similarity to HDLC given in Chapter 3.

Supervisory Format S. The S-format PDU is used to perform data-link supervisory control functions in Type 2 operation, such as acknowledging I-format PDUs, requesting retransmission of I-format PDUs, and requesting a temporary suspension of transmission of I-format PDUs. The functions of $N(R)$ and P/F are independent (i.e., each S-format PDU has an $N(R)$ sequence number that does or does not acknowledge additional I-format PDUs at the receiving LLC, and it also has a P/F bit that is set to "1" or "0").

Unnumbered Format U. The U-format PDUs are used in either Type 1 or Type 2 operation, depending upon the specific function utilized, to provide additional data-link control functions and to provide unsequenced information transfer. The U-format PDUs contain no sequence numbers but include a P/F bit that is set to "1" or "0."

9.4.2.3 *Control Field Parameters*

Type 1 Operation Parameters. The only parameter that exists in Type 1 operation is the P/F bit. The P/F bit set to "1" is only used in Type 1 operation with the XID and TEST command/response PDU functions. The poll (P) bit set to "1" is used to solicit (poll) a correspondent response PDU with the F bit set to "1" from the addressed LLC. The final (F) bit set to "1" is used to indicate that the response PDU which is sent by the LLC is the result of a soliciting (poll) command PDU (P bit set to "1").

Type 2 Operation Parameters

MODULUS. Each I PDU is sequentially numbered with a number that has a value between 0 and MODULUS minus ONE (where MODOLUS is the modulus of the sequence numbers). The modulus is equal to 128 for Type 2 LLC control field format. The sequence numbers cycle through the entire range.

The maximum number of sequentially numbered I PDUs that may be outstanding (i.e., unacknowledged) in a given direction on a data-link connection at any given time never exceeds one less than the modulus of the sequence numbers. This restriction prevents any ambiguity in the association of sent I PDUs with sequence numbers during normal operation and/or error recovery action.

LLC PDU STATE VARIABLES AND SEQUENCE NUMBERS. A station LLC maintains a send state variable $V(S)$ for the I PDUs it sends and a receive state variable $V(R)$ for the I PDUs it receives on each data-link connection. The operation of $V(S)$ is independent of the operation of $V(R)$.

The send state variable denotes the sequence number of the next in-sequence I PDU to be sent on a specific data-link connection. The send state variable takes on a value between 0 and MODULUS minus ONE (where MODULUS equals 128 and the numbers cycle through the entire range). The value of the send state variable is incremented by one with each successive I PDU transmission on the associated data-link connection, but will not exceed $N(R)$ of the last received PDU by more than MODULUS minus ONE. Again the similarity to HDLC should be noted.

Only I PDUs contain $N(S)$, the send sequence number of the sent PDU. Prior to sending an I PDU, the value of $N(S)$ is set equal to the value of the send state variable for that data-link connection.

The receive state variable denotes the sequence number of the next in-sequence I PDU to be received on a specific data-link connection. The receive state variable takes on a value between 0 and MODULUS minus ONE (where MODULUS equals 128 and the numbers cycle through the entire range). The value of the receive state variable associated with a specific data-link connection is incremented by one whenever an error-free, in-sequence I PDU is received whose send sequence number $N(S)$ equals the value of the receive state variable for the data-link connection.

All I-format PDUs and S-format PDUs contain $N(R)$, the expected sequence number of the next received I PDU on the specified data-link connection. Prior to sending an I-format PDU or S-format PDU, the value of $N(R)$ is set equal to the current value of the associated receive state variable for that data-link connection. $N(R)$ indicates that the station sending the $N(R)$ has received correctly all I PDUs numbered up through $N(R) - 1$ on the specified data-link connection.

POLL/FINAL (P/F) BIT. The poll (P) bit is used to solicit (poll) a response from the addressed LLC. The final (F) bit is used to indicate the response PDU sent as the result of a soliciting (poll) command.

The poll/final (P/F) bit serves as a function in Type 2 operation in both command PDUs and response PDUs. In command PDUs the P/F bit is referred to as the P bit. In response PDUs it is referred to as the F bit. P/F bit exchange provides a distinct command/response linkage that is useful during normal operation and recovery operations.

A command PDU with the P bit set to "1" is used on a data-link connection to solicit a response PDU with the F bit set to "1" from the addressed LLC on that data-link connection.

Only one PDU with a P bit set to "1" will be outstanding in a given direction at a given time on the data-link connection between any specific pair of LLCs. Before an LLC issues another PDU on the same data-link connection with the P bit set to "1," the LLC will have received a response PDU with the F bit set to "1" from the addressed LLC. If no valid response PDU is received within a system-defined P-bit timer timeout period, the (re)sending of a command PDU with the P bit set to "1" is permitted for error recovery purposes.

A response PDU with the F bit set to "1" is used to acknowledge receipt of a command PDU with the P bit set to "1."

Following the receipt of a command PDU with the P bit set to "1," the LLC sends a response PDU with the F bit set to "1" on the appropriate data-link connection at the earliest possible opportunity. The LLC is permitted to send appropriate response PDUs with the F bit set to "0" at any medium access opportunity on an asynchronous basis (without the need for a command PDU).

9.4.2.4 *Commands and Responses.*

The command/response (C/R) bit, located as the low-order bit in the SSAP field, is used to distinguish between commands and responses. Table 9.1 lists appropriate commands and related responses.

Type 1 Operation Commands and Responses. The Type 1 commands and responses are all U-format PDUs. The U-format PDU command encodings for Type 1 operation are listed in Figure 9.9.

The unnumbered information (UI) command PDU is used to send information to one or more LLCs. Use of the UI command PDU is not dependent on the existence of a data-link connection between the destination and the source LLCs, and its use will not affect the $V(S)$ and $V(R)$ variables associated with any data-link connections. There is no LLC response PDU to the UI command PDU.

TABLE 9.1   LLC Commands and Related Responses

Information Transfer Format Commands	Information Transfer Format Responses
I—Information	I—Information
Supervisory Format Commands	*Supervisory Format Responses*
RR—receive ready	RR—receive ready
RNR—receive not ready	RNR—receive not ready
REJ—reject	REJ—reject
Unnumbered Format Commands	Unnumbered Format Responses
UI—unnumbered information	UA—unnumbered acknowledgement
DISC—disconnect	DM—disconnected mode
SABME—set asynchronous balanced mode extended	FRMR—frame reject
XID—exchange identification	XID—exchange identification
TEST—test	TEST—test

FIRST CONTROL FIELD BIT DELIVERED
TO/RECEIVED FROM THE MAC SUBLAYER

1	2	3	4	5	6	7	8	
1	1	0	0	P	0	0	0	UI COMMAND
1	1	1	1	P	1	0	1	XID COMMAND
1	1	0	0	P	1	1	1	TEST COMMAND

Figure 9.9. Type 1 operation command control field bit assignments.

Reception of the UI command PDU is not acknowledged nor is the sequence number verified; therefore, the data contained in a UI PDU may be lost if a logical data-link exception (such as a transmission error or a receiver-busy condition) occurs during the sending of the command PDU.

A UI command PDU has either an individual, group, global, or null address as the destination DSAP address and has the originator's individual address as the SSAP address.

The XID command PDU is used to (a) convey the types of LLC services supported (for all LLC services) and the receive window size on a per-data-link connection basis to the destination LLC and (b) cause the destination LLC to respond with the XID response PDU at the earliest opportunity. The XID command PDU has no effect on any mode or sequence numbers maintained by the remote LLC. An XID command PDU has either an individual, group, global, or null address as the destination DSAP address and has the originator's individual address as the SSAP address.

The information field of an XID basic format command PDU consists of an 8-bit XID format identifier field plus a 16-bit parameter field that is encoded to identify the LLC services supported plus the receive window size, as shown in Figure 9.10. The receive window size (k) is the maximum number that the send state variable $V(S)$ can exceed the $N(R)$ of the last received PDU.

TEST COMMAND. The TEST command PDU is used to cause the destination LLC to respond with the TEST response PDU at the earliest opportunity, thus performing a basic test of the LLC to LLC transmission path. An information field is optional with the TEST command PDU. If present, however, the received information field is returned, if possible, by the addressed LLC in the TEST response PDU. The TEST command PDU has no effect on any mode or sequence numbers maintained by the remote LLC and may be used with an individual, group, global, or null DSAP address and with an individual, group, or global DA address.

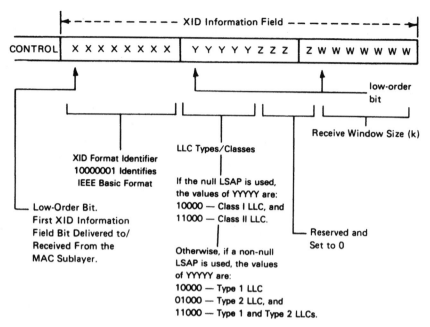

Figure 9.10. XID information field basic format.

TYPE 1 OPERATION RESPONSES. The U-format PDU response encodings for Type 1 operation are shown in Figure 9.11.

Type 2 Operation Commands and Responses. Type 2 commands and responses consist of I-format, S-format, and U-format PDUs.

The function of the information, I, command and response is to transfer sequentially numbered PDUs containing an octet-oriented information field across a data-link connection. The encoding of the I PDU control field for Type 2 operation is shown in Figure 9.12.

The I PDU control field contains two sequence numbers: $N(S)$, send sequence number, which indicates the sequence number associated with the I PDU; and $N(R)$, receive sequence number, which indicates the sequence

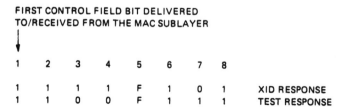

Figure 9.11. Type 1 operation response control field bit assignments.

FIRST CONTROL FIELD BIT DELIVERED
TO/RECEIVED FROM THE MAC SUBLAYER

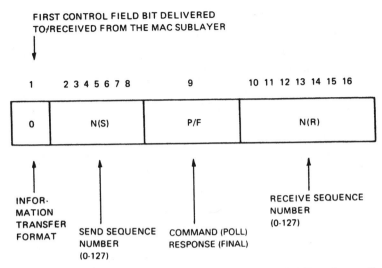

Figure 9.12. Information transfer format control field bits, I PDU, Type 2 operation.

number (as of the time the PDU is sent) of the next expected I PDU to be received and consequently indicates that the I PDUs numbered up through $N(R) - 1$ have been received correctly. Supervisory, S, PDUs are used to perform numbered supervisory functions such as acknowledgments, temporary suspension of information transfer, or error recovery.

PDUs with the S format do not contain an information field and, therefore, do not increment the send state variable at the sender or the receive state variable at the receiver. The encoding of the S-format PDU control field for Type 2 operation is shown in Figure 9.13.

An S-format PDU contains an $N(R)$, receive sequence number, which indicates, at the time of sending, the sequence number of the next expected I PDU to be received and consequently indicates that all received I PDUs numbered up through $N(R) - 1$ have been received correctly.

When sent, an RR (receive ready) or REJ PDU indicates the clearance of any busy condition at the sending LLC that was indicated by the earlier sending of an RNR (receive not ready) PDU.

The RR PDU is used by an LLC to indicate that it is ready to receive an I PDU(s). I PDUs numbered up through $N(R) - 1$ are considered to be acknowledged.

The reject (REJ) PDU is used by an LLC to request resending of I PDUs starting with the PDU numbered $N(R)$. I PDUs numbered up through $N(R) - 1$ are considered as acknowledged. It is possible to send additional I PDUs awaiting initial sending after the re-sent I PDUs.

With respect to each direction of sending on a data-link connection, only one "sent REJ" condition is established at any given time. The "sent REJ"

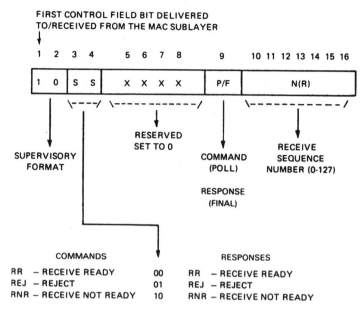

Figure 9.13. Supervisory format control field bits.

condition will be cleared upon receipt of an I PDU with an $N(S)$ equal to $N(R)$ of the REJ PDU.

The receive not ready (RNR) PDU is used by an LLC to indicate a busy condition (i.e., a temporary inability to accept subsequent I PDUs). I PDUs numbered up through $N(R) - 1$ are considered as acknowledged. I PDUs numbered $N(R)$ and any subsequent I PDUs received, if any, are not considered as acknowledged; the acceptance status of these PDUs is indicated in subsequent exchanges.

UNNUMBERED FORMAT COMMANDS AND RESPONSES. Unnumbered, U, commands and responses are used in Type 2 operation to extend the number of data-link control functions. PDUs sent with the U format do not increment the state variables on the data-link connection at either sending or receiving LLC. The encoding of the U-format command/response PDU control field is given in Figures 9.14 and 9.15.

Set Asynchronous Balanced Mode Extended (SABME) Command. The SABME command is used to establish a data-link connection to the destination LLC in the asynchronous balanced mode. No information is permitted with the SABME command PDU. The destination LLC confirms receipt of the SABME command PDU by sending a UA (unnumbered acknowledgement) response PDU on the data-link connection at the earliest opportunity according to whether a DL-CONNECT (primitive) response or a DL-

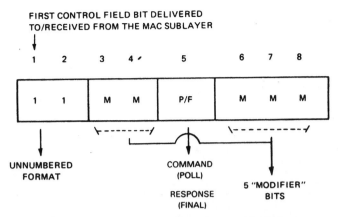

FIRST CONTROL FIELD BIT DELIVERED
TO/RECEIVED FROM THE MAC SUBLAYER

Figure 9.14. Unnumbered format control field bits.

FIRST CONTROL FIELD BIT DELIVERED
TO/RECEIVED FROM THE MAC SUBLAYER

1	2	3	4	5	6	7	8	
1	1	1	1	P	1	1	0	SABME COMMAND
1	1	0	0	P	0	1	0	DISC COMMAND
1	1	0	0	F	1	1	0	UA RESPONSE
1	1	1	1	F	0	0	0	DM RESPONSE
1	1	1	0	F	0	0	1	FRMR RESPONSE

Figure 9.15. Unnumbered command and response control field bit assignments.

DISCONNECT request primitive is passed from the network layer to the
LLC sublayer. Upon acceptance of the SABME command PDU, the destination LLCs send and receive state variables which are set to zero. If the UA
response PDU is received correctly, then the initiating LLC also assumes the
asynchronous balanced mode with its corresponding send and receive state
variables set to zero.

Previously sent I PDUs that are unacknowledged when this command is
actioned remain unacknowledged. Whether or not an LLC resends the
contents of the information field of unacknowledged outstanding I PDUs is
decided at a higher (OSI) layer.

Disconnect (DISC) Command. The DISC command PDU is used to terminate an asynchronous balanced mode previously set by a SABME command
PDU. It is used to inform the destination LLC that the source LLC is
suspending operation of the data-link connection and that the destination
LLC should assume the logically disconnected mode. No information field is

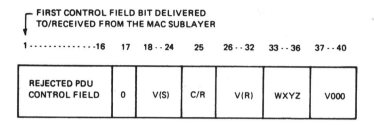

(1) Rejected PDU control field shall be the control field of the received PDU which caused the FRMR exception condition on the data link connection. When the rejected PDU is a U-format PDU, the control field of the rejected PDU shall be positioned in bit positions 1-8, with 9-16 set to 0.

(2) V(S) shall be the current send state variable value for this data link connection at the rejecting LLC (bit 18 = low-order bit).

(3) C/R set to "1" shall indicate that the PDU which caused the FRMR was a response PDU, and C/R set to "0" shall indicate that the PDU which caused the FRMR was a command PDU.

(4) V(R) shall be the current receive state variable value for this data link connection at the rejecting LLC (bit 26 = low-order bit).

(5) W set to "1" shall indicate that the control field received and returned in bits 1 through 16 was invalid or not implemented. Examples of invalid PDU are defined as
    (a) the receipt of a supervisory or unnumbered PDU with an information field which is not permitted;
    (b) the receipt of an unsolicited F bit set to "1"; and
    (c) the receipt of an unexpected UA response PDU.

(6) X set to "1" shall indicate that the control field received and returned in bits 1 through 16 was considered invalid because the PDU contained an information field which is not permitted with this command or response. Bit W shall be set to "1" in conjunction with this bit.

(7) Y set to "1" shall indicate that the information field received exceeded the established maximum information field length which can be accommodated by the rejecting LLC on that data link connection.

(8) Z set to "1" shall indicate that the control field received and returned in bits 1 through 16 contained an invalid N(R).

(9) V set to "1" shall indicate that the control field received and returned in bits 1 through 16 contained an invalid N(S). Bit W shall be set to "1" in conjunction with this bit.

Figure 9.16. FRMR information field format.

permitted with the DISC command PDU. Prior to actioning the command, the destination LLC confirms the acceptance of the DISC command PDU by sending a UA response PDU on that data-link connection.

Previously sent I PDUs that are unacknowledged when this command is actioned remain unacknowledged. Whether or not the LLC resends the contents of the information field of unacknowledged outstanding I PDUs is decided at a higher (OSI) layer.

The frame reject response (FRMR) information field is shown in Figure 9.16.

Section 9.4 consists of abstracted and edited material from Ref. 2.

9.5 MEDIUM ACCESS CONTROL

9.5.1 Introduction

With baseband LAN transmission, one medium interconnects all participating stations. In such a scheme, only one LAN station can transmit at a time. If two stations transmit at the same time, they will interfere with each other, each corrupting the other's traffic. There are three approaches to control medium access: random access or contention (first come, first served), token passing, and time-slot assignment schemes. In this chapter and Chapter 10 we deal with the first two. A typical time-slot assignment scheme is distributed queue dual bus, which is described in Ref. 4.

9.5.2 Carrier Sense Multiple Access with Collision Detection (CSMA/CD)—Basic Concepts

9.5.2.1 Introduction to CSMA/CD. Carrier sense multiple access with collision detection (CSMA/CD) (Ref. 3) is a random access method for LANs. Conceptually, this is a baseband technique for LAN transmission. As with other baseband LANs, two or more LAN stations share a common transmission medium. To transmit, a station waits (defers) until a quiet period on the medium (i.e., no other station is transmitting) and then sends the message in bit serial form. If, after initiating a transmission, the message collides with that of another station, then each transmitting station intentionally sends a few additional bytes to ensure propagation of the collision throughout the system. The station remains silent for a random amount of time (called *backoff*) before attempting to transmit again.

CSMA/CD relationship to OSI is shown in Figure 9.17. This section deals with the lowest two OSI layers: the data-link layer and the physical layer. These encompass the MAC sublayer and the physical layer. The OSI data-link layer in this case consists of the MAC and the LLC (Section 9.4).

9.5.2.2 MAC Frame Structure for CSMA/CD. Figure 9.18 illustrates the MAC frame format. There are nine fields: preamble, start frame delimiter (SFD), destination and source addresses, length/type, MAC Client Data, PAD (when required), frame check sequence, and extension. The extension field is required for 1-Gbps operation in the half-duplex mode only. Of these nine fields, all are of fixed length except the client data, PAD, and extension fields. The nine fields may contain an integer number of octets between the minimum and maximum values determined by the specific implementation of the CSMA/CD MAC (see Section 9.6.5).

In the tables in Section 9.6.5, the minimum and maximum frame size limits refer to that portion of the frame from the destination address field through the frame check sequence inclusive. When dealing with Figure 9.18, the octets of a frame are transmitted from top to bottom and the bits of each octet are transmitted from left to right.

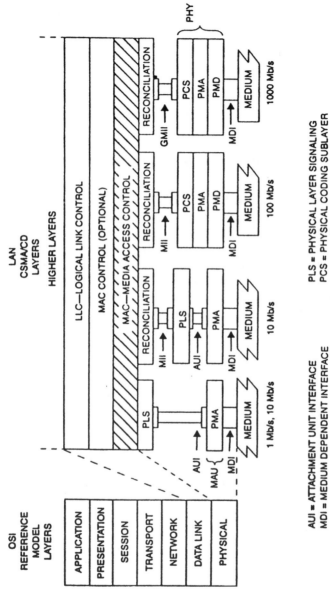

AUI = ATTACHMENT UNIT INTERFACE
MDI = MEDIUM DEPENDENT INTERFACE
MII = MEDIA INDEPENDENT INTERFACE
GMII = GIGABIT MEDIA INDEPENDENT INTERFACE
MAU = MEDIUM ATTACHMENT UNIT

PLS = PHYSICAL LAYER SIGNALING
PCS = PHYSICAL CODING SUBLAYER
PMA = PHYSICAL MEDIUM ATTACHMENT
PHY = PHYSICAL LAYER DEVICE
PMD = PHYSICAL MEDIUM DEPENDENT

Figure 9.17. CSMA/CD relationship with the OSI reference model, MAC sublayer partitioning. (From Figure 4-1, page 42, IEEE Std. 802.3, Ref. 3.)

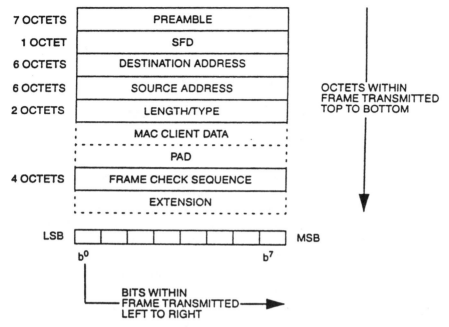

Figure 9.18. MAC frame format.

Elements of a MAC Frame

PREAMBLE FIELD. A preamble field contains seven octets and is used to allow the PLS (physical layer signaling) circuitry to reach its steady-state synchronization with received frame timing. The preamble pattern is alternating binary 1s and 0s (i.e., 10101010). The nature of the pattern is such that, for Manchester coding, it appears as a periodic waveform on the medium that enables bit synchronization. It should be noted that the pattern ends with a binary "0."

Start Frame Delimiter (SFD) Field. The SFD field is the sequence 10101011. It immediately follows the preamble and indicates the start of frame.

ADDRESS FIELDS. Each MAC frame contains two address fields: destination address field and the source address field, in that order (see Figure 9.19). The destination address field specifies the destination addressee(s) for which the frame is intended. The source address field identifies the station from which the frame was initiated.

(a) Each address field is 48 bits long.
(b) The first bit (least significant bit, LSB) is used in the destination address field as an address-type designation bit to identify the destina-

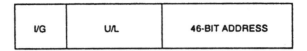

I/G = 0 INDIVIDUAL ADDRESS
I/G = 1 GROUP ADDRESS
U/L = 0 GLOBALLY ADMINISTERED ADDRESS
U/L = 1 LOCALLY ADMINISTERED ADDRESS

Figure 9.19. Address field format.

tion address either as an individual or as a group address. If this bit is 0, it indicates that the address field contains an individual address. If the bit is 1, it indicates that the address field contains a group address that identifies none, one or more, or all of the stations connected to the LAN. In the source address field, the first bit is reserved and set to 0.

(c) The second bit is used to distinguish between locally or globally administered addresses. For globally administered (or U, universal) addresses, the bit is set to 0. If an address is to be assigned locally, this bit is set to 1. Note that for the broadcast address, this bit is also set to 0.

(d) Each octet of each address field is transmitted least significant bit first. If there are all 1s in the destination address field, it is predefined to be the broadcast address.

LENGTH/TYPE FIELD. This is a two-octet field that takes on one of two meanings depending on its numeric value. For numerical evaluation, the first octet is the most significant octet in this field.

(a) If the value of this field is less than or equal to the value of maxValidFrame, then the length/type field indicates the number of MAC client data octets contained in the subsequent data field of the frame (length interpretation).

MaxValidFrame = maxFrameSize

$$- (2 \text{XaddressSize} + \text{LengthOrTypeSize} + \text{crcSize})/8$$

Substitute the following in the above equation: MaxFrameSize (in bits) uses data from tables in Section 9.6.5 (e.g., for 10 Mbps, 1518 octets ×8).

$$2 \text{XaddressSize} = 32 \text{ bits}$$

$$\text{LengthOrTypeSize} = 16 \text{ bits}$$

$$\text{CrcSize} = 32 \text{ bits}$$

(b) If the value of this field is greater than or equal to 1536 decimal (equal to 0600 hexadecimal), then the length/type fields indicates the nature of the MAC client protocol (Type interpretation).* The length and type interpretations of this field are mutually exclusive.

Regardless of the interpretation of the length/type field, if the length of the data field is less than the minimum required for proper operation of the protocol, a PAD field (a sequence of octets) is added at the end of the data field but prior to the FCS field, as discussed below. The procedure that determines the size of the PAD field is specified in Section 4.2.8 of Ref. 3. The length/type field is transmitted and received with the high-order octet first.

DATA AND PAD FIELDS. The data field (MAC client data) contains a sequence of n octets. Full data transparency is provided in the sense that any arbitrary sequence of octet values may appear in the data field up to the maximum number specified by the implementation of the standard that is used. Keep in mind that there is a minimum frame size required (i.e., 64 octets) for proper operation of the protocol.

The length of the PAD field required for MAC client data that is n octets long is max [0, minFrameSize − (8xn + 2XaddressSize + 48)] bits. The maximum possible size of the data field is maxFrameSize − (2XaddressSize + 48)/8 octets.

FRAME CHECK SEQUENCE (FCS) FIELD. A CRC is used by the transmit and receive algorithms to generate a CRC value for the FCS field. The FCS field contains a 32-bit CRC value. The value is computed as a function of the contents of the source address, destination address, length, MAC client data, and PAD (that is, all fields except the preamble, SFD, FCS, and extension). The encoding is defined by the following generating polynomial:

$$G(x) = x^{32} + x^{26} + x^{23} + x^{22} + x^{16} + x^{12} + x^{11} + x^{10}$$
$$+ x^8 + x^7 + x^5 + x^4 + x^2 + x + 1$$

Mathematically, the CRC value corresponding to a given frame is defined by the following procedure:

(a) The first 32 bits of the frame are complemented.
(b) The n bits of the frame are then considered to be the coefficients of a polynomial $M(x)$ of degree $n - 1$. (The first bit of the destination address field corresponds to the $x^{(n-1)}$ term, and the last bit of the data field corresponds to the x^0 term.)

---

*Type field assignments are administered by the Registration Authority, IEEE Standards Dept., PO Box 1331, Piscataway, NJ 08855-1331.

(c) $M(x)$ is multiplied by x^{32} and divided by $G(x)$, producing a remainder $R(x)$ of degree ≤ 31.

(d) The coefficients of $R(x)$ are considered to be a 32-bit sequence.

(e) The bit sequence is complemented and the result is the CRC.

The 32 bits of the CRC value are placed in the frame check sequence field so that the x^{31} term is the leftmost bit of the first octet, and the x^0 term is the rightmost bit of the last octet. (The bits of the CRC are thus transmitted in the order $x^{31}, x^{30}, \ldots, x^1, x^0$.)

EXTENSION FIELD. The extension field follows the FCS field and is made up of a sequence of extension bits, which are readily distinguished from data bits. The length of the field is in the range of zero to (slotTime-minFrame-Size) bits, inclusive. The contents of the extension field are not included in the FCS computation.

Carrier extension is employed in the half-duplex mode only and at operating speeds above 100 Mbps. The slotTime employed at speeds of 100 Mbps and below is inadequate to accommodate network topologies of the desired physical extent. Carrier extension provides a means by which the slotTime can be increased to a sufficient value for the desired topologies, without increasing the minFrameSize parameter, because this would have deleterious effects. Nondata bits, referred to as extension bits, are appended to frames that are less than slotTime bits in length so that the resulting transmission is at least one slotTime in duration. Carrier extension can be performed only if the underlying physical layer is capable of sending and receiving symbols that are readily distinguished from data symbols, as is the case in most physical layers that use a block encoding/decoding scheme. The maximum length of the extension is equal to the quantity (slotTime − minFrameSize).

The MAC continues to monitor for collisions while it is transmitting extension bits, and it will treat any collision that occurs after the threshold (slotTime) as a late collision.

Order of Bit Transmission. Each octet of the MAC frame, with the exception of the FCS, is transmitted low-order bit first.

Invalid MAC Frame. An invalid MAC frame is defined as one that meets at least one of the following conditions:

1. The frame length is inconsistent with the length field.
2. It is not an integral of octets in length.
3. The bits of the incoming frame (exclusive of the FCS field itself) do not generate a CRC value identical to the one received. The contents of invalid MAC frames are not passed to the LLC.

It should be noted that the LLC PDU is inserted into the MAC client data field as shown in Figure 9.18.

9.5.2.3 CSMA/CD Basic Operation

Transmission Without Contention. When an LLC sublayer requests the transmission of a frame, the transmit data encapsulation component of the MAC sublayer constructs the frame from the LLC-supplied data. It appends a preamble and an SFD to the beginning of a frame. A PAD is also appended, if required. It also appends destination and source addresses, a length count field, and an FCS. The frame is then handed to the transmit media access management component in the MAC sublayer for transmission. The transmit media access management then attempts to avoid contention with other traffic on the medium by monitoring the carrier sense signal provided by the PLS and deferring to passing traffic. When the medium is clear, frame transmission starts. The MAC sublayer then provides a serial stream of bits to the PLS interface for transmission.

The PLS performs the task of actually generating the electrical signals on the medium that represent the bits of the frame. Simultaneously, it monitors the medium and generates the collision detect signal, which, in the contention-free case, remains off for the duration of the frame.

Reception Without Contention. At each LAN receiving station, the arrival of a frame is first detected by the PLS, which responds by synchronizing with the incoming preamble and by turning on the carrier sense signal. As the encoded bits arrive from the medium, they are decoded and translated back into binary data. The PLS passes subsequent bits up to the MAC sublayer, where the leading bits are discarded, up to and including the end of the preamble and SFD.

Meanwhile, the receive media access management component of the MAC sublayer, having observed carrier sense, has been waiting for the incoming bits to be delivered. Receive media access management collects bits from the PLS as long as the carrier sense signal remains on. When the carrier sense signal is removed, the frame is truncated to the octet boundary, if necessary, and passed to receive data decapsulation for processing.

Receive data decapsulation checks the frame's destination address field to decide whether the frame should be received by this station. If so, it passes the destination address (DA), the source address (SA), and the LLC data unit (LLCDU) to the LLC sublayer along with the appropriate status code indicating "reception complete" or "reception too long." It also checks for invalid MAC frames by inspecting the FCS to detect any damage to the frame en route and by checking for proper octet boundary alignment of the frame. Frames with a valid FCS may also be checked for proper octet boundary alignment.

Access Interference and Recovery. If multiple LAN stations attempt to transmit at the same time, it is possible for them to interfere with each other's transmissions, in spite of their attempts to avoid this by deferring.

When transmissions from two stations overlap, the resulting contention is called *collision*. A given station can experience collision during the initial part of its transmission (the collision window) before its transmitted signal has had time to propagate to all stations on the medium. Once the collision window has passed, a transmitting station is said to have acquired the medium. Subsequent collisions are avoided because all other (properly functioning) LAN stations can be assumed to have noticed the signal (by means of carrier sense) and to be deferring to it. The time to acquire the medium is thus based on the round-trip propagation time of the physical layer whose elements include the PLS, the physical medium attachment (PMA), and the physical medium itself.

In the event of a collision, the transmitting station's physical layer initially notices the interference on the medium and then turns on the collision detect signal. This is noticed in turn by the transmit media access management component of the MAC sublayer, and the collision handling begins. First, the transmit media access management enforces the collision by transmitting a bit sequence called a *jam*. This ensures that the duration of the collision is sufficient to be noticed by the other transmitting station(s) involved in the collision. After the jam is sent, the transmit media access management terminates the transmission and schedules another transmission attempt after a randomly selected time interval. Retransmission is attempted again in the face of repeated collisions. Because repeated collisions indicate a busy medium, however, the transmit media access management attempts to adjust to the medium load by backing off (voluntarily delaying its own retransmissions to reduce its load on the medium). This is accomplished by expanding the interval from which the random retransmission time is selected on each successive transmit attempt. Eventually, either the transmission succeeds, or the attempt is abandoned on the assumption that the medium has failed or has become overloaded.

At the receiving end, the bits resulting from the collision are received and decoded by the PLS just as are the bits of a valid frame. Fragmentary frames received during collisions are distinguished from valid transmissions by the MAC sublayer's receive media access management component.

Carrier Deference. Even when it has nothing to transmit, the CSMA/CD MAC sublayer monitors the physical medium for traffic by watching the carrier sense signal provided by the PLS. Whenever the medium is busy, the CSMA/CD MAC sublayer defers to the passing frame by delaying any pending transmissions of its own. After the last bit of a passing frame (i.e., when carrier sense changes from true to false), the CSMA/CD MAC sublayer continues to defer for a specified interframe spacing period.

If, at the end of the interframe spacing, a frame is waiting to be transmitted, transmission is initiated independently of the value of carrier sense. When transmission has completed (or immediately, if there was nothing to transmit), the CSMA/CD MAC sublayer resumes its original monitoring of carrier sense.

When a frame is submitted by the LLC sublayer for transmission, the transmission is initiated as soon as possible, but in conformance with the rules of deference stated above.

Collision Handling. Once a CSMA/CD sublayer has finished deferring and has started transmission, it is still possible for it to experience contention for the medium. Collisions can occur until acquisition of the network has been accomplished through the deference of all other stations' CSMA/CD sublayers.

The dynamics of collision handling are largely determined by a single parameter called the *slot time*. This single parameter describes three important aspects of collision handling:

1. It is an upper bound on the acquisition time of the medium.
2. It is an upper bound on the length of a frame fragment generated by a collision.
3. It is the scheduling quantum for retransmission.

To fulfill all three functions, the slot time must be larger than the sum of the physical layer round-trip propagation time and the media access layer maximum jam time. The slot time is determined by the parameters of the implementation.

Collision Detection and Enforcement. Collisions are detected by monitoring the collision detect signal provided by the physical layer. When a collision is detected during a frame transmission, the transmission is not terminated immediately. Instead, the transmission continues until additional bits specified by jam size have been transmitted (counting from the time collision detect went on). This collision enforcement or jam guarantees that the duration of the collision is sufficient to ensure its detection by all transmitting stations on the network. The content of the jam is unspecified; it may be any fixed or variable pattern convenient to the media access implementation, but the implementation will not be intentionally designed to the 32-bit CRC value corresponding to the (partial) frame transmitted prior to the jam.

Collision Backoff and Retransmission. When a transmission attempt has terminated due to a collision, it is retried by the transmitting CSMA/CD sublayer until it is successful or a maximum number of attempts (attempt limit) have been made and all have terminated due to collisions. The scheduling of the retransmissions is determined by a controlled randomization process called *truncated binary exponential backoff*. At the end of enforcing a collision (jamming), the CSMA/CD sublayer delays before attempting to retransmit the frame. The delay is an integer multiple of the slotTime. The number of slot times to delay before the nth retransmission attempt is chosen as a uniformly distributed random integer r in the range

$$0 \leq r \leq 2^k, \qquad \text{where } k = \min(n, 10)$$

If all limit attempts fail, this event is reported as an error. Algorithms used to generate the integer r are designed to minimize the correlation between the numbers generated by any two stations at any given time.

It is noted that the values given above define the most aggressive behavior that a station may exhibit in attempting to retransmit after a collision. In the course of implementing the retransmission scheduling procedure, a station may introduce extra delays that will degrade its own throughput, but in no case may a station's retransmission scheduling result in a lower average delay between retransmission attempts than the procedure defined above.

9.6   CSMA/CD—CURRENT STATUS AND ADVANCED OPERATION

9.6.1   General

The original CSMA/CD system operated at a nominal 10-Mbps data rate. Subsequently, the system has undergone modifications for 100-Mbps and 1000-Mbps capabilities. Numerous variants on transmission media have also evolved. Table 9.2 lists 22 varieties of CSMA/CD, variously called *Ethernet*.

9.6.2   Half-Duplex and Full-Duplex

Half-duplex operation is where two or more stations share a common transmission medium. Full-duplex operation, on the other hand, allows simultaneous communication between a pair of stations using point-to-point means (dedicated channel). Full-duplex operation does not require that transmitters defer, nor do they monitor or react to receive activity, as there is no contention for a shared medium in this mode. Full-duplex mode can only be used when all of the following are true:

(a) The physical medium is capable of supporting simultaneous transmission and reception without interference.

(b) There are exactly two stations connected with a full-duplex point-to-point link. Since there is no contention for use of a shared medium, the multiple access (i.e., CSMA/CD) algorithms are unnecessary.

(c) Both stations on the LAN are capable of, and have been configured to use, full-duplex operation.

The most common configuration envisioned for full-duplex operation consists of a central bridge (also known as a switch) with a dedicated LAN connecting each bridge port to a single device. Repeaters as defined in the reference standard are outside the scope of full-duplex operation.

9.6.3   Compatibility Interfaces—Physical Layer

In this section five important compatibility interfaces are defined in what is architecturally the physical layer. These are shown in Figure 9.20.

TABLE 9.2 Variants of CSMA/CD

Type Name	Bit Rate	Transmission Medium	Comments
10BASE2	10 Mbps	RG 58 coaxial cable	
10BASE5	10 Mbps	Coaxial cable	Thicknet
10BASE-F	10 Mbps	Fiber-optic cable	
10BASE-FB	10 Mbps	Fiber-optic cable	Has MAU port on one repeater that can connect to another repeater
10BASE-FB segment	10 Mbps	Fiber-optic cable	Fiber-optic link segment provides connectivity between two 10BASE-FB ports on repeaters
10BASE-FL segment	10 Mbps	Fiber-optic cable	Fiber-optic link segment provides connectivity between two 10BASE-FL MAUs
10BASE-FP segment	10 Mbps	Fiber-optic cable	Fiber-optic mixing segment, including one 10BASE-FP star and all attached fiber pairs
10BASE-FP star	10 Mbps	Fiber-optic cable / star	Passive device couples fiber pairs together to form 10BASE-FP segment
10BASE-T	10 Mbps	Two wire pairs	Ordinary telephone pair cable
100BASE-FX	100 Mbps	Two optical fibers	
100BASE-T	10 & 100 Mbps	Shielded wire pair; twisted pair	Limited length shielded cable, twisted pair
100BASE-T2	100 Mbps	Two twisted pairs	Category 3 or better balanced cabling
100BASE-T4	100 Mbps	Four twisted pairs	Category 3, 4, or 5 UTP
100BASE-TX	100 Mbps	Two twisted pairs	Category 5 UTP or STP
100BASE-X	100 Mbps		General reference to 100BASE-TX and 100BASE-FX
1000BASE-CX	1000 Mbps	Specialty shielded copper cable	Specialty shielded copper balanced jumper cable assembles
1000BASE-LX	1000 Mbps	Fiber-optic cable	Long wavelength, multimode or single mode
1000BASE-SX	1000 Mbps	Fiber-optic cable	Short wavelength over multimode fiber
1000BASE-T	1000 Mbps	Four wire pairs	Balanced category 5
1000BA SE-X	1000 Mbps	General category	Embodies three types of transmission media: 1000BASE-CX, 1000BASE-SX, and 1000BASE-LX
10BROAD36	10 Mbps	Single broadband cable	
1BASE5	1 Mbps	Two twisted pairs	Telephone wire pair

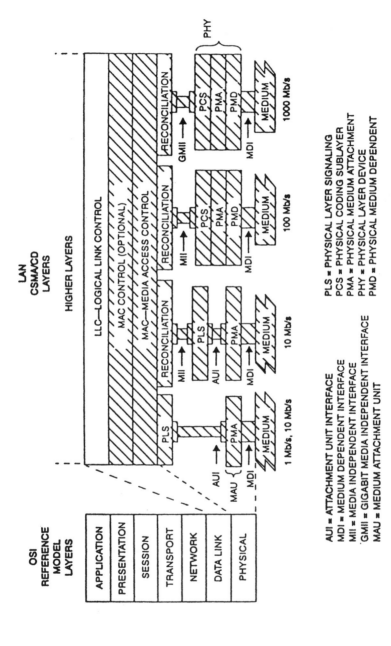

Figure 9.20. Compatibility interfaces illustrated, physical layer, with reference to the OSI reference model. (From Figure 1-1, page 3, IEEE Std. 802.3, Ref. 3.)

AUI = ATTACHMENT UNIT INTERFACE
MDI = MEDIUM DEPENDENT INTERFACE
MII = MEDIA INDEPENDENT INTERFACE
GMII = GIGABIT MEDIA INDEPENDENT INTERFACE
MAU = MEDIUM ATTACHMENT UNIT

PLS = PHYSICAL LAYER SIGNALING
PCS = PHYSICAL CODING SUBLAYER
PMA = PHYSICAL MEDIUM ATTACHMENT
PHY = PHYSICAL LAYER DEVICE
PMD = PHYSICAL MEDIUM DEPENDENT

(a) *Medium-Dependent Interfaces (MDI)*. To communicate in a compatible manner, all stations shall adhere rigidly to the exact specification of physical media signals defined in clause 8 (and beyond) in IEEE Std. 802.3 1998 Edition, (Ref. 3), and to the procedures that define correct behavior of a station. The medium-independent aspects of the LLC sublayer and the MAC sublayer should not be taken as detracting from this point; communication by way of the ISO/IEC 8802-3 [ANSI/IEEE Std 802.3] local area network requires complete compatibility at the physical medium interface (that is, the physical cable interface).

(b) *Attachment Unit Interface (AUI)*. It is anticipated that most DTEs will be located some distance from their connection to the physical cable. A small amount of circuitry will exist in the medium attachment unit (MAU) directly adjacent to the physical cable, while the majority of the hardware and all of the software will be placed within the DTE. The AUI is defined as a second compatibility interface. While conformance with this interface is not strictly necessary to ensure communication, it is highly recommended, since it allows maximum flexibility in intermixing MAUs and DTEs. The AUI may be optional or not specified for some implementations of this standard that are expected to be connected directly to the medium and so do not use a separate MAU or its interconnecting AUI cable. The PLS and PMA are then part of a single unit, and no explicit AUI implementation is required.

(c) *Media Independent Interface (MII)*. It is anticipated that some DTEs will be connected to a remote PHY, and/or to different medium-dependent PHYs. The MII is defined as a third compatibility interface. While conformance with implementation of this interface is not strictly necessary to ensure communication, it is highly recommended, since it allows maximum flexibility in intermixing PHYs and DTs. The MII is optional.

(d) *Gigabit Media Independent Interface (GMII)*. The GMII is designed to connect a gigabit-capable MAC or repeater unit to a gigabit PHY. While conformance with implementation of this interface is not strictly necessary to ensure communication, it is highly recommended, since it allows maximum flexibility in intermixing PHYs and DTEs at gigabit speeds. The GMII is intended for use as a chip-to-chip interface. No mechanical connector is specified for use with the GMII. The GMII is optional.

(e) *Ten-Bit Interface (TBI)*. The TBI is provided by the 1000BASE-X PMA sublayer as a physical instantiation of the PMA service interface. The TBI is highly recommended for 1000BASE-X systems, since it provides a convenient partition between the high-frequency circuitry associated with the PMA sublayer and the logic functions associated with the PCS and MAC sublayers. The TBI is intended for use as a chip-to-chip interface. No mechanical connector is specified for use with the TBI. The TBI is optional.

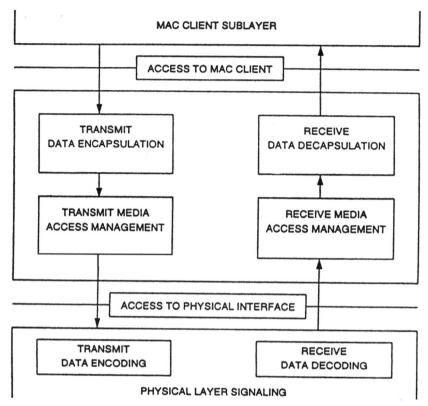

Figure 9.21. CSMA / CD MAC functions.

9.6.4 Layer Interface Requirements

The interface requirements for CSMA/CD are as follows:

(a) The interface between the MAC sublayer and its client includes facilities for transmitting and receiving frames, and it provides per-operation status information for use by higher-layer error recovery procedures.

(b) The interface between the MAC sublayer and the physical layer includes signals for framing (carrier sense, receive data valid, transmit initiation) and contention resolution (collision detect), facilities for passing a pair of serial bit streams (transmit, receive) between the two layers, and a wait function for timing.

9.6.4.1 Functional Capabilities of the MAC Sublayer. Figure 9.21 outlines the CSMA/CD media access control functions. This is followed by the function listing below (from Ref. 3):

(a) For Frame Transmission

    (1) Accepts data from the MAC client and constructs a frame

    (2) Presents a bit-serial data stream to the physical layer for transmission on the medium

    Note: Assumes data passed from the client sublayer are octet multiples.

(b) For Frame Reception

    (1) Receives a bit-serial data stream from the physical layer

    (2) Presents to the MAC client sublayer frames that are either broadcast frames or directly addressed to the local station

    (3) Discards or passes to network management all frames not addressed to the receiving station

(c) In half-duplex mode, defers transmission of a bit-serial stream whenever the physical medium is busy

(d) Appends proper FCS value to outgoing frames and verifies full octet boundary alignment

(e) Checks incoming frames for transmission errors by way of FCS and verifies octet boundary alignment

(f) Delays transmission of frame bit stream for specified interframe gap period

(g) In half-duplex mode, halts transmission when collision is detected

(h) In half-duplex mode, schedules retransmission after a collision until a specified retry limit is reached

(i) In half-duplex mode, enforces collision to ensure propagation throughout network by sending jam message

(j) Discards received transmissions that are less than a minimum length

(k) Appends preamble, start frame delimiter, DA, SA, length/type field, and FCS to all frames, and inserts PAD field for frames whose data length is less than a minimum value

(l) Removes preamble, start frame delimiter, DA, SA, length/type field, FCS, and PAD field (if necessary) from received frames

(m) Appends extension bits to the first (or only) frame of a burst if it is less than slotTime bits in length when in half-duplex mode at speeds above 100 Mbps

(n) Strips extension bits from received frames when in half-duplex mode at speeds above 100 Mbps

9.6.5 Allowable Implementations—Parameterized Values

Table 9.3 identifies the parameter values used with the 10-Mbps implementation of CSMA/CD. Table 9.4 gives the parameter values of 1BASE5. Table 9.5 provides the parameter values for 100-Mbps implementations, and Table 9.6 gives the parameter values for 1000-Mbps implementations.

TABLE 9.3 Parameters for 10-Mbps Implementations

Parameters	Values
slotTime	512 bit times
interFrameGap	9.6 μs
attemptLimit	16
backoffLimit	10
jamSize	32 bits
maxFrameSize	1518 octets
minFrameSize	512 bits (64 octets)
burstLimit	Not applicable

Note: The spacing between two successive noncolliding packets, from start to idle at the end of the first packet to start of preamble of the subsequent packet, can have a minimum value of 47 bit times, at the AUI receive line of the DTE. This interFrameGap shrinkage is caused by variable network delays, added preamble bits, and clock skew.

Source: Section 4.4.2 of Ref. 3.

TABLE 9.4 Parameters for 1BASE5 Implementations

Parameters	Values
slotTime	512 bits times
interFrameGap	96 μs
attemptLimit	16
backoffLimit	10
jamSize	32 bits
maxFrameSize	1518 octets
minFrameSize	512 bits (64 octets)
burstLimit	Not applicable

Source: Section 4.4.2 of Ref. 3.

TABLE 9.5 Parameters for 100-Mbps Implementations

Parameters	Values
slotTime	512 bit times
interFrameGap	0.96 μs
attemptLimit	16
backoffLimit	10
jamSize	32 bits
maxFrameSize	1518 octets
minFrameSize	512 bits (64 octets)
burstLimit	Not applicable

Source: Section 4.4.2 of Ref. 3.

TABLE 9.6 Parameters for 1000-Mbps Implementations

Parameters	Values
slotTime	4096 bit times
interFrameGap	0.096 μs
attemptLimit	16
backoffLimit	10
jamSize	32 bits
maxFrameSize	1518 octets
minFrameSize	512 bits (64 octets)
burstLimit	65, 536 bits

Note: The spacing between two noncolliding packets, from the last bit of the FCS field of the first packet to the first bit of the preamble of the second packet, can have a minimum value of 64 BT (bit times), as measured at the GMII receive signals at the DTE. This inter-FrameGap shrinkage may be caused by variable network delays, added preamble bits, and clock tolerances.

Source: Section 4.4.2 of Ref. 3.

9.6.5.1 Definition of Slot Time

The dynamics of collision handling are largely determined by a single parameter called the *slot time*. This single parameter describes three important aspects of collision handling:

1. It is an upper bound on the acquisition time of the medium.
2. It is an upper bound on the length of a frame fragment generated by a collision.
3. It is the scheduling quantum for retransmission.

To fulfill all three functions, the slot time must be larger than the sum of the physical layer round-trip propagation time and the media access layer maximum jam time. The slot time is determined by the parameters of a specific implementation.

9.6.5.2 *CSMA/CD System Designation.*
There are a number of different implementations of CSMA/CD. These are designated by a nomenclature made up of three parts: A number indicating the bit rate in Mbps, followed by either the words "BASE" or "BROAD," standing for baseband (BASE) or broadband (BROAD). The third element is either a number or letter(s). If a number is used, it describes a segment length in hundreds of meters. For example, in 10BASE5, the "10" indicates that it is a 10-Mbps CSMA/CD system, "BASE" tells us that it is a baseband system (versus broadband), and the "5" indicates that a segment length is 500 m. With some other implementations, the third element may be a letter or a letter-number. Consider the

following three examples:

100BASE-T4. This is a 100-Mbps baseband CSMA/CD system using 4 pairs of Category 3, 4, or 5 UTP (unshielded twisted pair).

1000BASE-LX. This is 1000-Mbps baseband CSMA/CD system using long-wavelength laser devices over fiber-optic cable.

1BASE5. This is a 1-Mbps CSMA/CD system with 500-m segments using two pairs of twisted-pair telephone wire.

9.6.6 Physical Signaling (PLS) and Attachment Unit Interface (AUI)—Selected Discussion

Each direction of data transfer is serviced with two (making a total of four) balanced circuits: "Data" and "Control." The Data and Control circuits are independently self-clocked, thereby eliminating the need for separate timing circuits. This is accomplished by encoding of all signals. The Control circuit signaling rate is nominally (but not of necessity exactly) equal to the Data circuit signaling rate.

The data circuits are used only for data transfer. No control signals associated with the interface are passed on these circuits. Likewise, the Control circuits are used only for control message transfer. No data signals associated with the interface are passed on these circuits.

Using such an arrangement (see Figure 9.20) allows the DTE full media independence for baseband coax, baseband twisted pair, broadband coax, and baseband fiber media so that identical PLS, MAC, and MAC clients may be used with any of these media.

9.6.6.1 *Frame Structure Across the AUI.* Frames transmitted on the AUI have the following structure:

$$\langle \text{silence} \rangle \langle \text{preamble} \rangle \langle \text{sfd} \rangle \langle \text{data} \rangle \langle \text{edt} \rangle \langle \text{silence} \rangle$$

These frame elements in sequence have the following characteristics:

Element	Characteristics
⟨silence⟩	= no transitions
⟨preamble⟩	= alternating (CD1) and (CD0) 56 bit times (ending in CD0)
⟨sfd⟩	= (CD1)(CD0)(CD1)(CD0)(CD1)(CD0) (CD1)(CD1)
⟨data⟩	= 8XN instances of CD0 or CD1
⟨etd⟩	= IDL

These elements are defined as follows:

Silence. The ⟨silence⟩ delimiter provides an observation window for an unspecified period of time during which no transitions occur on the AUI. The minimum length of this period is specified by the access procedure.

Preamble. The ⟨preamble⟩ delimiter begins a frame transmission and provides a signal for receiver synchronization. The signal is an alternating pattern of (CD1) and (CD0). This pattern is transmitted on the Data Out circuit by the DTE to the MAU for a minimum of 56 bit times* at the beginning of each frame. The last bit of the preamble [that is, the final bit of preamble before the start of frame delimiter (sfd)] is a CD0. The DTE is required to supply at least 56 bits of preamble to satisfy system requirements. System components consume preamble bits in order to perform their functions. The number of preamble bits sourced ensures an adequate number of bits are provided to each system component to correctly implement its function.

Start of Frame Delimiter (*sfd*). The ⟨sfd⟩ indicates the start of a frame, and follows the preamble. The ⟨sfd⟩ element is

$$(CD1)(CD0)(CD1)(CD0)(CD1)(CD0)(CD1)(CD1)$$

Data. The ⟨data⟩ in a transmission is in multiples of eight encoded data bits (CD0s and CD1s).

End of Transmission Delimiter (*etd*). The ⟨etd⟩ delimiter indicates the end of a transmission and serves to turn off the transmitter. The signal starts the IDL.

9.6.6.2 *Signal Characteristics*

Signal Encoding. Two different signal encoding mechanisms may be used by the AUI (attachment unit interface). One of the mechanisms is used to encode data and the other to encode control.

Data Encoding. Manchester encoding is used for the transmission of data across the AUI. Manchester encoding is a binary signaling mechanism that combines data and clock into "bit-symbols." Each bit-symbol is split into two halves, with the second half containing the binary inverse of the first half; a transition always occurs in the middle of each bit-symbol. During the first half of the bit-symbol, the encoded signal is the logical complement of the bit value being encoded. During the second half of the bit-symbol, the encoded

---

*A bit time is the duration of one bit. It is equal to the inverse of the bit rate.

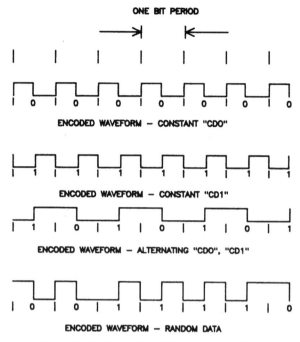

Figure 9.22. Examples of Manchester coding.

signal is the uncomplemented value of the bit being encoded. Thus, a CD0 is encoded as a bit-symbol in which the first half is HI and the second half is LO. A CD1 is encoded as a bit-symbol in which the first half is LO and the second half is HI. Examples of Manchester coding are shown in Figure 9.22.

The line condition IDL is also used as an encoded signal. An IDL always starts with a HI signal level. Since IDL always starts with a HI signal, an additional transition will be added to the data stream if the last bit sent was a zero. This transition cannot be confused with clocked data (CD0 or CD1) since the transition will occur at the start of a bit cell. There will be no transition in the middle of the bit cell. The IDL condition, as sent by a driver, shall be maintained for a minimum of 2 bit times. The IDL condition shall be detected within 1.6 bit times at the receiving device.

(a) System jitter considerations make detection of IDL (etd, end transmission delimiter) earlier than 1.3 bit times impractical. The specific implementation of the phase-locked loop or equivalent clock recovery mechanism determines the lower bound on the actual IDL detection time. Adequate margin between lower bound and 1.6 bit times should be considered.

(b) Recovery of timing implicit in the data is easily accomplished at the receiving side of the interface because of the wealth of binary transitions guaranteed to be in the encoded waveform, independent of the data sequence. A phase-locked loop or equivalent mechanism maintains continuous tracking of the phase of the information on the Data circuit.

Control Encoding. A simpler encoding mechanism is used for control signaling than for data signaling. The encoded symbols used in this signaling mechanism are CS0, CS1, and IDL. The CS0 signal is a signal stream of frequency equal to the bit rate (BR). The CS1 signal is a signal stream of frequency equal to half of the bit rate (BR/2). If the interface supports more than one bit rate, the bit rate in use on the data circuits is the one to which the control signals are referenced. The IDL signal used on the control circuits is the same as the IDL signal defined for data circuits. The Control Out circuit is optional (O) as is the one message on Control In (CI).

The frequency tolerance of the CS1 and CS0 signals on the CO circuit is $\pm 5\%$, and that of the CS1 signal on the CI circuit is $\pm 15\%$. The duty cycle of the above signals is nominally 50%/50% and is no worse than 60%/40%. The CS0 signal on the CI circuits has a frequency tolerance of BR + 25%, −15% with the pulse widths no less than 35 ns and no greater than 70 ns at the zero crossing points.

The meanings of the signals on the Control Out circuit (DTE to MAU—medium attachment unit) are as follows:

Signal	Message	Description
IDL	*normal*	Instructs the MAU to enter (remain in) normal mode
CS1	*mau_request* (O)	Requests that the MAU should be made available
CS0	isolate (O)	Instructs the MAU to enter (remain in) monitor mode

The meaning of the signals on the Control In circuit (MAU to DTE) are as follows:

Signal	Message	Description
IDL	*mau_available*	Indicates MAU is ready to output data
CS1	*mau_not_available*	Indicates MAU is not ready to output data
CS0	*signal_quality_error*	Indicates MAU has detected an error output data

SIGNALING RATE. The CSMA/CD specification covers multiple signaling rates. The signaling specified for these sections is 10 Mbps $\pm 0.01\%$.

It is intended that a given MDI (medium-dependent interface) operate at a single data rate. It is not precluded that specific DTE and MAU designs be manually switched or set to alternate rates. A given local network operates at a single signaling rate. To facilitate the configuration of operational systems, DTE and MAU devices are labeled with the actual signaling rate used with that device.

9.6.6.3 *Medium Attachment Unit (MAU)*—10BASE5. The MAU has the following general characteristics:

(a) Enables coupling the PLS by way of the AUI to the explicit baseband coaxial transmission system defined in this clause of the standard (Ref. 3).

(b) Supports message traffic at a data rate of 10 Mbps (alternative data rates may be considered in future additions to the standard).

(c) Provides for driving up to 500 m (1640 ft) of coaxial trunk cable without the use of a repeater.

(d) Permits the DTE to test the MAU and the medium itself.

(e) Supports system configurations using the CSMA/CD access mechanism defined with baseband signaling.

(f) Supports a bus topology interconnection means.

(g) Provides a communication channel with a mean BER at the PLS of better than 1×10^{-8}.

9.6.6.4 *Broadband CSMA/CD, 10BROAD36*—Overview. Commonly, broadband LANs use community antenna television (CATV) techniques to provide a multichannel LAN. In this case, each channel is governed by the CSMA/CD access method. The medium, of course, comprises CATV-type cable, taps, connectors, and broadband amplifiers. A coaxial broadband system permits the assignment of different frequency bands to multiple applications.

The purpose of the MAU (in this application) is to provide a means of attaching devices to a broadband local network medium. The physical tap (of the coaxial cable) is a passive directional device such that the MAU transmission is directed toward the head-end location (reverse direction). On a single-cable system the transmission from the MAU is at a carrier frequency f1. A frequency translator (or remodulator) located at the head-end up-converts to a carrier frequency f2, which is sent in the forward direction to the taps (receiver inputs). On a dual-cable system the transmit and receive carrier frequencies are identical (both f1) and the MAU connects to the medium via two taps, one on the receive cable and the other on the transmit cable. The transmit and receive cables are connected to each other at the head-end location. Figure 9.23 is a conceptual block diagram of broadband single- and dual-cable systems.

The broadband MAU operates by accepting data from the attached data termination equipment (DTE) and transmitting a modulated radio-frequency (RF) data signal in a data band on the broadband coaxial cable system. All MAUs attached to the cable system receive and demodulate this RF signal to recover the DTE data. The broadband MAU emulates a baseband MAU

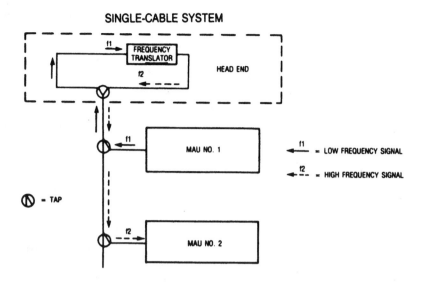

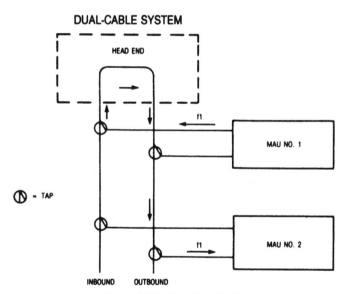

Figure 9.23. Broadband cable systems.

except for delay between transmission and reception, which is inherent in the broadband cable system. A transmitting MAU logically compares the beginning of the received data with the data transmitted. Any difference between them, which may be due to either (a) errors caused by colliding transmissions, (b) reception of an earlier transmission from another MAU, or (c) a bit error on the channel, is interpreted as a collision.

When a collision is recognized, the MAU stops transmission in the data band and begins transmission of an RF collision enforcement (CE) signal in a separate CE band adjacent to the data band. The CE signal is detected by all MAUs and informs them that a collision has occurred. All MAUs signal to their attached MAC controllers the presence of a collision. The transmitting MACs then begin the collision handling process.

Collision enforcement is necessary because RF data signals from different MAUs on the broadband cable system may be received at different power levels. During a collision between RF data signals at different levels, the MAU with the higher received power may see no errors in the detected data stream. However, the MAU with the lower RF signal will see a difference between transmitted and received data; this MAU then transmits the CE signal to force recognition of the collision by all transmitting MAUs.

Broadband MAU Functional Requirements. The MAU component provides the means by which signals on the physically separate AUI signal circuits to and from the DTE and their associated interlayer messages are coupled to the broadband coaxial medium. To achieve this basic objective, the MAU component contains the following capabilities to handle message flow between the DTE and the broadband medium:

(a) *Transmit Function.* The ability to transmit serial data bit streams originating at the local DTE in a band-limited modulated RF carrier form, to one or more remote DTEs on the same network.

(b) *Receive Function.* The ability to receive a modulated RF data signal in the band of interest from the broadband coaxial medium and demodulate it into a serial bit stream.

(c) *Collision Presence Function.* The ability to detect the presence of two or more stations' concurrent transmissions.

(d) *Jabber Function.* The ability of the MAU itself to interrupt the Transmit function and inhibit an abnormally long output data stream.

MAU TRANSMIT FUNCTION REQUIREMENTS. The transmit function requirements are shown in Figure 9.24. These requirements include the following capabilities:

(a) Receive Manchester-encoded data sent by the local DTE to the attached MAU on circuit DO (transmit data pair).

(b) Decode the Manchester-encoded data received on circuit DO to produce NRZ data and recovered clock signal.

(c) Scramble NRZ data using CCITT Rec. V.29 (Chapter 8) scrambler with seed changed on each transmitted packet.

(d) Transform the incoming bits (prior to modulation) to provide an unscrambled alternating zero–one pattern terminated by an Unscrambled Mode Delimiter (UMD), scramble the remainder of the incoming

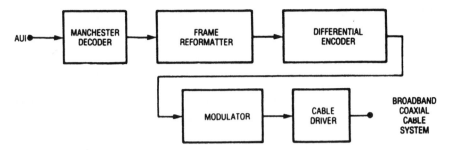

Figure 9.24. Transmit function requirements block diagram.

preamble, Start Frame Delimiter (SFD), and data frame; and append an unscrambled post-amble (Broadband End of Frame Delimiter [BEOFD]).

(e) Differentially encode the packet generated previously.

(f) Produce a band-limited double-sideband, suppressed carrier, binary PSK modulated RF signal representing the generated differentially encoded packet.

(g) Drive the coaxial cable with the modulated RF signal.

MAU RECEIVE FUNCTION REQUIREMENTS. The receive functions include the following:

(a) Receive the differentially encoded binary PSK modulated RF signal from the broadband coaxial medium.

(b) Receive the data band RF signals and reject signals in bands other than the data band (rejection of signals in the adjacent collision enforcement band is optional).

(c) Demodulate and differentially decode the incoming RF data signal from the coaxial medium to provide a receive bit stream that represents the scrambled bit stream at the transmitter.

(d) Descramble the receive bit stream using a self-synchronizing descrambler.

(e) Manchester encode the descrambled bit stream.

(f) Send to the DTE, using Circuit DI (receive data pair), an additional, locally generated, Manchester-encoded preamble equal to the number of preamble bits lost in the receive data path (plus or minus one bit), followed by Manchester-encoded bit stream. No more than 6 preamble bits may be lost from the preamble presented to Circuit DO at the transmitting MAU.

(g) Detect end of frame, using the postamble (BEOFD), and ensure that no extraneous bits are sent to the DTE on Circuit DI.

(h) Receive signals in the collision enforcement band and reject signals in the data band and all other bands on the broadband medium.

MAU COLLISION DETECTION FUNCTION REQUIREMENTS. The MAU shall perform the following functions to meet the collision detection requirements:

(a) Store the scrambled bits (not differentially encoded) in the transmit section through to the last bit in the source address.

(b) Detect the UMD in the transmit and receive paths.

(c) Compare received scrambled bits after the received UMD with transmitted scrambled bits after the transmit UMD through to the last bit in the source address.

(d) A Receive UMD Timer function shall be performed by the MAU. The timer shall be as long as the time required from initial detection of RF data signal presence to detection of a UMD in a normally received (no collision) packet.

(e) Enter a LOCAL COLLISION DETection state if one of the following occurs:

   (1) A bit error is found in the bit compare process through the last bit in the source address.

   (2) The Receive UMD Timer expires before a UMD is detected in the received bit stream.

   (3) The MAU receives the *output* (that is, transmit) signal from the AUI AFTER having received an RF signal from the coaxial cable.

(f) Upon entering the LOCAL COLLISION DET state, cease transmission in the data band and commence transmission in the collision enforcement band for as long as the DTE continues to send data to the MAU.

(g) Upon entering the LOCAL COLLISION DET state send the *signal_quality_error* (SQE) message on Circuit CI (collision presence pair) using the CS0 signal for as long as RF signals are detected on the broadband coaxial medium in either the data or collision enforcement bands.

(h) Detect power in the collision enforcement band and send the SQE message on Circuit CI using the CS0 signal. Send the SQE message for as long as energy is detected in the collision enforcement band.

(i) Ensure that during collisions, due to phase cancellations of the colliding carriers, Circuit DI does not become inactive before Circuit CI becomes active.

(j) Test the collision detection circuitry following every transmission that does not encounter a collision. This test consists of transmitting a burst of collision enforcement RF signal after the end of the postamble transmission and detecting this burst on the receive side. If the burst is detected, the CS0 (BR) signal is sent on Circuit CI of the transmitting MAU.

TABLE 9.7   Broadband Cable System Electrical Requirements

Impedance	75 Ω
Return loss	14 dB min
Transmit level	+50 dBmV ± 2 dB
Receive level	+6 dBmV ± 10 dB
Maximum receive noise level	−30 dBmV / 14 MHz
Loss variation[a] (per 18-MHz band)	2 dB min, 52 dB max
Path loss (between any transmit port and receive port, including loss variation)	36 dB min, 52 dB max
Group delay variation	
—around data carrier	20 ns / 10 MHz max
—over 18 MHz band	34 ns max

[a]Not including head-end.

Source: Table 11.5-1, page 246, Ref. 3.

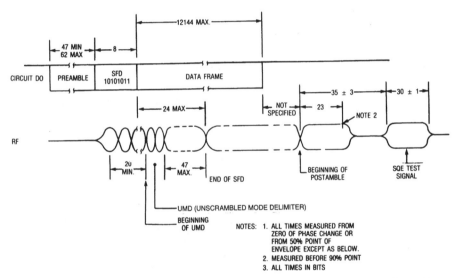

Figure 9.25. Packet format and timing diagram, AUI to coaxial cable interface. (From Figure 11-6, page 233, IEEE Std. 802.3, Ref. 3.)

Selected Coaxial Cable Electrical Interface Requirements. These requirements are listed in Table 9.7.

Figure 9.25 shows the packet format and timing diagram at the coaxial cable interface. Tables 9.8 and 9.9 show the single cable and dual-cable frequency plans.

9.6.7   System Considerations for Multisegment 10-Mbps Baseband Networks

9.6.7.1   Overview. Two transmission models are presented in Section 13 of Ref. 3: Model 1 and Model 2. These models require that the network size be limited to control round-trip propagation delay, and that the number of

TABLE 9.8   Single-Cable Frequency Allocations (MHz)[a]

			Receiver			
	Transmitter		Translation 156.25 MHz		Translation 192.25 Mhz	
Data Carrier	Coll Enf[b] Center Freq	Transmit Band	Head-End Local Osc	Receive Band	Head-End Local Osc	Receive Band
43	52	35.75–53.75	245.75	192–210	192.25	228–246
49	58	41.75–59.75	257.75	198–216	192.25	234–252
55	64	47.75–65.75	269.75	204–222	192.25	240–258
61[c]	70	53.75–71.75	281.75	210–228	192.25	246–264
67	76	59.75–77.75	293.75	216–234	192.25	252–270
73	82	65.75–83.75	305.75	222–240	192.25	258–276

[a]Some of these optional bands are overlapping. Frequency tolerance of the data carrier and head-end local oscillator shall each be ±25 kHz.
[b]Coll Enf, collision enforcement.
[c]Preferred frequency allocation.
Source: Table 11.2-1, page 237, Ref. 3.

TABLE 9.9   Dual-Cable Frequency Allocations (MHz)

Data Carrier[a]	Coll Enf[b] Center Freq	Data Band	Coll Enf[b] Band
43	52	36–50	50–54
49	58	42–56	56–60
55	64	48–62	62–66
61[c]	70	54–68	68–72
67	76	69–74	74–78
73	82	66–80	80–84
235.25	244.25	228–242	242–246
241.25	250.25	234–248	248–252
247.25	256.25	240–254	254–258
253.25[c]	262.25	246–260	260–264
259.25	268.25	252–266	266–270
265.25	274.25	258–272	272–276

[a]Some of these optional bands are overlapping. Frequency tolerance of the data carrier shall be ±25 kHz.
[b]Coll Enf, collision enforcement.
[c]Preferred frequency allocations.
Source: Table 11.2-2, page 237, Ref. 3.

repeaters between any two DTEs be limited in order to limit the shrinkage of the interpacket gap as it travels through the network. Model 1 is the more conservative that meets the two basic design criteria set forth above. The Model 2 transmission system model validates an additional broad set of topologies that are fully functional and do not fit within the simpler but more restrictive rules of Model 1.

The physical size of a CSMA/CD network is limited by the characteristics of individual network components. These characteristics include the

TABLE 9.10   Delays for Network Media Segments

Media Type	Maximum Number of MAUs per Segment	Maximum Segment Length (m)	Maximum Medium Delay per Segment (ns)
Mixing segment			
10BASE5	100	500	2165
10BASE2	30	185	950
10BASE-FP	33[a]	1000[b]	5000
Link segment			
FOIRL	2	1000	5000
10BASE-T	2	100[c]	1000
10BASE-FB	2	1000	10 000
10BASE-FL	2	2000	10 000
AUI[d] 1 DTE / 1 MAU		50	257

[a]Actual number depends on the passive-star characteristics.
[b]In addition, a MAU to passive-star link will not exceed 500 m.
[c]Actual maximum segment length depends on cable characteristics.
[d]AUI is not a segment.
Source: Table 13-1, page 284, Ref. 3.

following:

(a) Media lengths and their associated propagation time delay
(b) Delay of repeater units (start-up and steady-state)
(c) Delay of MAUs (start-up and steady-state)
(d) Interpacket gap shrinkage due to repeater units
(e) Delays within the DTE associated with the CSMA/CD access method
(f) Collision detect and deassertion times associated with MAUs

Table 9.10 summarizes the delays for the various network media segments. For more specific delays of the various MAUs, the reader should consult the reference specification, Sections 8–18.

9.6.7.2   Shrinkage of the Interpacket Gap (IPG). The worst-case variabilities of transmission elements in the network plus some of the signal reconstruction facilities required in the 10-Mbps baseband repeater specification combine in such a way that the gap between two packets travelling across the network may be reduced below the interFrameGap specified in Tables 9.3 through 9.6. This parameter limits the equipment (i.e., number of repeaters) between any two DTEs. Again the limit applies to all combinations of DTEs on any network, but the worst case is apparent from an inspection of a map or schematic representation of the topology in question.

9.6.7.3   Transmission System Model 1. The following network topology constraints apply to networks Transmission System Model 1. If no segment length constraints are given for a segment type, the maximum segment length, as defined in the relevant MAU clause, applies.

(a) Repeater sets are required for all segment interconnection.

(b) MAUs that are part of repeater sets count toward the maximum number of MAUs on a segment.

(c) The transmission path permitted between any two DTEs may consist of up to five segments, four repeater sets (including optional AUIs), two MAUs, and two AUIs.

(d) AUI cables for 10BASE-FP and 10BASE-FL shall not exceed 25 m. (Since two MAUs per segment are required, 25 m per MAU results in a total AUI cable length of 50 m per segment.)

(e) When a transmission path consists of four repeater sets and five segments, up to three of the segments may be mixing and the remainder must be link segments (Figures 9.26, 9.27, and 9.30). When five segments are present, each fiber-optic link segment (FOIRL, 10BASE-FB, or 10BASE-FL) shall not exceed 500 m, and each 10BASE-FP segment shall not exceed 300 m.

(f) When a transmission path consists of three repeater sets and four segments (Figures 9.28 and 9.29), the following restrictions apply:

(1) The maximum allowable length of any inter-repeater fiber segment shall not exceed 1000 m for FOIRL, 10BASE-FB, and 10BASE-FL segments and shall not exceed 700 m for 10BASE-FP segments.

(2) The maximum allowable length of any repeater to DTE fiber segment shall not exceed 400 m for 10BASE-FL segments and shall not exceed 300 m for 10BASE-FP segments and 400 m for segments terminated in a 10BASE-FL MAU.

(3) There is no restriction on the number of mixing segments in this case.

9.6.7.4 *Typical Topologies Applicable to Transmission Model 1.* Figures 9.26 through 9.30 show several of the network topologies consistent with Transmission Model 1.

9.7   100-Mbps CSMA / CD BASEBAND NETWORKS

9.7.1   Overview—Key Points

Figure 9.31 shows the relationship between 100BASE-T, the existing IEEE 802.3 (CSMA/CD MAC) and the ISO OSI reference model.

100BASE-T uses existing IEEE 802.3 MAC layer interface, connected through a media-independent interface layer to a physical layer entity (PHY) sublayer such as 100BASE-T4, 100BASE-TX, or 100BASE-FX.

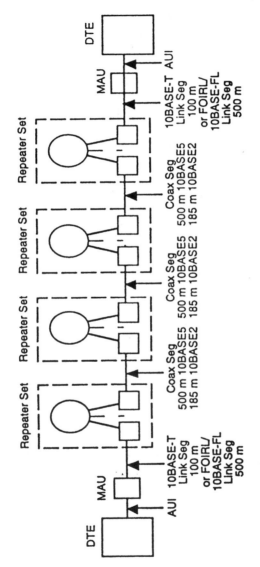

Figure 9.26. Maximum transmission path with three coaxial cable segments and two link segments.

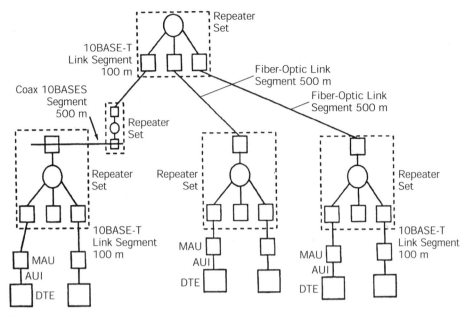

Figure 9.27. Example of maximum transmission path using coaxial cable segments, 10BASE-T link segments, and fiber-optic link segments.

100BASE-T extends the IEEE 802.3 MAC to 100-Mbps operation. The following key points should be considered:

- The bit rate is faster.
- Bit times are shorter.
- Packet transmission times are reduced.
- Cable delay budgets are smaller.

All of these factors are in proportion to the change in bit rate (i.e., 10 Mbps vs. 100 Mbps). This means that the ratio of packet duration to network propagation delay for 100BASE-T is the same as 10BASE-T.

9.7.1.1 Reconciliation Sublayer (RS) and Media-Independent Interface (MII)—Introduction.

The media-independent interface (MII) provides an interconnection between the media access control (MAC) sublayer and the physical layer entities (PHY) and between PHY layer and station management (STA) entities. This MII is capable of supporting both 10-Mbps and 100-Mbps data rates through 4-bit wide (nibble wide) transmit and receive paths. The reconciliation sublayer provides a mapping between signals provided at the MII and the MAC/PLS service definition.

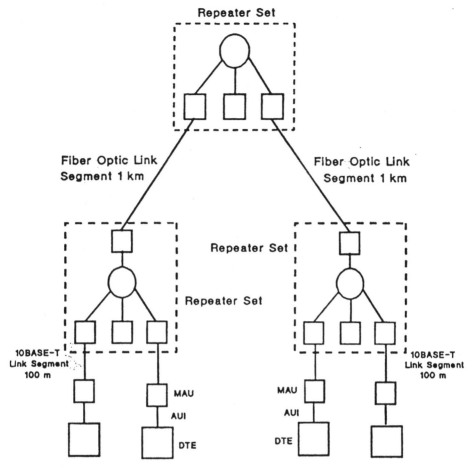

Figure 9.28. Example of maximum transmission path with three repeater sets, four link segments (two are 100-m 10BASE-T and two are 1-km fiber).

9.7.1.2 *Physical Layer Signaling Systems.* The MII attaches to any one of a number of physical layer implementations as expressed in the following table:

Designation	Physical Layer Description
100BASE-T4	Four pairs Cat. 3, 4, or 5 balanced cabling or 150-ohm shielded balanced cabling
100BASE-FX	Two multimode fibers (FDDI/ANSI X3T12)
100BASE-TX	See 100BASE-FX, both defined under 100BASE-X
100BASE-T2	Two pairs of Cat. 3, 4, or 5 balanced cabling

Note: Category 3, 4, and 5 cabling is described in Chapter 14.

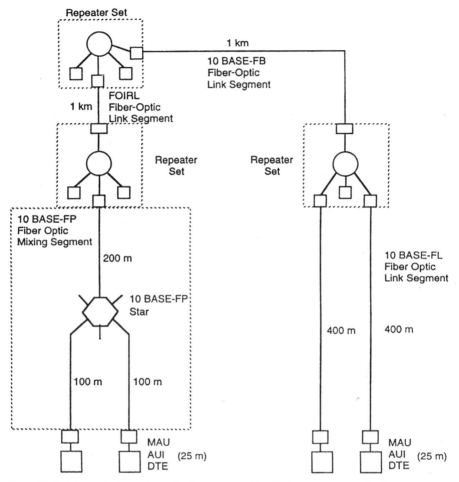

Figure 9.29. Example of maximum transmission path with three repeater sets, four segments (one 1-km 10BASE-FB, one 1-km FOIRL, one 400-m 10BASE-FL, and one 300-m 10BASE-FP).

9.7.1.3 Repeater. Repeater sets are an integral part of any 100BASE-T network with more than two DTEs in a collision domain. They extend the physical system topology by coupling two or more segments. Multiple repeaters are permitted within a single collision domain to provide the maximum path length.

9.7.2 Reconciliation Sublayer (RS) and Media Independent-Interface (MII)—Description Details

Figure 9.32 illustrates the relationship of the reconciliation sublayer and MII to the OSI reference model.

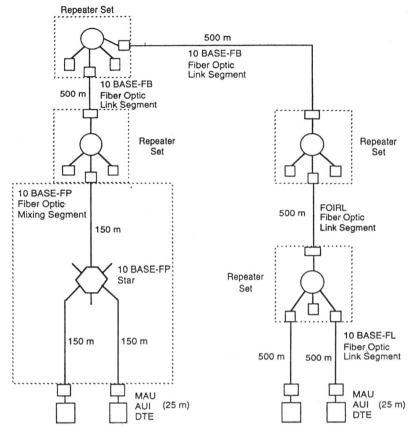

Figure 9.30. Example of maximum transmission path with four repeater sets, five segments (two 500-m 10BASE-FB, one 500-m FOIRL, one 500-m 10BASE-FL, and one 300-m 10BASE-FP).

Summary of Major Concepts

(a) Each direction of data transfer is serviced with 7 (making a total of 14) signals: data (a 4-bit bundle), delimiter error, and check.

(b) Two media status signals are provided. One indicates the presence of carrier, and the other indicates the occurrence of a collision.

(c) A management interface comprised of two signals provides access to management parameters and services.

(d) The reconciliation sublayer maps the signal set provided at the MII to the PLS service definition specified in Section 6 of the reference publication.

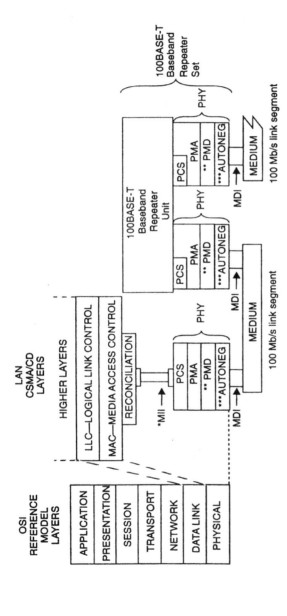

MDI = MEDIUM DEPENDENT INTERFACE
MII = MEDIA INDEPENDENT INTERFACE

PCS = PHYSICAL CODING SUBLAYER
PMA = PHYSICAL MEDIUM ATTACHMENT
PHY = PHYSICAL LAYER DEVICE
PMD = PHYSICAL MEDIUM DEPENDENT

* MII is optional for 10 Mb/s DTEs and for 100 Mb/s systems and is not specified for 1 Mb/s systems.

** PMD is specified for 100BASE-X only; 100BASE-T4 does not use this layer.

   Use of MII between PCS and Baseband Repeater Unit is optional.

*** AUTONEG is optional.

Figure 9.31. Architectural positioning of 100BASE-T. (From Figure 21-1, page 480, IEEE Std. 802.3, Ref. 3.)

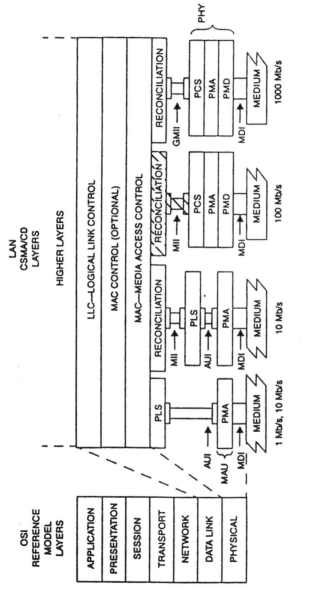

AUI = ATTACHMENT UNIT INTERFACE
MDI = MEDIUM DEPENDENT INTERFACE
MII = MEDIA INDEPENDENT INTERFACE
GMII = GIGABIT MEDIA INDEPENDENT INTERFACE
MAU = MEDIUM ATTACHMENT UNIT

PLS = PHYSICAL LAYER SIGNALING
PCS = PHYSICAL CODING SUBLAYER
PMA = PHYSICAL MEDIUM ATTACHMENT
PHY = PHYSICAL LAYER DEVICE
PMD = PHYSICAL MEDIUM DEPENDENT

Figure 9.32. MII location in the protocol stack. (From Figure 22-1, page 488, IEEE Std. 802.3, Ref. 3.)

259

$$\langle\texttt{interframe}\rangle\langle\texttt{preamble}\rangle\langle\texttt{sfd}\rangle\langle\texttt{data}\rangle\langle\texttt{efd}\rangle$$

Figure 9.33. MII frame format. Sfd = start frame delimiter; efd = end frame delimiter.

The interface through the MII connector is used to provide media independence for various forms of unshielded twisted-pair wiring, shielded twisted-pair wiring, fiber-optic cabling, and potentially other media, so that identical media access controllers may be used with any of these media.

Rates of Operation. The MII can support two specific data rates, 10 Mbps and 100 Mbps. The functionality is identical at both data rates, as are the signal timing relationships. The only difference between 10-Mbps and 100-Mbps operation is the nominal clock frequency.

Relationship of MII and GMII (gigabit media-independent interface). The GMII is similar to the MII. The GMII uses MII management interface and register set specified in Section 22.2.4 of the reference document. These common elements of operation allow station management to determine PHY capabilities for any supported speed of operation and configure the station based on those capabilities. In a station supporting both MII and GMII operation, configuration of the station would include enabling either the MII or GMII operation as appropriate for the data rate of the selected PHY.

Most of the MII and GMII signals use the same names, but the width of the RXD and TXD data bundles and the semantics of the associated control signals differ between MII and GMII operation. The GMII transmit path clocking differs significantly from MII clocking. MII operation of these signals and clocks is specified in Section 22 and GMII operation is specified in Section 35 of the reference publication.

9.7.2.1 MII Frame Structure. The data frames transmitted through the MII have the format as shown in Figure 9.33.

For MII transmission and reception of each octet of data is done a nibble* at a time with the order of nibble transmission and reception as shown in Figure 9.34. The bits of each octet are transmitted and received as two nibbles: Bits 0 through 3 correspond to bits 0 through 3 of the first nibble transmitted and received, and bits 4 through 7 of the octet correspond to bits 0 through 3 of the second nibble.

Interframe. The interframe period provides an observation window for an unspecified amount of time during which no data activity occurs on the MII. The absence of data activity is indicated by the de-assertion of the RX_DV (receive data valid) signal on the receive path, and the de-assertion of the TX_EN (transmit enable) signal on the transmit path. The MAC interFrameSpacing parameter defined in Tables 9.3 through 9.6 is measured from

*A *nibble* is a sequence of 4 bits.

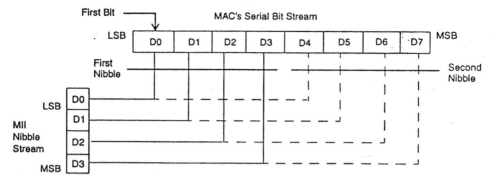

Figure 9.34. Octet / nibble transmit and receive order. (From Figure 22-11, page 500, IEEE Std. 802.3, Ref. 3.)

the de-assertion of the CRS (carrier sense) signal to the assertion of the CRS signal.

Preamble. The preamble begins a frame transmission and consists of at least seven octets of alternating 1s and 0s transmitted serially.

Start of Frame Delimiter (sfd). The sdf indicates the start of a frame and follows the preamble. It consists of the octet:

$$10101011$$

Data. The data in a well-formed frame consists of N octets of data transmitted as $2N$ nibbles. For each octet of data, the transmit order of each nibble is as specified in Figure 9.34. Data in a collision fragment may consist of an odd number of nibbles.

9.7.3 100BASE-T4, Its PMA and PCS

9.7.3.1 Introduction. 100BASE-T4 and baseband medium specifications are aimed at users who want 100-Mbps performance, but would like to retain the benefits of using voice-grade twisted-pair cable. 100BASE-T4 requires four pairs of Category 3 or better cables.

9.7.3.2 Objectives. The following are the objectives of 100BASE-T4:

(a) To support the CSMA/CD MAC in the half duplex mode of operation.
(b) To support the 100BASE-T MII, repeater, and optional autonegotiation.
(c) To provide 100-Mbps data rate at the MII.

(d) To provide for operating over unshielded twisted pairs of Category 3, 4, or 5 cable, installed as horizontal runs in accordance with ISO/IEC 11801: 1995, as specified in 23.6 of the reference document, at distances up to 100 m (328 ft).

(e) To allow for a nominal network extent of 200 m, including:

   (1) Unshielded twisted-pair links of 100 m.

   (2) Two-repeater networks of approximately a 200-m span.

(f) To provide a communication channel with a mean ternary symbol error rate, at the PMA service interface, of less than one part in 10^8.

9.7.3.3 Description. The PCS transmit function accepts data nibbles from the MII. The PCS transmit function encodes these nibbles using an 8B6T coding scheme described below. It passes the resulting ternary symbols to the PMA. In the reverse direction, the PMA conveys received ternary symbols to the PCS receive function. The PCS receive function decodes them into octets, and then it passes the octets one nibble at a time up to the MII. The PCS also contains a PCS carrier sense function, a PCS error sense function, a

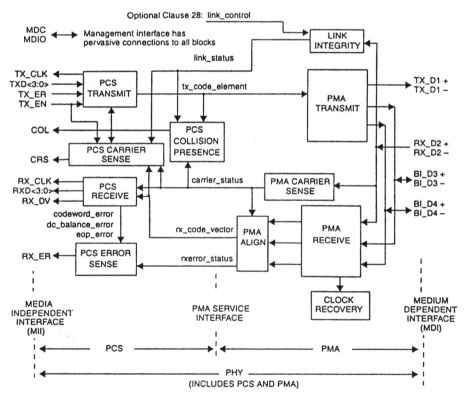

Figure 9.35. Division of responsibilities between 100BASE-T4 PCS and PMA. (From Figure 23-2, page 537, IEEE Std. 802.3, Ref. 3.)

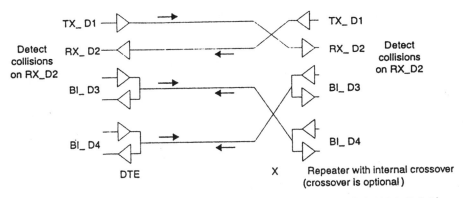

Figure 9.36. Use of wire pairs. (From Figure 23-3, page 537, IEEE Std. 802.3, Ref. 3.)

PCS collision presence function, and a management interface. Figure 9.35 shows the division of responsibilities between PCS, PMA and MDI layers.

Physical level communication between PHY entities takes place over four twisted pairs using Category 3, 4, or 5 UTPs. Figure 9.36 shows how the PHY manages the four twisted pairs at its disposal.

The 100BASE-T4 transmission algorithm always leaves one pair open for detecting carrier from the far end (see Figure 9.36). Leaving one pair open for carrier detection in each direction greatly simplifies media access control. All collision detection functions are accomplished using only the unidirectional pairs TX_D1 and RX_D2, in a manner similar to 10BASE-T. This collision detection strategy leaves three pairs in each direction free for data transmission, which uses an 8B6T block code, schematically represented in Figure 9.37.

8B6T coding, as used with 100BASE-T4 signaling, maps data octets into ternary symbols. Each octet is mapped to a pattern of 6 ternary symbols, called a 6T code group. The 6T code groups are fanned out to three independent serial channels. The effective data rate carried on each pair is one third of 100 Mbps, which is 33.333 ... Mbps. The ternary symbol transmission rate on each pair is 6/8 times 33.33 Mbps, or precisely 25.000 MHz.

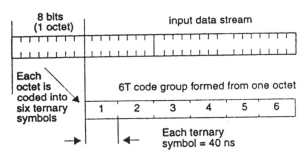

Figure 9.37. 8B6T coding.

9.7.3.4 *Summary of Physical Medium Attachment (PMA) Specification.* The PMA couples messages from the PMA service interface onto the twisted-pair physical medium. The PMA provides communications, at 100 Mbps, over four pairs of twisted-pair wiring up to 100 m in length.

The PMA transmit function, shown in Figure 9.35, comprises three independent ternary data transmitters. Upon receipt of a PMA_UNITDATA.request message, the PMA synthesizes one ternary symbol on each of the three output channels (TX_D1, BI_D3, and BI_D4). Each output driver has a *ternary* output, meaning that the output waveform can assume any of three values, corresponding to the transmission of ternary symbols CS0, CS1, or CS-1. (A CS0 is a ternary symbol 0 with an output voltage of 0 volts, CS1 is a ternary symbol 1 with a nominal peak voltage of $+3.5$ V, and a CS-1 conveys the ternary symbol -1 and has a nominal peak voltage of -3.5 volts), on each of the twisted pairs.

The PMA receive function comprises three independent ternary data receivers. The receivers are responsible for acquiring clock, decoding the start of stream delimiter (SSD) on each channel, and providing data to the PCS in the synchronous fashion defined by the PMA_UNIDATA.indicate message. The PMA also contains functions for PMA carrier sense and link integrity.

9.7.4 Physical Coding Sublayer (PCS) and Physical Medium Attachment (PMA) Sublayer, Type 100BASE-X

9.7.4.1 *Objectives.* The following are the objectives of 100BASE-X:

(a) Support the CSMA/CD MAC in the half-duplex and the full-duplex modes of operation.

(b) Support the 100BASE-T MII, repeater, and optional autonegotiation.

(c) Provide 100-Mbps data rate at the MII.

(d) Support cable plants using Category 5 UTP, 150 Ω STP, or optical fiber, compliant with ISO/IEC 11801.

(e) Allow for a nominal network extent of 200–400 m, including:

    (1) Unshielded twisted-pair links of 100 m

    (2) Two repeater networks of approximately 200 m span

    (3) One repeater network of approximately 300 m span (using fiber)

    (4) DTE/DTE links of approximately 400 m (half-duplex mode using fiber) and 2 km (full-duplex mode using multimode fiber)

(f) Preserve full-duplex behavior of underlying PMD channels.

9.7.4.2 *Functional Requirements.* The PCS comprises the transmit, receive, and carrier sense functions for 100BASE-T. In addition, the collision-Detect signal required by the MAC (COL on the MII) is derived from the

PMA code-bit stream. The PCS shields the reconciliation sublayer (and MAC) from the specific nature of the underlying channel. Specifically for receiving, the 100BASE-X PCS passes to the MII a sequence of data nibbles derived from incoming code groups, each comprised of five code bits, received from the medium. Code-group alignment and MAC packet delimiting is performed by embedding special nondata code groups. The MII uses a nibble-wide, synchronous data path, with packet delimiting being provided by separate TX_EN and RX_DV signals. The PCS provides the functions necessary to map these two views of the exchanged data. The process is reversed for transmit.

The following provides a detailed specification of the functions performed by the PCS, which comprise five parallel processes (transmit, transmit bits, receive, receive bits, and carrier sense). Figure 9.38 includes a functional block diagram of the PCS.

The receive bits process accepts continuous code bits via the PMA_UNITDATA.indicate primitive. Receive monitors these bits and generates RXD $\langle 3:0 \rangle$, RX_DV and RX_ER on the MII, and the internal flag, receiving, used by the carrier sense and transmit processes.

The transmit process generates continuous code groups based upon the TXD $\langle 3:0 \rangle$, TX_EN, and TX_ER signals on the MII. These code groups are transmitted by transmit bits via the PMA_UNITDATA.request primitive. The transmit process generates the MII signal COL based on whether a reception is occurring simultaneously with transmission. Additionally, it generates the internal flag, transmitting, for use by the carrier sense process.

The carrier sense process asserts the MII signal CRS when either transmitting or receiving is TRUE. Both the transmit and receive processes monitor link_status via the PMA_LINK.indicate primitive, to account for potential link failure conditions.

Code Groups. The PCS maps 4-bit nibbles from the MII into 5-bit code groups, and vice versa, using a 4B/5B block coding scheme. A code group is a consecutive sequence of five code-bits interpreted and mapped by the PCS. Implicit in the definition of a code-group is an establishment of code group boundaries by an alignment function within the PCS receive process. It is important to note that, with the sole exception of the SSD, which is used to achieve alignment, code groups are undetectable and have no meaning outside the 100BASE-X physical protocol data unit, called a "stream."

The coding method used provides the following:

(a) Adequate codes (32) to provide for all data code groups (16) plus necessary control code-groups

(b) Appropriate coding efficiency (4 data bits per 5 code bits; 80%) to effect a 100-Mbps physical layer interface on a 125-Mbps physical channel as provided by FDDI PMDs

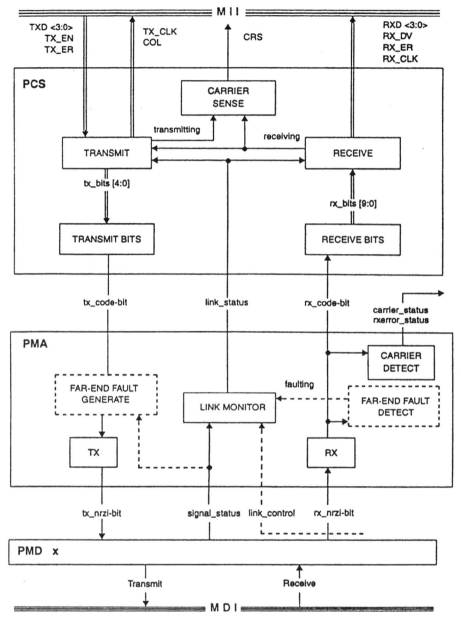

Figure 9.38. Functional block diagram, 100BASE-X. (From Figure 24-4, page 615, IEEE Std. 802.3, Ref. 3.)

   (c) Sufficient transition density to facilitate clock recovery (when not scrambled)

   Table 9.11 specifies the interpretation assigned to each 5-bit code group, including the mapping to the nibble-wide (TXD or RXD) Data signals on

TABLE 9.11   4B / 5B Code Groups

	PCS Code Group [4:0] 4 3 2 1 0	Name	MII (TXD / RXD) ⟨3:0⟩ 3 2 1 0	Interpretation
D	1 1 1 1 0	0	0 0 0 0	Data 0
A	0 1 0 0 1	1	0 0 0 1	Data 1
T	1 0 1 0 0	2	0 0 1 0	Data 2
A	1 0 1 0 1	3	0 0 1 1	Data 3
	0 1 0 1 0	4	0 1 0 0	Data 4
	0 1 0 1 1	5	0 1 0 1	Data 5
	0 1 1 1 0	6	0 1 1 0	Data 6
	0 1 1 1 1	7	0 1 1 1	Data 7
	1 0 0 1 0	8	1 0 0 0	Data 8
	1 0 0 1 1	9	1 0 0 1	Data 9
	1 0 1 1 0	A	1 0 1 0	Data A
	1 0 1 1 1	B	1 0 1 1	Data B
	1 1 0 1 0	C	1 1 0 0	Data C
	1 1 0 1 1	D	1 1 0 1	Data D
	1 1 1 0 0	E	1 1 1 0	Data E
	1 1 1 0 1	F	1 1 1 1	Data F
	1 1 1 1 1	I	Undefined	IDLE; used as interstream fill code
C	1 1 0 0 0	J	0 1 0 1	Start-of-stream delimiter, Part 1 of 2; always used in pairs with K
O	1 0 0 0 1	K	0 1 0 1	Start-of-stream delimiter, Part 2 of 2; always used in pairs with J
N	0 1 1 0 1	T	Undefined	End-of-stream delimiter, Part 1 of 2; always used in pairs with R
T				
R	0 0 1 1 1	R	Undefined	End-of-stream delimiter, Part 2 of 2; always used in pairs with T
O				
L				
I	0 0 1 0 0	H	Undefined	Transmit error; used to force signaling errors
N	0 0 0 0 0	V	Undefined	Invalid code
V	0 0 0 0 1	V	Undefined	Invalid code
A	0 0 0 1 0	V	Undefined	Invalid code
L	0 0 0 1 1	V	Undefined	Invalid code
I	0 0 1 0 1	V	Undefined	Invalid code
D	0 0 1 1 0	V	Undefined	Invalid code
	0 1 0 0 0	V	Undefined	Invalid code
	0 1 1 0 0	V	Undefined	Invalid code
	1 0 0 0 0	V	Undefined	Invalid code
	1 1 0 0 1	V	Undefined	Invalid code

Source: Table 24-1, page 618, Ref. 3.

the MII. The 32 code groups are divided into four categories, as shown in Table 9.11.

DATA CODE GROUPS. A data code group conveys one nibble of arbitrary data between the MII and the PCS. The sequence of data code groups is arbitrary, where any data code group can be followed by any other data code group. Data code groups are coded and decoded but not interpreted by the PCS. Successful decoding of data code groups depends on proper receipt of the start-of-stream delimiter sequence as defined in Table 9.11.

IDLE CODE GROUPS. The idle code group (/I/) is transferred between streams. It provides a continuous fill pattern to establish and maintain clock synchronization. Idle code groups are emitted from, and interpreted by, the PCS.

CONTROL CODE GROUPS. The control code groups are used in pairs (/J/K/, /T/R/) to delimit MAC packets. Control code groups are emitted from, and interpreted by, the PCS.

START-OF-STREAM DELIMITER (/J / K /). A start-of-stream delimiter (SSD) is used to delineate the boundary of a data transmission sequence and to authenticate carrier events. The SSD is unique in that it may be recognized independently of previously established code-group boundaries. The receive function within the PCS uses the SSD to establish code group boundaries. An SSD consists of the sequence /J/K/.

END-OF-STREAM DELIMITER (/T / R /). An end-of-stream delimiter (ESD) terminates all normal data transmissions. Unlike the SSD, an ESD cannot be recognized independent of previously established code group boundaries. An ESD consists of the sequence /T/R/.

INVALID CODE GROUPS. The /H/ code group indicates that the PCS's client wishes to indicate a transmit error to its peer entity. The normal use of this indicator is for repeaters to propagate received errors. Transmit error code groups are emitted from the PCS, at the request of the PCS's client through the use of the TX_ER signal.

The presence of any invalid code group on the medium, including /H/, denotes a collision artifact or an error condition. Invalid code groups are not intentionally transmitted onto the medium by DTEs. The PCS indicates the reception of an invalid code group on the MII through the use of the RX_ER signal.

Encapsulation. The 100BASE-X PCS accepts frames from the MAC through the reconciliation sublayer and MII. Due to the continuously signaled nature of the underlying PMA, and the encoding performed by the PCS, the

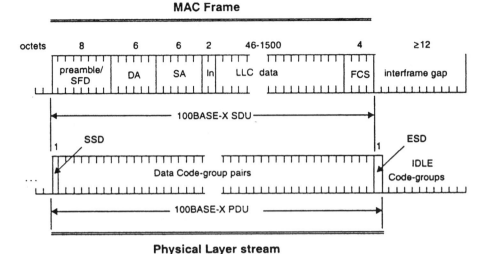

Figure 9.39. PCS encapsulation. (From Figure 24-5, page 619, IEEE Std. 802.3, Ref. 3.)

100BASE-X PCS encapsulates the MAC frame (100BASE-X service data unit, SDU) into a physical layer stream (100BASE-X protocol data unit, PDU).

Except for the two code-group SSD, data nibbles within the SDU (including the non-SSD portions of the MAC preamble and SFD) are not interpreted by the 100BASE-X PHY. The conversion from a MAC frame to a physical layer stream and back to a MAC frame is transparent to the MAC.

Figure 9.39 depicts the mapping between MAC frames and physical layer streams.

A properly formed stream can be viewed as comprising three elements:

(a) *Start-of-Stream Delimiter* (*SSD*). The start of a physical layer stream is indicated by an SSD, as defined in Section 9.7.4.2. The SSD replaces the first octet of the preamble from the MAC frame and vice versa.

(b) *Data Code Groups*. Between delimiters (SSD and ESD), the PCS conveys data code groups corresponding to the data nibbled of the MII. Data code groups comprise the 100BASE-X service data unit (SDU). Data nibbles within the SDU (including those corresponding to the MAC preamble and SFD) are not interpreted by the 100BASE-X PCS.

(c) *End-of-Stream Delimiter* (*ESD*). The end of a properly formed stream is indicated by an ESD, as defined in Section 9.7.4.2. The ESD is transmitted by the PCS following the de-assertion of TX_EN on the MII, which corresponds to the last data nibble composing the FCS

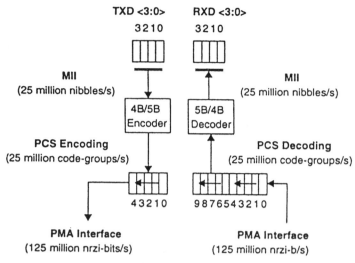

Figure 9.40. PCS reference diagram. (From Figure 24-6, page 620, IEEE Std. 802.3, Ref. 3.)

from the MAC. It is transmitted during the period considered by the MAC to be the interframe gap (IFG). On reception, ESD is interpreted by the PCS as terminating the SDU.

Between streams, IDLE code groups are conveyed between the PCS and PMA.

Data Delay. The PCS maps a nonaligned code bit data path from the PMA to be an aligned, nibble-wide data path on the MII, for both transmit and receive. Logically, received bits must be buffered to facilitate SSD detection and alignment, code translation, and ESD detection. These functions necessitate even longer delays of the incoming code-bit stream.

When the MII is present as an exposed interface, the MII signals TX_CLK and RX_CLK, not depicted in the following state diagrams, are generated by the PCS in accordance with Section 22 of the reference document.

Mapping Between MII and PMA. Figure 9.40 depicts the mapping of the nibble-wide data path of the MII to the 5-bit-wide code groups (internal to the PCS) and the code-bit path of the PMA interface.

Upon receipt of a nibble from the MII, the PCS encodes it into a 5-bit code group in accordance with the code group description paragraphs. Code groups are serialized into code bits and passed to the PMA for transmission on the underlying medium, according to Figure 9.40. The first transmitted code bit of a code group is bit 4, and the last code bit transmitted is bit 0. There is no numerical significance ascribed to the bits within a code group; that is, the code group is simply a 5-bit pattern that has some predefined interpretation.

Similarly, the PCS deserializes code bits received from the PMA, in accordance with Figure 9.40. After alignment is achieved, based on SSD detection, the PCS converts code groups into MII data nibbles, according to the code group description paragraphs.

9.7.5 System Considerations for Multisegment 100BASE-T Networks

This section provides a brief review on building 100BASE-T networks. The 100BASE-T technology is designed to be deployed in both homogeneous 100-Mbps networks and heterogeneous 10/100-Mbps mixed CSMA/CD networks. Network topologies can be developed within a single 100BASE-T collision domain, but maximum flexibility is achieved by designing multiple-collision-domain networks that are joined by bridges and/or routers configured to provide a range of service levels to DTEs. For example, a combined 100BASE-T/10BASE-T system built with repeaters and bridges can deliver dedicated 100-Mbps, shared 100-Mbps, dedicated 10-Mbps, and shared 10-Mbps service to DTEs, as shown in Figure 9.41. The effective bit rate capacity of shared service is controlled by the number of DTEs that share the service.

Linking multiple 100BASE-T collision domains with bridges maximizes flexibility. Bridged topology design can provide single bit rate capacity as shown in Figure 9.41 or multiple bit rate capacity services as shown in Figure 9.42.

Individual collision domains can be linked by single devices, as shown in Figures 9.41 and 9.42, or by multiple devices from any of several transmission systems. The design of multiple-collision-domain networks is governed by the rules defining each of the transmission systems incorporated into the design.

The design of shared bit rate capacity 10-Mbps collision domains is defined in Section 9.6.7 of this document. The design of shared bit rate capacity 100-Mbps CSMA/CD collision domains is defined in the present section.

9.7.5.1 *Physical Size of the Network.* The physical size of a CSMA/CD network is limited by the characteristics of individual network components. These characteristics include the following:

(a) Media lengths and their associated propagation delay
(b) Delay of repeater units (start-up, steady-state, and end of event)
(c) Delay of MAUs and PHYs (start-up, steady-state, and end of event)
(d) Interpacket gap shrinkage due to repeater units
(e) Delays within the DTE associated with the CSMA/CD access method
(f) Collision detect and de-assertion times associated with the MAUs and PHYs

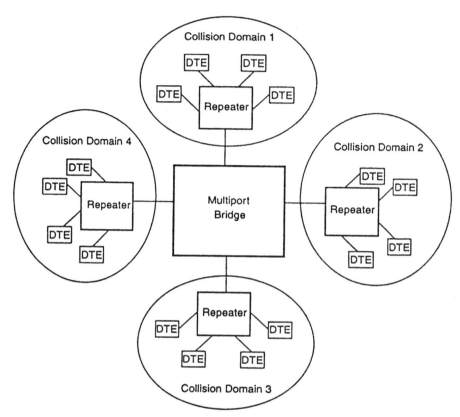

Figure 9.41. 100-Mbps multiple-collision-domain topology using multiport bridge. (From Figure 29-1, page 735, IEEE Std. 802.3, Ref. 3.)

Table 9.12 summarizes the delays for 100BASE-T media segments.

9.7.5.2 Transmission Models. Two different transmission system models are used to determine maximum network size and complexity:

Transmission System Model 1. The following network topology constraints apply to networks using Transmission System Model 1.

(a) All balanced cable (copper) segments less than or equal to 100 m each.
(b) Fiber segments less than or equal to 412 m each.
(c) MII cables for 100BASE-T shall not exceed 0.5 m each. When evaluating system topology, MII cable delays need not be accounted for separately. Delays attributable to the MII are incorporated into DTE and repeater component delays.

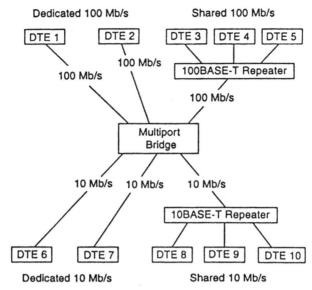

Figure 9.42. Multiple bit rate capacity, multiple-collision-domain topology using a multiport bridge.

TABLE 9.12 Delays for Network Media Segments Model 1

Media Type	Maximum Number of PHYs per Segment	Maximum Segment Length (m)	Maximum Medium Round-Trip Delay per Segment (BT)[a]
Balanced Cable Link Segment 100BASE-T	2	100	114
Fiber Link Segment	2	412	412

[a]BT, bit times.
Source: Table 29-1, page 737, Ref. 3.

Transmission System Model 2. The physical size and number of topological elements in a 100BASE-T network is limited primarily by round-trip collision delay. A network configuration must be validated against collision delay using a network model. Since there are a limited number of topology models for any 100BASE-T collision domain, the modeling process is quite straightforward and can easily be done either manually or with a spreadsheet.

This model derives from Model 2 presented in Section 9.6.7. Modifications have been made to accommodate adjustments for DTE, repeater, and cable speeds.

9.7.5.3 *Round-Trip Collision Delay*—Key to Network Operation and Maximum Size.
For a network to be valid, it must be possible for any two DTEs on the network to contend for the network at the same time. Each station attempting to transmit must be notified of the contention by the

TABLE 9.13   Network Component Delays, Transmission System Model 2

Component	Round-Trip Delay in Bit Times per Meter	Maximum Round-Trip Delay in Bit Times
Two TX / FX DTEs		100
Two T4 DTEs		138
Two T2 DTEs		96
One T2 or T4 and one TX / FX DTE[a]		127
Cat. 3 cabling segment	1.14	114 (100 m)
Cat. 4 cabling segment	1.14	114 (100 m)
Cat. 5 cabling segment	1.112	111.2 (100 m)
STP cabling segment	1.112	111.2 (100 m)
Fiber-optic cabling segment	1.0	412 (412 m)
Class I repeater		140
Class II repeater with all ports TX / FX		92
Class II repeater with any port T4		67
Class II repeater with any port T2		90

[a]Worst-case values are used (TX / FX values for MAC transmit start and MDI input to collision detect; T4 value for MDI input to MDI output).

Source: Table 29-3, page 741, Ref. 3.

returned "collision" signal within the "collision window." Additionally, the maximum length fragment created must contain less than 512 bits after the start-of-frame delimiter (SFD). These requirements limit the physical diameter (maximum distance between DTEs) of a network. The maximum round-trip delay must be qualified between all pairs of DTEs in the network. In practice this means that the qualification must be done between those that, by inspection of the topology, are candidates for the longest delay. The following network modeling methodology is provided to assist that calculation.

Table 9.13 is presented for guidelines on various system component delays, Transmission System Model 2.

9.8   1000-Mbps CSMA / CD NETWORKS

9.8.1   Overview

Gigabit Ethernet (CSMA/CD) uses the extended IEEE 802.3 MAC layer interface, connected through a gigabit media-independent interface layer to physical layer entities (PHY sublayers) such as 1000BASE-LX, 1000BASE-SX, 1000BASE-CX, and 1000BASE-T. The architectural positioning of gigabit Ethernet with the total CSMA/CD LAN structure and the OSI reference model is shown in Figure 9.43.

Gigabit Ethernet extends the IEEE 802-3 MAC beyond 100 Mbps to 1000 Mbps. The bit rate is faster, and the bit times are shorter—both in proportion to the change in bandwidth. In full-duplex mode, the minimum packet

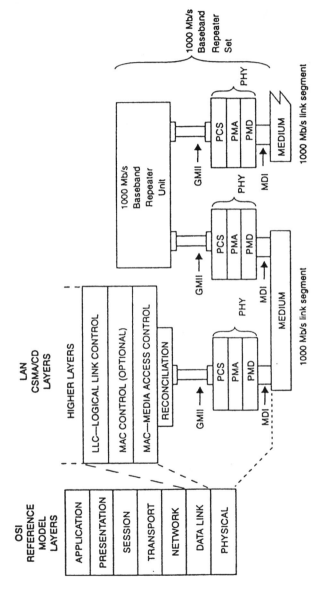

Figure 9.43. Architectural positioning of gigabit Ethernet. (From Figure 34-1, page 893, IEEE Std. 802.3, Ref. 3.)

MDI = MEDIUM DEPENDENT INTERFACE
GMII = GIGABIT MEDIA INDEPENDENT INTERFACE

PCS = PHYSICAL CODING SUBLAYER
PMA = PHYSICAL MEDIUM ATTACHMENT
PHY = PHYSICAL LAYER DEVICE
PMD = PHYSICAL MEDIUM DEPENDENT

transmission time has been reduced by a factor of ten. Achievable topologies for 1000-Mbps full-duplex operation are comparable to those found in 100BASE-T full-duplex mode. In half-duplex mode, the minimum packet transmission time has been reduced, but not by a factor of ten. Cable delay budgets are similar to those in 100BASE-T. The resulting achievable topologies for the half-duplex 1000-Mbps CSMA/CD MAC are similar to those found in half-duplex 100BASE-T.

9.8.2 Reconciliation Sublayer (RS) and Gigabit Media-Independent Interface (GMII)

9.8.2.1 Overview. Figure 9.43 shows the relationship of the reconciliation sublayer and the GMII with the OSI reference model. The purpose of this interface is to provide a simple, inexpensive, and easy-to-implement interconnection between media access control (MAC) sublayer and PHYs and between PHYs and station management. This interface has the following characteristics:

(a) It is capable of supporting 1000-Mbps operation.
(b) Data and delimiters are synchronous to clock references.
(c) It provides independent 8-bit-wide transmit and receive data paths.
(d) It provides a simple management interface.
(e) It uses signal levels, compatible with common CMOS digital ASIC processes and some bipolar processes.
(f) It provides for full-duplex operation.

9.8.2.2 GMII Data Stream. Data frames transmitted through the GMII are transferred within the data stream as shown in Figure 9.44.

For the GMII, transmission and reception of each octet of data is as shown in Figure 9.45.

Interframe ⟨*interframe*⟩. The interframe ⟨interframe⟩ period on a GMII transmit or receive path is an interval during which no data activity occurs on the path. Between bursts or single frame transmission, the absence of data activity on the receive path is indicated by the de-assertion of both RX_DV (receive data valid) and RX_ER (receive error) or the de-assertion of the

<inter-frame><preamble><sfd><data><efd><extend>

Figure 9.44. GMII data stream.

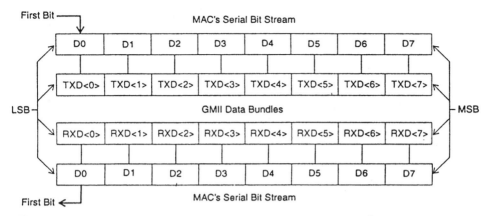

Figure 9.45. Relationship of data bundles to MAC serial bit stream. (From Figure 35-16, page 909, IEEE Std. 802.3, Ref. 3.)

RX_DV signal with the RXD⟨7:0⟩* value of 100 hexadecimal. On the transmit path the absence of data activity is indicated by the de-assertion of both TX_EN (transmit enable) and TX_ER. Between frames within a burst, the interframe period is signaled as carrier extend on the GMII.

9.8.3   Physical Coding Sublayer (PCS) and Physical Medium Attachment (PMA) Sublayer, 1000BASE-X

9.8.3.1   Overview. There are currently three embodiments within this family: 1000BASE-CX, 1000BASE-LX, and 1000BASE-SX. 1000BASE-CX covers operation over a single copper medium: two pairs of 150-Ω balanced copper cabling. 1000BASE-LX covers operation over a pair of optical fibers using long-wavelength optical transmission. 1000BASE-SX covers operation over a pair of optical fibers using short-wavelength optical transmission. The use of the term 1000BASE-X is to cover issues common to any of these subvariants.

1000BASE-X is based on the physical layer standards developed by ANSI X3.230-1994 (fiber channel physical and signaling interface). In particular, this standard uses the same 8B/10B coding as fiber channel, a PMA sublayer compatible with speed-enhanced versions of the ANSI 10-bit serializer chip, and similar optical and electrical specifications.

1000BASE-X PCS and PMA sublayers map the interface characteristics of the PMD layer (including MDI) to the services expected by the reconciliation sublayer. 1000BASE-X can be extended to support any other full-duplex medium requiring only that the medium be compliant at the PMD level.

*RXD = a bundle of 8 data signals (RXD⟨7:0⟩).

9.8.3.2 *Objectives.* The following are the objectives of 1000BASE-X:

(a) To support the CSMA/CD MAC
(b) To support the 1000 Mbps repeater
(c) To provide for autonegotiation among like 1000-Mbps PMDs
(d) To provide 1000-Mbps data rate at the GMII
(e) To support cable plants using 150-Ω balanced copper cabling, or optical fiber compliant with ISO/IEC 11801: 1995
(f) To allow for a nominal network extent of up to 3 km, including
  (1) 150-Ω balanced links of 25-m span
  (2) one-repeater networks of 50-m span (using all 150-Ω balanced copper cabling)
  (3) one-repeater networks of 200-m span (using fiber)
  (4) DTE/DTE links of 3000 m (using fiber)
(g) To preserve full-duplex behavior of underlying PMD channels
(h) To support a BER objective of 10^{-12}

9.8.3.3 *Summary of 1000BASE-X Sublayers**

Physical Coding Sublayer. The PCS interface is the gigabit media-independent interface (GMII) that provides a uniform interface to the reconciliation sublayer for all 1000-Mbps PHY implementations (e.g., not only 1000BASE-X but also other possible types of gigabit PHY entities). 1000BASE-X provides services to the GMII in a manner analogous to how 100BASE-X provides services to the 100-Mbps MII.

The 1000BASE-X PCS provides all services required by the GMII, including

(a) Encoding (decoding) of GMII data octets to (from) 10-bit code groups (8B/10B) for communication with the underlying PMA
(b) Generating carrier sense and collision detect indications for use by PHY's half-duplex clients
(c) Managing the autonegotiation process, and informing the management entity via the GMII when the PHY is ready for use

Physical Medium Attachment (PMA) Sublayer. The PMA provides a medium-independent means for the PCS to support the use of a range of serial-bit-oriented physical media. The 1000BASE-X PMA performs the

---

*1000BASE-X PHY consists of that portion of the physical layer between the MDI and GMII consisting of the PCS, PMA, and PMD sublayers. 1000BASE-X PHY is roughly analogous to the 100BASE-X PHY.

following functions:

(a) Mapping of transmit and receive code groups between the PCS and PMA via the PMA service interface

(b) Serialization (deserialization) of code groups for transmission (reception) on the underlying serial PMD

(c) Recovery of clock from the 8B/10B-coded data supplied by the PMD

(d) Mapping of transmit and receive bits between the PMA and PMD via the PMD service interface

(e) Data loopback at the PMD service interface

Physical Medium-Dependent (PMD Sublayer. 1000BASE-X physical layer signaling for fiber and copper media is adapted from ANSI X.320-1994 (FC and PH), Clauses 6 and 7, respectively. These clauses define 1062.5-Mbps, full-duplex signaling systems that accommodate single-mode optical fiber, multimode optical fiber, and 150-Ω balanced copper cabling. 1000BASE-X adapts these basic physical layer specifications for use with the PMD sublayer and mediums specified in Sections 38 and 39 of the reference document.

The MDI,* logically subsumed within each PMD subsection, is the actual medium attachment, including connectors, for the various supported media.

Figure 9.46 depicts the relationship between 1000BASE-X and its associated PMD sublayers. Figure 9.47 is a functional block diagram of the 1000BASE-X PHY.

9.8.3.4 *Physical Coding Sublayer (PCS).*

The PCS service interface allows the 1000BASE-X PCS to transfer information to and from a PCS client. PCS clients include the MAC (via the reconciliation sublayer) and repeater. In the reference specification the GMII variables are set to "true" or "false." This is equivalent to, respectively, "asserting" or "de-asserting" them.

Functions Within the PCS. The PCS comprises the PCS transmit, carrier sense, synchronization, PCS receive, and autonegotiation processes for 1000BASE-X. The PCS shields the reconciliation sublayer (and MAC) from the specific nature of the underlying channel. When communicating with the GMII, the PCS uses an octet-wide, synchronous data path, with packet delimiting being provided by separate transmit control signals (TX_EN and TX_ER) and receive control signals (RX_DV and RX_ER). When communicating with the PMA, the PCS uses a 10-bit-wide, synchronous data path, which conveys 10-bit code groups. At the PMA service interface, code-group alignment and MAC packet delimiting are made possible by embedding special nondata code groups in the transmitted code-group stream. The PCS provides the functions necessary to map packets between the GMII format and the PMA service interface format.

*MDI, medium-dependent interface.

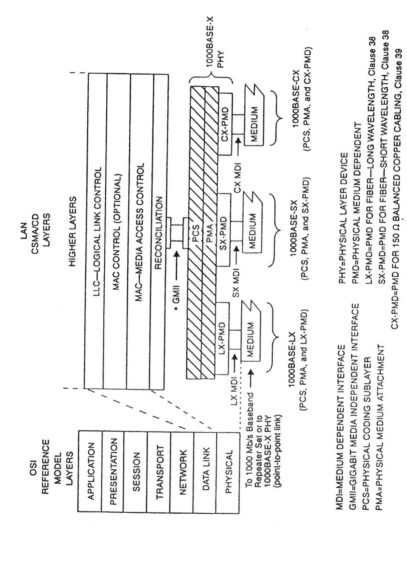

Figure 9.46. Relationship of 1000BASE-X and the PMDs. (From Figure 36-1, page 925, IEEE Std. 802.3, Ref. 3.)

MDI=MEDIUM DEPENDENT INTERFACE
GMII=GIGABIT MEDIA INDEPENDENT INTERFACE
PCS=PHYSICAL CODING SUBLAYER
PMA=PHYSICAL MEDIUM ATTACHMENT

PHY=PHYSICAL LAYER DEVICE
PMD=PHYSICAL MEDIUM DEPENDENT
LX-PMD=PMD FOR FIBER—LONG WAVELENGTH, Clause 38
SX-PMD=PMD FOR FIBER—SHORT WAVELENGTH, Clause 38
CX-PMD=PMD FOR 150 Ω BALANCED COPPER CABLING, Clause 39

NOTE—The PMD sublayers are mutually independent.

• GMII is optional.

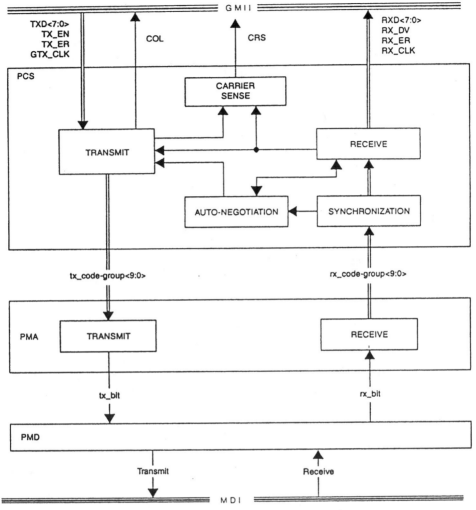

Figure 9.47. Functional block diagram of the 1000BASE-X PHY. (From Figure 36-2, page 926, IEEE Std. 802.3, Ref. 3.)

The PCS transmit process continuously generates code groups based upon the TXD⟨7:0⟩, TX_EN, and TX_ER signals on the GMII, sending them immediately to the PMA service interface via the PMA_UNITDATA.request primitive. The PCS transmit process generates the GMII signal COL based on whether a reception is occurring simultaneously with transmission. Additionally, it generates the internal flag, transmitting, for use by the carrier sense process. The PCS transmit process monitors the autonegotiation process xmit flag to determine whether to transmit data or reconfigure the link.

The carrier sense process controls the GMII signal CRS (see Figure 36-8 of the reference document).

The PCS synchronization process continuously accepts code groups via the PMA_UNITDATA.indicate primitive and conveys received code groups to the PCS receive process via the SYNC_UNITDATA.indicate primitive. The PCS synchronization process sets the sync_status flag to indicate whether the PMA is functioning dependably (as well as can be determined without exhaustive error-rate analysis).

The PCS receive process continuously accepts code groups via the SYNC_UNITDATA.indicate primitive. The PCS receive process monitors these code groups and generates RXD⟨7 : 0⟩, RX_DV, and RX_ER on the GMII and generates the internal flag, receiving, used by the carrier sense and transmit processes.

The PCS autonegotiation process sets the xmit flag to inform the PCS transmit process to either transmit normal idles interspersed with packets as requested by the GMII or to reconfigure the link. The PCS autonegotiation process is specified in Clause 37 of the reference document.

8B / 10B Transmission Code. The PCS uses a transmission code to improve the transmission characteristics of information to be transferred across the link. The encodings defined by the transmission code ensure that sufficient transitions are present in the PHY bit stream to make clock recovery possible at the receiver. Such encoding also greatly increases the likelihood of detecting any single or multiple bit errors that may occur during transmission and reception of information. In addition, some of the special code groups of the transmission code contain a distinct and easily recognizable bit pattern that assists a receiver in achieving code-group alignment on the incoming PHY bit stream. The 8B/10B transmission code specified for use in this standard has a high transition density, is a run-length-limited code, and is dc-balanced. The transition density of the 8B/10B symbols ranges from 3 to 8 transitions per symbol.

The definition of the 8B/10B transmission code is identical to that specified in ANSI X3.230-1994, Clause 11. The relationship of code-group bit positions to PMA and other PCS constructs is illustrated in Figure 9.48.

Notation Conventions. 8B/10B transmission code uses letter notation for describing the bits of an unencoded information octet and a single control variable. Each bit of the unencoded information octet contains either a binary zero or a binary one. A control variable, Z, has either the value D or the value K. When the control variable associated with an unencoded information octet contains the value D, the associated encoded code group is referred to as a data code group. When the control variable associated with an unencoded information octet contains the value K, the associated encoded code group is referred to as a special code group.

The bit notation of A, B, C, D, E, F, G, H for an unencoded information octet is used in the description of the 8B/10B transmission code. The bits A, B, C, D, E, F, G, H are translated to bits a, b, c, d, e, f, g, h, j of 10-bit transmission code groups. 8B/10B code-group bit assignments are illustrated in

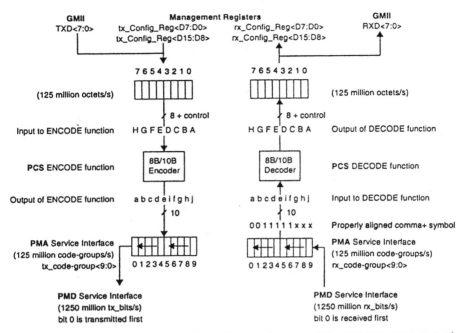

Figure 9.48. PCS reference diagram. (From Figure 36-3, page 928, IEEE Std. 802.3, Ref. 3.)

Figure 9.48. Each valid code group has been given a name using the following convention: /Dx.y/ for the 256 valid data code groups and /Kx.y/ for special control code groups, where x is the decimal value of bits EDCBA and y is the decimal value of bits HGF.

Transmission Order. Code group bit transmission order is shown in Figure 9.48.

Code groups within multi-code-group ordered_sets (as specified in Table 36-3 of the reference document) are transmitted sequentially beginning with the special code group used to distinguish the ordered_set (e.g., /K28.5/) and proceeding code group by code group from left to right within the definition of the ordered_set until all code groups of the ordered_set are transmitted.

9.8.4 System Considerations for Multisegment 1000-Mbps Networks

9.8.4.1 Overview. The 1000-Mbps technology is designed to be deployed in both homogeneous 1000-Mbps networks and 10/100/1000 Mbps mixed networks using bridges and/or routers. Network topologies can be developed within a single 1000-Mbps collision domain, but maximum flexibility is achieved by designing multiple-collision-domain networks that are joined by bridges and/or routers configured to provide a range of service levels to

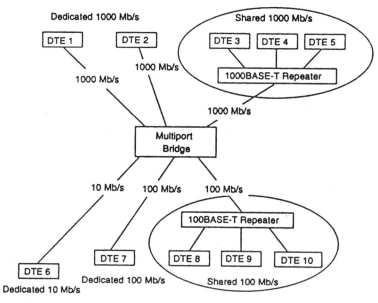

Figure 9.49. Multiple data rate, multiple-collision-domain topology using multiport bridge. (From Figure 42-2, page 1067, IEEE Std. 802.3, Ref. 3.)

DTEs. For example, a combined 1000BASE-T/100BASE-T/10BASE-T system built with repeaters and bridges can deliver dedicated or shared service to DTEs at 1000 Mbps, 100 Mbps, or 10 Mbps.

Figure 9.41 outlines a 100-Mbps multiple-collision-domain topology using a multiport bridge. This figure serves equally as well for a 1000-Mbps multiple-collision-domain topology, and it shows how individual collision domains can be linked by single devices or by multiple devices from any of several transmission systems. (Also see Figure 9.49.) The design of multiple-collision-domain networks is governed by the rules defining each of the transmission systems incorporated into the design.

9.8.4.2 Single-Collision-Domain Multisegment Networks.

The proper operation of a CSMA/CD network requires the physical size of the collision domain to be limited in order to meet the round-trip propagation delay requirements defined by the slot time discussed in Section 9.6.5.1. It also requires that the number of repeaters be limited so as not to exceed the interframe shrinkage given in Tables 9.3 through 9.6 and discussed in Section 9.6.7.2.

As in the previous section, there are two transmission models: Transmission System Model 1 and Model 2. Model 1 is the more conservative, whereas Model 2 provides a set of calculation aids that allow those configuring a network to test a proposed configuration against a simple set of criteria that allows it to be qualified. Transmission System Model 2 validates an

TABLE 9.14   Delays for Network Media Segments—Model 1

Media Type	Maximum Number of PHYs per Segment	Maximum Segment Length (m)	Maximum Medium Round-Trip Delay per Segment (BT)
Category 5 UTP link segment (1000BASE-T)	2	100	1112
Shielded jumper cable link segment (1000BASE-CX)	2	25	253
Optical fiber link segment (1000BASE-SX, 1000BASE-LX)	2	316[a]	3192

[a]May be limited by the maximum transmission distance of the link.
Source: Table 42-1, page 1068, Ref. 3.

additional broad set of topologies that are fully functional and do not fit within the simpler, but more restrictive rules of Model 1.

The physical size of a CSMA/CD network is discussed in Section 9.7.5.1. Table 9.14 summarizes the delays, measured in bit times (BTs), for 1000-Mbps media segments.

Repeater Usage. Repeaters are the means used to connect segments of a network medium together into single collision domain. Different physical signaling systems (e.g., 1000BASE-CX, 1000BASE-SX, 1000BASE-LX, 1000BASE-T) can be joined into a common collision domain using a repeater. Bridges can also be used to connect different signaling systems. However, if a bridge is so used, each LAN connected to the bridge will comprise a separate collision domain.

9.8.4.3   *Transmission System Model 1.*
The following network topology constraints apply to networks using Transmission System Model 1:

(a) Single repeater topology maximum
(b) Link distances not to exceed the lesser of 316 m or the maximum transmission distance of the link

9.8.4.4   *Transmission System Model 2.*
Transmission System Model 2 is a single repeater topology with the physical size limited primarily by round-trip collision delay. A network configuration must be validated against collision delay using a network model for a 1000-Mbps collision domain. The modeling process is quite straightforward and can easily be done either manually or with a spreadsheet.

Key elements in the calculation are two: round-trip collision delay and interpacket gap shrinkage. The validation for each of these must be done separately. The model consists of a series of segments consisting of a left-end segment, mid-segments, and a right-end segment. The model consists of the worst-case path through the network for that particular calculation.

Round-Trip Collision Delay. For a network to be valid, it must be possible for any two DTEs on that network to properly arbitrate for the network. When two or more stations attempt to transmit within the slot time interval,

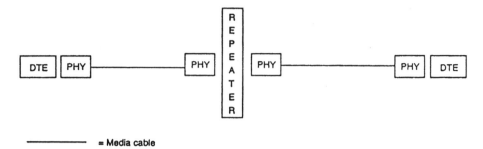

= Media cable

Figure 9.50. System model 2: single repeater.

each station must be notified of the contention by the returned "collision" signal within the "collision window."* Additionally, the maximum length fragment created on a 1000-Mbps network must contain less than 512 bytes after the start frame delimiter (SFD). These requirements limit the physical diameter (maximum distance between DTEs) of a network. The maximum round-trip delay must be qualified between all pairs of DTEs in the network. In practice, this means that the qualification must be done between those that, by inspection of the topology, are candidates for the longest delay. The following network modeling methodology is provided to assist that calculation.

WORST-CASE PATH DELAY VALUE (PDV) SELECTION. The worst-case path through a network to be validated is identified by examination of the aggregate DTE delays, cable delays, and repeater delay. The worst case consists of the path between the two DTEs at opposite ends of the network that have the longest round-trip time. Figure 9.50 shows a schematic representation of a one-repeater path.

WORST-CASE PDV CALCULATION. Once a set of paths is selected for calculation, each is checked for validity against the following formula:

$$\text{PDV} = \Sigma \text{link delays(LSDV)} + \text{repeater delay}$$
$$+ \text{DTE delays} + \text{safety margin}$$

Values for the formula variables are determined by the following method:

(a) Determine the delay for each link segment (link segment delay value, or LSDV), using the formula

$$\text{LSDV} = 2 \text{ (for round-trip delay)} \times \text{segment length}$$
$$\times \text{cable delay for this segment}$$

---

*A given station can experience a collision during the initial part of its transmission ("collision window") before its transmitted signal has had time to propagate to all stations on the CSMA/CD medium.

TABLE 9.15   Network Component Delays, Transmission System Model 2

Component	Round-Trip Delay in Bit Times per Meter (BT/m)	Maximum Round-Trip Delay in Bit Times (BT)
Two DTEs		864
Category 5 UTP cable segment	11.12	1112 (100 m)
Shielded jumper cable segment	10.10	253 (25 m)
Optical fiber cable segment	10.10	1111 (110 m)
Repeater		976

Source: Table 42-3, page 1071, Ref. 3.

TABLE 9.16   Conversion Table for Cable Delays

Speed Relative to c	ns/m	BT/m	Speed Relative to c	ns/m	BT/m
0.4	8.34	8.34	0.62	5.38	5.38
0.5	6.67	6.67	0.63	5.29	5.29
0.51	6.54	6.54	0.64	5.21	5.21
0.52	6.41	6.41	0.65	5.13	5.13
0.53	6.29	6.29	0.654	5.10	5.10
0.54	6.18	6.18	0.66	5.05	5.05
0.55	6.06	6.06	0.666	5.01	5.01
0.56	5.96	5.96	0.67	4.98	4.98
0.57	5.85	5.85	0.68	4.91	4.91
0.58	5.75	5.75	0.69	4.83	4.83
0.5852	5.70	5.70	0.7	4.77	4.77
0.59	5.65	5.65	0.8	4.17	4.17
0.6	5.56	5.56	0.9	3.71	3.71
0.61	5.47	5.47			

Source: Table 42-4, pages 1072 and 1073, Ref. 3.

Note 1: Length is the sum of the cable lengths between the PHY interfaces at the repeater and PHY interfaces at the farthest DTE. All measurements are in meters.

Note 2: Cable delay is the delay specified by the manufacturer or the maximum value for the type of cable used as shown in Table 9.15. For this calculation, cable delay must be specified in bit times per meter (BT/m). Table 9.16 can be used to convert values specified relative to the speed of light (%c) or nanoseconds per meter (ns/m).

Note 3: When actual cable lengths or propagation delays are not known, use the Max delay in bit times as specified in Table 9.15 for copper cables. Delays for fiber should be calculated, as the value found in Table 9.15 will be too large for most applications.

Note 4: The value found in Table 9.15 for shielded jumper cable is the maximum delay for cable with solid dielectric. Cables with foam dielectric may have a significantly smaller delay.

(b) Sum together the LSDVs for all segments in the path.

(c) Determine the delay for the repeater. If model specific data is not available from the manufacturer, enter the appropriate default value from Table 9.15.

(d) Use the DTE delay value shown in Table 9.15 unless the equipment manufacturer defines a different value. If the manufacturer's supplied values are used, the DTE delays of both ends of the worst-case path should be summed together.

(e) Decide on appropriate safety margin—0 to 40 bit times—for the PDV calculation. Safety margin is used to provide additional margin to accommodate unanticipated delay elements, such as extra-long connecting cable runs between wall jacks and DTEs. (A safety margin of 32 BT is recommended.)

(f) Insert the values obtained through these calculations into the following formula to calculate the PDV. (Some configurations may not use all the elements of the formula.)

$$PDV = \Sigma \text{link delays (LSDV)} + \text{repeater delay}$$
$$+ \text{ DTE delay} + \text{safety margin}$$

(g) If the PDV is less than 4096, the path is qualified in terms of worst-case delay.

(h) Late collisions and/or CRC errors may be indications that path delays exceed 4096 BT.

REFERENCES

1. *The IEEE Standard Dictionary of Electrical and Electronic Terms*, 6th ed., IEEE Std. 100-1996, IEEE, New York, 1996.

2. *Information Processing Systems—Local Area Networks, Part 2, Logical Link Control*, ANSI/IEEE Std. 802.2, 1998 edition, IEEE, New York, 1998.

3. *Information Technology—Telecommunication and Information Exchange Between Systems—Local and Metropolitan Area Networks—Specific Requirements, Part 3: Carrier Sense Multiple Access with Collision Detection (CSMA / CD) Access Method and Physical Layer Specification*, IEEE Std. 802.3, 1998 edition, IEEE, New York, 1998.

4. *Information Technology—Telecommunications and Information Exchange Between Systems—Local and Metropolitan Area Networks—Specific Requirements, Part 6: Distributed Queue Dual Bus (DQDB) Access Method and Physical Layer Specifications*, ANSI/IEEE 802.6, 1994 edition, IEEE, New York, 1994.

10

DATA COMMUNICATIONS IN THE OFFICE ENVIRONMENT, PART 2

10.1 MEDIUM ACCESS CONTROL—TOKEN-PASSING SCHEMES

10.1.1 Introduction

Whereas CSMA/CD (Section 9.5) was a random access/contention method of local area network (LAN) access, token-passing LAN methods are disciplined access techniques. The following lists the four major points of the token-passing method:

1. A token controls the right of access to the physical medium. The LAN station that holds (possesses) the token has control over the medium for a short time period.
2. The token is passed by LAN stations residing on the medium. As the token is passed from station to station, a logical ring is formed.
3. Steady-state operation consists of a data transfer phase and a token-passing phase.
4. Ring maintenance functions within the LAN stations provide for ring initialization, lost token recovery, new station addition to the logical ring, and general housekeeping of the logical ring. The ring maintenance functions are replicated among all the token-using LAN stations on the network.

Shared media generally can be categorized into two major types: broadcast and sequential. The token-bus overview that follows deals with the broadcast type. On a broadcast medium, every station may receive signals transmitted. Media of the broadcast type are usually configured as a physical bus.

This section covers three token-passing methods:

- Token bus
- Token ring
- Fiber distributed data interface (FDDI)

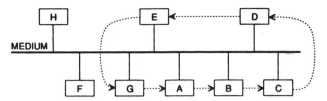

Figure 10.1. Logical ring on a physical bus.

The token-bus method is covered by only a brief overview because its market penetration has dropped to just several percent of the total LAN market.* Token ring and FDDI are discussed in greater detail. The IEEE 802.4 (i.e., token-bus) committee has a current specification available. Token ring was backed by IBM and is covered by the IEEE 802.5 committee. FDDI is an ANSI initiative.

10.1.2 Token-Passing Bus

Figure 1.1 shows a simple example of a bus network. Figure 10.1 illustrates a *logical* ring on a physical bus. It should be noted that the token medium access method is always sequential in a logical sense. That is, during normal steady-state operation, the right to access the medium passes from station to station in a sequential manner (i.e., station #1 to station #2, to station #3 ... to station #n in the logical ring). In addition, the physical connectivity has little impact on the order of the logical ring, and stations can respond to the token holder even without being part of the logical ring. For example, Stations H and F in Figure 10.1 can receive frames and could respond but cannot initiate a transmission because they will never be sent the token.

The MAC sublayer provides sequential access to the shared bus medium by passing control of the medium from station to station in a logically circular fashion. The MAC sublayer determines when the LAN station has the right to access the shared medium by recognizing and accepting the token from the predecessor station, and it determines when the token will be passed to the successor station.

The MAC sublayer carries out the following general functions:

1. Lost token timer
2. Distributed initialization
3. Token-holding timer
4. Limited data buffering
5. Node address recognition

*In contrast, Ethernet (Section 9.5) has over 75% market share (year 2000).

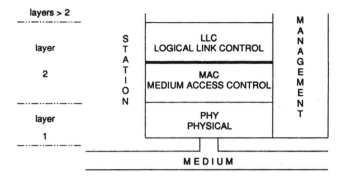

Figure 10.2. Token-passing bus architecture relates to OSI layers.

6. Frame encapsulation (including token preparation)
7. Frame check sequences (FCS) generation and checking

Figure 10.2 relates the token bus architecture to the OSI reference model. Token-passing bus is a 1-, 5-, 10-, or 20-Mbps system.

10.1.2.1 Token-Bus Frame Formats. All frames sent or received by the MAC sublayer conform to the format shown in Figure 10.3. In a MAC frame the number of octets between SD and ED is 8191 or less. The components of a frame are detailed in the legend under the figure.

Figure 10.4 shows the format of a token. Here the destination address (DA) has the value of the next station in sequence following the station issuing the token. The frame control (FC) field has the unique binary sequence: 00001000.

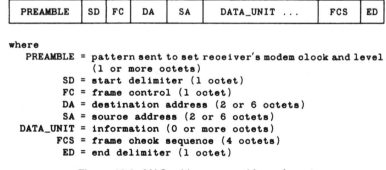

| PREAMBLE | SD | FC | DA | SA | DATA_UNIT ... | FCS | ED |

where
    PREAMBLE = pattern sent to set receiver's modem clock and level
               (1 or more octets)
          SD = start delimiter (1 octet)
          FC = frame control (1 octet)
          DA = destination address (2 or 6 octets)
          SA = source address (2 or 6 octets)
   DATA_UNIT = information (0 or more octets)
         FCS = frame check sequence (4 octets)
          ED = end delimiter (1 octet)

Figure 10.3. MAC sublayer general frame format.

| PRE | SD | 00001000 | DA | SA | FCS | ED |

Figure 10.4. The token format.

10.1.2.2   *Operational Aspects*

SLOT_TIME. Slot_time is the maximum time any station need wait for an immediate MAC level response from another station. Slot_time is measured in octet times and is defined as

$$\text{slot\_time} = \text{INTEGER}(\{[(2*\text{transission\_path\_delay} + \text{station\_delay}$$

$$+\text{safety\_margin})/\text{MAC-symbol\_time}] + 7\}/8)$$

The slot_time, along with the station's address and several other management parameters, is known to the station before it attempts to transmit on the network. If all stations on the network are not using the same value for slot_time, the MAC protocol may not operate properly.

RIGHT TO TRANSMIT. The token (right to transmit) is passed from station to station in descending order of station address. When a station which belongs to the logical ring hears a token frame addressed to itself, it "has the token" and may transmit data frames. When a station has completed transmitting data frames, it passes the token to the next station in the logical ring.

When a LAN station has the token, it may temporarily delegate to another station its right to transmit by sending a request_with_response data frame. When a station hears a request_with_response data frame addressed to itself, it responds with a response data frame, if the request_with_response option is implemented. The response data frame causes the right to transmit to revert back to the station which sent the request_with_response data frame.

TOKEN PASSING. After each station has completed transmitting any data frames it may have and has completed other maintenance functions, the station passes the token to its successor by sending a "token" MAC_control frame.

After sending the token frame, the station listens for evidence that its successor has heard the token frame and is active. If the sender hears a valid frame following the token within one slot_time, it assumes that its successor has the token and is transmitting. Otherwise the token-sending station attempts to assess the state of the network.

If the token-sending station hears a noise_burst (e.g., an unidentifiable sequence or a frame with an incorrect FCS), it cannot be sure which station sent the transmission. The MAC protocol treats this condition in a way which minimizes the chance of the station causing a serious error.

Because a station on a broadband network should always hear its own frames, if the token-sending station hears a single noise_burst without hearing its own token frame, the station assumes that it heard its own token that had been garbled and continues to listen. If a second noise_burst is heard or if a noise_burst is heard after the station hears its own token frame, the token-sending station continues to listen in the CHECK_TOKEN_PASS state for up to four more slot_times. If nothing more is heard, the station assumes that the noise_burst it heard was not a garbled frame from the successor station and so it repeats the token transmission. If anything is heard during the four-slot_time delay, the station assumes that its successor successfully received the token.

RESPONSE WINDOW. New stations are added to the logical ring through a controlled contention process using "response windows." A response window is a controlled interval of time (equal to one slot_time) after transmission of a MAC_control frame in which the station sending the frame pauses and listens for a response. If a station hears a transmission start during the response window, the station continues to listen to the transmission, even after the response window time expires, until the transmission is complete. Thus, the response windows define the time interval during which a station will hear the beginning of a response from another station (Ref. 1).

10.1.3 Token-Passing Ring

*10.1.3.1 **Overview.*** A token ring consists of a set of LAN stations serially connected by a transmission medium as shown in Figure 10.5, with the last station and first station folded back connecting one to the other. Thus a ring is formed. Information is transferred sequentially bit by bit from one LAN station to the next around the ring. Each LAN station regenerates and repeats each bit and serves as the means of attaching one or more data devices (terminals, workstations, computers, print servers, etc.) to the ring for

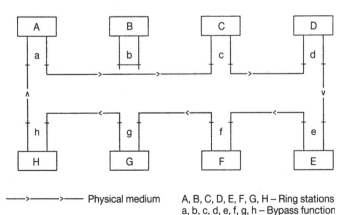

——>———>—— Physical medium   A, B, C, D, E, F, G, H – Ring stations
a, b, c, d, e, f, g, h – Bypass function

All stations are active except B (b illustrated in bypass mode)

Figure 10.5. Typical token-passing ring configuration.

the purpose of communicating with other devices on the network. A given LAN station has access to the medium for a given time (token-holding time). During this time a station may transfer information onto the ring, where the information circulates from one station to the next. The addressed destination station(s) copies the information as it passes. Finally, the station that transmitted the information effectively removes the information from the ring as it is returned around to that station again.

A LAN station gains the right to transmit its information onto the medium when it detects a token passing on the medium. The token is a control signal comprising a unique binary sequence that circulates on the medium following each information transfer. Any LAN station detecting the token may capture the token by modifying it to a start-of-frame sequence (SFS) and by appending the appropriate control and status fields, address fields, information field, frame check sequence (FCS), and end-of-frame sequence (EFS). On completion of its information transfer and after appropriate checking for proper operation, the LAN station initiates a new token. This provides other stations the opportunity to gain access to the ring.

A LAN station may only hold the token for a maximum amount of time before it must pass the token onwards.

There are multiple levels of priority available for independent and dynamic assignment depending on the relative class of service required for any given message, such as synchronous (real-time voice), asynchronous (interactive), or immediate (network recovery). The allocation of priorities is by mutual agreement among users of the network.

Error detection and recovery mechanisms are provided to restore network operation in the event that transmission errors or medium transients cause the access method to deviate from normal operation. Detection and recovery for these cases utilize a network monitoring function that is performed in a specific station with backup capability in all other stations that are attached to the ring.

Figure 10.6 shows a token ring with server stations (servers) that are the means through which the system manager manages the stations in such a system. Servers are data collection and distribution points on each ring where reports from data stations are gathered. Servers then communicate the necessary information to the system manager for the purpose of managing a LAN token ring system.

Data stations communicate with servers by reporting errors that are detected, such as lost token, FCS error, or lost frames, and requesting operating parameters when inserting into the ring, reporting changes in configuration due to insertion or removal of stations, responding to requests for various status information, and removal from the ring when requested.

Figure 10.7 shows the token ring's relationship with the open system interconnection (OSI) reference model. Essentially the LLC and MAC sublayers relate to OSI layer 2, the data-link layer, and PHY (the physical layer) relates to OSI layer 1.

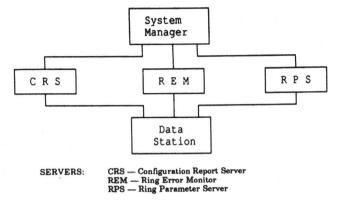

SERVERS: CRS — Configuration Report Server
REM — Ring Error Monitor
RPS — Ring Parameter Server

Figure 10.6. Relationship between data stations, servers, and system manager.

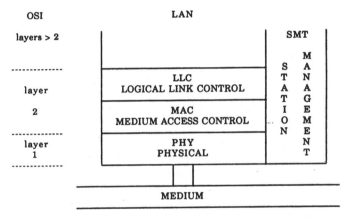

Figure 10.7. Relating the OSI reference model to the LAN model.

10.1.3.2 *Token-Ring Frame Formats.* Figure 10.8 shows the token format, and Figure 10.9 illustrates the frame format. The frame format is used for transmitting both MAC and LLC messages to the destination station(s). A frame may or may not have an information field.

Abort Sequence. The abort sequence consists of two fields: the starting delimiter (1 octet) and the ending delimiter (1 octet). This sequence is not used for the purpose of terminating the transmission of a frame prematurely.

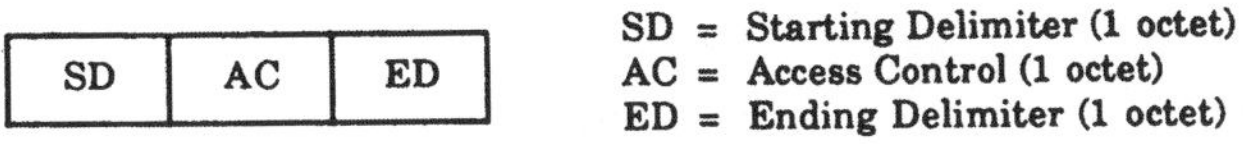

SD = Starting Delimiter (1 octet)
AC = Access Control (1 octet)
ED = Ending Delimiter (1 octet)

Figure 10.8. Token format.

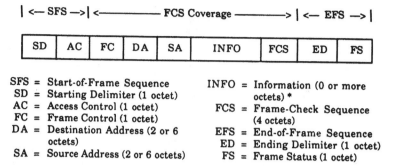

Figure 10.9. Token-ring frame format. *Maximum length only limited by the token-holding time.

The abort sequence may occur anywhere in the bit stream; that is, receiving stations are able to detect an abort sequence even if it does not occur on octet boundaries.

Fill. When a LAN station is transmitting (as opposed to repeating), it transmits fill preceding or following frames, tokens, or abort sequences to avoid what would otherwise be an inactive or indeterminate transmitter state. Fill may be either "1s" or "0s" or any combination thereof and may be any *number* of bits in length.

Starting Delimiter (SD). A frame or token is started with these eight symbols. If otherwise, the sequence is not considered valid.

$$J\ K\ 0\ J\ K\ 0\ 0\ 0$$

where

$$J = \text{nondata J}$$
$$K = \text{nondata K}$$
$$0 = \text{zero bit}$$

Access Control (AC). The access control sequence is one octet and consists of the following bits:

$$PPP\ T\ M\ RRR$$

where

$$PPP = \text{priority bits}$$
$$T = \text{token bit}$$
$$M = \text{monitor bit}$$
$$RRR = \text{reservation bits}$$

PRIORITY BITS. Priority bits indicate the priority of a token and, therefore, which stations are allowed to use the token. In a multiple priority system, stations use different priorities depending on the priority of the PDU (protocol data unit) to be transmitted.

There are eight levels of priority increasing from the lowest (000) to the highest (111) priority. For the purposes of comparing priority values, the priority is transmitted most significant bit first. For example, 110 has higher priority than 011, with leftmost bit transmitted first.

TOKEN BIT. The token bit is a 0 in a token and 1 in a frame. When a station with a PDU to transmit detects a token that has a priority equal to or less than the PDU to be transmitted, it may change the token to a start-of-frame sequence (SFS) and transmit the PDU.

MONITOR (M) BIT. The M bit is used to prevent a token that has a priority greater than 0 or any frame from continuously circulating on the ring. If an active monitor detects a frame or a high-priority token with the M bit equal to 1, the frame or token is aborted. The M bit is transmitted as 0 in all frames and tokens. The active monitor inspects and modifies this bit. All other LAN stations repeat this bit as received.

RESERVATION BITS. These allow stations with high-priority PDUs to request (in frames or tokens as they are repeated) that the next token be issued at the requested priority. The eight levels of reservation increase from 000 to 111, and these are operated upon in a similar fashion as the priority bits.

Frame Control (FC). The FC field defines the type of the frame and certain MAC and information frame functions. It consists of one octet as follows:

FF ZZZZZZ

where

FF = frame-type bits

ZZZZZZ = control bits

Frame-type bits indicate the type of frame as follows:

00 = MAC frame (i.e., contains a MAC PDU)

01 = LLC frame (i.e., contains an LLC PDU)

1X = undefined format (reserved for future use)

MAC FRAMES. If the frame-type bits indicate a MAC frame, all stations on the ring shall interpret and, based on the finite state of the system, act on the ZZZZZZ control bits.

LLC FRAMES. If the frame-type bits indicate an LLC frame, the ZZZZZZ bits are designated as rrrYYY. The rrr bits are reserved and are transmitted as 0s in all designated frames and ignored upon reception. The YYY bits may be used to carry the priority (Pm) of the PDU from the source LLC entity to the target LLC entity or entities. Note that P, the priority of the AC field of a frame, is less than or equal to Pm when the frame is transmitted onto the ring.

UNDEFINED FORMAT. The value of "1x" is reserved for frame types that may be defined in the future. However, although currently undefined, any future frame formats will adhere to the following conditions:

1. The format is delimited by the 2-octet SFS (start-of-frame sequence) field and the 2-octet EFS (end-of-frame sequence) field, as they are defined above. Additional fields may follow the EFS field.

2. The position of the FC (frame control) field is unchanged.

3. The SFS and EFS of the format are separated by an integral number of octets. This number is at least 1 (i.e., the FC field), and the maximum length is subject to the constraints of the token-holding timer (THT).

4. All symbols between the SFS and EFS are 0 and 1 bits.

5. All stations on the ring are to check for data symbols and an integral number of octets between the SFS and EFS fields. The error-detected (E) bit of formats that are repeated shall be set to 1 when a nondata symbol or nonintegral number of octets is detected between the SFS and EFS fields.

6. All bit errors having a Hamming distance* of less than four must be detectable by stations using the format and shall not be accepted by any other stations conforming to this standard.

Destination and Source Address Fields. Each frame contains two address fields: the destination (LAN station) address and the source (LAN station) address, in that order. Addresses may be either 2 or 6 octets in length. However, all stations of a specific LAN will have addresses of uniform length.

Destination address (DA) field identifies the LAN station(s) for which the information field of the frame is intended. Included in the DA is a bit to indicate whether the DA is an individual or group address; and, for 6-octet addresses only, the second bit indicates whether it is a universally or locally administered address.

---

*Hamming distance is described in Section 4.5.

INDIVIDUAL AND GROUP ADDRESSES. The first bit transmitted of the DA distinguishes individual from group addresses as follows:

$$0 = \text{individual address}$$

$$1 = \text{group address}$$

Individual addresses identify a particular station on the LAN and are distinct from all other individual station addresses on the same LAN (in the case of local administration), or from individual addresses of other LAN stations on a global basis (in the case of universal administration).

A group address is used to address a frame to multiple destination stations. Group addresses may be associated with zero or more stations on a given LAN. In general, a group address is an address associated by convention with a group of logically related stations.

BROADCAST ADDRESS. The group address consisting of 16 or 48 1s (for 2- or 6-octet addressing, respectively) constitutes a broadcast address, denoting the set of stations on a given LAN. Stations using 48-bit addressing must also recognize X'C000FFFFFFFF' as a broadcast address in MAC frames.

NULL ADDRESS. An address of 16 or 48 0s (for 2- or 6-octet addressing, respectively) is considered a null address. It means that a frame is not addressed to any particular station.

Source address (SA) field identifies the station originating the frame and has the same format and length as the DA in a given frame. The individual/group bit is 0.

Information (INFO) Field. The INFO field contains zero, one, or more octets that are intended for MAC, SMT, or LLC. Although there is no maximum length specified for the information field, the time required to transmit a frame may be no greater than the token-holding time that has been established for that LAN station. The format of the INFO field is indicated in the frame-type bits of the FC field. There are two frame types: MAC and LLC.

MAC FRAME FORMAT. Figure 10.10 shows the format of the information field, when present, for MAC frames.

The vector is the fundamental unit of MAC and SMT information. A vector contains its length, an identifier of its function, and zero or more subvectors. Only one vector is permitted per MAC frame.

LLC FRAME FORMAT. This is not specified in the IEEE 802.5 standard (Ref. 2). However, all LAN stations should be capable of receiving frames with information fields up to and including 133 octets in length.

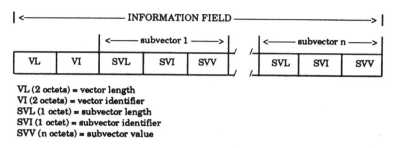

VL (2 octets) = vector length
VI (2 octets) = vector identifier
SVL (1 octet) = subvector length
SVI (1 octet) = subvector identifier
SVV (n octets) = subvector value

Figure 10.10. MAC frame information field structure.

FRAME CHECK SEQUENCE (FCS). The FCS is a 32-bit sequence based on the following standard generating polynomial of degree 32:

$$G(X) = X^{32} + x^{26} + X^{23} + X^{22} + X^{16} + X^{12} + X^{11} + X^{10} + X^8 + X^7$$
$$+ X^5 + X^4 + X^2 + X + 1$$

The FCS is the 1s complement of the sum (modulo 2) of the following: (a) the remainder of $X^k \times (X^{31} + X^{30} + X^{29} + \cdots + X^2 + X + 1)$ divided (modulo 2) by $G(X)$, where k is the number of bits in the FC, DA, SA, and INFO fields, and (b) the remainder after multiplication by X^{32} and then division (modulo 2) by $G(X)$ of the content (treated as a polynomial) of the FC, DA, SA, and INFO fields. The FCS is transmitted commencing with the coefficient of the highest term.

Ending Delimiter (ED)

$$J \ K \ 1 \ J \ K \ 1 \ I \ E$$

where

$$J = \text{nondata J}$$
$$K = \text{nondata K}$$
$$1 = \text{binary 1}$$
$$I = \text{intermediate frame bit}$$
$$E = \text{error-detected bit}$$

The transmitting station transmits the delimiter as shown. Receiving stations consider the ED valid if the first six symbols J K 1 J K 1 are received correctly.

INTERMEDIATE FRAME BIT (I BIT). If the I flag is used to determine the end of a station's transmission, the I bit is transmitted as a 1 in intermediate (or first) frames of a multiple-frame transmission. The I bit in the last or only frame of the transmission is sent as a 0.

ERROR-DETECTED BIT (E BIT). The E bit is transmitted as 0 by the station that originates the token, abort sequence or frame. All stations on the ring, as they repeat frames, check tokens and frames for errors. The E bit of tokens and frames that are repeated are set to 1 when a frame with error is detected; otherwise the E bit is repeated as received.

Frame Status (FS). The frame status (FS) field consists of one octet as follows:

$$A \ C \ r \ r \ A \ C \ r \ r$$

where

A = address-recognized bits

C = frame-copied bits

r = reserved bits

The reserved bits are reserved for future standardization. They are transmitted as 0s; however, their value is ignored by receivers.

ADDRESS-RECOGNIZED (A) BITS AND FRAME-COPIED (C) BITS. The A and C bits are transmitted as 0 by the LAN station originating the frame. If another station recognizes the DA as its own address or relevant group address, it sets the A bits to 1. If it copies the frame into its receiver buffer, it also sets the C bits to 1. This allows the originating station to differentiate among three conditions:

1. Station nonexisting or nonactive on the ring
2. Station exists but frame not copied
3. Frame copied

Notes. After transmission of frame(s) has been completed, the station checks to see if the station's address has returned in the SA field, as indicated by the MA_FLAG. If it has not been seen, the station transmits fill until the MA_FLAG is set, at which time the station transmits a token.

After transmission of the token, the station will remain in the transmit state until all the frames that the station originated are removed from the ring. This is called *stripping* and is done to avoid unnecessary recovery action that would be caused if a frame were allowed to continuously circulate on the ring.

10.1.3.3 Token-Ring Symbol Transmission Characteristics

PHY Serial Stream of Symbols. Each symbol is one of the following:

0 = binary zero
1 = binary one
J = nondata J
K = nondata K

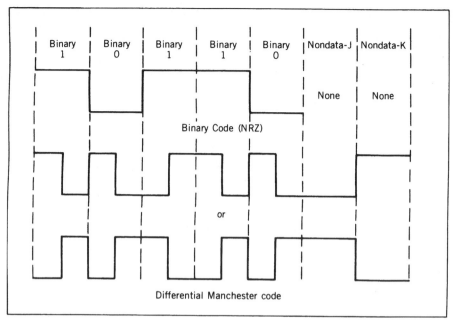

Figure 10.11. Symbol encoding using differential Manchester coding.

The symbols are transmitted to the medium in the form of differential Manchester encoding that is characterized by the transmission of two signal elements per symbol as shown in Figure 10.11.

In the case of two data symbols, binary 1 and binary 0, a signal element of one polarity is transmitted for one-half the duration of the symbol to be transmitted, followed by the contiguous transmission of a signal element of the opposite polarity for the remainder of the symbol duration. This provides two distinct advantages:

1. The resulting signal has no DC component and can readily be inductively or capacitively coupled.
2. The forced *mid-bit* transition conveys inherent timing information on the channel.

In the case of differential Manchester coding, the sequence of signal element polarities is completely dependent on the polarity of the trailing signal element of the previously transmitted data or nondata bit symbol. If the symbol to be transmitted is a binary zero, the polarity of the leading signal element of the sequence is opposite to that of the trailing element of the previous symbol, and, consequently, a transition occurs at the bit symbol boundary as well as at mid-bit. If the symbol to be transmitted is a binary one, the algorithm is reversed and the polarity of the leading signal element is the same as that of the trailing signal element of the previous bit. Here there is no transition at the bit symbol boundary.

The nondata symbols J and K depart from this rule in that a signal element of the same polarity is transmitted for both signal elements of the symbol and there is, therefore, no mid-bit transition. A J symbol has the same polarity as the trailing element of the preceding symbol. The transmission of only one nondata symbol introduces a DC component, and thus nondata symbols are normally transmitted as a pair of J or K symbols. (By its nature, a K symbol is opposite polarity of the preceding symbol.)

Data Signaling Rates. The data signaling rates should be within $\pm 0.01\%$ of 4 and 16 Mbps.

Symbol Timing. The PHY recovers the symbol timing information inherent in the transitions between levels of the received signal. It minimizes the phase jitter in this recovered timing signal to provide suitable timing at the data signaling rate for internal use and for the transmission of symbols on the ring. The rate at which symbols are transmitted is adjusted continuously in order to remain in phase with the received signal.

In normal operation there is one station on the ring that is the active monitor. All other stations on the ring are frequency- and phase-locked to this station. They extract timing from received data by means of a phase-locked loop design based on the requirement to accommodate a combined total of at least 250 stations and repeaters on the ring.

10.1.3.4 *Symbol Decoder.* Received symbols are decoded using an algorithm that is the inverse of that used for symbol encoding (Section 10.1.3.3). The station monitors the received data for the signal element pattern corresponding to a starting delimiter (SD) and ending delimiter (ED) as shown in Figure 10.12.

Starting delimiters (SD) and ending delimiters (ED) shall be detected even if they do not occur on a symbol boundary. If an ending delimiter is detected that does not occur on the symbol boundary established by the previous SD, then an ED_alignment_error is indicated to the MAC (the MAC uses this signal to verify that the ending delimiter falls exactly on an octet boundary). Once a delimiter has been detected, it shall establish the symbol and octet boundary for the decoding of the received signal elements until another delimiter is detected.

10.1.3.5 *Latency.* Latency is the time, expressed in number of symbols, it takes for a signal element to pass through a ring component. Two latency buffers are specified: a fixed buffer and an elastic buffer, which are inserted in the ring signal path as requested by the MAC protocol. To ensure that stations do not introduce excessive latency, the specification recommends

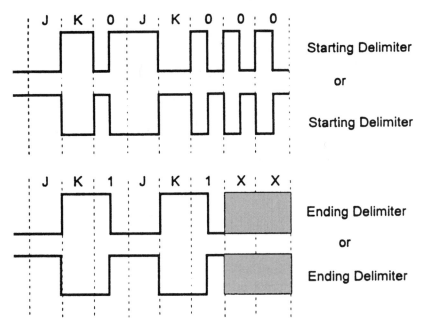

Figure 10.12. Starting and ending delimiter patterns. (From Figure 28, page 110, ANSI/IEEE Std. 802.5, Ref. 2.)

that the average PHY latency should not exceed 6.25 ms to prevent the expiration of the *timer, return to repeat* (*TRR*), which is set between 4.0 and 4.1 ms. It is further recommended that this time should not exceed 2.5 ms for performance/access reasons.

Ring Latency. Ring latency, defined as the time it takes for a signal element to travel around the entire ring, is equal to the cumulative latency of the ring plus the latency of the active monitor. Cumulative latency is the time it takes for a signal element to travel from the active monitor's transmitter output to its receiver input. The latency of the active monitor consists of two buffers, elastic buffer and fixed latency buffer, in addition to the normal station latency. The active monitor station uses the elastic buffer to compensate for variations in cumulative latency and uses the fixed latency buffer to provide the required latency to ensure ring operation even when only one station is on the ring.

When the ring is in a normal operating state, the MAC protocol ensures that only one station, the active monitor, will have Crystal_transmit asserted, which inserts the fixed latency and elastic latency buffers in the ring path. The active monitor transmits data to the ring timed from its local clock, which provides the master timing for the ring. Although the mean data signaling rate around the ring is established by the active monitor, segments of the ring can, for short periods of time, operate at a slightly higher or lower signaling rate than the signaling rate of the active monitor. The cumulative effect of these signaling rate variations may cause variations in the cumulative latency.

The elastic buffer in the active monitor compensates for the variation in the cumulative latency as follows. When the frequency of the received signal at the active monitor is slightly faster than the crystal clock, the elastic buffer will expand, as required, to maintain a constant ring latency. Conversely, if the frequency of the received signal is slightly slower, the elastic buffer will contract to maintain a constant ring latency. Constant ring latency is a requirement to avoid adding or dropping signal elements from the data stream. The elastic buffer compensates for dynamic variations in latency due to jitter. The elastic buffer does not compensate for some changes such as step changes in latency caused by stations becoming or ceasing to be stacking stations.

The fixed latency buffer is provided by the active monitor to provide the latency required for token circulation.

In order for the token to continuously circulate around the ring, the ring must have a minimum ring latency of at least 24 symbols (the number of symbols in a token). To ensure this minimum ring latency, a fixed latency buffer of at least 24 symbols is inserted into the data path of the station when MAC asserts Crystal_transmit.

Cumulative latency variations are handled by the elastic buffer. These variations are due to jitter. The elastic buffer accommodates without error a minimum cumulative latency variation of positive or negative B symbols where

$$\text{Data rate} \quad = \quad 4 \quad \quad 16 \quad \quad \text{Mbps}$$

$$B \quad = \quad 3 \quad \quad 15 \quad \quad \text{symbols}$$

The elastic buffer in the active monitor is initialized to the center of the elastic latency range whenever the MAC indicates Token_received (station is repeating token).

10.1.3.6 Station Organization and Basic PMC Requirements. Figure 10.13 shows station functional organization and data flow. The PMC (physical medium components) feeds the TCU (trunk coupling unit) concentrator. Two transmission parameters are specified for both active and passive channels. These are attenuation and differential impedance. Figure 10.14 gives an example of station-to-TCU connection.

The channel attenuation of a passive channel consists of the loss of all channel components between a ring station medium interface connector (MIC) and the MIC of the next downstream station. The channel attenuation of an active channel consists of the loss between a station MIC and its respective active concentrator port MIC, or the loss between an active RO

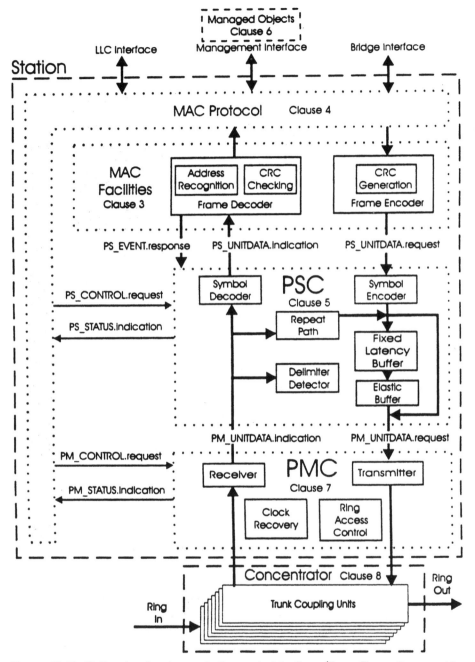

Figure 10.13. Station functional organization and data flow. (From Figure 2, page 16, ANSI/IEEE Std. 802.5, Ref. 2.)

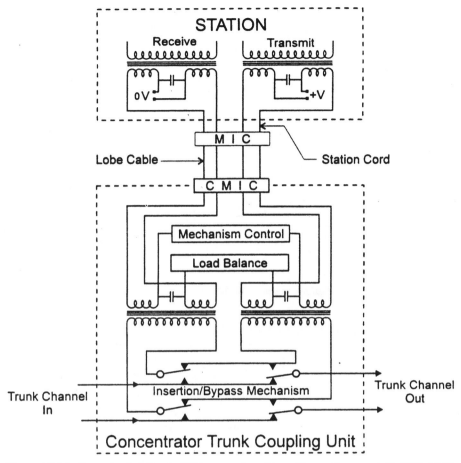

Figure 10.14. Example of station-to-TCU connection. (From Figure 29, page 145, ANSI/IEEE Std. 802.5, Ref. 2.)

(ring out) port MIC of one concentrator and the active RI (ring in) port MIC of a second concentrator. Table 10.1 gives the maximum attenuation values. These loss values are based on the channel terminated in 100 ohms for UTP cable and 150 ohms for STP cable.

The magnitude of the channel differential input and output impedance for both passive and active channels using STP and UTP media will meet the specifications given in Table 10.2.

Error Performance. The token-ring transmission system provides a communication channel that consists of a transmitter, receiver and channel with an equivalent BER of less than or equal to 1×10^{-9} with a ring equivalent BER of 1×10^{-8}.

Section 10.1.3 is based on ANSI/IEEE Standard 802.5, 2nd ed., December 29, 1995, Ref. 2.

TABLE 10.1   Channel Attenuation Specifications

Channel Type	Channel Attenuation	
	Only 4 Mbps	16 / 4 Mbps
Passive	< 19 dB @ 4 MHz	< 19 dB @ 16 MHz
Active	< 19 dB @ 4 MHz	< 16 dB @ 16 MHz

Source: Table 28, page 160, Ref. 2.

TABLE 10.2   Channel Differential Impedance Specifications

Media Type	Differential Impedance	Frequency Range	
		Only 4 Mbps	16/4 Mbps
UTP	100 ± 15 Ω	1–12 MHz	1–25 MHz[a]
STP	150 ± 15 Ω	1–12 MHz	1–25 MHz

[a]Category 3 is only specified to 16 MHz but it is assumed that it can be extrapolated to 25 MHz.
Source: Table 29, page 160, Ref. 2.

10.1.4   Fiber Distributed Data Interface (FDDI)

10.1.4.1   Introduction and Overview. Fiber distributed data interface (FDDI) is a LAN protocol that uses fiber-optic cable or other transmission medium. Its peak data rate is 100 Mbps and a sustained data rate of 80 Mbps. It uses a 4-out-of-5 code on the medium such that the line modulation rate is 125 Mbaud. FDDI can support 500 stations linked by up to 100 km of dual fiber cable. Larger networks can be supported by increasing timer values.

The FDDI operates in a token-ring format and is an outgrowth of IEEE 802.5 standard (Section 10.1.3). Table 10.3 compares FDDI with IEEE 802.5 standard.

FDDI consists of a set of nodes connected by a counterrotating ring topology, which is illustrated in Figure 10.15. There are two classes of nodes

TABLE 10.3   Token-Ring Comparison

FDDI	Token Ring
Half-duplex architecture and symbol (or byte)-level manipulation	Full-duplex architecture and bit-level manipulation
Token sent immediately behind packet	Token sent only after source address has returned
Traffic regulated through timed token	Traffic regulated through priority and reservation bits on each packet
Uses 4-out-of-5 group coding, up to 10% DC component	Uses differential Manchester coding, no DC component
Ring is decentralized; individual clocks limit packet size	Centralized control with active monitor clock, allowing very long packets
Fiber-optic medium (wire-pair option)	Wire-pair medium

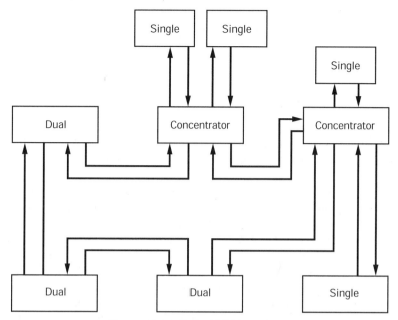

Figure 10.15. Typical FDDI topology.

or stations: single attachment and dual attachment. FDDI trunk rings are composed of dual attachment stations and have two physical-layer medium-dependent (PMD) entities. Single attachment stations are connected to the trunk ring via concentrators.

Two kinds of data service can be provided in a logical ring: packet service and circuit-connected service. The more common application today is the packet service, where a given station (or node) transmits information on to the ring as a series of data packets or frames, where each packet circulates from one station to the next. Each station copies the packet or frame header as the packet passes. The addressed station(s) pass(es) the packet internally to the SMT (station management) or LLC. Finally, the station that transmitted the packets effectively removes them from the ring. For circuit service, some of the logical bit rate capacity is allocated to independent channels. Two or more stations can simultaneously communicate via each channel.

FDDI provides packet service via a token ring in a similar manner as described in Section 10.1.3. With FDDI a station gains the right to transmit frames or packets on to the medium when it detects a token passing on the medium. A token is a unique sequence consisting of four fields, which is illustrated in Figure 10.16. The token follows each series of transmitted packets. Any station, upon detection of a token, may capture the token by removing it from the ring. That station may then transmit one or more data packets that it has in queue.

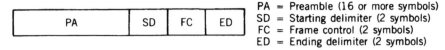

Figure 10.16. The FDDI token format.

The time a station may hold a token is controlled by the *token-holding timer*. At the expiration of its preset time, the station must release the token and return it to the ring. This permits other stations to gain access to the ring.

Multiple levels of priority are available for independent and dynamic assignment depending upon the relative class of service required. The classes of service may be synchronous, which is typically used for such applications as real-time packet voice. They may be asynchronous such as for interactive applications, or they may be immediate, which is used for extraordinary applications such as ring recovery. The allocation of ring bit rate capacity occurs by mutual agreement among users of the ring.

Error detection and recovery mechanisms are provided to restore ring operation in the event that transmission errors or medium transients (such as those resulting from station insertion or removal) cause the access method to deviate from normal operation. Detection and recovery for these cases utilize a recovery function that is distributed among the stations attached to the ring.

In the hybrid mode both token-ring operation and isochronous data transfer are multiplexed on the same medium. Hybrid operation requires the existence of a hybrid ring control (HRC) entity. HRC passes all packet data to the packet MAC entity. HRC does not pass isochronous data to the packet MAC.

10.1.4.2 *Ring Transmission.* The FDDI MAC transmits *symbols*. The symbol is the smallest signal element used by the MAC. A symbol provides three types of information:

1. Line states, such as the *halt line state* or the *idle line state*.
2. Data quartets, where each quartet represents four ordered data bits.
3. Control sequences, such as the *starting delimiter* (SD), *ending delimiter* (ED), or *control indicator* sequences.

Table 10.4 provides the FDDI symbol set and line coding. Each data quartet shown in the table conveys a hexadecimal digit (0–F). There are seven 5-bit invalid code groups, which are not transmitted by the MAC because they can

TABLE 10.4 FDDI Code Symbol Set

Code Group		Symbol	
Decimal	Binary	Name	Assignment

Line State Symbols

00	00000	Q	Quiet
04	00100	H	Halt
31	11111	I	Idle

Starting Delimiter

24	11000	J	First symbol of JK pair
17	10001	K	Second symbol of JK pair

Embedded Delimiter

05	00101	L	Second symbol of IL pair

Data Quartets

Decimal	Binary	Name	**Hexadecimal**	**Binary**
30	11110	0	0	0000
09	01001	1	1	0001
20	10100	2	2	0010
21	10101	3	3	0011
10	01010	4	4	0100
11	01011	5	5	0101
14	01110	6	6	0110
15	01111	7	7	0111
18	10010	8	8	1000
19	10011	9	9	1001
22	10110	A	A	1010
23	10111	B	B	1011
26	11010	C	C	1100
27	11011	D	D	1101
28	11100	E	E	1110
29	11101	F	F	1111

Ending Delimiter

13	01101	T	Terminate

Control Indicators

07	00111	R	Reset (logical Zero or Off)
25	11001	S	Set (logical One or On)

Note the 4B/5B code.
Source: Extracted from Ref. 3. X3.231-1994.

generate patterns that can violate run length or duty cycle requirements. The invalid code points are 01, 02, 03, 06, 08, 12, and 16.

In FDDI, information is signaled as a stream of fixed-length code groups, each containing 5 bits (see Table 10.4). Each valid code group represents a symbol. The first level code performed by PHY is the conversion of symbols from the data-link layer (MAC or HRC) to code groups. Each symbol is mapped to its corresponding code group and consists of a sequence of five NRZ code bits. The second level of coding performed by the PHY is the

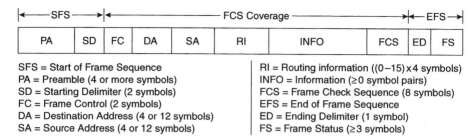

SFS = Start of Frame Sequence
PA = Preamble (4 or more symbols)
SD = Starting Delimiter (2 symbols)
FC = Frame Control (2 symbols)
DA = Destination Address (4 or 12 symbols)
SA = Source Address (4 or 12 symbols)

RI = Routing information ((0–15) x 4 symbols)
INFO = Information (≥0 symbol pairs)
FCS = Frame Check Sequence (8 symbols)
EFS = End of Frame Sequence
ED = Ending Delimiter (1 symbol)
FS = Frame Status (≥3 symbols)

Figure 10.17. FDDI frame format.

conversion of NRZ code bits to NRZI* code bits. Each NRZ code bit 1 generates a transition in polarity of the NRZI signal, while an NRZ code bit 0 maintains the previous NRZI signal polarity. For the incoming pulse stream, NRZI code bits are first decoded to equivalent NRZ code bits and then decoded to symbols for delivery to the data-link layer (DLL).

10.1.4.3 Protocol Data Units. Two protocol data unit (PDU) formats are used by the FDDI MAC: tokens and frames. Figure 10.16 illustrates the token and Figure 10.17 shows the FDDI frame. The token is the means by which a station gains the right to transmit MAC SDUs[†] (as opposed to the normal process of repeating). The token is passed from one MAC to another.

The frame format is used for transmitting MAC SDUs between peer MAC entities and for transmitting MAC recovery information.

Frame Length. In the basic mode (as opposed to the HRC), the physical layer (PHY) of FDDI requires limiting the maximum frame length to 9000 symbols, including four symbols of the preamble. The hybrid mode does not have a limit on frame length, but it is recommended that it also limit its frame length to 9000 symbols.

Frame Format. Refer to Figures 10.16 and 10.17. The preamble (PA) of a frame is transmitted by the frame originator as a minimum of 16 symbols of idle pattern (only 4 symbols in the hybrid mode). This sequence allows for station synchronization and frame alignment. PHYs of subsequent repeating stations may change the length of the idle pattern consistent with PHY clocking requirements. Thus, repeating stations may see a variable-length preamble that may be shorter or longer than the originally transmitted preamble. Stations in the basic mode are not required to be capable of copying frames with preambles of less than 12 symbols.

The starting delimiter (SD) consists of a J and K symbol (see Table 10.4). No frame or token is considered valid unless it starts with this explicit

---

*NRZI stands for non-return-to-zero, invert on ones.
[†]SDU stands for service data unit.

sequence. After the SD there is a frame control (FC) field which defines the type of frame and associated control functions. It contains a class ⓒ bit, an address length (L) bit, format bits (FF), and four control bits (ZZZZ). For example, the C bit tells us whether the operation is asynchronous or synchronous; the address length (L) bit tells us whether the address is 16 or 48 bits in length; frame format bits indicate whether a token is restricted or nonrestricted, whether the frame is a station management frame, a MAC frame, or an LLC frame, among others.

Destination and source addresses (DA and SA) follow the FC. These addresses may be 16 or 48 bits long. However, all stations must have a 16-bit address capability as a minimum. A station with only a 16-bit address capacity must be capable of functioning in a ring with stations concurrently operating with 48-bit addresses.

The routing information (RI) field is included in the frame when there is a 48-bit address field. In this case the RII bit is set to 1. The RI field contains 2 to 30 octets (symbol pairs) whose format and meaning are specified in the ISO/IEC 10038 standard on bridging.

The information (INFO) field contains zero, one, or more data symbol pairs whose meaning is determined by the FC field and whose interpretation is made by the destination entity (e.g., MAC, LLC, or SMT*). The length of the INFO field is variable. It does have a maximum length restriction based on a maximum frame length of 9000 symbols, including the four symbols of the preamble.

The frame check sequence (FCS) is used to detect bits in error within the frame as well as erroneous addition or deletion of bits to the frame. The fields covered by the FCS include FC, DA, SA, INFO, and FCS fields. The FCS generating polynomial is the ANSI 32-bit standard algorithm given in Section 10.1.3.2.

The ending delimiter (ED) is the symbol T, which ends tokens and frames. Ending delimiters and optional control indicators form a balanced symbol sequence. These are transmitted in pairs so as to maintain octet boundaries. This is accomplished by adding a trailing symbol T as required. The ending delimiter of a token consists of two consecutive T symbols; the ending delimiter of a frame has a single T symbol.

The frame status (FS) field consists of an arbitrary length sequence of control indicator symbols (R and S). The FS field follows the ending delimiter of a frame. It ends if any symbol other than R and S is received. A trailing T symbol, if present, is repeated as part of the FS field. The first three control indicators of the FS field are mandatory. These indicators indicate error detected (E), address recognized (A), and frame copied (C). The use of additional trailing control indicators in the FS field is optional and may be defined by the user.

---

*SMT stands for station management.

10.1.4.4 FDDI Timers and Timing Budget. There are three timers used to regulate the operation of the FDDI ring: token-holding timer (THT), valid transmission timer (TVX), and the token rotation timer (TRT). The values of these timers are locally administered. They may vary from station to station on the ring, provided that the applicable ring limits are not violated.

Definitions. Token time is the time required to transmit a token (6 symbols) and a nominal length preamble (16 symbols). Token time (token length) in basic mode: 0.00088 ms; in hybrid mode, 0.00089 ms to 0.11458 ms.

D_max is the maximum latency (circulation delay) for a starting delimiter to travel around the logical ring. It includes node delay and propagation delay.

D_max = 1.7745 ms (default) = maximum ring latency with hybrid mode disabled.

D_max = 2.8325 ms (default) = maximum ring latency with hybrid mode enabled.

10.1.4.5 FDDI Ring Operation. Access to the ring is controlled by passing the token around the ring as in standard token-ring practice. The token gives the downstream station (the receiving station relative to the station passing the token) the opportunity to transmit a frame or sequence of frames. If a station has frame(s) in queue for transmission, it strips the token from the ring before the frame control field of the token is repeated. After the captured token is completely received, the station begins transmitting the eligible queued frames. After completion of transmission, the station issues a new token for use by a downstream station.

Stations that have nothing to transmit at a particular time merely repeat the incoming symbol stream. While in the process of repeating the incoming symbol stream, the station determines whether the information is intended for that station. This is done by matching the downstream address to its own address or a relevant group address. If a match occurs, subsequent received symbols up to the FCS are processed by the MAC or sent to the LLC.

Frame stripping is an important concept. Each transmitting station is responsible for stripping frames from the ring that it originated. This is accomplished by stripping the remainder of each frame whose source address matches the station's own address from the ring and replacing it with idle symbols (symbol I Table 10.4).

It will be noted that the process of stripping leaves remnants of frames consisting of the PA, SD, FC, DA, and SA fields, followed by idle symbols. This happens because the decision to strip a frame is based on the recognition of the station's own address in the SA field. This cannot occur until after the initial part of the frame has already been repeated and passed on to the next downstream station. These remnants cause no ill effects. This is because of the various specified criteria, including recognition of an ending delimiter

(ED), which must be met to indicate that a frame is valid. To the level of accuracy required by statistical purposes, the remnants cannot be distinguished from errored or lost frames because they are always followed by an idle (I) symbol pattern. Such remnants are removed from the ring when they encounter the first transmitting station.

The FDDI specification distinguishes *asynchronous* from *synchronous* transmission. In this context, asynchronous transmission is a class of data transmission service whereby all requests for service contend for a pool of dynamically allocated ring bandwidth and response time. Synchronous transmission is a class of data service whereby each requester is preallocated a maximum bandwidth and guaranteed a response time not to exceed a specific delay.

Clocking. A local clock is used to synchronize both the internal operation of the PHY and its interface to the data-link layer (DLL). The clock is derived from a fixed frequency reference. This reference may be created internally within the PHY implementation or supplied to the PHY. Often a crystal oscillator is used for this purpose.

Characteristics of the local clock are as follows:

(a) Nominal symbol time (UI) = 40 ns (1/UI = 25 MHz)

(b) Nominal code bit cell time (UI) = 8 ns (1/UI = 125 MHz)

(c) Frequency accuracy $< \pm 0.005\%$ (± 50 ppm)

(d) Harmonic content (above 125.02 MHz) < -20 dB

(e) Phase jitter (above 20 kHz) $< \pm 8°$ (0.044 UI pp)

(f) Phase jitter (below 20 kHz in hybrid mode) $< \pm 270°$ (1.5 UI pp)

The receive function derives a clock by recovering the timing information from the incoming serial bit stream. This clock is locked in frequency and phase to the transmit clock of the upstream node. The maximum difference between the received bit frequency and the local bit frequency is 0.01% of the nominal frequency. The received frequency can be either slower or faster than the local frequency, resulting in an excess or a deficiency of bits unless some compensation is included. The elasticity buffer function provides this compensation by adding or dropping Idle bits in the preamble between DLL PDUs.

The operation of the elasticity buffer function produces variations in the lengths of the preambles between DLL PDUs as they circulate around the logical ring. The cumulative effect on preamble size of PDU propagation through many elasticity buffers can result in excessive preamble erosion and, in hybrid mode, excessive cycle clock jitter. The smoothing function serves to filter out these undesirable effects.

Section 10.1.4 consists of abstracts of Refs. 3, 4, and 5.

10.1.5 Wireless LANS (WLANs)

10.1.5.1 Introduction. A wireless LAN (WLAN) offers a flexible alternative to the wired LAN. In many cases a WLAN is an extension to a wired LAN, augmenting the resources of an enterprise network. Using radio frequency (RF) technology, wireless LANs transmit and receive data by means of radiated RF or light signals, minimizing or negating the need for wired connections. WLANs are particularly attractive in the factory and office environment where repeated rearrangement of communication facilities can be expected.

The following are some of the advantages of wireless LANs:

- Ideal when many rearrangements are expected (moves and changes).
- *Mobility.* Allows a LAN user to set up anywhere within range of radio/wireless link.
- *Installation speed.* Rapid installation of system. No requirement to stretch wires.
- *Economic.* Less expensive than its wire counterpart.
- *Scalability.* Changes in topology; added users.

A wireless LAN consists of interconnected access points which are tied together by means of either frequency-hop or direct sequence spread spectrum radio on 2.5 GHz ISM band, or by means of infrared radiated light connectivity. The bit rate of these systems is either 1 Mbps or 2 Mbps when based on the IEEE 802.11 standard.

10.1.5.2 Definitions

Access Point (AP). Any entity that has station functionality and provides access to the distribution services via the wireless medium (WM) for associated stations.

Association. The service used to establish AP/STA mapping and enable STA invocation of the distribution system services.

Basic Service Area (BSA). The conceptual area within which members of a basic service set may communicate.

Basic Service Set (BSS). A set of stations controlled by a single coordination function.

Coordination Function (CF). The logical function that determines when a station operating within a BSS is permitted to transmit and may be able to receive PDUs via the wireless medium. The CF within a BSS may have one PCF and shall have one DCF.

Distributed Coordination Function (DCF). A class of coordination function where the same coordination function logic is active in every station in the BSS when ever the network is in operation.

Distribution. The service which (by using the association information) delivers MSDUs within the DS.

Distribution System (DS). A system used to interconnect a set of basic service sets and integrated LANs to create an extended service set.

Distribution System Medium (DSM). The medium or set of media used by a distribution system for communications between access points and portals of an ESS.

Distribution System Services (DSS). The set of services provided by the distribution system which enable the MAC to transport MSDUs between stations that are not in direction communication with each other over a single instance of the WM. These services include transport of MSDUs between the APs of BSSs within an ESS, transport of MSDUs between portals and BSSs within an ESS, and the transport of MSDUs between stations in the same BSS in cases where the MSDU has a multicast or broadcast destination address or where the destination is an individual address, but the station sending the MSDU chooses to involve DSS. DS services are provided between pairs of 802.11 MACs.

Extended Service Area (ESA). The conceptual area within which members of an extended service set may communicate. An extended service area is larger or equal to a basic service area and may involve BSSs in overlapping, disjoint, or both configurations.

Extended Service Set (ESS). A set of one or more interconnected basic service sets and integrated LANs which appear as a single basic service set to the LLC layer at any station associated with one of those BSSs.

Infrastructure. The infrastructure includes the distribution system medium, access point, and portal entities, as well as being the logical location of distribution and integration service functions of an ESS. An infrastructure contains one or more access points and zero or more portals in addition to the distribution system.

MAC Management Protocol Data Unit (MMPDU). The unit of data exchanged between two peer MAC entities using the services of the PHY.

MAC Protocol Data Unit (MPDU). The unit of data exchanged between two peer MAC entities using the services of the PHY.

MAC Service Data Unit (MSDU). The MAC service data unit is information that is delivered as a unit between MAC service access points.

Net Allocation Vector (NAV). An indicator, maintained by each station, of time periods when transmission onto the WM shall not be initiated by the station whether or not the station's CCA function senses the WM as being busy.

Point Coordination Function_(PCF). A class of possible coordination functions where the coordination function logic is active in only one station in a BSS at any given time that the network is in operation.

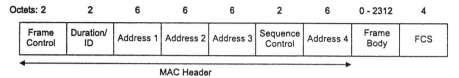

Figure 10.18. WLAN MAC frame format. (From Figure 12, page 34, ANSI/IEEE Std. 802.11, Ref. 8.)

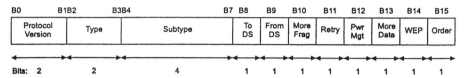

Figure 10.19. Frame control field. (From Figure 13, page 35, IEEE Std. 802.11-1997, Ref. 6.)

Portal. The logical point at which MSDUs from an integrated, non-802.11 LAN enter the distribution system of an ESS.

Wireless Medium (WM). The medium used to implement the transfer of PDUs between peer PHY entities of a wireless LAN.

10.1.5.3 Frame Formats. Figure 10.18 shows the general MAC frame format. In the figure the following fields are only present for certain frame types: Address 2, Address 3, Sequence Control, Address 4, and Frame Body. A selected group of fields is briefly described below.

Frame Control Field. The format of the frame control field is illustrated in Figure 10.19.

PROTOCOL VERSION. The Protocol Version subfield consists of 2 bits. For the present version of IEEE 802.11, its value is 0. All other values are reserved. The revision level will be incremented only when a fundamental incompatibility exists between a new revision and this revision of the standard.

TYPE AND SUBTYPE FIELDS. The Type field consists of 2 bits, and the Subtype field is 4 bits long. These two fields identify the function of the frame. There are three frame types: control, data, and management. Each of the frame types has several defined subtypes. Type value binary 00 indicates a management frame, type value binary 01 indicates the frame is a control frame, and binary value 10 for Type indicates that the frame is a data frame. For the Subtype field, an example is the binary value 0100 which indicates a probe request.

MORE FRAGMENTS. The More Fragments field is 1 bit in length and is set to binary 1 in all frames which have another fragment of the current MSDU or current MMPDU to follow. It is set to binary 0 in all other frames.

RETRY. The Retry subfield is 1 bit in length and is set to binary 1 in a Data or Management type frame that is a retransmission of an earlier frame. A receiving station uses this indication to aid in the process of eliminating duplicate frames.

WEP FIELD. The WEP (Wired Equivalent Privacy) field is 1 bit in length. It is set to binary 1 if the Frame Body field contains information that has been processed by the WEP algorithm. The WEP field is only set to 1 within frames of Type Data and frames of Type Management, Subtype Authentication. The WEP field is set to binary 0 in all other frames.

ORDER FIELD. The Order field is 1 bit in length and is set to 1 in any Data Type frame that contains an MSDU, or fragment thereof, which is being transferred using the Strictly Ordered service class. This field is set to 0 in all other frames.

ADDRESS FIELDS. There are four address fields in the MAC frame format. These fields are used to indicate the BSSID (Basic Service Set Identification) source address, destination address, transmitting station address and receiving station address. The usage of the four address fields in each frame type is indicated by the abbreviations BSSID, DA, SA, RA, and TA indicating BSS Identifier, Destination Address, Source Address, Receiver Address, and Transmitter Address, respectively. Certain frames may not contain some of the address fields.

Certain address field usage is specified by the relative position of the address field (1–4) within the MAC header, independent of the type of address present in that field. For example, receiver address matching is always performed on the contents of the Address 1 field in received frames, and the receiver address of CTS (clear to send) and ACK frames is always obtained from the Address 2 field in the RTS (request to send) frame, or from the frame being acknowledged.

Each Address field contains a 48-bit address.

Address Designation. A MAC sublayer address is one of the following two types:

(a) *Individual Address.* An address associated with a particular station on the network.

(b) *Group Address.* A multidestination address, associated with one or more stations on a given network.

The two kinds of group addresses are as follows:

1. *Multicast-Group Address*. An address associated by higher-level convention with a group of logically related stations.
2. *Broadcast Address*. A distinguished, predefined multicast address that always denotes the set of all stations on a given LAN. All 1s in the Destination Address field are interpreted to be the broadcast address. This group is predefined for each communication medium to consist of all stations actively connected to that medium; it is used to broadcast to all active stations on that medium. All stations are able to recognize the broadcast address. It is not necessary that a station be capable of generating the broadcast address.

The address space is also partitioned into locally administered and universal (globally administered) addresses. The nature of a body and the procedures by which it administers these universal (globally administered) addresses is beyond the scope of the reference standard.

BSSID Field. The BSSID field is a 48-bit field of the same format as an IEEE 802 MAC address. This field uniquely identifies each BSS (basic service set). The value of this field, in an infrastructure BSS, is the MAC currently in use by the STA (station) in the AP (access point) of the BSS.

The value of this field in an IBSS (independent basic service set) is a locally administered IEEE MAC address formed from a 46-bit random number generated according to the procedure defined in Section 11.1.3 of the reference document. The individual/group bit of the address is set to 0. The universal/local bit of the address is set to 1. This mechanism is used to provide a high probability of selecting a unique BSSID.

The value of all 1s is used to indicate the broadcast BSSID. A broadcast BSSID may only be used in the BSSID field of management frames of subtype probe request.

Destination Address (DA) Field. The DA field contains an IEEE MAC individual or group address that identifies the MAC entity or entities intended as the final recipient(s) of the MSDU (or fragment thereof) contained in the frame body.

Source Address (SA) Field. The SA field contains an IEEE MAC individual address that identifies the MAC entity from which the transfer of the MSDU (or fragment thereof) contained in the frame body field was initiated. The individual/group bit is always transmitted as a zero in the source address.

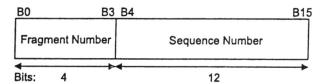

Figure 10.20. Sequence control field.

Receiver Address (RA) Field. The RA field contains an IEEE MAC individual or group address that identifies the intended immediate recipient [STS(s), on the WM (wireless medium)] for the information contained in the frame body field.

Transmitter Address (TA) Field. The TA field contains an IEEE MAC individual address that identifies the STA that has transmitted, onto the WM, the MPDU contained in the frame body field. The Individual/Group bit is always transmitted as a zero in the transmitter address.

SEQUENCE CONTROL AND SEQUENCE NUMBERS FIELDS. The Sequence Control field is 16 bits long and consists of two subfields: the Sequence Number and the Fragment Number. The format of the Sequence Control field is illustrated in Figure 10.20. The Sequence Number field is a 12-bit field indicating the sequence number of the MSDU or MMPDU. Each MSDU or MMPDU transmitted by an STA is assigned a sequence number. Sequence numbers are assigned from a single modulo 4096 counter, starting at 0 and incrementing by 1 for each MSDU or MMPDU. Each fragment of an MSDU or MMPDU contains the assigned sequence number. The sequence number remains constant in all retransmissions of an MSDU, MMPDU, or fragment thereof.

Fragment Number Field. The Fragment Number field is a 4-bit field indicating the number of each fragment of an MSDU or MMPDU. The fragment number is set to zero in the first or only fragment of an MSDU or MMPDU and is incremented by one for each successive fragment of the MSDU or MMPDU. The fragment number remains constant in all retransmissions of the fragment.

FRAME BODY FIELD. The frame body is a variable length field that contains information specific to individual frame types and subtypes. The minimum frame body is 0 octets. The maximum-length frame body is defined by the maximum length (MSDU + ICV + IV), where ICV and IV are the WEP fields defined in Section 8.2.5 of the reference publication.

THE FCS FIELD. The FCS field is a 32-bit field containing the standard 32-bit CRC defined in Section 10.1.3.2.

Octets: 2   2   6   6   4

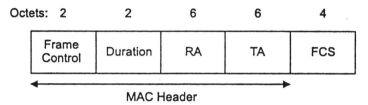

MAC Header

Figure 10.21. RTS frame. (From Figure 16, page 41, ANSI/IEEE Std. 802.11, Ref. 8.)

RTS (REQUEST TO SEND) AND CTS (CLEAR TO SEND) FRAME FORMATS. The format for the RTS frame is defined in Figure 10.21.

The RA (receiver address) of the RTS frame is the address of the station, on the wireless medium, that is the intended immediate recipient of the pending directed Data or Management frame. The TA is the address of the station transmitting the RTS frame.

The Duration value is the time, in microseconds, required to transmit the pending Data or Management frame, plus one CTS frame, plus one ACK frame, plus three SIFS (short interframe space) intervals. If the calculated duration includes a fractional microsecond, that value is rounded up to the next higher integer.

The format of the CTS frame is shown in Figure 10.22.

The RA of the CTS frame is copied from the Transmitter Address (TA) field of the immediately previous RTS frame to which the CTS is a response.

The Duration value is the value obtained from the Duration field of the Immediately previous RTS frame, minus the time, in microseconds, required to transmit the CTS frame and its SIFS interval. If the calculated duration includes a fractional microsecond, it is rounded up to the next higher integer.

ACK FRAME FORMAT. The frame format of the ACK frame is given in Figure 10.23.

The Receiver Address (RA) of the ACK frame is copied from the Address 2 field of the immediately previous direct Data, Management, or PS-Poll Control frame.

Octets: 2   2   6   4

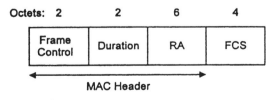

MAC Header

Figure 10.22. CTS frame.

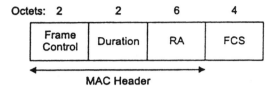

Figure 10.23. ACK frame.

If the More Fragment bit was set to 0 in the Frame Control field of the immediately previous directed data or management frame, the duration value is set to 0. If the More Fragment bit was set to 1 in the Frame Control field of the immediately previous directed data or management frame, the duration value is the value obtained from the Duration field of the immediately previous data or management frame, minus the time, in microseconds, required to transmit the ACK frame and its SIFS interval. If the calculated duration includes a fractional microsecond, that value is rounded up to the next higher integer.

10.1.5.4 Overview of the WEP. Eavesdropping is a familiar problem of other types of "wireless" technology. IEEE 802.11 specifies a wired LAN equivalent data confidentiality algorithm. *Wired equivalent privacy* (WEP) is defined as protecting authorized users of a wireless LAN from casual eavesdropping. This service is intended to provide functionality for the wireless LAN equivalent to that provided by the physical security attributes inherent to a wired medium. The use of the WEP algorithm is optional.

10.1.5.5 MAC Access

MAC Architecture. One way of illustrating the WLAN MAC architecture is illustrated in Figure 10.24. It is shown as providing the PCF (point coordination function) through the services of the DCF (distributed coordination function).

DISTRIBUTED COORDINATION FUNCTION (DCF). The principal underlying access method is a DCF known as *carrier sense multiple access with collision avoidance* (CSMA/CA). The DCF is implemented at all stations (STAs) for use without both the IBSS and infrastructure network configurations.

For a station to transmit, it uses the medium to determine if another station is transmitting. If the medium is not determined to be busy (see "Carrier Sense Mechanism,"), the transmission may proceed. The CSMA/CA distributed algorithm mandates that a gap of a minimum specified duration exist between contiguous frame sequences. A transmitting station ensures that the medium is idle for this required duration before attempting to transmit. If the medium is determined to be busy, the station defers until the

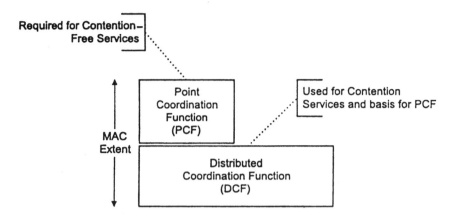

Figure 10.24. MAC architecture. (From Figure 47, page 70, ANSI/IEEE Std. 802.11, Ref. 8.)

end of the current transmission. After deferral, or prior to attempting to transmit again immediately after a successful transmission, the station selects a random backoff interval and decrements the backoff interval counter while the medium is idle. A refinement of the method may be used under various circumstances to further minimize collisions; here the transmitting and receiving stations exchange short control sequences [request to send (RTS) and clear to send (CTS) frames] after determining that the medium is idle and after deferrals or backoffs, prior to data transmission. "Deferrals" and "Backoffs" are described in the following text.

POINT COORDINATION FUNCTION (PCF). The IEEE 802.11 MAC may also incorporate an optional access method called a PCF, which is only usable on infrastructure network configurations. This access method uses a point coordinator (PC), which operates at the access point of the BSS to determine which station currently has the right to transmit. The operation is essentially that of polling, with the PC performing the roll of polling master. The operation of the PCF may require additional coordination, not specified in the reference standard, to permit efficient operation in cases where multiple point-coordinated BSSs are operating on the same channel in overlapping physical space.

The PCF uses a virtual carrier-sense mechanism aided by an access priority mechanism. The PCF distributes information within Beacon management frames to gain control of the medium by setting the network allocation vector (NAV) in stations. In addition, all frame transmission under the PCF may use an interframe space (IFS) that is smaller than the IFS for frames transmitted via the DCF. The use of a smaller IFS implies that point-coordinated traffic has priority access to the medium over stations in overlapping BSSs operating under the DCF access method.

The access priority provided by a PCF may be utilized to create a *contention-free* (CF) access method. The PC controls the frame transmissions of the stations so as to eliminate contention for a limited period of time.

COEXISTENCE OF DCF AND PCF. The DCF and PCF coexist in a manner that permits both to operate concurrently within the same BSS. When a PC is operating in a BSS, the two access methods alternate, with a contention-free period (CFP) followed by a contention period (CP).

FRAGMENTATION / DEFRAGMENTATION OVERVIEW. The process of partitioning a MAC service data unit (MSDU) or a MAC management protocol data unit (MMPDU) into smaller MAC level frames, MAC protocol data units (MPDUs), is called fragmentation. Fragmentation creates MPDUs smaller than the original MSDU or MMPDU length to increase reliability, by increasing the probability of successful transmission of the MSDU or MM-PDU in cases where channel characteristics limit reception reliability for longer frames. Fragmentation is accomplished at each immediate transmitter. The process of recombining MPDUs into a single MSDU or MMPDU is defined as defragmentation. Defragmentation is accomplished at each immediate recipient.

Only MPDUs with a unicast receiver address shall be fragmented. Broadcast/multicast frames shall not be fragmented even if their length exceeds aFragmentationThreshold.

When a directed MSDU is received from the LLC or a directed MMPDU is received from the MAC sublayer management entity (MLME) with a length greater than aFragmentationThreshold, the MSDU or MMPDU shall be fragmented. The MSDU or MMPDU is divided into MPDUs. Each fragment is a frame no larger than aFragmentationThreshold. It is possible that any fragment may be a frame smaller than aFragmentationThreshold. An illustration of fragmentation is shown in Figure 10.25.

The MPDUs resulting from the fragmentation of an MSDU or MMPDU are sent as independent transmissions, each of which is separately acknowledged. This permits transmission retries to occur per fragment, rather than per MSDU or MMPDU. Unless interrupted due to medium occupation limitations for a given PHY, the fragments of a single MSDU or MMPDU

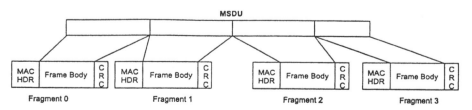

Figure 10.25. Fragmentation. (From Figure 48, page 71, IEEE Std. 802.11-1997, Ref. 6.)

are sent as a burst during the CP, using a single invocation of the DCF medium access procedure. The fragments of a single MSDU or MMPDU are sent during a CFP as individual frames obeying the rules of the PC medium access procedure.

MAC DATA SERVICE. The MAC data service translates MAC service requests from an LLC into input signals utilized by the MAC state machines. The MAC data services also translates output signals from the MAC state machines into service indications to an LLC. The translations are given in the MAC data service state machine defined in Annex C of the reference document.

DCF. The basic medium access protocol is a DCF that allows for automatic medium sharing between compatible PHYs through the use of CSMA/CA and a random backoff time following a busy medium condition. In addition, all directed traffic uses immediate positive acknowledgment (ACK frame) where retransmission is scheduled by the sender if no ACK is received.

The CSMA/CA protocol is designed to reduce the collision probability between multiple STAs accessing a medium, at the point where collisions would most likely occur. Just after the medium becomes idle following a busy medium (as indicated by the CS function) is when the highest probability of a collision exists. This is because multiple STAs could have been waiting for the medium to become available again. This is the situation that necessitates a random backoff procedure to resolve medium contention conflicts.

Carrier sense shall be performed both through physical and virtual mechanisms.

The virtual carrier-sense mechanism is achieved by distributing reservation information announcing the impending use of the medium. The exchange of RTS and CTS frames prior to the actual data frame is one means of distribution of this medium reservation information. The RTS and CTS frames contain a Duration/ID field that defines the period of time that the medium is to be reserved to transmit the actual data frame and the returning ACK frame. All STAs within the reception range of either the originating STA (which transmits the RTS) or the destination STA (which transmits the CTS) shall learn of the medium reservation. Thus an STA can be unable to receive from the originating STA, yet still know about the impending use of the medium to transmit a data frame.

Another means of distributing the medium reservation information is the Duration/ID field in directed frames. This field gives the time that the medium is reserved, either to the end of the immediately following ACK, or in the case of a fragment sequence, to the end of the ACK following the next fragment.

The RTS/CTS exchange also performs both a type of fast collision inference and a transmission path check. If the return CTS is not detected by the STA originating the RTS, the originating STA may repeat the process (after observing the other medium-use rules) more quickly than if the long data frame had been transmitted and a return ACK frame had not been detected.

Another advantage of the RTS/CTS mechanism occurs where multiple BSSs utilizing the same channel overlap. The medium reservation mechanism works across the BSA boundaries. The RTS/CTS mechanism may also improve operation in a typical situation where all STAs can receive from the AP, but cannot receive from all other STAs in the BSA.

The RTS/CTS mechanism cannot be used for MPDUs with broadcast and multicast immediate address because there are multiple destinations for the RTS, and thus potentially multiple concurrent senders of the CTS in response. The RTS/CTS mechanism need not be used for every data frame transmission. Because the additional RTS and CTS frames add overhead inefficiency, the mechanism is not always justified, especially for short data frames.

The use of the RTS/CTS mechanism is under control of the dot11RTSThreshold attribute. This attribute may be set on a per-STA basis. This mechanism allows STAs to be configured to use RTS/CTS either always, never, or only on frames longer than a specified length.

An STA configured not to initiate the RTS/CTS mechanism shall still update its virtual carrier-sense mechanism with the duration information contained in a received RTS or CTS frame, and shall always respond to an RTS addressed to it with a CTS.

The medium access protocol allows for STAs to support different sets of data rates. All STAs shall receive all the data rates in aBasicRateSet and transmit at one or more of the aBasicRateSet data rates. To support the proper operation of the RTS/CTS and the virtual carrier-sense mechanism, all STAs shall be able to detect the RTS and CTS frames. For this reason the RTS and CTS frames shall be transmitted at one of the aBasicRateSet rates. (See Section 9.6 of the reference document for a description of multirate operation.)

Data frames sent under the DCF shall use the frame type Data and subtype Data or Null Function. STAs receiving Data type frames shall only consider the frame body as the basis of a possible indication to LLC.

CARRIER SENSE MECHANISM. Physical and virtual carrier-sense functions are used to determine the state of the medium. When either function indicates a busy medium, the medium is considered busy; otherwise, it is considered idle.

A physical carrier-sense mechanism is provided by the PHY. See Section 12 of the reference document for how this information is conveyed to the MAC. The details of physical carrier sense are provided in the individual PHY specifications.

A virtual carrier-sense is provided by the MAC. The mechanism is referred to as the *network allocation vector* (NAV). The NAV maintains a prediction of future traffic on the medium based on duration information that is announced in the RTS/CTS frames prior to actual exchange of data.

The duration information is also available on the MAC headers of all frames sent during the CP other than PS-POLL Control frames.

The carrier-sense mechanism combines the NAV state and the station's transmitter status with the physical carrier sense to determine the busy/idle state of the medium. The NAV may be thought of as a counter which counts down to zero at a uniform rate. When the counter is zero, the virtual carrier-sense indication is that the medium is idle; when nonzero, the indication is busy. The medium is determined to be busy whenever the station is transmitting.

MAC-LEVEL ACKNOWLEDGMENTS. The reception of some frames requires the receiving station to respond with an acknowledgment, generally an ACK frame, if the FCS of the received frame is correct. This technique is known as positive acknowledgment.

Lack of reception of an expected ACK frame indicates to the source STA that an error has occurred. Note, however, that the destination station may have received the frame correctly and that the error may have occurred in the reception of the ACK frame. To the initiator of the frame exchange, this condition is indistinguishable from an error occurring in the initial frame.

INTERFRAME SPACE (IFS). The time interval between frames is called the IFS. A station determines that the medium is idle through the use of the carrier-sense function for the interval specified. Four different IFSs are defined to provide priority levels for access to the wireless media; they are listed in order, from the shortest to the longest. Figure 10.26 illustrates some of these relationships.

1. SIFS   short interframe space
2. PIFS   PCF interframe space
3. DIFS   DCF interframe space
4. EIFS   extended interframe space

The different IFSs shall be independent of the STA bit rate. The IFS timings shall be defined as time gaps on the medium, and they shall be fixed

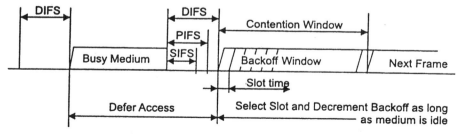

Immediate access when medium is free >= DIFS

Figure 10.26. Some IFS relationships. (From Figure 49, page 74, ANSI/IEEE Std. 802.11, Ref. 8.)

for each PHY (even in multirate-capable PHYs). The IFS values are determined from attributes specified by the PHY.

Short IFS (SIFS). The SIFS is used for an ACK frame, a CTS frame, and the second or subsequent MPDU of a fragment burst, as well as by a station responding to any polling by the PCM. It may also be used by a PC for any types of frames during the CFP. The SIFS is the time from the end of the last symbol of the previous frame to the beginning of the first symbol of the preamble of the subsequent frame as seen at the air interface.

The SIFS timing is achieved when the transmission of the subsequent frame is started at the TxSIFS Slot boundary as specified in Section 9.2.10 of the reference document. An IEEE 802.11 implementation does not allow the space between frames that are defined to be separated by a SIFS time, as measured on the medium, to vary from the nominal SIFS value by more than ±10% of aSlotTime for the PHY in use.

SIFS is the shortest of the interframe spaces. SIFS is used when stations have seized the medium and need to keep it for the duration of the frame exchange sequence to be performed. Using the smallest gap between transmissions within the frame exchange sequence prevents other stations (which are required to wait for the medium to be idle for a longer gap) from attempting to use the medium, thus giving priority to completion of the frame exchange sequence in progress.

PCF IFS (PIFS). The PIFS is used only by stations under the PCF to gain priority access to the medium at the start of the CFP. A station using the PCF is allowed to transmit contention-free traffic after its carrier-sense mechanism determines that the medium is idle at the TxPIFS slot boundary as defined in Section 9.2.10 of the reference document. Section 9.3 of this document describes the use of the PIFS by stations operating under the PCF.

DCF IFS (DIFS). The DIFS is used by stations operating under the DCF to transmit data frames (MPDUs) and management frames (MMPDUs). A station using the DCF is allowed to transmit if its carrier-sense mechanism determines the medium is idle at the TxDIFS slot boundary as defined in Section 9.2.10 of the reference document after a correctly received frame, and its backoff time has expired. A station using the DCF does not transmit within an EIFS after it determines that the medium is idle following reception of a frame for which the PHYRXEND.indication primitive contained an error or a frame for which the MAC FCS value was not correct. A station may transmit after subsequent reception of an error-free frame, resynchronizing the station. This allows the station to transmit using the DIFS following that frame.

Extended IFS (EIFS). The EIFS is used by the DCF whenever the PHY has indicated to the MAC that a frame transmission has begun that did not result in the correct reception of a complete MAC frame with a correct FCS value. The duration of an EIFS is defined in Section 9.2.10 of the reference document. The EIFS interval begins following indication by the PHY that the medium is idle after detection of the erroneous frame, without regard to the virtual carrier-sense mechanism. The EIFS is defined to provide enough time for another station to acknowledge what was, to this station, an incorrectly received frame before this station commences transmission. Reception of an error-free frame during the EIFS resynchronizes the station to the actual busy/idle state of the medium, so the EIFS is terminated and normal medium access (using DIFS and, if necessary, backoff) continues following reception of that frame.

RANDOM BACKOFF TIME. A station desiring to initiate transfer of data MPDUs and/or management MMPDUs shall invoke the carrier-sense mechanism to determine the busy/idle state of the medium. If the medium is busy, the station shall defer until the medium is determined to be idle without interruption for a period of time equal to DIFS when the last frame detected on the medium was received correctly, or after the medium is determined to be idle without interruption for a period of time equal to EIFS when the last frame detected on the medium was not received correctly. After this DIFS or EIFS medium idle time, the station shall then generate a random backoff period for an additional deferral time before transmitting, unless the backoff timer already contains a nonzero value, in which case the selection of a random number is not needed and not performed. This process minimizes collisions during contention between multiple stations that have been deferring to the same event.

$$\text{Backoff Time} = \text{Random( )} \times \text{aSlotTime}$$

where

Random( ) = Pseudorandom integer drawn from a uniform distribution over the interval $[0, \text{CW}]$, where CW is an integer within the range of values of the PHY characteristics aCWmin and aCWmax, $\text{aCWmin} \leq \text{CW} \leq \text{aCWmax}$. It is important that designers recognize the need for statistical independence among the random number streams among stations

aSlotTime = Value of the correspondingly named PHY characteristic.

The contention window (CW) parameter shall take an initial value of aCWmin. Every STA shall maintain an STA short retry count (SSRC) as well as an STA long retry count (SLRC), both of which shall take an initial value of zero. The SSRC shall be incremented whenever any short retry count associated with any MSDU is incremented. The SLRC shall be incremented whenever any long retry count associated with any MSDU is incremented. The CW shall take the next value in the series every time an unsuccessful attempt to transmit an MPDU causes either station retry counter to increment, until the CW reaches the value of aCWmax. A retry is defined as the

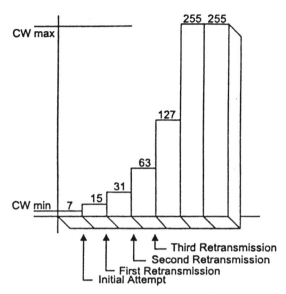

Figure 10.27. An example of exponential increase of CW. (From Figure 50, page 76, IEEE Std. 802.11-1997, Ref. 6.)

entire sequence of frames sent, separated by SIFS intervals, in an attempt to deliver an MPDU, as described in Section 9.7 of the reference document. Once it reaches aCWmax, the CW remains at the value of aCWmax until it is reset. This improves stability of the access protocol under high-load conditions. See Figure 10.27.

The CW shall be reset to aCWmin after every successful attempt to transmit an MSDU or MMPDU, when SLRC reaches aLongRetryLimit, or when SSRC reaches dot11ShortRetryLimit. The SSRC shall be reset to 0 whenever a CTS frame is received in response to an RTS frame, whenever an ACK frame is received in response to an MPDU or MMPDU transmission, or whenever a frame with a group address in the Address1 field is transmitted. The SLRC shall be reset to 0 whenever an ACK frame is received in response to transmission of an MPDU or MMPDU of length greater than dot11RTSThreshold, or whenever a frame with a group address in the Address1 field is transmitted.

The set of CW values shall be sequentially ascending integer powers of 2, minus 1, beginning with a PHY-specific aCWmin value, and continuing up to and including a PHY-specific aCWmax value.

DCF Basic Access Procedure. The CSMA/CA access method is the foundation of the DCF. Operational rules vary slightly between the DCF and the PC.

Basic access refers to the core mechanism a station uses to determine whether it may transmit. In general, a station may transmit a pending MPDU when it is operating under the DCF access method, either in the absence of a

PC or in the CP of the PCF access method, when the station determines that the medium is idle for greater than or equal to a DIFS period, or an EIFS period if the immediately preceding medium-busy event was caused by detection of a frame that was not received at this station with a correct MAC FCS value. If, under these conditions, the medium is determined by the carrier-sense mechanism to be busy when a station desires to initiate the initial frame of one of the frame exchanges described in Section 9.7 of the reference document, exclusive of the CF period, the random backoff algorithm described in Section 9.2.5.2 of the reference document is followed. There are conditions, specified in Sections 9.2.5.2 and 9.2.5.5 of the reference document, where the random backoff algorithm is followed even for the first attempt to initiate a frame exchange sequence.

In a station having a frequency hop (FH) PHY, control of the channel is lost at the dwell time boundary and the station has to contend for the channel after that dwell boundary. It is required that the stations having an FH PHY complete transmission of the entire MPDU and associated acknowledgment (if required) before the dwell time boundary. If, when transmitting or retransmitting an MPDU, there is not enough time remaining in the dwell to allow transmission of the MPDU plus acknowledgment (if required), the station defers the transmission by selecting a random backoff time, using the present CW (without advancing to the next value in the series). The short retry counter and long retry counter for the MSDU are not affected. The basic access mechanism is illustrated in Figure 10.28.

10.1.5.6 Overview of the Air Interface. The data rates presently offered by the 802.11 specification are 1.0 and 2.0 Mbps.

There are three different modulation plans: Two of these plans use a form of spread spectrum, and the third uses radiated light in the IR region. The spread spectrum types are frequency hop (FH) and direct sequence (DS). The hopping rate for FH systems is 2.5 or more hops/s.

Frequency Hopping Spread Spectrum Operation. Section 15.247 of the FCC Rules and Regulations (CFR 47, Ref. 7) states that the dwell time is not

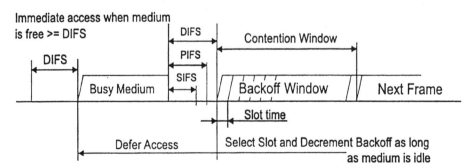

Figure 10.28. Basic access method. (From Figure 51, page 77, IEEE Std. 802.11-1997, Ref. 6.)

greater than 0.4 hops/s in any 30 hops/s period. This type of operation is permitted in two frequency bands: 2400 to 2483.5 MHz and 5725 to 5850 MHz. The IEEE 802.11 specification only allows operation in the former. The FCC has assigned 79 specific hop frequencies in the 2.4-GHz band beginning at 2402 MHz and extending to 2483 MHz. These frequencies are valid for North America and Europe, not including France and Spain.

Maximum transmitter output power (FCC Part 15.247) is 1.0 watt ($+30$ dBm). Transmit antennas may have gains as high as $+6$ dBi before there is a power penalty. For example, if a transmit antenna has a gain of $+10$ dBi, the transmit output power must be reduced 4 dB. Thus, the maximum EIRP is $+36$ dBm or $+6$ dBW.

If we were to allow a receive threshold of -80 dBm (Section 14.6.15.4 of IEEE 802.11, Ref. 8), what kind of line-of-sight range could we expect?

Based on *Radio System Design for Telecommunications*, (Ref. 9), the free space loss (FSL) equation is

$$FSL = 36.58 + 20 \log D + 20 \log F \text{ (in dB)}$$

where D is in miles and F is in MHz.

Let $F = 2483$ MHz, what is D?

$$FSL = 36.58 + 20 \log(2483) + 20 \log D$$
$$= 36.58 + 67.9 + 20 \log D$$

At 1 mile, FSL = 104.48 dB.

With $+36$-dBm EIRP, the isotropic receive level (IRL) would be

$$IRL = +36 \text{ dBm} - 104.48 \text{ dB}$$
$$= -68.48 \text{ dBm}$$

If we assume that the gain of the receive antenna is 0 dBi and that the transmission line to the receiver input is lossless, then there is a margin on the link of -80 dBM* $- (-68.48$ dBm) or 11.52 dB. This is at 1-mile distance. At half a mile, the margin will be 6 dB greater, or 17.52 dB; at one quarter-mile, there is another 6-dB improvement giving a 23.52-dB margin. Because of the complex propagation conditions inside a building, link margin is at a premium. Remember, every time we reduce to half of its value, the free space loss decreases 6 dB (Ref. 9).

The modulation employed for the radio systems is a form of frequency shift keying (FSK) called Gaussian FSK. For 1-Mbps systems it is binary GFSK, and for the 2-Mbps systems it is 4-ary GFSK with a theoretical channel occupancy of 2 bits/Hz of bandwidth. The channel occupancy for either 1 or 2 Mbps is 1 MHz.

One concern of a system designer is the maximum permissible input level to the receiver and its dynamic range. The maximum input signal is -20 dBm, and the dynamic range should extend to -8 dBm.

---

*Input threshold for a frame error ratio (FER) 8×10^{-2} at 2-Mbps operation.

Direct Sequence Spread Spectrum (DSSS) Operation. The DSSS system provides a wireless LAN with both 1-Mbps and 2-Mbps data payload communication capability. Based on US FCC Part 15.247, the DSSS system must provide a processing gain of at least 10 dB. This is accomplished by chipping the baseband signal at 11 MHz with an 11-bit PN code. The baseband modulation employed is differential binary phase shift keying (DBPSK) for 1-Mbps operation and differential quadrature phase shift keying (DQPSK) for the 2-Mbps data rate.

The North American and ETSI (European, less France and Spain) provide 11 channels starting with 2412 MHz and extending up to 2462 MHz with 5-MHz spacing. ETSI adds two more 5-MHz channels to the group, 2467 and 2472 MHz. Transmit power and antenna gain values are the same as for FHSS systems given previously.

Transmit RF power control is provided for transmitted power greater than 100 mW. A maximum of four power levels may be provided. At a minimum, a radio capable of transmission greater than 100 mW must be capable of switching power back to 100 mW or less.

At the receiver, the frame error ratio (FER) must be less than 8×10^{-2} at an MPDU length of 1024 bytes for an input level of -80 dBm measured at the antenna connector. The FER is specified for 2-Mbps DQPSK modulation. A similar FER must be achieved with an input level as high as -4 dBm measured at the antenna.

Infrared (IR) Transmission. The physical layer for the infrared system uses near-visible light in the 850- to 950-nm range. This is similar to the spectral usage of both common consumer devices such as infrared remote controls, as well as other data communications equipment, such as Infrared Data Association (IrDA) devices.

Unlike many other infrared devices, however, the IR PHY is not directed. That is, the receiver and transmitter do not have to be aimed at each other and do not need a clear line-of-sight. This permits the construction of a true LAN system, whereas with an aimed system, it would be difficult or impossible to install a LAN because of physical constraints.

A pair of conformant infrared devices would be able to communicate in a typical environment at a range up to about 10 m. This standard allows conformant devices to have more sensitive receivers, and this may increase range up to about 20 m.

The IR PHY relies on both reflected infrared energy as well as line-of-sight infrared energy for communications. Most designs anticipate that *all* of the energy at the receiver is reflected energy. This reliance on reflected infrared energy is called *diffuse infrared* transmission.

This standard specifies the transmitter and receiver in such a way that a conformant design will operate well in most environments where there is no line-of-sight path from the transmitter to the receiver. However, in an environment that has few or no reflecting surfaces and where there is no line-of-sight, an IR PHY system may suffer reduced range.

The IR PHY will operate only in indoor environments. Infrared radiation does not pass through walls, and it is significantly attenuated passing through most exterior windows. This characteristic can be used to "contain" an IR PHY in a single physical room, like a classroom or conference room. Different LANs using the IR PHY can operate in adjacent rooms separated only by a wall without interference and without the possibility of eaves-dropping.

At the time of this standard's preparation, the only known regulatory standards that apply to the use of infrared radiation are safety regulations, such as IEC 60825-1: 1998 [B2] and ANSI Z136.1-1993 [B]. While a conformant IR PHY device can be designed to also comply with these safety standards, conformance with this standard does not ensure conformance with other standards.

Worldwide, there are currently no frequency allocation or bandwidth allocation regulatory restrictions on infrared emissions.

10.2 REPEATERS, BRIDGES, ROUTERS, AND HUBS

10.2.1 Definitions and Interactions

In these types of applications, a repeater extends a LAN's range (see Figure 10.29). It simply regenerates the baseband signal being carried on the LAN. A repeater operates on the physical layer. A bridge may be used to segment a LAN. Users on a segment have a high community of interest. They do this based on the MAC destination address(es) (the work-station hardware address). Routers connect dissimilar LANs or provide LAN-to-WAN connectivity by operating at OSI layer 3, the network layer. Repeaters, bridges, and routers are compared in Figure 10.29.

10.3 LAN BRIDGES—OVERVIEW

There are a number of types of bridges. We'll start with the transparent bridge, the "learning" bridge, and then discuss the *spanning tree algorithm*.

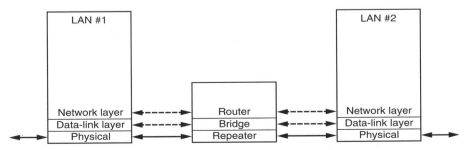

Figure 10.29. Conceptual drawing illustrating the functions of repeaters, bridges, and routers.

The *transparent bridge* simply relays frames/packets. Such a bridge has several ports. Frames/packets received on a particular port are stored and then forwarded on all other ports. At first blush, it would appear that this type of bridge is the same as a repeater. However, this is not the case. A LAN repeater simply repeats an incoming bit stream bit-by-bit. A transparent bridge, on the other hand, receives an entire frame and delivers the frame to an outgoing port(s). In the case of CSMA/CD, an outgoing port is just another station on the LAN to which it is connected. The port must wait until the LAN is free of transmissions before it sends the frame. Such a frame is sent unaltered. Thus the bridge is transparent.

If a number of LANs were connected to the bridge, traffic peaks could possibly occur. If the bridge has sufficient storage capacity, there would be no traffic peak problem. In this case the bridge would hold the frames in queue and the frames would be transmitted on each port (other than the incoming port) as their turn arises at each port of interest. If the LAN has insufficient storage capacity, some frames would have to be dropped and would then be lost.

Adding Intelligence to a Bridge. Suppose a processor is incorporated in the bridge design so that ports are associated with destination and source fields. Knowing the source field assures that a frame is not returned to the originator's LAN. Knowledge of destination fields and their association with ports cuts down on unnecessary traffic on LANs. What we mean here is that a particular frame will only travel to LAN(s) containing stations with the pertinent MAC destination addresses, whereas before, with the "dumb" bridges described above, frames went out every port but the port associated with that incoming frame. All frames received were relayed out all bridge ports but the port of origin of a particular frame.

Radia Perlman (Ref. 11) describes the strategy of such a "smart" bridge as follows:

1. The bridge receives every packet transmitted.
2. For each packet received, the bridge reads the source address and along with it the port on which the packet was received. This information is stored in a *cache,** which can be called the "station cache."
3. For each received packet, the station cache is reviewed by the bridge for a frame/packet destination address.
   (a) If no destination address is found in the cache, the frame/packet is forwarded to each of the bridge's ports less the port on which the frame/packet was received.

*The IEEE defines *cache* as follows: (Ref. 12) "A buffer inserted between one or more processors and the bus, used to hold currently active copies of blocks from main memory. (2) A small portion of high-speed memory used for temporary storage of frequently used data, instructions, or operands."

(b) If the address is located in the station cache, the bridge forwards the frame/packet only to the port indicated in the cache table. If the specified interface is the one from which the frame/packet was received, the packet is discarded (*filtered* in IEEE 802.1 terminology).

4. The bridge ages each entry in the station cache and deletes it after a period of time (a parameter known as *aging time*) in which no traffic is received with that address as the source address.

Drawbacks of LAN Bridges: Proliferation and Loops. When transparent bridges have not "learned" on which port a particular destination address resides, an incoming frame/packet is submitted to each port on the bridge except the port on which the frame was received. Consider the situation where there is more than one bridge in the network, and none of the bridges have learned a certain destination port for a particular frame/packet. The frame/packet tends to proliferate with useless capacity occupation on every LAN but the real destination LAN.

Loops can occur when a network has more than one path to a LAN station destination address exists. This is shown in Figure 10.30. Traffic in the loop can circulate around it indefinitely as specific traffic is routed from one bridge to the next on the loop so formed. This phenomenon can cause a network to crash or tie up the paths involved in such a way that they cannot handle real traffic. To avoid such loops and proliferation, a *spanning tree algorithm* is employed along with the use of *configuration messages*.

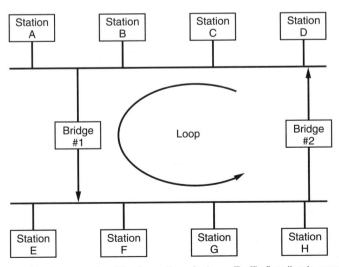

Figure 10.30. An example of the formation of a loop. Traffic flooding is assumed.

Spanning Tree Algorithm. The spanning tree algorithm allows bridges to dynamically discover a subset of the network topology that is loop-free and yet be able to provide sufficient connectivity so that every pair of LAN stations on the network can communicate. To carry out this process, bridges exchange configuration messages called *configuration bridge protocol data units* or configuration BPDUs.

The configuration message contains enough information such that a single bridge can be nominated a *root bridge*. Other bridges on the network can then calculate the shortest path from themselves to the root bridge.

On each LAN, a *designated bridge* is selected from the bridges residing on that LAN. The designated bridge is the one that is closest to the root bridge. The designated bridge forwards frame/packets on the LAN toward the root bridge. A port is then selected that gives the best path from themselves to the root bridge. This port is called the *root port*.

Ports are selected on a bridge to be included in the spanning tree. The ports selected include the root port plus any ports on which "self" has been selected as designated bridge.

Ports selected for inclusion in the spanning tree forward data traffic to/from their connected LANs. Data traffic is never forwarded to ports that are not selected for inclusion in the spanning tree. This traffic is discarded.

See Section 10.3.1.3 for specifics on the spanning tree algorithm.

10.3.1   IEEE 802.1 Bridges

10.3.1.1   *Application of the 802.1 Bridge.* IEEE 802 LANs of all types, including FDDI, may be connected together with MAC bridges. Each individual LAN has its own independent MAC. The bridged local area network created allows the interconnection of stations attached to separate LANs as if they were attached to a single LAN. A MAC bridge operates below the MAC service boundary, and is transparent to protocols operating above this boundary, in the logical link control (LLC) sublayer or the network layer. A bridged local area network may provide the following:

(a) The interconnection of stations attached to 802 LANs of different MAC types.
(b) An effective increase in the physical extent, the number of permissible attachments, or the total performance of a LAN.
(c) Partitioning of the physical LAN support for administrative or maintenance reasons.

This section describes the operation of the IEEE 802.1 MAC bridge and highlights a selected group of its operational parameters (Ref. 10).

10.3.1.2   *Principles of Operation.* The principal elements of 802.1 bridge operation are the relay and filtering of frames. It includes the maintenance of the information required to make frame filtering and relaying decisions. A typical 802.1 MAC bridge is shown in Figure 10.31.

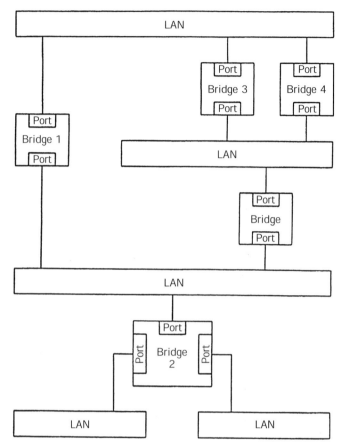

Figure 10.31. An IEEE Std. 802.1 bridge showing active topology.

A MAC bridge relays individual MAC user data frames between the separate MACs of the bridged local area networks connected to its ports. The order of frames of given user_priority received on one port and transmitted on another is preserved.

An IEEE 802.1 bridge carries out the following functions: It forwards received frames to other bridge ports. It discards frames in error and discards frames to ensure that a maximum bridge transit delay is not exceeded. It selects outbound access priority and discards frames following the application of filtering information. The bridge maps service data units and recalculates the frame check sequence of a frame.

The bridge carries out a filtering function. It eliminates the possibility of "end-less routing loops." This is done by eliminating frame duplication. Thus frames transmitted between a pair of end stations can be confined to LANs that form a path between those end stations.

The bridge calculates and configures a bridged local area network topology. It automatically learns dynamic filtering information through observation of bridged LAN traffic. Filtering information that has been automatically learned is deleted. It explicitly configures static filtering information and permanently configures reserved addresses. The bridge "ages out" filtering information that has been automatically learned.

The bridge protocol entity and other higher layer protocol users, such as bridge management, make use of the logical link control procedures. These procedures are provided separately for each port, and they use the MAC service provided by the individual MAC entities.

Model of Operation. Frames are accepted for transmission and delivered on reception to and from processes and entities that model the operation of the MAC relay entity in a bridge. These are as follows:

1. The forwarding process, which forwards received frames that are to be relayed to other bridge ports, filtering frames on the basis of information contained in the filtering database and on the state of the bridge ports.
2. The learning process, which by observing the source addresses of frames received on each port, updates the filtering database conditionally on the state of the port on which frames are observed.
3. The filtering database, which holds filtering information either explicitly configured by management action or automatically entered by the learning process, and which supports queries by the forwarding process as to whether frames with given values of the destination MAC address field should be forwarded to a given port.

Each bridge port also functions as an end station providing the MAC service to LLC that supports:

1. The bridge protocol entity, which operates a MAC sublayer configuration protocol between bridges, which determines, in part, the state of each bridge port and its participation in the active topology of the bridged local area network.
2. Other users of LLC, such as protocols providing bridge management.

Figure 10.32 illustrates the relaying of a single frame between ports of a bridge with two ports. Figure 10.33 shows the inclusion of information carried by a single frame, received on one of the ports of a bridge with two ports, in the filtering database. The bridged protocol entity operation is illustrated in Figure 10.34. This entity operates the spanning tree algorithm and protocol. It includes its modification of port state information as part of determining the active topology of the bridged local area network.

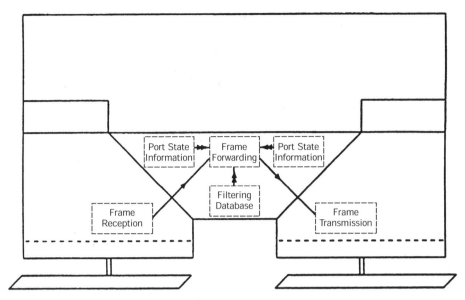

Figure 10.32. Relaying MAC frames. (From Figure 3-4, page 33, ANSI/IEEE Std. 802.1D, Ref. 10.)

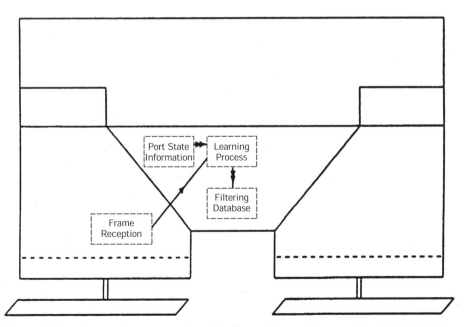

Figure 10.33. Observation of network traffic. (From Figure 3-5, page 33, ANSI/IEEE Std. 802.1D, Ref. 10.)

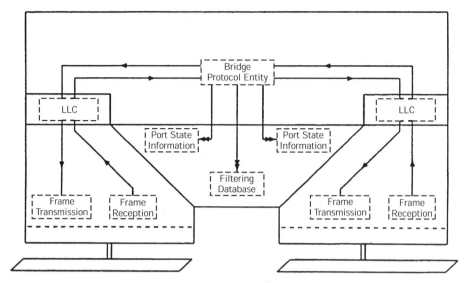

Figure 10.34. Operation of the interbridge protocol. (From Figure 3-6, page 34, ANSL/IEEE Std. 802.1D, Ref. 10.)

Learning Process. The learning process observes the source addresses of frames received on each port and update the filtering database conditionally on the state of the receiving port. Frames are submitted to the learning process by the individual MAC entities associated with each bridge port.

The learning process may deduce the path through the bridged local area network to particular stations by inspection of the source address field of received frames. It creates or updates a dynamic entry in the filtering database, associating the port on which the frame was received with the MAC address in the source address field of the frame, if and only if:

1. The port on which the frame was received is in a state that allows learning.
2. The source address field of the frame denotes a specific end station (i.e., is not a group address).
3. A static entry for the associated MAC address does not already exist.
4. The resulting number of entries would not exceed the capacity of the filtering database.

Figure 10.32 illustrates the operation of the learning process.

It should be noted that *static entries* are added or removed from the filtering database under explicit management control. They are not automatically removed by any timeout mechanism.

Static entries specify the following:

1. The MAC address for which filtering is specified.
2. For each inbound port on which frames are received, a port map that specifies for each outbound port on which frames may be transmitted, whether frames will be filtered or forwarded to that port.

These MAC addresses that can be specified include group and broadcast addresses.

Dynamic addresses, on the other hand, are created and updated by the learning process. They are automatically removed after a specified time (timeout) has elapsed since the entry was created or last updated. This timing out of entries ensures that end stations that have been moved to a different part of the bridged local area network will not be permanently prevented from receiving frames. It also takes into account the changes on the active topology of the bridged local area network which can cause end stations to appear to move from the point of view of a bridge (i.e., the path to those end stations subsequently lies through a different bridge port).

The timeout value, also called *aging time*, after which a dynamic entry is automatically removed, may be set by management. The recommended default value is 300 s, and the range is 10.0 to 1.0×10^6 s.

The spanning tree algorithm (discussed in the next subsection) includes a procedure for notifying all bridges in the bridged local area network of topology changes and specifies a short value for the timeout value which is enforced for a period after any topology change.

Dynamic entries specify the following:

1. The MAC address for which filtering is specified
2. A port number

Frames with the specified destination MAC address are forwarded only to the specified port. A dynamic entry acts like a static entry with a single port selected in the port map.

It should be pointed out that the filtering database contains a permanent database that provides fixed storage for static entries. The filtering database is initialized with the static entries contained in this fixed data store. Static entries may be added to and removed from the permanent database under explicit management control.

Addressing. All MAC entities communicating across a bridged local area network use 48-bit addresses, which may be universally administered or locally administered addresses or both. Bridges may also use 16-bit locally administered addresses.

UNIQUE IDENTIFICATION OF A BRIDGE. It is necessary for the operation of the Bridge Protocol that a single unique identifier be associated with each bridge. This identifier is derived from a unique address for the bridge, known as the "bridge address."

It is recommended that this address be the specific MAC address of the lowest numbered bridge port (Port 1).

10.3.1.3 The Spanning Tree Algorithm

Key Requirements

(a) The algorithm configures the active topology of a bridged local area network or arbitrary topology into a single spanning tree,* such that there is at most one data route between any two end stations, eliminating data loops.

(b) The entire active topology will stabilize in any sized bridged local area network. It will, with a high probability, stabilize within a short, known bounded interval in order to minimize the time where service is unavailable for communication between any pair of end stations.

(c) The active topology is predictable and reproducible, and may be selected by management of the parameters of the algorithm, thus allowing the application of configuration management, following traffic analysis to meet performance objectives.

(d) It will operate transparently to the end stations such that they are unaware of their attachment to a single LAN or a bridged local area network regardless of the total number of bridges or LANs.

Overview

THE ACTIVE TOPOLOGY AND ITS COMPUTATION. The spanning tree algorithm and protocol configure a simply connected active topology from the arbitrarily connected components of a bridged local area network. Frames are forwarded through some of the bridge ports in the bridged local area network and not through others, which are held in a blocking state. At any time, bridges effectively connect just the LANs to which ports in a forwarding state are attached. Frames are forwarded in both directions through bridge ports that are in a forwarding state. Ports that are in a blocking state do not forward frames in either direction but may be included in the active topology (i.e., be put into a forwarding state if components fail, are removed, or are added). Figure 10.31 shows the active topology (i.e., the logical connectivity), of the same bridged local area network following configuration.

*The IEEE (Ref. 12) defines *Spanning tree* as: "A bridging technique where a network of randomly interconnected bridges can automatically build a logical tree structure so as to guarantee a unique path between any pair of stations on the network. In this scheme, the transmitter does not have to know how to route the frame to the destination; that is the job of the bridges."

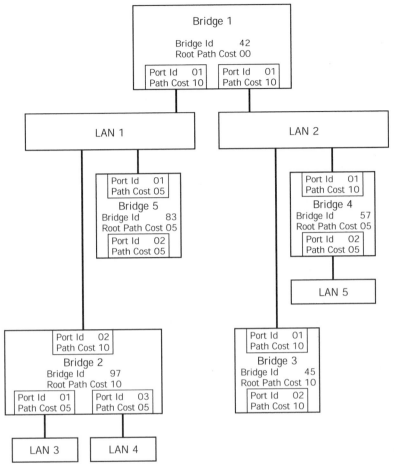

Figure 10.35. Spanning tree (example). (From Figure 4-2, page 46, ANSI/IEEE Std. 802.1D, Ref. 10.)

One of the bridges of such a configured network is known as the "root" or the root bridge. Each individual LAN has a bridge port connected to it that forwards frames from the direction of the root onto that LAN. This port is known as the *designated port* for that LAN, and the bridge of which it is part is the *designated bridge* for the LAN. The root is the designated bridge for all the LANs to which it is connected. The ports on each bridge that are in the forwarding states are the root port (the closest to the root) and the designated ports (if there ary any).

In Figure 10.35, Bridge 1 has been selected as the root (although one cannot tell simply by looking at the topology which bridge is the root) and is the designated bridge for LAN 1 and LAN 2, Bridge 2 is the designated bridge for LAN 3 and LAN 4, and Bridge 4 is the designated bridge for LAN 5. Figure 10.35 shows the logical tree topology of this configuration of the bridged local area network.

The stable active topology of a bridged local area network is determined by the following:

1. The unique bridge identifiers with each bridge port
2. The path cost* associated with each bridge port
3. The port identifier associated with each bridge port

The bridge with the highest priority bridge identifier is the root (for convenience of calculation, this is the identifier with the lowest numerical value). Each bridge port in the bridge local area network has a root path cost associated with it. This is the sum of the path costs for each bridge port receiving frames forwarded from the root on the least-cost path to the bridge. The designated port for each LAN is the bridge port for which the value of the root path cost is lowest: If two or more ports have the same value of root path cost, then the first bridge identifier of their bridges and then their port identifiers are used as tie-breakers. Thus, a single bridge port is selected as the designated port for each LAN, the same computation selects the root port of a bridge amongst the bridge's own ports, and the active topology of the bridged local area network is completely determined.

A component of the bridge identifier of each bridge, and the path cost and port identifier of each bridge port, can be managed, thus allowing a manager to select the active topology of the bridged local area network.

PROPAGATING TOPOLOGY INFORMATION. Bridges send a type of bridge protocol data unit known as a configuration BPDU to each other to communicate and compute the above information. A MAC frame conveying a BPDU carries the bridge group address in the destination address field and is received by all bridges connected to the LAN on which the frame is transmitted.

BPDUs are not directly forwarded by bridges, but the information in them may be used by a bridge in calculating its own BPDU to transmit and may stimulate the transmission. The configuration BPDU, which is conveyed between bridge ports attached to a single LAN, is distinguished from the notion of a configuration message, which expresses the propagation of the information carried throughout the bridged local area network.

Each configuration BPDU contains, among other parameters, the unique identifier of the bridge that the transmitting bridge believes to be the root, the cost of the path to the root from the transmitting port, the identifier of the transmitting bridge, and the identifier of the transmitting port. This information is sufficient to allow a receiving bridge to (a) determine whether the transmitting port has a better claim to be the designed port on the LAN on which the configuration BPDU was received than the port currently

---

*We call *cost* a routing parameter that determines shortest path route.

believed to be the designated port and (b) determine whether the receiving port should become the root port for the bridge if it is not already.

Timely propagation throughout the bridged local area network of the necessary information to allow all bridge ports to determine their state (i.e., blocking or forwarding) is achieved through three basic mechanisms:

1. A bridge that believes itself to be the root (all bridges start believing themselves to be the root until they discover otherwise) will originate configuration messages (by transmitting configuration BPDUs) on all LANs to which it is attached, at regular intervals.
2. A bridge that receives a configuration BPDU on what it decides is its root port conveying better information (i.e., highest priority root identifier, lowest root path cost, highest priority transmitting bridge and port) will pass that information on to all the LANs for which it believes itself to be the designated port.
3. A bridge that receives inferior information on a port it considers to be the designated port on the LAN to which it is attached will transmit its own information in reply for all other bridges to the LAN to hear.

Hence, spanning tree paths to the bridge with highest priority root identifier are quickly learned throughout the bridged local area network, with inferior information about other potential roots and paths being contradicted.

Figure 10.36 shows the various port states of a bridge and the transitions between the port states.

SELECTED KEY BRIDGE PARAMETERS

Root Path Cost. The cost of the path to the root from this bridge. It is equal to the sum of the values of the designated cost and path cost parameters held for the root port. When the bridge is the root, this parameter has the value zero. The parameter is used to test the value of the root path cost parameter conveyed in received configuration BPDUs; it is also used as the value of the root path cost parameter offered in all configuration BPDUs transmitted by the bridge.

Root Port. The port identifier of the port that offers the lowest cost path to the root (i.e., that port for which the sum of the values of the designated cost and path cost parameters held for the port is lowest).

Max Age. The maximum age of received protocol information before it is discarded.

Hello Time. The time interval between the transmission of configuration BPDUs by a bridge that is attempting to become the root or is the root.

Bridge Hello Time. The value of the hello time parameter when the bridge is the root or is attempting to become the root. The time interval between

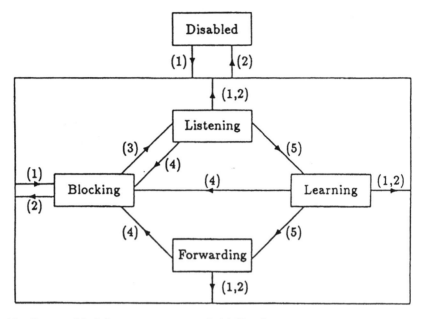

(1) Port enabled, by management or initialization
(2) Port disabled, by management or failure
(3) Algorithm selects as Designated or Root Port
(4) Algorithm selects as not Designated or Root Port
(5) Protocol timer expiry (Forwarding Timer)

Figure 10.36. Port states. (From Figure 4-3, page 48, ANSI/IEEE Std. 802.1D, Ref. 10.)

transmissions of topology change notification BPDUs toward the root when the bridge is attempting to notify the designated bridge on the LAN to which its root port is attached of a topology change. This parameter may be updated by management action.

Port Parameter—Path Cost. The contribution of the path through this port, when the port is the root port, to the total cost of the path to the root for this bridge. The parameter is used, added to the value of the designated cost parameter for the root port, as the value of the root path cost parameter offered in all configuration BPDUs transmitted by the bridge, when it is not the root. This parameter may be updated by management action.

Performance. The IEEE 802.1D specification recommends default operational values for performance parameters. These have been specified in order to avoid the need to set values prior to operation, and they have been selected with a view of optimizing the ease with which bridged local area network components interoperate. It specifies absolute maximum values for performance parameters. The ranges of applicable values specified to assist in the choice of operational values and to provide guidance to implementors.

REQUIREMENTS. For correct operation, the parameters and configuration of bridges in the bridge local area network ensure the following:

1. Bridges do not initiate reconfiguration if none is needed. This means that a bridge protocol message is not timed out before its successor arrives, unless a failure has occurred.
2. Following reconfiguration frames are not forwarded on the new active topology, while frames that were initially forwarded on the previous active topology are still in the network. This ensures that frames are not duplicated.

These requirements are met through placing restrictions on the following:

Maximum bridge diameter expresses the maximum number of bridges between any two points of attachment of end stations. The recommended value is 7.

Maximum bridge transit delay (in seconds) is the maximum time elapsing between reception and transmission by a bridge of a forwarded frame, frames that would otherwise exceed this limit being discarded. It consists of three components for which we express the recommended value and the maximum value:

Maximum bridge transit delay:	1.0 s	4.0 s
Maximum BPDU transmission delay:	1.0 s	4.0 s
Maximum message age increment overestimate:	1.0 s	4.0 s

Spanning Tree Algorithm Timer Values:

Bridge hello time:	2.0 s	(range 1.0–10.0)
Bridge max age:	20.0 s	(range 6.0–40.0)
Bridge forward delay:	15.0 s	(range 4.0–30.0)
Hold time:	Fixed value of 1.0	

Priority Parameter Values (recommended or default values)

Bridge priority:	32,768	(range 0–65535)
Port priority:	128	(range 0–255)

Path Cost Parameter Values

Path cost = 100/(attached LAN speed in Mbps)   for each bridge port

Example: 10 Mbps LAN, path cost = 100

Absolute minimum path cost = 1 (range 1–65535)

Granularities are in 1-s increments.

A bridge must enforce the following relationships:

$$2 \times (\text{bridge\_forward\_delay} - 1.0 \text{ s}) \geq \text{Bridge\_Max\_Age}$$

$$\text{Bridge\_Max\_Age} \geq 2 \times (\text{Bridge\_Hello\_Time} + 1.0 \text{ s})$$

10.3.1.4   Bridge Protocol Data Unit (BPDU). Figure 10.37 illustrates the format of a configuration BPDU.

Each transmitted configuration contains the following parameters (and no others):

1. The protocol identifier is encoded in octets 1 and 2 of the BPDU. It takes the value 0000 0000 0000 0000, which identifies the spanning tree algorithm and protocol as specified in Section 4 of this standard.

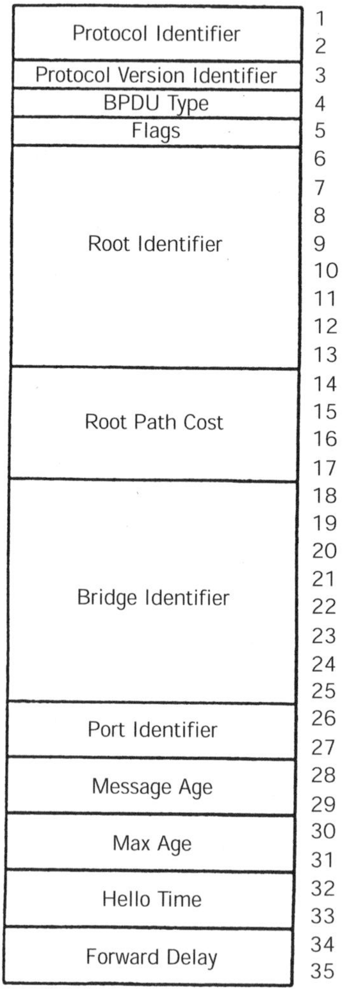

Field	Octet
Protocol Identifier	1
	2
Protocol Version Identifier	3
BPDU Type	4
Flags	5
Root Identifier	6
	7
	8
	9
	10
	11
	12
	13
Root Path Cost	14
	15
	16
	17
Bridge Identifier	18
	19
	20
	21
	22
	23
	24
	25
Port Identifier	26
	27
Message Age	28
	29
Max Age	30
	31
Hello Time	32
	33
Forward Delay	34
	35

Figure 10.37. Configuration BPDU, parameters and format. (From Figure 5-1, page 103, ANSI/IEEE Std. 802.1D, Ref. 10.)

2. The protocol version identifier is encoded in octet 3 of the BPDU. It takes the value 0000 0000.

3. The BPDU type is encoded in octet 4 of the BPDU. This field shall take the value 0000 0000. This denotes a configuration BPDU.

4. The topology change acknowledgment flag is encoded in bit 8 of octet 5 of the BPDU.

5. The topology change flag is encoded in bit 1 of octet 5 of the BPDU.

6. The root identifier is encoded in octets 6 through 13 of the BPDU.

7. The root path cost is encoded in octets 14 through 17 of the BPDU.

8. The bridge identifier is encoded in octets 18 through 25 of the BPDU.

9. The port identifier is encoded in octets 26 and 27 of the BPDU.

10. The message age timer value is encoded in octets 28 and 29 of the BPDU.

11. The max age timer value is encoded in octets 30 and 31 of the BPDU.

12. The hello time timer value is encoded in octets 32 and 33 of the BPDU.

13. The forward delay timer value is encoded in octets 34 and 35 of the BPDU.

ENCODING OF BRIDGE IDENTIFIERS. A bridge identifier is encoded as 8 octets, taken to represent an unsigned binary number. Two bridge identifiers may be numerically compared; the lesser number denotes the bridge of the higher priority.

The two most significant octets of a bridge identifier comprise a settable priority component that permits the relative priority of bridges to be managed. The six least significant octets ensure uniqueness of the bridge identified. They are derived from the globally unique bridge address according to the following procedure.

The third most significant octet is derived from the initial octet of the MAC address, the least significant bit of the octet (Bit 1) is assigned the value of the first bit of the bridge address, the next most significant bit is assigned the value of the second bit of the bridge address, and so on. In a bridged local area network utilizing 48-bit MAC address, the fourth through the eighth octets are similarly assigned the values of the second to the sixth octets of the bridge address.

In a bridged local area network utilizing 16-bit MAC addresses, the fourth octet is assigned the value of the second octet of the bridge address, and the fifth through eight octets are assigned the value 0000 0000.

ENCODING OF ROOT PATH COST. Root path cost is encoded as four octets, taken to represent an unsigned binary number, a multiple of arbitrary cost units. Recommendations for these values are given in Section 10.3.1.3 herein.

ENCODING OF PORT IDENTIFIERS. A port identifier is encoded as 2 octets, taken to represent an unsigned binary number. If two port identifiers are numerically compared, the lesser number denotes the port of higher priority. The more significant octet of a port identifier is a settable priority component that permits the relative priority of ports on the same bridge to be managed. The less significant octet is the port number expressed as an unsigned binary number. The value 0 is not used as a port number.

ENCODING OF TIMER VALUES. Timer values are encoded in 2 octets, taken to represent an unsigned binary number multiplied by a unit of time of 1/256 of a second. This permits times in the range 0 to, but not including, 256 s to be represented.

Sections 10.3.1 through 10.3.4 consist of abstracts of IEEE Std. 802.1D, 1993, edition, Ref. 10.

10.3.2 Source Routing Bridges

The basic concept behind source routing is that a frame header contains routing information that has been inserted by the source of the frame. How must a source end station know its route? It does not. So it must *discover* a route. It can do this by transmitting a special kind of frame. This frame replicates itself at each node where the route splits (i.e., where there are two or more routing choices for onward delivery). As a frame travels the route toward its destination, it accumulates routing information such that upon its arrival, future routing can be selected. This routing information is stored in cache so it can serve subsequent frames.

Source Routing Header and the Routing Information (RI) Field. The generic MAC frame for a LAN contains, as a minimum, a destination address field, a source address field, and a payload, variously called info field, text, data, and so on. This simplified header is

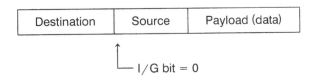

The object of this exercise is to show how an RI field may be added to a MAC header. To tell a receiving bridge that the rules have been changed, we take advantage of the first bit in the source address field (see Section 10.3.2). This bit has been defined as the individual/group address field. In standard MAC frames, this bit is always carried as a binary 0. To tell a receiving bridge that a frame carries an RI field, the bit is set to binary 1 as follows:

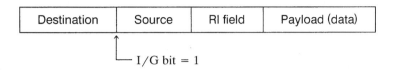

Destination	Source	RI field	Payload (data)

            └─ I/G bit = 1

The RI field contains the following source routing information (Ref. 11):

Type	Length	Direction	Largest frame	Route
3 bits	5 bits	1 bit	3 bits	16 bits

- *Type* (3 bits). The type of source routing may be one of the following:
  - (a) *Specifically routed* (the route appears in the header)
  - (b) *All paths explorer* (the route information accumulates as a frame traverses the network)
  - (c) *Spanning tree explorer* (the route information accumulates as frame copies traverse the network just as with all path explorers, but the frame only travels along branches of the spanning tree)
- *Length* (5 bits). Specifies the number of octets in the RI field.
- *Direction* (1 bit). Specifies whether the route should travel from right to left or left to right.
- *Largest Frame* (LF) (3 bits). A value representing one of a few popular frame sizes (516, 1500, 2052, 4472, 8144, 11407, 17800, or 65535 octets).
- *Route*. A sequence of 2-octet-long fields which are called *route designators*, each of which consists of a 12-bit LAN number followed by a 4-bit bridge number as illustrated:

LAN	Bridge
12 bits	4 bits

Bridge numbers are required to resolve any ambiguity of multiple frames. Within an extended LAN, each LAN must be assigned a unique LAN number. An extended LAN is assumed to have multiple bridges. As a frame travels across the network, it replicates itself at every bridge outlet port and the number of copies of a frame grows exponentially.

Each bridge is assigned a 4-bit identifying number to distinguish it from other bridges that connect the same pair of LANs. Thus there is a LAN number for each port plus a bridge number. The assignment of a bridge number must follow the rule that no two bridges interconnecting the same two LANs be allowed to have the same bridge number.

That important route is just an alternating sequence of LAN numbers and bridge numbers, always starting and ending with a LAN number.

From the viewpoint of a bridge, there are four types of frames to be handled:

1. Frames without an RI field. These are known as *transparent frames*.
2. Specifically routed frames.
3. All paths explorer frames.
4. Spanning tree explorer frames.

10.3.3 Remote Bridges

Remote bridges are used to connect two geographically distant LANs. In this case, a bridge is installed on each LAN and a wide area network (WAN) connects the two bridges. It is generally assumed that these two bridges will be bought in pairs to avoid interface problems.

The simplest way of implementing a remote bridge configuration is to entirely encapsulate the LAN data frame with the WAN header (and trailer, often an FCS).* The remote bridge encapsulation concept is shown in Figure 10.38. Here an entire CSMA/CD frame is encapsulated in a frame relay leader. Note that the CSMA/CD preamble has been stripped off. It is added back on at the remote bridge at the other side of the WAN link.

The spanning tree works well with the point-to-point WAN, and the two bridges on each side consider the WAN link as another LAN. One of the two bridges will be the designated bridge on the point-to-point link. The second bridge, then, considers the port onto that link as its root port.

10.4 HUBS AND SWITCHING HUBS

10.4.1 Rationale and Function

The original idea of a hub was to centralize equipment, usually in the wiring closet on a floor. We think of a LAN as a set of operational workstations (PCs) connected to some sort of wire or cable stretching around the floor. It

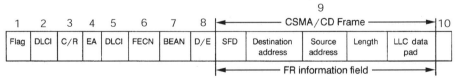

Figure 10.38. Remote bridging concept, WAN encapsulation. CSMA / CD encapsulated in frame relay.

Note: RFC 1483 discusses the application of a remote bridge by means of ATM; RFC 1490, by means of frame relay.

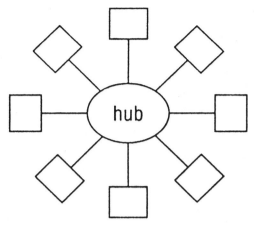

Figure 10.39. The star topology idea of a hub. The hub, of course, sits in the center. The lines connecting workstations to the hub represent wire twisted pairs or a continuous length of coaxial cable tapped at the workstation.

was found to be more practical to steal a page from the book on telephony in that each telephone is terminated on a floor's mainframe located in the wiring closet. Why couldn't we do the same for each computer or workstation? Thus we ended up with a star topology illustrated in Figure 10.39.

One advantage of the hub concept was centralization; another was that LAN equipment could be locked up in a room, away from unauthorized hands. To improve performance, each hub port termination was a repeater. In the case of CSMA/CD, the problem of collisions remained unchanged.

10.4.2 Improvements in Hub Technology

An early improvement in hubs was that of filtering frames based on sensing of the destination address. Here a frame would only be transmitted out a hub spoke if that frame was addressed to the LAN station attached to that "spoke." Of course the hub must learn which stations attach to which spoke based on the source address of frames emanating from a particular spoke. Store-and-forward techniques were employed. Note that if there are multiple hubs or hubs plus bridges in the LAN network, the spanning tree algorithm must be employed to avoid disastrous traffic loops.

With this type of CSMA/CD setup, the chance of collision is eliminated because the active link is isolated (i.e., there are only two stations on the link in the case of unicast—the hub port processor/repeater and the receiving LAN station). Another advantage is that ports on a hub can be at different speeds (e.g., 10 Mbps and 100 Mbps).

The introspective reader will ask, What is the difference between a hub of this type and a learning bridge? None. In fact, many in the industry call this type of hub a *layer 2 switch*.

One also may note the delay incurred in the "store" of store and forward switching. One cure for the delay is to use *cut-through* forwarding. All this means is that the layer 2 switch can forward as soon as it has made a forwarding decision. This happens sometime after all the bits of the destination address are received. Another name for a layer 2 switch is a *smart hub*.

10.5 ROUTERS

As illustrated in Figure 10.40, a router bases its routing decisions on the addressing found in OSI layer 3, the network layer. So much of this discussion deals with addressing, particularly in layer 3. Because the routing is based on OSI layer 3, we'll just call it *layer 3 switching*.

As we have learned in data communications, we can have connection-oriented service or connectionless service. Using real-life analogies: A telephone call can be related to connection-oriented service. To make a call, a circuit must first be set up between the call initiator and its recipient. The circuit setup is based on the telephone number of the called party. Then conversation is exchanged and carried on until one or the other party hangs up; the call is terminated and the circuit is taken down. The analogy of connectionless service is the mailing of a letter through the postal service. A letter is placed in the outgoing mail slot in the Post Office. Some days later (we hope) the letter arrives at its destination. It relies on the address on the envelope for its proper delivery.

A router works on the "connectionless" service concept. A frame is launched into the system. We hope that by means of its destination address, it will arrive at the recipient uncorrupted and in some reasonable length of time.

When discussing addressing, ask "Just what does an address identify?": a node, an interface, a switch, a LAN station, an end-user, or a service access point (SAP) of an end-user?

10.5.1 Addressing with Hierarchical Significance

As with telephone numbering, the addressing at layer 3 can have geographical (or other) significance. In other words, the addressing scheme is hierarchical. All this means that a network address is broken down in identifiable chunks. This can mean that a "chunk" or piece of the address tells us where the address resides.

Worldwide telephone numbering is based on pieces. In North America, telephone numbering is a mixed bag. Certainly, area codes have geographical significance. Exchange codes (the 3-digit number group following the area code) do not; nor do the 4-digit subscriber codes, the last 4 digits of a telephone number that identify an individual subscriber. In other countries, the exchange code can have geographical significance.

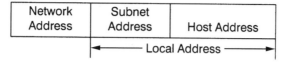

Figure 10.40. The subnet address field in relation to the total IP address.

All network layer addresses are hierarchical. However, for the purpose of routing, IEEE 802 addresses are not hierarchical, even though they seem so. The IEEE uses a block of 3 octets in its 802 series addressing for the OUI (organizationally unique identifier). It is not geographically hierarchical.

Certain network types, such as ISDN and frame relay, mix voice calls with data calls. They may be governed by one of two ITU-T Recommendations: E-164, which is more inclined to the voice telephone world, and X.121 for the data world. For example, X.121 addressing (numbering) consists of two chunks or pieces: the Data Network Identification Code (DNIC), where the first three digits of the DNIC is the DCC (data country code). The second part consists of the remaining 10 or 11 digits (maximum) which contains the network terminal number or the national number which is part of the national numbering scheme. Of course, this addressing will be used for wide area networks.

IP has its own addressing scheme, and there is a scheme for IP4* and another for IP6. For this discussion we will stick with IP4 or just IP. It has a source address and a destination address, and each is 32 bits long (4 octets). Its address, whether source or destination, consists of two pieces or chunks: network address and host address. Some texts refer to the host address as the *local address*. Refer to this simple drawing:

32 bits

The "link" in the field indicates a link number, and the second subfield indicates a system on the link. IP people call the second piece or chunk a *host*. The actual number of bits in the link and host fields is not actually fixed (Ref. 11). However, the total address length is 32 bits (IP4). See Section 11.4.3.

Another way of looking at an IP address is shown in Figure 10.40. This shows the subnet address structure in relation to the total address field.

Assignments for the local address is left up to the local network administrator. For instance, for the 16 bits assigned to the local address, half may be assigned to the subnetwork and half to the host; or 12 bits can be assigned to the subnetwork and only 4 bits assigned to the host.

---

*IP4 is currently the most widely used version of IP. It is so widely used that many people just call it IP, where they assume everybody knows it is IP4.

10.5.2 Subnet Masks

For each link in an IP network, the network planner decides the value of the link number and which bits in the address will correspond to the link number. A link (a LAN, for example) has a link number and a *mask* indicating which bits correspond to the link number. The purpose of the mask is to determine which part of the IP address pertains to the subnet and which pertains to the host.

According to Black (Ref. 13), the convention used for subnet masking is to use a 32-bit field in addition to the IP address. The contents of the mask field are set as follows:

Binary 1s: Identify the network address portion of the IP address.

Binary 0s: Identify the host address portion of the IP address.

A *bitwise* AND function is used to extract the fields of the IP address as follows.

A bitwise AND is performed on the IP address and subnet mask. The results of this operation are matched against a destination address in a routing table. If the results are equal, the next hop IP address (relative to the destination address) is used to determine the next hop on the route. Therefore, the last octet (all 0s) identifies the host on the subnet.

The mask becomes part of the routing algorithm's conditional statement: If the destination IP address and the subnet mask equal my IP address and subnet mask, then send the datagram to local network; otherwise, send datagram to gateway corresponding to destination address. Black comments that, indeed, the use of masks handles routes to direct conventions, host-specified routes, and default routes. The reference suggests that the implementor should note the following guidelines for subnetting:

- The IP algorithm must be implemented on all machines in a subnet.
- Subnet masks should be the same for all machines.
- If one or more machines do not support masks, proxy ARP (address resolution protocol; see Section 11.4.4.1) can be used to achieve subnetting.

10.6 VIRTUAL LOCAL AREA NETWORKS (VLANs)

VLANs are not well-defined by the industry. Reference 11 defines a VLAN as the territory over which a broadcast or multicast frame (packet) is delivered (also known as a *broadcast domain*). The difference between a VLAN and a LAN, if any, is in packaging. A virtual LAN allows the use of separate LANs among ports on the same switch. It breaks a switch up into two (or more bridges) where ports 1–41 belong to VLAN X and ports 42–73 to VLAN Y. In this case, the switch would be broken up into two separate bridges: one that forwarded between the first 41 ports, and one that covers the remainder of the ports. A router would be employed to forward traffic from one VLAN to the other. Of course the separate router connecting the two VLANs requires one port from each bridge. Some switches are capable of acting as routers between VLANs.

When two switches are involved in a VLAN environment with a connecting link between the switches, frames can belong to either VLAN (if there are only two). To solve this situation, additional information is provided called a VLAN *tag*, which is added to the frame (packet) so that switches can know to which VLAN frame or other frame is intended. A station would be confused if it received the frame with a VLAN tag. So switches must be configured to know which ports connect to switches and which ports connect to stations. A switch removes the VLAN tag from the frame before forwarding the frame to a non-switch-neighbor port.

A VLAN tag contains 2 octets where 3 bits are assigned a *priority* function, 12 bits for a VLAN ID, and 1 bit indicating the addresses are in canonical format. In IEEE 802.3 LANs, the VLAN tag is inserted just before the length field.

10.7 SERVERS AND INTRANETS

10.7.1 Servers

A *server* is a data device that *facilitates*. The IEEE (Ref. 12) calls it "a device or computer system that is dedicated to providing specific facilities to other devices attached to the network." There are many examples. A *network server* provides special network applications to all users at the network level; that is, it allows all users to share files more easily and have larger file storage areas available for their use. Commonly, network servers also carry out a policing function where all users check in through the server that validates passwords and establishes firewall. It often is the heart of a LAN, such that when the server is down, the network is down.

Other servers are more specialized—for example, *client server*, where the client is the requesting device, and the server is the supplying device. Another example, is where the user interface could reside in the client workstation while the storage and retrieval functions could reside in the server database. Another case is the *terminal server* which is a device that permits users to access a central mainframe computer. Still another example is the *print server*. This is a device that is dedicated to queueing and sending printer output from the networked computers to a shared printer.

We also have database servers, disk servers, file servers, and mail servers.

10.7.2 Intranet and Extranet

An *intranet* is an enterprise's internal network. Intranets can communicate with each other via the *intranet*, which provides the backbone communication. However, an internet does not need an outside connection to the internet in order to operate. It simply employs TCP/IP protocols and applications to provide a "private" internet.

When a business exposes part of its internal network to the outside community, it is known as an *extranet*. Some of this internet exposure may be a company's own web page. Outsiders (i.e., the general public) will not have access to the entire corporate network, but merely that part that the company wants the public to have access to. The enterprise blocks access in the networks routers and places *firewalls* on resources the enterprise may consider sensitive, forcing the public to have access only to a subset of its intranet.

REFERENCES

1. *Information Processing Systems—Local Area Networks—Part 4: Token-Passing Bus Access Method and Physical Layer Specification*, ANSI/IEEE Std. 802.4-1990, IEEE, New York, 1990.

2. *Information Technology—Telecommunications and Information Exchange Between Systems—Local and Metropolitan Area Networks—Specific Requirements—Part 5: Token Ring Access Method and Physical Layer Specifications*, ANSI/IEEE Std. 802.5, 1995 edition, IEEE, New York, 1995.

3. *Fiber Distributed Data Interface (FDDI)—Physical Layer Protocol (PHY-2)*, ANSI X3.231-1994, ANSI, New York, 1994.

4. *Fiber Distributed Data Interface (FDDI)—Token Ring Physical Layer Medium Dependent*, ANSI X3.166-1990, ANSI, New York, 1990.

5. *Fiber Distributed Data Interface (FDDI)—Token Ring Media Access Control—2 (MAC-2)*, ANSI X.239-1004, ANSI, New York, 1994.

6. *Information Technology—Telecommunications and Information Exchange Between Systems—Local and Metropolitan Area Networks—Specific Requirements—Part 11: Wireless LAN Medium Access Control (MAC) and Physical Layer (PHY) Specifications*, IEEE Std. 802.11-1997, IEEE, New York, 1997.

7. *Code of Federal Regulations: CFR 47—Telecommunication*, Part 15, U.S. Government Printing Office, Washington DC, October 1996.

8. *Information Technology—Telecommunication and Information Exchange Between Systems—Local and Metropolitan Area Networks—Specific Requirements: Part 11:*

Wireless LAN Medium Access Control (MAC) and Physical Layer (PHY) Specifications, ANSI/IEEE Std. 802.11, IEEE, New York, August 1999.

9. Roger L. Freeman, *Radio System Design for Telecommunications*, 2nd edition, John Wiley, New York, 1997.

10. *Information Technology—Telecommunications and Information Exchange Between Systems—Local Area Networks—Media Access Control (MAC) Bridges*, ANSI/IEEE Std. 802.1D, 1st edition, IEEE, New York, July 1993.

11. Radia Perlman, *Interconnections*, 2nd ed., Addison-Wesley, Longman, Reading, MA, 2000.

12. *The IEEE Standard Dictionary of Electrical and Electronic Terms*, 6th ed., IEEE Std. 100, IEEE, New York, 1996.

13. Uyless Black, *TCP/IP and Related Protocols*, McGraw-Hill, New York, 1992.

11

WIDE AREA NETWORKS (WANs)

11.1 BACKGROUND AND SCOPE

Early data networks consisted of point-to-point or multipoint circuits. A star topology with a computer at the star's hub was also widely employed. When the personal computer (PC) gained wide acceptance, distributed processing became the rule. As a result, data networks and data networking became much more complex. We now see network architectures such as intricate tree, multiple star and mesh topologies, and nodes with multiple inlets and outlets.

Data rates were based on the limitations of the analog network or roughly limited to 9600 bps (ca. 1985). To efficiently transport data in such an environment, packet switching/transmission came about. Meanwhile, in the late 1990s, data rates notably increased to 56 kbs or thereabouts on a subscriber loop, DSL (digital subscriber line) lines displayed rates well in excess of 1 Mbps, and trunk rates were (and are) based on the 64-kbps channel, T1/E1 rates, and very high SONET/SDH data transmission rates.

The preponderance of packet* networks is based on the ITU-T Rec. X.25 protocol. However, X.25 networks have greater acceptance in Europe than in the United States and Canada. As we will see, packet networks gave a perception of being slow and tedious, especially considering the demand for LAN interconnectivity where data rates are well into the megabit/second range, whereas X.25 supported transmission rates in the kilobit region.

This chapter provides design data and brief descriptions of the comparatively mature technology involved in wide area (data) networking as a whole. A brief review is given on the concept of statmultiplexers[†] and the X.25 protocol. Much of the chapter deals with the ever popular TCP/IP. It was

---

*We must be careful of language here. There seems to be a general underlying carelessness of language with data communications. Certainly, we can interpret TCP/IP circuits as carrying packets, although we may call them datagrams. There are other examples.
[†]Statmultiplexer is the shortened term for statistical multiplexer.

the author's decision to cover TCP/IP under wide area networks. It could have just as well been included in Chapter 10. IP version 6 is then reviewed, followed by two transport protocols: TCP and User Datagram Protocol (UDP). This is followed by a short discussion of CLNP (Connectionless Network Protocol). The concept of VSAT (very small aperture terminal) satellite data networks is covered, as well as a short discussion of the ITU-T Hypothetical Reference Connection.

11.2 BASIC APPROACHES

11.2.1 Point-to-Point Links

11.2.1.1 Economy Versus Data Perishability. *Perishability* means how soon must the distant end-user have the data. A credit check for a credit card purchase is very perishable. The purchaser wants an immediate reply. Many payrolls, on the other hand, can take much longer time (i.e., hours or days), so long as payday is on time.

The most economical way of shipping data that is not perishable is by U.S. mail. If overnight is required, U.S. mail, Federal Express, UPS, and others are there to do the job.

When perishability increases, some form of electrical data communications is probably in order, say when the data are needed in under 12 hours. The 12 hours drop to seconds for interactive data, such as the credit card credit check. Often this can be accomplished with a dial-up connection using an acoustic coupler or other dial-up device. Such a concept is shown in Figure 11.1.

This form of point-to-point data connectivity can be very cost-effective if there are only a few transactions per day of fairly short duration—on the order of minutes, not hours. Often the connection is half-duplex and we are relegated to using stop-and-wait ARQ. This is a cost-effective and efficient approach if there are not many transactions.

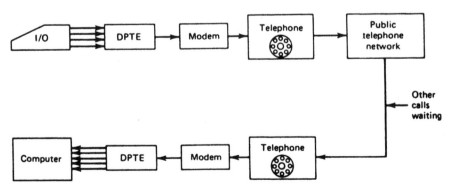

Figure 11.1. A dial-up data connection.

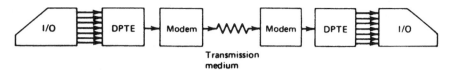

Figure 11.2. A typical dedicated (leased) point-to-point data circuit. I/O, input/output device. DPTE, data processing terminal equipment.

As the number of transactions increases and possibly the duration of each transaction increases, a dedicated circuit may be more effective. This should be justified by a simple cost analysis. A call to the business office of the local telephone company would be helpful; however, if the transactions will be interstate or international, a call to ATT/Sprint/MCI or others would be in order.

11.2.1.2 Dedicated Circuits. Figure 11.2 illustrates the concept of a point-to-point dedicated connection. Expect it to be a leased facility. It is nearly always 4-wire. This means that we can expect equal service in either direction simultaneously. It is ours 24 hours a day, 365 days a year. It can be quite costly. Can we justify the cost?

The demand for its use will ordinarily be through the conventional workday, 5 days a week. This investment may lay dormant after 5 P.M. until 8 A.M. and on weekends and holidays. There may be ways of prioritizing service so that less perishable data can wait until these dormant periods. In many instances, we may not keep this leased connection occupied even during normal working hours.

One way to increase utilization is to share the line with others in our own organization, locally or elsewhere. Convert the service from point-to-point to multipoint. Now we have to set out some rules for who gets to use the service and when. A representative multipoint (multidrop) arrangement is shown in Figure 11.3.

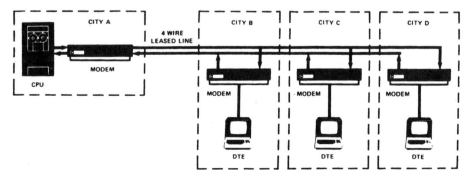

Figure 11.3. A multipoint arrangement for a leased (dedicated) line (4-wire).

If there are several users on one line, contention might serve using CSMA (carrier sense multiple access, meaning "listen before use.") Polling* is another alternative. Some scheme of error control would also probably be in order. A link-layer protocol such as HDLC described in Chapter 3 would solve most of these issues.

Frame relay, which is covered in Chapter 12, might be another alternative if we were concerned with speeding this up. Such an alternative become viable if the leased service provides 64 kbps (and multiples thereof) channelization. This type of channelization is described in Chapter 7. Care must be taken here in that T1 (DS1) PCM in North America may only provide a clear 56-kbps channel due to the signaling bit appearing in every sixth frame. Often clear 64-kbps channels must be specially ordered. European E1 systems do not suffer this shortcoming because signaling for an entire 30-channel group appears on a separate channel (i.e., Channel 31).

11.2.2 Data Multiplexers and Statmultiplexers

11.2.2.1 Time Division Multiplexing. We are dealing here with multiple users sharing a single channel. In the case of the analog channel, there are two approaches. The first, which predates World War II, involves a voice channel that is broken down into frequency slots each 120 Hz wide. Using frequency shift keying (FSK) modulation, each frequency slot can support 75-bps operation. In theory, 25 such channels can be accommodated in the analog voice channel (300–3400 Hz), but more often only 24 channels are carried. The aggregate bit rate would be $24 \times 75 = 1800$ bps.

The other approach is to use time division (rather than frequency division) multiplexing. Here each user would have a fixed time slot. The bit rate that is divided up among the users is 2400, 4800, or more probably 9600 bits per second. Assume again 24 users, each with a fixed time slot. Thus $9600/24 = 400$ bits, or each user would have a 400-bit time slot each second. Of course in this simple calculation, no overhead bits are added. Some framing bits would be needed as well as some way of identifying each user's channel so reconstitution could be carried out at the other end of the circuit. So from a gross bit rate standpoint, time division multiplex is much more efficient than multiplexing in the frequency domain. The TDM data concept is illustrated in Figure 11.4.

11.2.2.2 Statistical Time Division Multiplexing. In a data TDM configuration, a time slot is allocated to each user. In many typical applications, data traffic is bursty. Here we mean that there is a burst of traffic as in a credit card verification system, followed by a quiescent period of some duration. In

---

*Polling is a method of establishing discipline on a multiuser circuit. Each station is polled in sequence allowing them to have their turn transmitting traffic. Token passing is an outgrowth of polling.

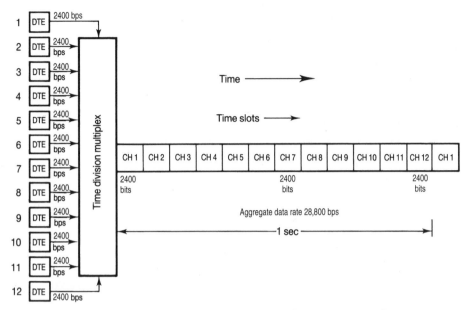

Figure 11.4. Time division multiplexing of data. Only 12 users are shown.

fact, Doll (Ref. 1) states that often individual time-slot utilization is not much more than 10%.

A statistical multiplexer (statmux) improves utilization by dynamically assigning time-slot duration to active users. As one would imagine, some overhead is required to provide the distant end with status information and time-slot assignment in nearly real time. A notable increase is required in buffering to avoid overflows, especially when several incoming traffic bursts coincide. A queueing regime is also required.

Doll (Ref. 1) reports the number of users can be increased from two to four times compared to the equivalent fixed time-slot TDM device. It is assumed with such increases that the typical user has a bursty traffic profile.

11.3   PACKET NETWORKS FOR DATA COMMUNICATIONS

11.3.1   Introduction to Packet Data

In Section 11.2 we found that point-to-point and often multipoint links have periods of inactivity and that they tended to be underutilized. The circuit-switched public telephone network is a prime example of a network that is underutilized and inefficient. It is sized or dimensioned for the *busy hour*, a 1-hour period, usually in the morning of a workday, when the network has the highest traffic activity. After the workday is over, traffic starts to really drop off, and at 2 A.M. it may be $1/500$ of what it was during the busy hour. The design criterion is a 99% probability that a user can successfully complete a call during the busy hour.

Store-and-forward systems, typically Telex, are very efficient with excellent utilization factors, even at 2 A.M. However, when a message is submitted there is no guarantee when it will arrive. Prioritization schemes help. Generally, links connecting store-and-forward nodes can boast nearly 100% utilization, even in off-peak hours.

Packet networks attempt to reach a happy medium: to be efficient with a high utilization factor and have a good delivery time record, providing nearly as good service as circuit switching can provide. They were also designed to better accommodate bursty traffic, typical of a data connectivity.

With the packet concept, data messages are broken down into fixed-length packets through the process of segmentation on one end of the circuit and reassembly on the other end. On a particular circuit, packets of many different users may be transported. Each packet is individually addressed and transits the network in a store-and-forward fashion. When a packet is transmitted, the circuit is immediately available for allocation of packets from other users.

Each node is a packet network has quasi-real-time information on the status of the network and, in theory, will route a packet on an optimum route toward its destination. Packets are routed around failed nodes and sectors of congestion.

Virtual Circuits and Logical Channels. When dealing with a packet network, we must understand the terms *virtual circuit* and *logical channel.* When a request is received to enter packets into the network, a virtual call is established. It is something like a dialed switched connection in the telephone network. The connection exists only for the duration of the call. For that period the network behaves as though there were a fixed path from the source to the destination. However, in an ideal packet network, each packet may follow a different route through the network.

For users with high traffic volumes, a permanent virtual circuit (PVC) can be established. This path is similar to a dedicated circuit, eliminating repeated call setup and release procedures.

The concept of a *virtual* circuit (versus a *real* circuit) we could say derives from the digital network which is a TDM network. For example, in a DS1 configuration (Chapter 7), a caller has use of a circuit for only $1/24$th of the time; for an E1 configuration, $1/32$nd of the time. In the analog network, a user has a physical connectivity 100% of the time.

For packet networks and for ATM networks, *logical channel numbers* are used as a quasi-routing tool. Each packet entering the network is assigned a logical channel number that indicates the *session* or *conversation* to which it belongs. The network can then associate the logical channel number of one end user and the logical channel number of another. CCITT notes in their definition that a number of logical channels may be derived from a data link by packet interleaving.

Certainly the most well-known and popular specification for a packet data network is based on ITU-T Rec. X.25. A general overview of the Recommendation is given in Section 11.3.2.

11.3.2   Packet-Switched and Virtual Connection Based on ITU-T Rec. X.25

11.3.2.1   Introduction. The X.25 protocol (Ref. 2) specifies the rules for communication between the data user (the DTE) and the entry to the public switched network (serving node). In practice, the electrical interface between the user and the network is located between the DTE and DCE. The DCE may be a modem or digital service unit (DSU) which connects to the nearest network node by a transmission line (or other media). The intelligent part of the X.25 protocol is handled by the DCE and the DSE (data switching equipment) of the node to which the DTE is connected.

The X.25 protocol does not define the internal operation of the data network (i.e., how the packets are routed between various nodes of the network). X.25 *does* precisely define the interface between the network and the user in such a way as to permit all X.25 DTEs to be connected to any public data network that conforms to the protocol. X.25 defines three protocol layers that correspond approximately to the first three layers of the open system interconnection (OSI) reference model. The X.25 architecture is shown in Figure 11.5.

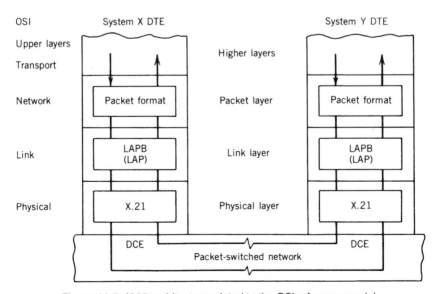

Figure 11.5. X.25 architecture related to the OSI reference model.

Layer 1 is defined by CCITT Rec. X.21 (Ref. 3) (Section 11.3.2.2) for a digital network, by X.21 bis for an analog entry point, and by X.31 (Ref. 4) for ISDN connectivity. Bit rates are supported all the way up to 2.048 Mbps.

Layer 2 of X.25 defines the point-to-point transfer of frames between the DTE and DCE. This is the data-link layer and is specified by the LAPB (link access procedure balanced). It is almost identical to HDLC as described in Chapter 3.

The third layer of X.25 concerns the structure of text and control information in packets. This may be called the *packet layer* in contrast to the OSI network layer. This "packet layer protocol" defines the format of the data field of the frames exchanged at layer 2. Thus, the data text transmitted by the user is first augmented at layer 3 with a packet header that specifies, for example, the logical channel number or the type of packet. The packet is then transferred to layer 2 (LAPB), which incorporates it into the information field of a data-link frame. This frame is then transferred to the network by means of the physical layer (layer 1).

11.3.2.2 X.25 Packets Are Contained in LAPB Frames.
Figure 11.6 shows the conceptual relationship of X.25 packets embedded in an LAPB frame. It shows the X.25 packet occupying the LAPB frame information field. Figure 11.7 shows the actual LAPB frame for modulo-8, modulo-128, and modulo 32,768 operation. All frame types have 8-bit address fields

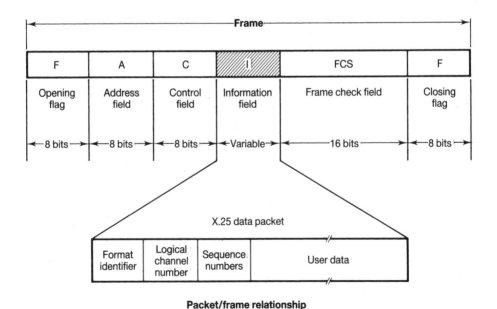

Packet/frame relationship

Figure 11.6. X.25 packet embedded in the LAPB frame information field.

Bit order of transmission	12345678	12345678	12345678	16 to 1	12345678
	Flag	Address	Control	FCS	Flag
	F 01111110	A 8 bits	C 8 bits	FCS 16 bits	F 01111110
Bit order of transmission	12345678	12345678	12345678		16 to 1	12345678
	Flag	Address	Control	Information	FCS	Flag
	F 01111110	A 8 bits	C 8 bits	Info N bits	FCS 16 bits	F 01111110

FCS   Frame check sequence

(a)

Bit order of transmission	12345678	12345678	1 to a)	16 to 1	12345678
	Flag	Address	Control	FCS	Flag
	F 01111110	A 8 bits	C bits a)	FCS 16 bits	F 01111110
Bit order of transmission	12345678	12345678	1 to a)		16 to 1	12345678
	Flag	Address	Control	Information	FCS	Flag
	F 01111110	A 8 bits	C bits a)	Info N bits	FCS 16 bits	F 01111110

FCS   Frame check sequence
a)   16 for frame formats that contain sequence numbers; 8 for frame formats that do not contain sequence numbers.

(b)

Bit order of transmission	12345678	12345678	1 to a)	16 to 1	12345678
	Flag	Address	Control	FCS	Flag
	F 01111110	A 8 bits	C · bits a)	FCS 16 bits	F 01111110
Bit order of transmission	12345678	12345678	1 to a)		16 to 1	12345678
	Flag	Address	Control	Information	FCS	Flag
	F 01111110	A 8 bits	C bits a)	Info N bits	FCS 16 bits	F 01111110

FCS   Frame Check Sequence
a)   32 for frame formats that contain sequence numbers; 8 for frame formats that do not contain sequence numbers.

(c)

Figure 11.7. LAPB frame formats. (a) Basic (modulo 8) operation, (b) extended (modulo 128) operation, and (c) super (modulo 32,768) operation. (From Tables 2-1, 2-2, and 2-3/X.25, page 8, ITU-T Rec. X.25, Ref. 2.)

providing an addressing capacity of $2^8 = 256$ distinct addresses. It is the control field that varies for the different operational types. The control field for basic operation has 1 octet; for extended (modulo 128) operation it has 2 octets, and for Super (modulo 32768) operation, the control field has 4 octets. LAPB is almost identical to HDLC as described in Chapter 3.

LAPB Frame Discussion. The frame formats shown in Figure 11.7 are for single-link procedures. LAPB is basically a subset of HDLC as described in Chapter 3. If start–stop transmission is to be used, there must be eight information bits between the start element and the stop element. The information field must be octet-aligned. This means there must be an integer number of octets. If not, the DCE will pad the information field so that it will become octet-aligned.

11.3.2.3 *Single-Link and Multilink Procedures.*
The multilink procedure (MLP) exists as an added upper sublayer of the data-link layer, operating between the packet layer and a multiplicity of single data-link protocol functions (SLPs) in the data-link layer. The multilink concept in X.25 is illustrated in Figure 11.8. An MLP performs the functions of accepting packets from the packet layer, distributing those packets across the available DCE or DTE SLPs for transmission to the DTE or DCE SLPs, respectively, and resequencing the packets received from the DTE or DCE SLPs for delivery to the DTE or DCE packet layer, respectively.

Field of Application. A multilink is a group of one or more single links in parallel between a DCE and a DTE. The multilink procedure provides the following general features:

1. It achieves economy and reliability of service by providing multiple SLPs between a DCE and a DTE.
2. It permits addition and deletion of SLPs without interrupting the service provided by the multiple SLPs.

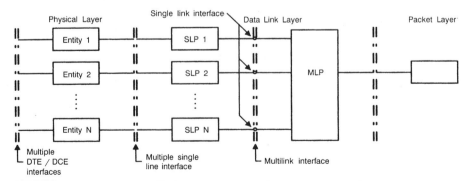

Figure 11.8. Multilink functional organization. SLP, single-link procedure; MLP, multilink procedure. (From Figure 2-2 / X.25, page 36, ITU-T Rec. X.25, Ref. 2.)

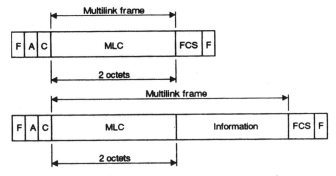

Figure 11.9. Multilink frame formats. MLC, multilink control field. (From Table 2-15/X.25, page 36, ITU-T Rec. X.25, Ref. 2.)

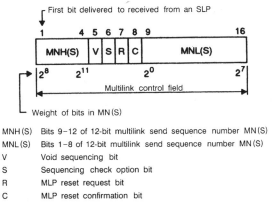

MNH(S)	Bits 9–12 of 12-bit multilink send sequence number MN(S)
MNL(S)	Bits 1–8 of 12-bit multilink send sequence number MN(S)
V	Void sequencing bit
S	Sequencing check option bit
R	MLP reset request bit
C	MLP reset confirmation bit

Figure 11.10. Multilink control field format. (From Table 2-16/X.25, page 37, ITU-T Rec. X.25, Ref. 2.)

3. It optimizes bandwidth utilization of a group of SLPs through load sharing.
4. It achieves graceful degradation of service when an SLP(s) fails.
5. It provides each multiple SLP group with a single logical data-link layer appearance to the packet layer.
6. It provides resequencing of the received packets prior to delivering them to the packet layer.

Multilink Frame Formats. The multilink frame formats are shown in Figure 11.9, and the multilink control field format is given in Figure 11.10.

11.3.2.4 Packet-Level DTE / DCE Interface. Each packet to be transferred across the DTE/DCE interface is contained within the data-link layer information field that delimits its length, and only one packet is contained in the information field.

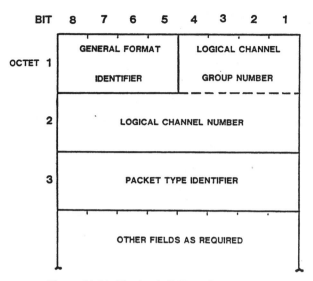

Figure 11.11. The basic X.25 packet structure.

Basic Structure of X.25 Packets. ITU-T Rec. X.25 describes an interface between the DTE and DCE for packet data communications. The DTE resides in the data terminal equipment, whereas the DCE resides on the network side. The DTE and DCE may be colocated physically or may be separated by some distance. The DCE, for example, may be in a public packet data switching center.

Every packet transferred across the DTE–DCE interface may consist of four or five fields: a protocol identifier octet, a general format identifier, a logical channel group number, a logical channel number, and a packet-type identifier. Other packet fields are appended as required. The basic packet structure is illustrated in Figure 11.11. This figure is valid for modulo 8 and 128 operation.

It should be noted that for modulo 8 and 128 operation, the protocol identifier octet is not present in any packet type. For modulo 32,768 operation, the protocol identifier octet is contained in the first octet of each packet.

Table 11.1 defines the general format identifier. This is a 4-bit binary coded field that is provided to indicate the general format of the rest of the header. For modulo 8 and 128 operation, the general format identifier is contained in the first octet of each packet. For modulo 32,768 operation, the general format identifier is contained in the second octet of each packet. The general format identifier field is located in bit positions 8, 7, 6, and 5 is the low-order bit as shown in Table 11.1. The *A bit*, which we discuss later, is in position 8. The A bit, when applied, deals with addressing. X in the Table means a bit that may be set to either binary 1 or binary 0.

TABLE 11.1   General Format Identifier

	General Format Identifier	Bit Position 8	7	6	5
Call setup packets (Note 1)	Sequence numbering scheme modulo 8	X	X	0	1
	Sequence numbering scheme modulo 128	X	X	1	0
	Sequence numbering scheme modulo 32,768	X	X	1	0
Clearing packets (Note 1)	Sequence numbering scheme modulo 8	X	0	0	1
	Sequence numbering scheme modulo 128	X	0	1	0
	Sequence numbering scheme modulo 32,768	X	0	1	1
Flow control, interrupt, reset, restart, and *diagnostic* packets	Sequence numbering scheme modulo 8	0	0	0	1
	Sequence numbering scheme modulo 128	0	0	1	0
	Sequence numbering scheme modulo 32,768	0	0	1	1
Data packets (Note 1)	Sequence numbering scheme modulo 8	X	X	0	1
	Sequence numbering scheme modulo 128	X	X	1	0
	Sequence numbering scheme modulo 32,768	X	X	1	1
Reserved format (Note 2)		[a]	[a]	0	0

[a]Undefined.

Note 1. A bit that is indicated as "X" may be set to either 0 or 1, as discussed in the text.

Note 2. When the general format identifier field is contained in the first octet of a packet, this value is reserved for other applications. When the first octet of a packet is the protocol identifier octet, then this value is reserved for general format identifier extension.

Source: From Table 5-1 / X.25, page 59, Ref. 2.

Logical Channel/Group Number. To enable simultaneous virtual calls and/or permanent virtual circuits, logical groups/channels are used. Each virtual call or permanent virtual circuit is assigned a logical channel group number (less than or equal to 15) and a logical channel number (less than or equal to 255).

The logical channel group number appears in every packet except *restart* and *diagnostic* packets. For modulo 8 and modulo 128 operation, the logical channel group number is contained in the first octet of each packet. In the case of modulo 32,786 operation, the logical channel group number is contained in the second octet of each packet. The logical channel group number is located in bit position 4, 3, 2, and 1. For each logical channel, this number has local significance at the DTE−DCE interface. This field is binary coded, and bit 1 is the low-order bit of the logical channel group number. In *restart* and *diagnostic* packets, this field is coded all zeros.

Logical Channel Number. The logical channel number appears in every packet except *restart* and *diagnostic* packets. For modulo 8 and modulo 128 operation, the logical channel number is contained in the second octet of each packet. For modulo 32,768 operation, the logical channel number is contained in the third octet of each packet. The logical channel group number is located in all bit positions of the octet. For each logical channel, this number has local significance at the DTE−DCE interface.

For virtual calls, a logical channel group number and a logical channel number are assigned during the call setup phase. The range of logical channels used for virtual calls is agreed upon with the administrations or service companies at the time of subscription of the service. A similar procedure is used for permanent virtual circuits.

Packet-Type Identifier. The packet-type identifier is carried in octet number 3 for modulo 8 and modulo 128 operation, and in octet number 4 for super modulo 32,678 operation. Each packet is identified in accordance with Table 11.2.

TABLE 11.2 Packet-Type Identifier

Packet Type		Bit Position							
From DCE to DTE	From DTE to DCE	8	7	6	5	4	3	2	1
Call Setup and Clearing									
Incoming call	Call request	0	0	0	0	1	0	1	1
Call connected	Call accepted	0	0	0	0	1	1	1	1
Clear indication	Clear request	0	0	0	1	0	0	1	1
DCE clear confirmation	DTE clear confirmation	0	0	0	1	0	1	1	1
Data and Interrupt									
DCE data	DTE data	X	X	X	X	X	X	X	0
DCE interrupt	DTE interrupt	0	0	1	0	0	0	1	1
DCE interrupt confirmation	DTE interrupt confirmation	0	0	1	0	0	1	1	1
Flow Control and Reset									
DCE RR (modulo 8)	DTE RR (modulo 8)	X	X	X	0	0	0	0	1
DCE RR (modulo 128)[a]	DTE RR (modulo 128)[a]	0	0	0	0	0	0	0	1
DCE RR (modulo 32,768)[a]	DTE RR (modulo 32,768)[a]	0	0	0	0	0	0	0	1
DCE RNR (modulo 8)	DTE RNR (modulo 8)	X	X	X	0	0	1	0	1
DCE RNR (modulo 128)[a]	DTE RNR (modulo 128)[a]	0	0	0	0	0	1	0	1
DCE RNR (modulo 32,768)[a]	DTE RNR (modulo 32,768)[a]	0	0	0	0	0	1	0	1
	DTE REJ (modulo 8)[a]	X	X	X	0	1	0	0	1
	DTE REJ (modulo 128)[a]	0	0	0	0	1	0	0	1
DCE REJ (modulo 32,768)[a]	DTE REJ (modulo 32,768)[a]	0	0	0	0	1	0	0	1
Reset indication	Reset request	0	0	0	1	1	0	1	1
DCE reset confirmation	DTE reset confirmation	0	0	0	1	1	1	1	1
Restart									
Restart indication	Restart request	1	1	1	1	1	0	1	1
DCE restart confirmation	DTE restart confirmation	1	1	1	1	1	1	1	1
Diagnostic									
Diagnostic[a]		1	1	1	1	0	0	0	1

[a]Not necessarily available on every network.
Note: A bit which is indicated as "X" may be set to either 0 or 1 as discussed in the text.
Source: Table 5-2/X.25, page 61, Ref. 2.

11.3.2.5   A Typical X.25 Packet. Figure 11.12 shows the format of a DTE
and DCE data packet. This is the packet that carries the revenue-bearing
data payload. All the other packet types would be listed as service packets.

Qualifier (Q) Bit. In some operational situations, an indicator bit may be
needed with the user data field to distinguish between two types of information—for example, user data and control information. When such a mechanism is required, the qualifier (Q) bit is used. When the Q bit is not needed,
it is set to binary 0.

Delivery Confirmation Bit. The setting of the delivery confirmation bit (D bit)
is used to indicate whether or not the DTE wishes to receive end-to-end
acknowledgment of delivery, for data it is transmitting, by means of the
packet receive sequence number P(R).

Sequence Numbering: P(R) and P(S). Each data packet transmitted at the
DTE/DCE interface for each direction of transmission in a virtual call or
permanent virtual circuit is sequentially numbered. The sequence numbering
scheme of the packets is performed using modulo 8. The packet sequence
numbers cycle through the entire range of 0 to 7. There are optional
extended packet sequence numbering facility providing a numbering scheme
in modulo 128 and 2^{15} in modulo 32,768. For example, for packet send
sequence number $P(S)$, bits 4, 3, and 2 of octet 3, or bits 8 through 2 of octet
3 when extended, or bits 8 through 2 of octet 4 and bits 8 through 1 of octet 5
when superextended are used for indicating the packet send sequence
number $P(S)$. $P(S)$ is binary coded and bit 2 is the low-order bit. When
superextended, bit 2 of octet 4 is the low-order bit, and bit 8 of octet 5 is the
high-order bit. Whatever the modulo scheme, the packet sequence numbering scheme is the same in both directions of transmission and is common for
all logical channels at the DTE–DCE interface.

Only data packets have the send sequence number $P(S)$. The send
sequence number $P(S)$ maintains the integrity of the data because it is a
basic mechanism for the detection of lost packets. The receive sequence
number $P(R)$ is used to acknowledge correct receipt of packets and autho-
rize transmission of additional packets.

Flow Control. There are three flow control packets: receive ready (RR),
receive not ready (RNR), and reject (REJ). Each of these packets is 3 octets
long in the modulo-8 case; 4 octets long when modulo-128 extended sequence
numbering is used; and six octets long when super modulo 32,768 is used.
The received sequence number $P(R)$ acknowledges receipt of packets numbered $P(R) - 1$.

There is also a *flow control window* defined for each direction of data
transmission of a logical channel used for a virtual call or a permanent virtual

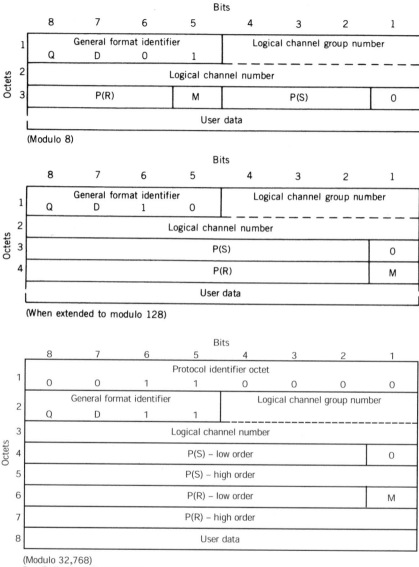

Figure 11.12. DTE and DCE data packet formats showing modulo 8 sequence numbering (upper panel), modulo 128 (middle panel), and modulo 32,678 (lower panel). (From Figure 5-8 / X.25, page 75, ITU-T Rec. X.25, Ref. 2.)

circuit. The window has a size, W, ranging from 1 to 7. The window size specifies the number of sequentially numbered packets that a DTE may have outstanding (i.e., unacknowledged). The standard window size W is 2 for each direction of transmission. Other window sizes can be negotiated through the *flow control parameter negotiation* facility.

When the sequence number $P(S)$ of the next data packet to be transmitted is within the window, the DCE is authorized to transmit this data packet to the DTE. When the sequence number $P(S)$ of the next data packet to be transmitted is outside of the window, the DCE will not transmit the data packet to the DTE. The DTE follows the same procedure.

When the sequence number $P(S)$ of the data packet received by the DCE is the next in sequence and is within the window, the DCE will accept this data packet. A received data packet containing a $P(S)$ that is out of sequence (i.e., there is a duplicate or gap in the $P(S)$ numbering), outside the window, or not equal to 0 for the first data packet after entering the *flow control ready* state (d1) is considered by the DCE as a local procedure error. The DCE will reset the virtual call or permanent virtual circuit. The DTE follows the same procedure.

The packet receive sequence number $P(R)$ conveys across the DTE/DCE interface information from the receiver for the transmission of data packets. When transmitted across the DTE/DCE interface, a $P(R)$ becomes the lower window edge. In this way, additional data packets may be authorized by the receiver to cross the DTE–DCE interface.

The packet receive sequence number $P(R)$ is conveyed in data, receive ready (RR), and receive not ready (RNR) packets.

The value of $P(R)$ received by the DCE must be within the range from the last $P(R)$ received by the DCE up to and including the packet send sequence number of the next data packet to be transmitted by the DCE. Otherwise, the DCE will consider the receipt of this $P(R)$ as a procedure error and will reset the virtual call or permanent virtual circuit. The DTE follows the same procedure.

The receive sequence number $P(R)$ is less than or equal to the sequence number of the next expected data packet and implies that the DTE or DCE transmitting $P(R)$ has accepted at least all data packets numbered up to and including $P(R) - 1$.

User Data Field Length. The standard maximum user data field length is 128 octets. In addition, other maximum user data field lengths may be offered by administrations from the following list: 16, 32, 64, 256, 512, 1026, 2048, and 4096 octets. The user data field of data packets transmitted by a DTE or DCE may contain any number of bits up to the agreed maximum. Rather than *bits*, some networks will use *octets*.

More Data Mark, the M Bit. If a DTE or DCE wishes to indicate a sequence of more than one packet, it uses a more data mark (M bit). The M bit can be

TABLE 11.3 Definition of Two Categories of Data Packets and Network Treatment of the M and D Bits

Data Packet Sent by Source DTE				Combining with Subsequent Packet(s) Is Performed by the Network When Possible	Data packet[a] Received by Destination DTE	
Category	M	D	Full		M	D
B	0 or 1	0	No	No	0 (Note 1)	0
B	0	1	No	No	0	1
B	1	1	No	No	1	1
B	0	0	Yes	No	0	0
B	0	1	Yes	No	0	1
A	1	0	Yes	Yes (Note 2)	1	0
B	1	1	Yes	No	1	1

[a]Refers to the delivered data packet whose last bit of user data corresponds to the last bit of user data, if any, that was present in the data packet sent by the source DTE.
Note 1. The originating network will force the M bit to 0.
Note 2 If the data packet sent by the source DTE is combined with other packets, up to and including a category B packet, the M and D bit settings in the data packet received by the destination DTE will be according to that given in the two right-hand columns for the last data packet sent by the source DTE that was part of the combination.
Source: Table 4-1 / X.25, page 51, Ref. 2.

set to 1 in any data packet. When it is set to 1 in a full data packet or in a partially full data packet also carrying the D bit set to 1, it indicates that more data are to follow. Recombination with the following data packet may only be performed within the network when the M bit is set to 1 in a full data packet which also has the D bit set to 0.

A sequence of data packets with every M bit set to 1 except for the last one will be delivered as a sequence of data packets with the M bit set to 1 except for the last one when the original packets having the M bit set to 1 are either full (irrespective of the setting of the D bit) or partially full but have the D bit set to 1.

Two categories of data packets, A and B, have been defined and are shown in Table 11.3. This table also illustrates the network's treatment of the M and D bits at both ends of a virtual call or permanent virtual circuit.

Complete Packet Sequence. A complete packet sequence is defined as being composed of a single *category B* packet and all contiguous preceding *category A* packets (if any). Category A packets have the exact maximum user data field length with the M bit set to 1 and the D bit set to 0. All other data packets are category B packets. When transmitted by a source DTE, a complete packet sequence is always delivered to the destination by a single complete packet sequence.

X.25 Octet Conventions. Bits of an octet are numbered 8 to 1, where 1 is the low-order bit and is transmitted first. Octets of a packet are consecutively numbered starting from 1 and are transmitted in this order.

11.3.2.6 Typical X.25 Transmissions Exchange of Packets. Figures 11.13a and 11.13b are X.25 call setup and call clearing state diagrams.

Ready State. If there is no call in existence, a logical channel is in the *ready* state.

Call Request/Incoming Call Packet. The call request is initiated by the DTE to indicate a request to establish a virtual circuit. The incoming call packet is identical but is initiated by the DCE. The packet carries the logical channel number, as well as the calling and called DTE addresses, and provides a means for carrying parameters used to define specific information such as billing information about the call. The packet also contains a user

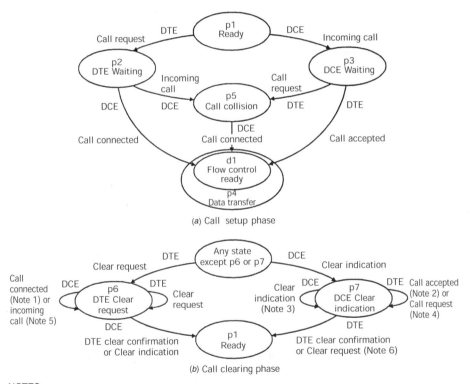

(a) Call setup phase

(b) Call clearing phase

NOTES
1 This transition is possible only if the previous state was *DTE Waiting* (p2).
2 This transition is possible only if the previous state was *DCE Waiting* (p3).
3 This transition takes place after time-out T13 expires the first time.
4 This transition is possible only if the previous state was *Ready* (p1) or *DCE Waiting* (p3).
5 This transition is possible only if the previous state was *Ready* (p1) or *DTE Waiting* (p2).
6 This transition also takes place after time-out T13 expires the second time (without transmission of any packet, except, possibly, a diagnostic packet).

Figure 11.13. X.25 call setup (upper) and call clearing (lower) state diagrams. (From Figure B.2/X.25, page 123, ITU-T Rec. X.25, Ref. 2.)

information field up to 16 octets long. In conjunction with the *fast select facility*, it can have 128 octets.

Call-Accepted / Call-Connected Packet. The call-accepted and call-connected packets are used to indicate that the called DTE has accepted a request to establish a call. Call accepted is initiated by the called DTE, and the network delivers a call-connected packet to the calling DTE. Data transfers can begin after the call is accepted. These packets include the general format identifier, the logical channel number, and the packet identifier. Address and facilities fields are optional.

Call Clearing. Three packets are used in the call-clearing phase. The clear request packet is issued by the DTE initiating the clearing (call take-down). It is delivered to the remote DTE as a clear indication packet. The clear confirmation packet is issued by both the DTE and DCE to acknowledge receipt of clear packets.

These packets contain the standard 3-octet header. In addition, the request and indication packets contain a clearing-cause field and an optional diagnostic code. On some networks, optional address and facilities fields are permitted.

The clearing-cause field indicates the reason for the clear, such as busy, network congestion, or invalid call.

11.3.2.7 *Addressing in X.25.* Figure 11.14 shows the relationship of X.25 address fields in typical X.25 frames. In this case, the frame group selected deals with call setup and call clearing, specifically the "call request and incoming call packet format."

The call setup and clearing packets contain an address block which can take on one of two possible formats. The first format, known as the non_TOA/NPI address format, can accommodate addresses conforming to the formats described in ITU-T Recs. X.121 and X.301 whose length (including possible prefixes and/or escape codes) is not greater than 15 digits. The second format, known as the TOA/NPI address format, can be used by networks and DTEs to accommodate addresses conforming to the formats described in ITU-T Recs. X.121 and X.301 whose length is greater than 15 digits and can also be used to carry an alternative address in the called DTE address field of the *call request* and *clear request* packet. The address block of the TOA/NPI format contains (in addition to the address itself) fields to specify the type of address (TOA) and numbering plan identification (NPI).

The non-TOA/NPI address format and the TOA/NPI address format are distinguished by bit 8 (A bit) of the general format identifier (Table 11.1). See Figures 11.11, 11.12, and 11.14. When the A bit is set to 0, the non-TOA/NPI address format is used. (See Figure 11.15.) When the A bit is set to 1, the TOA/NPI address format is used. (See Figure 11.16.) Note that the A-bit (address bit) is valid only for call setup and call clearing packets and is set to 0 for all other packets.

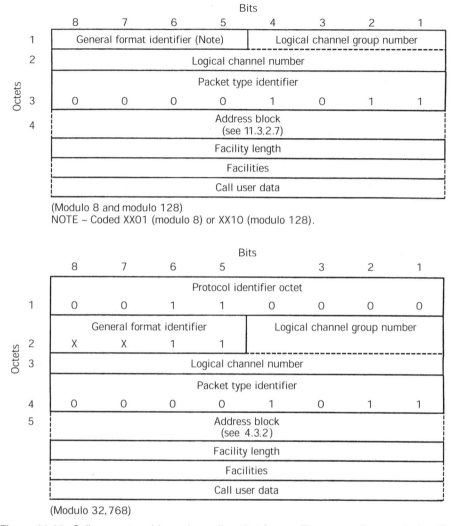

(Modulo 8 and modulo 128)
NOTE – Coded XX01 (modulo 8) or XX10 (modulo 128).

(Modulo 32,768)

Figure 11.14. Call request and incoming call packet format. The upper diagram deals with modulo 8 and modulo 128 operation; the lower diagram deals with modulo 32,768 operation. (From Figure 5-4 / X.25, page 67, ITU-T Rec. X.25, Ref. 2.)

The non-TOA/NPI address format is supported by all networks. The TOA/NPI address format may be supported by some networks and by some DTEs.

The use of the TOA/NPI address format is only possible if the TOA/NPI address format is supported by the network. When the address format used by one DTE in a call setup or call clearing packet is not the same as the address format used by the remote DTE, the network (if it supports the TOA/NPI address format) converts from one address format to the other.

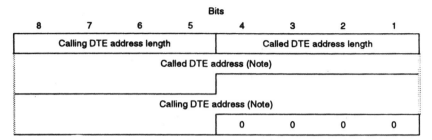

Figure 11.15. Format of the address block when the A bit is set to 0, non-TOA/NPI addressing. Note: The figure is drawn assuming the number of semi-octets present in the called DTE address field is odd and the number of semi-octets in the calling DTE address field is even. (From Figure 5-1/X.25, page 62, ITU-T Rec. X.25, Ref. 2.)

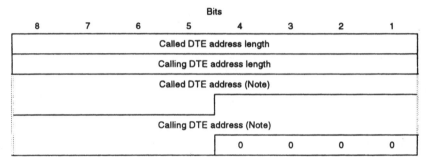

Figure 11.16. Format of the address block when the A bit is set to 1, TOA/NPI addressing. Note: The figure is drawn assuming the number of semi-octets present in the called DTE address field is odd and the number of semi-octets in the calling DTE address field is even. (From Figure 5-2/X.25, page 63, ITU-T Rec. X.25, Ref. 2.)

In either case, we see that there are four basic elements to the addressing broken down into two groups: address length fields and address fields. There is one each for the calling DTE and the called DTE. Address length fields are each 4 bits long, occupying the first octet in the address block (see Figures 11.15 and 11.16).

In these address formats, the ITU-T Organization has had to turn to using the term *semi-octet*, an octet of less than 8 bits, in this case 4 bits. The length indicators are binary encoded, and bit 1 is the low-order bit of the indicator.

The address fields themselves have three subfields: type of address (TOA) subfield, numbering plan identification (NPI) subfield, and the address digits subfield. The first two subfields are at the beginning of the address and are binary-encoded. These provide information such as whether the number is international, network-dependent, national, or alternative address and whether it is based on ITU-T Rec. E.164 or X.121.

When the type of address subfield indicates an address other than an alternative address, the other semi-octets of a DTE address are digits, coded

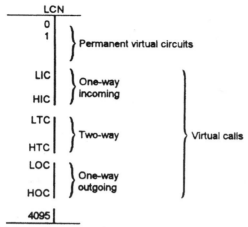

Logical channels 1 to LIC-1: Range of logical channels which may be assigned permanent virtual circuits.
Logical channels LIC to HIC: Range of logical channels which are assigned to one-way incoming logical channels for virtual calls.
Logical channels LTC to HTC: Range of logical channels which are assigned to two-way logical channels for virtual calls
Logical channels LOC to HOC: Range of logical channels which are assigned to one-way outgoing logical channels for virtual calls.
Logical channels HIC + 1 to LTC − 1, HTC + 1 to LOC − 1, and HOC + 1 to 4095 are non-assigned logical channels.

Figure 11.17. Logical channel assignment diagram. LCN, logical channel number; LIC, lowest incoming channel; HIC, highest incoming channel; LTC, lowest two-way channel; HTC, highest two-way channel; LOC, lowest outgoing channel; HOC, highest outgoing channel. (From Figure A.1/X.25, page 120, ITU-T Rec. X.25, Ref. 2.)

in binary-coded decimal with bit 5 or 1 being the low-order bit of the digit. Starting from the high-order digit, the address digits are coded in semi-octets. In each octet, the higher-order digit is coded in bits 8, 7, 6, and 5.

When present, the calling DTE address field starts on the first semi-octet following the end of the called DTE address field. Consequently, when the number of semi-octets of the called DTE address field is odd, the beginning of the calling DTE address field, when present, is not octet-aligned.

When the total number of semi-octets in the called and calling DTE address field is odd, a semi-octet with zeros in bits 4, 3, 2, and 1 is inserted after the calling DTE address field in order to maintain octet alignment.

11.3.2.8 Assignment of Logical Channels. Logical channel assignment is based on the coding of the second semi-octet of the first field of the X.25 packet format (see Figure 11.12) and the second octet, for a total of 12 binary digits or 4096 binary digit combinations. The assignment is carried out in a manner similar to that of telephone trunks in the way it is handled at each end of a connection. This is shown in Figure 11.17. One end of the connection selects channels from the low end of the channel assignment, starting with channel 0, then 1, 2, 3 and so forth. The other end of the circuit selects starting from the highest-numbered channel—in this case channel number 4095, then 4094, and then 4093 and so on.

11.4 TRANSMISSION CONTROL PROTOCOL/INTERNET PROTOCOL (TCP/IP)

11.4.1 Background and Application

TCP/IP protocol family was developed by the U.S. Department of Defense for the ARPANET (Advanced Research Projects Agency Network). ARPANET was one of the first large advanced packet networks. It dates back to 1968 and was well into existence before CCITT and ISO took interest in layered protocols.

The TCP/IP suite of protocols (Ref. 5) has wide acceptance today. These protocols are used on both local area networks (LANs) and wide area networks (WANs). They are particularly attractive for their internetworking capabilities. The internet protocol (IP) competes with CCITT Rec. X.75 protocol (Ref. 6).

The architectural model of the IP (Ref. 7) uses terminology that differs from the OSI reference model. (The IP predates OSI.) Figure 11.18 shows the relationship between TCP/IP and related Department of Defense (DoD) protocols and the OSI model. It traces data traffic from an originating host, which runs an applications program, to another host in another network. This may be a LAN-to-WAN-to-LAN, a LAN-to-LAN, or a WAN-to-WAN connectivity. The host enters its own network by means of a network access protocol such as LAPB (a derivative of HDLC, described in Section 11.3) or an IEEE 802 series protocol. A common LAN-to-WAN-to-LAN connectivity is shown in Figure 11.19.

OSI	TCP/IP and Related Protocols		
Application	File transfer	Electronic mail	Terminal emulation
Presentation	File transfer protocol (FTP)	Simple mail transfer protocol (SMTP)	Telnet protocol
Session			
Transport	Transmission control protocol (TCP)		User datagram protocol (UDP)
Network	Address resolution protocol (ARP)	Internet protocol (IP)	Internet control message protocol (ICMP)
Data link	——————Network interface cards—————— CSMA/CD (Ethernet), Token Ring, ARCNET, StarLan		
Physical	——————Transmission media—————— Wire pair, fiber optics, coaxial cable, radio		

Figure 11.18. How TCP/IP and associated DoD protocols relate to the OSI reference model.

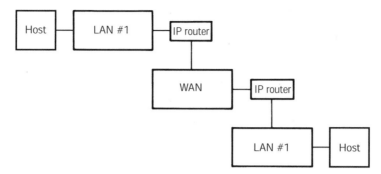

Figure 11.19. Connecting one LAN to another via a WAN, employing IP.

In the figure, the IP router will encapsulate* typically an 802.3 MAC frame derived from the LAN. The encapsulated information is called a *datagram* in IP language. IP operates in OSI layer 3 (see Figure 11.18). The datagram is now passed to the data-link layer (layer 2) where it is encapsulated still again. The resulting body of data is now called a *packet*, which is transmitted by the appropriate physical layer.

IP is transparent to the information contained in the datagram; it just adds some control information to the data received from an upper-layer protocol (often TCP or UDP) and then attempts to deliver it to a station on the network.

The term *host* is commonly used in the IP world meaning the lowest level of addressing and refers to a device. That device is often a computer or some other digital processor.

IP and related documentation are supported by RFCs (Request for Comment). We must remember that IP was a U.S. Department of Defense (DoD) initiative. RFC is typical language of the DoD. Before releasing a specification or request for proposal, the DoD would often issue an RFC document to industry. This allowed industry to shape DoD programs such that they would be more reasonable to bid, perhaps less costly. The RFC label stuck for the IP family of protocols (Figure 11.18). RFCs are issued sequentially, where RFC 791 defines the present IP4. As of this writing, the issuing agency is beyond RFC 4000.

11.4.2 TCP/IP and Data-Link Layers

TCP/IP is transparent to the type of data-link layer involved, and it is also transparent whether it is operating in a LAN or WAN regime or among them. There is wide document support available for interfacing a large

---

*"Encapsulate"—placing header information around a block or frame of data.

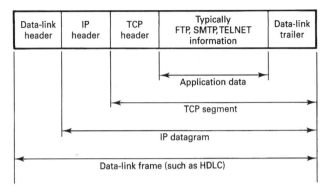

Figure 11.20. The incorporation of upper-layer PDUs into a data-link layer frame showing the relationship of TCP and IP.

number of popular data-link layer protocols. Among these we find the IEEE 802 series, Ethernet, ARCNET for LANs, and X.25 for WANs.

Typical IP encapsulation is shown in Figure 11.20. The figure illustrates how upper OSI layers are encapsulated with TCP and IP header information and then incorporated into the data-link layer frame.

For the case of IEEE 802 series LAN protocols, advantage is taken of the LLC common to all 802 protocols. The LLC extended header contains the subnetwork access protocol (SNAP) such that we have 3 octets for the LLC header and 5 octets in the SNAP. The LLC header has its fields fixed as follows:

DSAP = 10101010   (destination service access point)

SSAP = 10101010   (source service access point)

Control = 00000011   for unnumbered information (UI frame)

The 5 octets in the SNAP have three assigned for protocol ID or organizational code and two for "EtherType." Typical EtherType assignments are shown in Table 11.4. EtherType refers to the general class of LANs based on CSMA/CD.

Figure 11.21 shows how OSI relates to TCP/IP and IEEE 802, and Figure 11.22 illustrates an IEEE 802 frame incorporating TCP, IP, and LLC.

Addressing is resolved by a related TCP/IP protocol called *address resolution protocol* (ARP), which, typically, performs the mapping of the 32-bit IP4 internet address into a 48-bit IEEE 802 address.

Another interface problem may be the recommended minimum and absolute maximum IP datagram lengths. The total length is the length of the datagram, measured in octets, including internet header and data. The *total length* field in the IP header allows the length of a datagram to be up to

TABLE 11.4   EtherType Assignments

Ethernet Decimal	Hex	Description
512	0200	XEROX PUP
513	0201	PUP Address Translation
1,536	0600	XEROX NS IDP
2,048	0800	DOD Internet Protocol (IP)
2,049	0801	X.75 Internet
2,050	0802	NBS Internet
2,051	0803	ECMA Internet
2,052	0804	Chaosnet
2,053	0805	X.25 level 3
2,054	0806	Address Resolution Protocol (ARP)
2,055	0807	XNS Compatibility
4,096	1000	Berkeley Trailer
21,000	5208	BBN Simnet
24,577	6001	DEC MOP Dump / Load
24,578	6002	DEC MOP Remote Console
24,579	6003	DEC DECnet Phase IV
24,580	6004	DEC LAT
24,582	6005	DEC
24,583	6006	DEC
32,773	8005	HP Probe
32,784	8010	Excelan
32,821	8035	Reverse ARP
32,824	8038	DEC LANBridge
32,823	8098	Appletalk

Source: Reference 8.

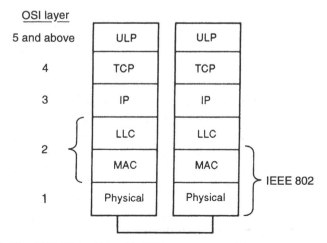

Figure 11.21. How TCP/IP working with IEEE 802 relates to OSI. ULP, upper-layer protocol.

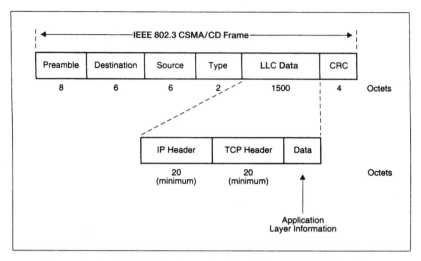

Figure 11.22. An IEEE 802 frame showing LLC and TCP/IP functions.

65,535 octets. Such long datagrams are impractical for most hosts and networks. All hosts should be prepared to accept datagrams up to 576 octets (whether they arrive whole or in fragments). It is recommended that hosts only send datagrams larger than 576 octets if they have assurance that the destination is prepared to accept larger datagrams.

The number 576 was selected to allow a reasonable-sized data block to be transmitted in addition to the required header information. For example, this size allows a data block of 512 octets plus 64 header octets to fit in a datagram. The maximum internet header is 60 octets, and a typical internet header is 20 octets, allowing a margin for headers of higher level protocols.

11.4.3   The IP Routing Function

IP offers typical connectionless service. In OSI, the network layer functions include routing and switching of a datagram through the telecommunications subnetwork. IP provides this essential function. It forwards the datagram based upon the network address contained within the IP header. Each datagram is independent and has no relationship with other datagrams. There is no guaranteed delivery of the datagram from the standpoint of IP. However, the next higher layer, the TCP layer, provides the reliability that IP lacks. It also carries out segmentation and reassembly functions of a datagram to match the frame sizes of the data-link layer protocols.

Addresses determine routing, and, at the far end, equipment (host hardware). Actual routing derives from the IP address, and equipment addresses derive from the data-link header, typically the 48-bit CSMA/CD MAC address. Some details of IP routing are covered in Section 10.5.1.

User data from upper-layer protocols are passed to the IP layer. The IP layer examines the network address (IP address) for a particular datagram and determines if the destination node is on its own local area network or some other network. If it is on the same network, the datagram is forwarded directly to the destination host. If it is on some other network, it is forwarded to the local IP router (gateway). The router, in turn, examines the IP address and forwards the datagram as appropriate. Routing is based on a lookup table residing in each router or gateway. (See Section 11.4.4.2 for a more detailed discussion of IP routing.)

11.4.4 Detailed IP Operation

The IP provides connectionless service, meaning that there is no call setup phase prior to exchange of traffic. There are no flow control nor error control capabilities incorporated in IP. These are left to the next higher layer, the TCP. The IP is transparent to subnetworks connecting at lower layers, and thus different types of networks can attach to an IP gateway.

Whereas prior to this discussion we have used the term *segmentation* to mean the breaking up of a data file into manageable segments, frames, packets, blocks, and so on, the IP specifications refer to this as *fragmentation*. IP messages are called *datagrams*. As we mentioned in Section 11.4.2, the minimum datagram length is 576 octets, and the maximum length is 65,535 octets. Fragmentation resolves protocol data unit (PDU) sizes of the different networks with which IP carries out an interface function. For example. X.25 packets typically have data fields of 128 octets, Ethernet limits the size of a PDU to 1500 octets, and so forth. Of course, IP does a reassembly of the "fragments" at the opposite end of its circuit.

11.4.4.1 *Frame Format of an IP Datagram; Description of Functions.*
The IP datagram frame format is illustrated in Figure 11.23. The format should be taken in context with Figures 11.21 and 11.22, showing how the IP datagram relates to TCP and the data-link layer. In Figure 11.23, we move from left to right and top down in our description.

Ver. Version field (4 bits) gives the release number of the IP version for which a particular router (or gateway) is equipped. IP has versions 4, 5, and 6. Several sources tell us that version 5 does not exist and that version 6 (IP6) is experimental (Ref. 9). Thus the version will be IP4, until IP6 gets the wrinkles worked out. (See RFC 791 for IP4 specification.) It is assigned decimal 4, and IP6 (RFC 1883) is assigned decimal 6.

IHL. Internet header length (4 bits) is measured in units of 32 bits.* Example: A header without certain options, such as QoS, typically has 20

---

*It is interesting to note that IP is based on 32-bit boundaries.

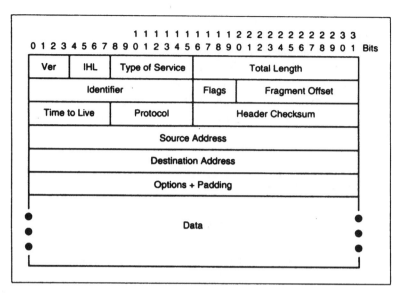

Figure 11.23. The IP datagram format.

octets. Thus its length is 5 or simply 20/4 (32 bits or 4 octets per unit; $5 \times 4 = 20$).

Type of Service. TOS is rarely used and usually set to 0 (8 bits). The TOS, when employed, is used to identify several QoS parameters provided by IP. The 8-bit field is broken down into four active groupings, and the last two bits are reserved. The first is *precedence,* which consists of three bits as follows:

000	Routine	001	Priority
010	Immediate	011	Flash
100	Flash override	101	CEITIC/ECP
110	Internetwork control	111	Network control

The three bits for precedence are followed by three 1-bit fields:

D delay:	0 = normal, 1 = low
T throughput:	0 = normal, 1 = high
R Reliability:	0 = normal, 1 = high

For example, if there is more than one route to a destination, the router could read this field to pick a route. This becomes important in open shortest path first (OSPF) routing protocol, which was the first IP protocol to take advantage of this.

Total Length. The total length field specifies the total length of an IP datagram (not packet) in octets. It includes the length of the header. There are 16 bits in the field. Therefore the maximum length of a datagram is $2^{16} - 1$ or 65,535 octets; the default length, as we know, is 576 octets. IP6 allows a concept known as the *jumbogram* to manage very long files.

The IP folks call segmentation *fragmentation*. It allows for the breaking up of long files into "fragments." The next three fields in the IP header deal with fragmentation. These are:

Identifier: 16 bits

Flags: 3 bits

Fragment offset: 3 bits

The term *fragment* suggests part of a whole; the *identifier* field identifies a fragment as part of a complete datagram, along with the source. It indicates which datagram fragments belong together so datagrams do not get mismatched. The receiving IP layer uses this field and the source IP address to identify which fragments belong together.

Flags (3 bits). Flags indicate whether more fragments are to arrive or no more data is to be sent for that particular datagram. They are coded in sequence as follows:

Bit 0 = reserved

Bit 1 = 0 = fragment; 1 = do not fragment

Bit 2 = (M bit) = 0 = last fragment; 1 = more fragments to follow

It should be noted that as a fragment datagram passes through the network it can be further fragmented. If a router receives a packet that must be fragmented and the don't fragment bit is set to 1, then it will discard the packet and send an error message via the ICMP protocol.

Fragment Offset (or just "offset"). The fragmentation offset field contains a value that specifies the relative position of the fragment in the original datagram. The value is initialized at 0 and subsequently set to the proper number when/if the router fragments the data. The offset value is measured in units of 8 octets (64 bits).

Using the total length and the fragment offset fields, IP can reconstruct a fragmented datagram and deliver it to the upper-layer software. The total length field indicates the total length of the original packet, and the offset field indicates to the node that is reassembling the packet the offset from the beginning of the packet. It is at this point that the data will be placed in the data segment to reconstruct the packet.

Time to Live (TTL) (8 bits). This indicates the amount of time that a datagram is allowed to stay in the network. Each router in the network through which the datagram passes is required to check this field and discard it if the TTL values equals 0. A router is also required to decrement this field in each datagram it processes. In actual implementations, the TTL field is a number of hops value. Therefore, when a datagram proceeds through a router, it has completed one hop and the field value is decremented by one. Implementation of IP may use a time counter in this field and decrement the value in 1-s decrements.

The TTL field can be used by routers to prevent endless loops. It can also be used by hosts to limit the lifetime that segments have in the network. Network management facilities employing, say, SNMP* might wish to set the TTL values for diagnostic purposes.

Protocol (8 bits). The protocol field identifies the next-level protocol above the IP that is to receive the datagram at the final host destination. It is similar to the "type field" found in Ethernet/CSMA/CD. Table 11.5, taken from RFC 1060 (Ref. 10) shows the protocol decimal numbering scheme and the corresponding protocol for each number. It identifies the most widely used upper-layer protocols. The most common protocols we would expect above the IP layer are TCP and UDP.

Header Checksum (16 bits). This detects errors in the header by means of cyclic redundancy check. CRC is described in Sections 4.4 and 4.5. If an error does occur in the header, the router notifies the originating host via an ICMP† message.

Source Address, Destination Address. Each of these fields is 32 bits long. There are two types of network addressing schemes used with IP (Ref. 9) (see RFC 1918):

Classless. The full address range (all 32 bits) can be used without regard to bit reservation classes. This type of addressing scheme is basically not used in direct host assignment. The scheme is directly applied to the routing tables used on the Internet and ISPs.

Classful. Where the 32-bit address is segmented into specific classes denoting networks and hosts.

The IP address structure is shown in Figure 11.24. It shows four address formats in which the lengths of the component fields making up the address field change with each of the different formats. These component fields are

*SNMP stands for simple network management protocol, a member of the TCP/IP family of protocols. See Chapter 18.
†ICMP stands for internet control message protocol. See Section 11.4.4.3.

TABLE 11.5   IP Protocol Field Numbering (Assigned Internet Protocol Numbers)

Decimal	Key Word	Protocol
0		Reserved
1	ICMP	Internet Control Message Protocol
2	IGMP	Internet Group Management Protocol
3	GGP	Gateway-to-Gateway Protocol
4		Unassigned
5	ST	Stream
6	TCP	Transmission Control Protocol
7	UCL	UCL
8	EGP	Exterior Gateway Protocol
9	IGP	Interior Gateway Protocol
10	BBN-MON	BBN-RCC Monitoring
11	NVP-II	Network Voice Protocol
12	PUP	PUP
13	ARGUS	ARGUS
14	EMCON	EMCON
15	XNET	Cross Net Debugger
16	CHAOS	Chaos
17	UDP	User Datagram Protocol
18	MUX	Multiplexing
19	DCN-MEAS	DCN Measurement Subsystems
20	HMP	Host Monitoring Protocol
21	PRM	Packet Radio Monitoring
22	XNS-IDP	XEROX NS IDP
23	TRUNK-1	Trunk-1
24	TRUNK-2	Trunk-2
25	LEAF-1	Leaf-1
26	LEAF-2	Leaf-2
27	RDP	Reliable Data Protocol
28	IRTP	Internet Reliable TP
29	ISO-TP4	ISO Transport Class 4
30	NETBLT	Bulk Data Transfer
31	MFE-NSP	MFE Network Services
32	MERIT-INP	MERIT Internodal Protocol
33	SEP	Sequential Exchange
34–60		Unassigned
61		Any host internal protocol
62	CFTP	CFTP
63		Any local network
64	SAT-EXPAK	SATNET and Backroom EXPAK
65	MIT-SUBN	MIT Subnet Support
66	RVD	MIT Remote Virtual Disk
67	IPPC	Internet Plur. Packet Core
68		Any distributed file system
69	SAT-MON	SATNET Monitoring
70		Unassigned
71	IPCV	Packet Core Utility
72–75		Unassigned
76	BRSAT-MON	Backroom SATNET Monitoring
77		Unassigned
78	WB-MON	Wideband Monitoring
79	WB-EXPAK	Wideband EXPAK
80–254		Unassigned
255		Reserved

Source: RFC 1060 (Ref. 10) and *Internet Protocol Transition Workbook* (Ref. 11).

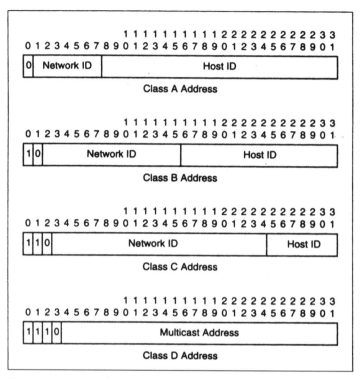

Figure 11.24. Internet protocol address formats. (RFC 791). Note that an IP address has three sections from left to right: Class (1, 2, 3, or 4 bits), Network ID, and Host ID. This holds for both destination and source addresses. (*Note:* This is basis of classful addressing.)

Class (A, B, C, or D; occupying the first 1, 2, 3, or 4 bits), "network address," and "local address." In Section 10.3.5.1, we break down the "local address" into two subfields: subnetwork address and host address.

The *Class A* addressing is for very large networks, such as ARPANET. The field starts with binary 0, indicating that it is a Class A format. In this case the local or host address component field is 24 bits long and has an address capacity of 2^{24}. *Class B* addressing is for medium-size networks, such as campus networks. The field begins with 10 to indicate it is a Class B format; the network component field is 14 bits long, and the host or local address component field is 16 bits in length. The *Class C* format is for small networks with a very large network ID field with an addressing capacity of 2^{24} and a considerably small host ID field of only 8 bits (2^{8} addressing capacity). The address field in this case starts with binary 110. The *Class D* format is for multicasting, a form of broadcasting. Its first 4 bits are the binary sequence 1110.

11.4.4.2 *IP Routing.* An IP router needs only the network ID portion of the address to perform its routing function. Each router or gateway has a routing table which consists of destination network addresses and a specified next-hop router (gateway).

Three types of routing are performed with the support of a routing table:

1. Direct routing to locally attached devices (hosts).
2. Routing to networks that are reached by one or more routers (gateways).
3. Default routing to destination network in case the first types of routing are unsuccessful.

A typical routing table is shown in Figure 11.25. Each row of the table contains data for each route that is known to the IP module storing the table.

Destination. Destination is self-explanatory and contains destination address. When the column is coded 0.0.0.0, the route is a default route.

IfIndex (Interface Index). Identifies the local physical port through which the next hop is routed.

Metric. Columns 3, 4, and 5 are labeled Metric. Actually five columns are assigned a Metric entry. Each column contains information about the cost metric used in determining the route. With most systems, the cost metric is the number of hops necessary to reach the destination. Ordinarily one metric would be used for route determination. But with

	1 Destination	2 If index	3 Metric 1	4 Metric 2	5 Metric 5	6 Next hop	7 Route type	8 Routing protocol	9 Route age	10 Routing mask	11 Route information	
Route 1												
Route 2												
Route 3												
Route 4												
Route n												

Figure 11.25. IP routing table.

new sophisticated algorithms [typically OSPF (open shortest path first)], more than one metric could be used in route calculation.

Next Hop (Ref. 6). This contains the IP address which identifies the next hop for the route selected.

Route Type (Ref. 7). There are four possibilities here: 1, none of the following; 2, an invalid route; 3, directly connected route meaning a directly connected subnet; 4, indirect meaning that an indirect connection is used to reach the destination.

Routing Protocol (Ref. 5). This tells the route discovery protocol by which the route was learned. This entry may be, for example, the RIP protocol where the column value would be 8, the Cisco internal group routing protocol where the value would be 11, the exterior gateway protocol (EGP) where the value would be 5, and so on.

Route Age (Ref. 10). Value in seconds since the route was verified or updated.

Routing Mask (Ref. 9). The mask is an overlay showing the valid binary digits. This is logical ANDed with the destination address in the IP datagram before being compared to column 1 of the table.

Route Information (Ref. 11). This is a reference to MIB (management information base) definition dealing with a particular protocol. Value depends on the type of protocol used.

Suppose a datagram (or datagrams) is (are) directed to a host that is not in the routing table resident in a particular gateway. Likewise, there is a possibility that the network address for that host is also unknown. These problems may be resolved with the *address resolution protocol* (ARP) (Ref. 11).

First the ARP searches a mapping table that relates IP addresses with corresponding physical addresses. If the address is found, it returns the correct address to the requester. If it cannot be found, the ARP broadcasts a request containing the IP target address in question. If a device recognizes the address, it will reply to the request where it will update its ARP cache with that information. The ARP cache contains the mapping tables maintained by the ARP module.

There is also a *reverse address resolution protocol* (RARP, Ref. 12). It works in a fashion similar to that of ARP, but in reverse order. RARP provides an IP address to a device when the device returns its own hardware address. This is particularly useful when certain devices are booted and only know their own hardware address.

Routing with IP involves a term called *hop*. A hop is defined as a link connecting adjacent nodes (routers or gateways) in a connectivity involving IP. A *hop count* indicates how many nodes (routers or gateways) must be traversed from source to destination.

There are two routing protocols used with IP. These are RIP (routing information protocol) and OSPF (open shortest path first).

RIP is known as a *distance vector* protocol. As one might imagine, it uses a database that we call a routing table. It contains two fields needed for routing: a vector (a known IP address) and distance, meaning how many routers away is the destination. There are other fields in the table, but these are the basic ones which provide the essential concept of the protocol. In other words, RIP simply builds a table (or database) in memory that contains all the routes that it has learned and the distance to that network. A router associates a cost with those interfaces. The cost is initially set to 1. As other routers report their topology and connectivity regarding locally attached subnets, all the information needed will be stored in each router's database (routing table). Routing will then be based on cost, which really equates to hop count.

OSPF (Ref. 13) is a *link state protocol*. It maintains the state of every link in the domain. Information is *flooded* to all routers in the domain. Flooding is discussed in Chapter 10. Flooding is the process of receiving information on one port and retransmitting to all other active ports of a router. This is how routers receive the same information. The information is stored in a router database that is called a *link-state* database. This database is identical on every router in an autonomous system. Based on information from the database, the Dykstra algorithm (Ref. 9) is employed which seeks the shortest-path tree based on metrics. It uses itself as the root of the tree. It is able to build a routing table based on shortest-path routing.

11.4.4.3 *Internet Control Message Protocol (ICMP).* ICMP (Ref. 14) is used as an adjunct to IP when there is an error in datagram processing. ICMP uses the basic support of IP as if it were a high-level protocol. However, ICMP is actually an integral part of IP and is implemented in every IP module.

ICMP messages are sent in several situations—for example, when a datagram cannot reach its destination, when a gateway does not have the buffering capacity to forward a datagram, and when the gateway can direct the host to send traffic on a shorter route.

ICMP messages typically report errors in the processing of datagrams. To avoid the possibility of infinite regress of messages about messages, and so on, no ICMP messages are sent about ICMP messages. Also ICMP messages are only sent about errors in handling fragment zero of fragmented datagrams. (*Note:* Fragment zero has the fragment offset equal to zero.)

Message Formats. ICMP messages are sent using the basic IP header (see Figure 11.23). The first octet of the data portion of the datagram is an ICMP-type field. The "data portion" is the last field at the bottom of Figure 11.23. The ICMP-type field determines the format of the remaining data. Any field labeled *unused* is reserved for later extensions and is fixed at zero when sent, but receivers should not use these fields (except to include them in the checksum). Unless otherwise noted under individual format descriptions, the values of the internet header fields are as follows:

- Version: 4.
- IHL (internet header length): length in 32-bit words.
- Type of service: 0.
- Total length: length of internet header and data in octets.
- Identification, flags, and fragment offset: used in fragmentation, as in basic IP described previously.
- Time to live: in seconds; as this field is decremented at each machine in which the datagram is processed, the value in this field should be at least as great as the number of gateways which this datagram will traverse.
- Protocol: ICMP = 1.
- Header checksum: The 16-bit one's complement of the one's complement sum of all 16-bit words in the header. For computing the checksum, the checksum field should be zero. The reference RFC (Ref. 14) (RFC 792) states that this checksum may be replaced in the future.
- Source address: The address of the gateway or host that composes the ICMP message. Unless otherwise noted, this is any of a gateway's addresses.
- Destination address: The address of the gateway or host to which the message should be sent.

There are eight distinct ICMP messages covered in RFC 792:

1. Destination unreachable message
2. Time exceeded message
3. Parameter problem message
4. Source quench message
5. Redirect message
6. Echo or echo reply message
7. Timestamp or timestamp reply message
8. Information request or information reply message

Example: Destination Unreachable Message. The ICMP fields in this case are shown in Figure 11.26.

IP Fields

Destination Address: The source network and address from the original datagram's data.

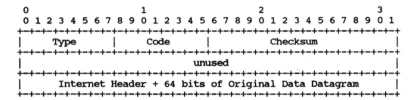

Figure 11.26. A typical ICMP message format: destination unreachable message. *Note:* The relevant RFC for ICMP is RFC 792 (Ref. 14).

ICMP Fields

Type 3
Code: 0 = net unreachable
         1 = host unreachable
         2 = protocol unreachable
         3 = port unreachable
         4 = fragmentation needed and DF set
         5 = source route failed
Checksum: as above
Internet header + 64 bits of data datagram

The Internet Header Plus the First 64 Bits of the Original Datagram's Data. These data are used by the host to match the message to the appropriate process. If a higher-level protocol uses port numbers, they are assumed to be in the first 64 data bits of the original datagram's data.

Description. If, according to the information in the gateway's routing tables, the network specified in the internet destination field of a datagram is unreachable (e.g., the distance to the network is infinity), the gateway may send a destination unreachable message to the internet source host of the datagram. In addition, in some networks the gateway may be able to determine if the internet destination host is unreachable. Gateways in these networks may send destination unreachable messages to the source host when the destination host is unreachable.

If, in the destination host, the IP module cannot deliver the datagram because the indicated protocol module or process port is not active, the destination host may send a destination unreachable message to the source host.

Another case is when a datagram must be fragmented to be forwarded by a gateway, yet the "Don't Fragment" flag is on. In this case the gateway must discard the datagram and may return a destination unreachable message.

It should be noted that Codes 0, 1, 4, and 5 may be received from a gateway; Codes 2 and 3 may be received from a host.

Section 11.4.4.3 is based on RFC 792 (Ref. 14).

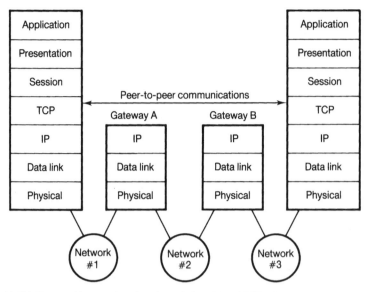

Figure 11.27. Protocol layers showing the relationship of TCP with other layered protocols.

11.4.5   The Transmission Control Protocol

11.4.5.1   TCP Defined. TCP (Refs. 15 and 16) was designed to provide reliable communication between pairs of processes in logically distinct hosts on networks and sets of interconnected networks. TCP operates successfully in an environment where network congestion and the loss, damage, duplication, or misorder of data can occur. This robustness in spite of unreliable communications media makes TCP well-suited to support commercial, military, and government applications. TCP appears at the transport layer of the protocol hierarchy. Here, TCP provides connection-oriented data transfer that is reliable, ordered, full-duplex, and flow-controlled. TCP is designed to support a wide range of ULPs.* The ULP can channel continuous streams of data through TCP for delivery to peer ULPs. The TCP breaks the streams into portions which are encapsulated together with appropriate addressing and control information to form a segment—the unit of exchange between TCPs. In turn, the TCP passes the segments to the network layer for transmission through the communication system to the peer TCP.

As shown in Figure 11.27, the layer below the TCP in the protocol hierarchy is commonly the IP layer. The IP layer provides a way for the TCP to send and receive variable-length segments of information enclosed in internet datagram "envelopes." The internet datagram provides a means for addressing source and destination TCPs in different networks. The IP also deals with fragmentation and reassembly of TCP segments required to

---

*ULP stands for upper layer protocol. Typically, there are three ULPs: file transfer protocol (FTP), simple mail transfer protocol (SMTP), telnet protocol. See Figure 11.18.

achieve transport and delivery through the multiple networks and interconnecting gateways. The IP also carries information on the precedence, security classification, and compartmentation of the TCP segments, so this information can be communicated end-to-end across multiple networks.

11.4.5.2 TCP Mechanisms. TCP builds its services on top of the network layer's potentially unreliable services with mechanisms such as error detection, positive acknowledgments, sequence numbers, and flow control. These mechanisms require certain addressing and control information to be initialized and maintained during data transfer. This collection of information is called a *TCP connection*. The following paragraphs describe the purpose and operation of the major TCP mechanisms.

PAR Mechanism. TCP uses a positive acknowledgment with retransmission (PAR) mechanism to recover from the loss of a segment by the lower layers. The strategy with PAR is for a sending TCP to retransmit a segment at timed intervals until a positive acknowledgment is returned. The choice of retransmission interval affects efficiency. An interval that is too long reduces data throughput, whereas one that is too short floods the transmission media with superfluous segments. In TCP, the timeout is expected to be dynamically adjusted to approximate the segment round-trip time plus a factor for internal processing; otherwise performance degradation may occur. TCP uses a simple checksum to detect segments damaged in transit. Such segments are discarded without being acknowledged. Hence, damaged segments are treated identically to lost segments and are compensated for by the PAR mechanism. TCP assigns sequence numbers to identify each octet of the data stream. These enable a receiving TCP to detect duplicate and out-of-order segments. Sequence numbers are also used to extend the PAR mechanism by allowing a single acknowledgment to cover many segments worth of data. Thus, a sending TCP can still send new data although previous data have not been acknowledged.

Flow Control Mechanism. TCP's flow control mechanism enables a receiving TCP to govern the amount of data dispatched by a sending TCP. The mechanism is based on a *window* which defines a contiguous interval of acceptable sequence-numbered data. As data are accepted, TCP slides the window upward in the sequence number space. This window is carried in every segment, enabling peer TCPs to maintain up-to-date window information.

Multiplexing Mechanism. TCP employs a multiplexing mechanism to allow multiple ULPs within a single host and multiple processes in a ULP to use TCP simultaneously. This mechanism associates identifiers, called *ports*, to ULP processes accessing TCP services. A ULP connection is uniquely identified with a *socket*, the concatenation of a port and an internet address. Each connection is uniquely named with a socket pair. This naming scheme allows a single ULP to support connections to multiple remote ULPs. ULPs which provide popular resources are assigned permanent sockets, called *well-known sockets*.

11.4.5.3 ULP Synchronization. When two ULPs wish to communicate, they instruct their TCPs to initialize and synchronize the mechanism of information flow on each to open the connection. However, the potentially unreliable network layer (i.e., internet protocol layer) can complicate the process of synchronization. Delayed or duplicate segments from previous connection attempts might be mistaken for new ones. A handshake proce-dure with clock-based sequence numbers is used in connection opening to reduce the possibility of such false connections. In the simplest handshake, the TCP pair synchronizes sequence numbers by exchanging three segments —thus the name *three-way handshake*.

11.4.5.4 ULP Modes. A ULP can open a connection in one of two modes: passive or active. With a passive open, a ULP instructs its TCP to be *receptive* to connections with other ULPs. With an active open, a ULP instructs its TCP to actually initiate a three-way handshake to connect to another ULP. Usually an active open is targeted to a passive open. This active/passive model supports server-oriented applications where a permanent resource, such as a database management process, can always be accessed by remote users. However, the three-way handshake also coordinates two simultaneous active opens to open a connection. Over an open connection, the ULP pair can exchange a continuous stream of data in both directions. Normally, TCP groups the data into TCP segments for transmission at its own convenience. However, a ULP can exercise a *push* service to force TCP to package and send data passed up to that point without waiting for additional data. This mechanism is intended to prevent possible deadlock situations where a ULP waits for data internally buffered by TCP. For example, an interactive editor might wait forever for a single input line from a terminal. A push will force data through the TCPs to the awaiting process. A TCP also provides the means for a sending ULP to indicate to a receiving ULP that "urgent" data appear in the upcoming data stream. This urgent mechanism can support, for example, interrupts or breaks. When a data exchange is complete, the connection can be closed by either ULP to free TCP resources for other connections. Connection closing can happen in two ways. The first, called a *graceful close*, is based on the three-way handshake procedure to complete data exchange and coordinate closure between the TCPs. The second, called an *abort*, does not allow coordination and may result in the loss of unac-knowledged data.

11.4.5.5 Scenario. The following scenario provides a walk-through of a connection opening, data exchange, and connection closing as might occur between the database management process and the user mentioned previ-ously. The scenario focuses more on (a) the three-way handshake mechanism in connection with opening and closing and (b) the positive acknowledgment with retransmission mechanism supporting reliable data transfer. Although not pictured, the network layer transfers the information between TCPs. For the purpose of this scenario, the network layer is assumed not to damage, lose, duplicate, or change the order of data unless explicitly noted. The scenario is organized into three parts:

1. A simple connection opening (steps 1–7)
2. Two-way data transfer (steps 8–17)
3. A graceful connection close (steps 18–24)

Scenario Notation. The following notation is used in Figures 11.28–11.30:

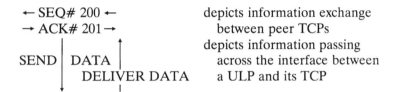

← SEQ# 200 ←	depicts information exchange
→ ACK# 201 →	between peer TCPs
SEND │ DATA │	depicts information passing across the interface between
DELIVER DATA	a ULP and its TCP

Start at Figure 11.28.

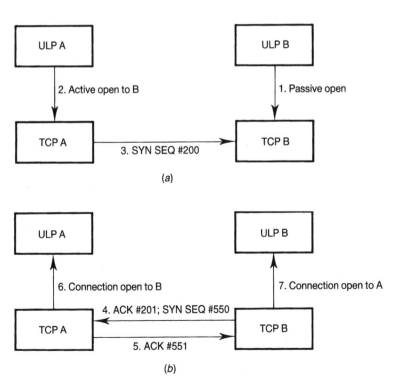

Figure 11.28. Two examples of a simple connection opening.

1. ULP B (the DB manager) issues a PASSIVE OPEN to TCP B to prepare for connection attempts from other ULPs in the system.

2. ULP A (the user) issues an ACTIVE OPEN to open a connection to ULP B.

3. TCP A sends a segment to TCP B with an OPEN control flag, called an SYN, carrying the first sequence number (shown as SEQ#200) it will use for data sent to B.

4. TCP B responds to the SYN by sending a positive acknowledgment, or ACK, marked with the next sequence number expected from TCP A. In the same segment, TCP B sends its own SYN with the first sequence number for its data (SEQ#550).

5. TCP A responds to TCP B's SYN with an ACK showing the next sequence number expected from B.

6. TCP A now informs ULP A that a connection is open to ULP B.

7. Upon receiving the ACK, TCP B informs ULP B that a connection has been opened to ULP A.

Turn to Figure 11.29.

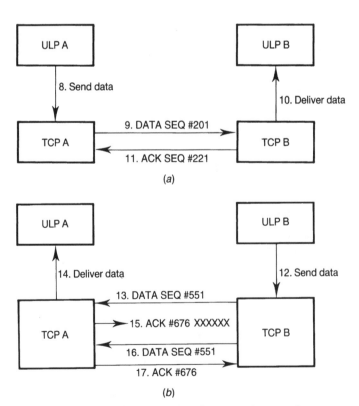

Figure 11.29. Two examples of two-way data transfer.

8. ULP A passes 20 octets of data to TCP A for transfer across the open connection to ULP B.

9. TCP A packages the data in a segment marked with current "A" sequence number.

10. After validating the sequence number, TCP B accepts the data and delivers it to ULP B.

11. TCP B acknowledges all 20 octets of data with the ACK set to the sequence number of the next data octet expected.

12. ULP B passes 125 bytes of data to TCP B for transfer to ULP A.

13. TCP B packages the data in a segment marked with the "B" sequence number.

14. TCP A accepts the segment and delivers the data to ULP A.

15. TCP A returns an ACK of the received data marked with the sequence number of the next expected data octet. However, the segment is lost by the network and never arrives at TCP B.

16. TCP B times out waiting for the lost ACK and retransmits the segment. TCP A receives the retransmitted segment, but discards it

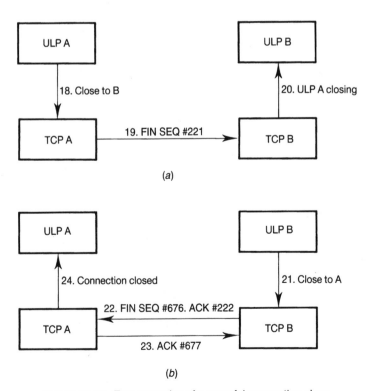

Figure 11.30. Two examples of a graceful connection close.

because the data from the original segment have already been accepted. However, TCP A re-sends the ACK.

17. TCP B gets the second ACK.

Turn to Figure 11.30.

18. ULP A closes its half of the connection by issuing a CLOSE to TCP A.
19. TCP A sends a segment marked with a CLOSE control flag, called a FIN, to inform TCP B that ULP A will send no more data.
20. TCP B gets the FIN and informs ULP B that ULP A is closing.
21. ULP B completes its data transfer and closes its half of the connection.
22. TCP B sends an ACK of the first FIN and its own FIN to TCP A to show ULP B's closing.
23. TCP A gets the FIN and the ACK, then responds with an ACK to TCP B.
24. TCP A informs ULP A that the connection is closed. (Not pictured.) TCP B receives the ACK from TCP A and informs ULP B that the connection is closed.

11.4.5.6 TCP Header Format. The TCP header format is shown in Figure 11.31. It should be noted that TCP works with 32-bit segments.

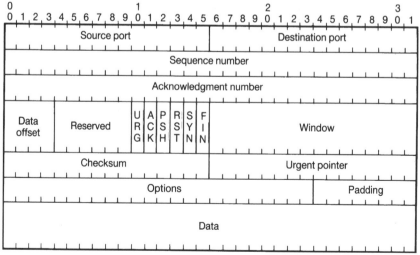

Figure 11.31. TCP header format. Note that one tick mark represents one bit position.

Source Port. The "port" represents the source ULP initiating the exchange. The field is 16 bits long.

Destination Port. This is the destination ULP at the other end of the connection. This field is also 16 bits long.

Sequence Number. Usually, this value represents the sequence number of the first data octet of a segment. However, if an SYN is present, the sequence number is the initial sequence number (ISN) covering the SYN; the first data octet is then numbered SYN + 1. The "SYN" is the *synchronize control flag*. It is the opening segment of a TCP connection. SYNs are exchanged from either end. When a connection is to be closed, there is a similar "FIN" sequence exchange.

Acknowledgment Number. If the ACK control bit* is set (bit 2 of the 6-bit control field), this field contains the value of the next sequence number that the sender of the segment is expecting to receive. This field is 32 bits long.

Data Offset. This field indicates the number of 32-bit words in the TCP header. From this value the beginning of the data can be computed. The TCP header is an integral number of 32 bits long. The field size is 4 bits.

Reserved. This is a field of 6 bits set aside for future assignment. It is set to zero.

Control Flags. The field size is 6 bits covering six items (1 bit per item):

1. URG: Urgent pointer field significant
2. ACK: Acknowledgment field significant
3. PSH: Push function
4. RST: Reset the connection
5. SYN: Synchronize sequence numbers
6. FIN: No more data from sender

Window. The number of data octets beginning with the one indicated in the acknowledgment field which the sender of this segment is willing to accept. The field is 2 octets in length.

Checksum. The checksum is the 16-bit one's complement of the one's complement sum of all 16-bit words in the header and text. The checksum also covers a 96-bit pseudo-header conceptually prefixed to the TCP header. This pseudo-header contains the source address, the destination address, the protocol, and TCP segment length.

---

*ACK is a control bit (acknowledge) occupying no sequence space, which indicates that the acknowledgment field of this segment specifies the next sequence number the sender of this segment is expecting to receive, thereby acknowledging receipt of all previous sequence numbers.

Urgent Pointer. This field indicates the current value of the urgent pointer as a positive offset from the sequence number in this segment. The urgent pointer points to the sequence number of the octet following the urgent data. This field is only to be interpreted in segments with the URG control bit set. The urgent point field is two octets long.

Options. This field is variable in size; and, if present, options occupy space at the end of the TCP header and are a multiple of 8 bits in length. All options are included in the checksum. An option may begin on any octet boundary. There are two cases of an option:

1. Single octet of option kind
2. An octet of option kind, an octet of option length, and the actual option data octets

Options include "end of option list," "no-operation," and "maximum segment size."

Padding. The field size is variable. The padding is used to ensure that the TCP header ends and data begin on a 32-bit boundary. The padding is composed of zeros.

11.4.5.7 TCP Entity State Diagram. Figure 11.32 summarizes TCP operation. It shows a TCP state diagram.

Section 11.4 is based on various RFCs, MIL-STD-1778, and Refs. 9, 15, and 16.

11.4.6 Internet Protocol Version 6 (IPv6)

IPv6 is an evolutionary step improvement on IPv4. The principal problem with IPv4 was its addressing, a 32-bit field. IPv6 has a 128-bit field, which supports more levels of addressing.

What happened to IPv5? It is covered by RFC 1819 and is called ST2 + . It serves as an adjunct to IP where a guaranteed QoS is required.

In IPv6 the header has been simplified and certain IPv4 fields have been dropped. The IPv6 header is a static 40 octets long.

A new capability has been added to enable labeling of packets belonging to particular traffic *flows* for which the sender requests special handling, such as nondefault QoS or *real-time* service.

There have also been added authentication and privacy capabilities and improved data integrity.

11.4.6.1 IPv6 Routing. *Prefix routing* is used with IPv6 where every address has an associated prefix. This is just a mask identifier which indicates how many bits, starting from the left are used for routing and how many

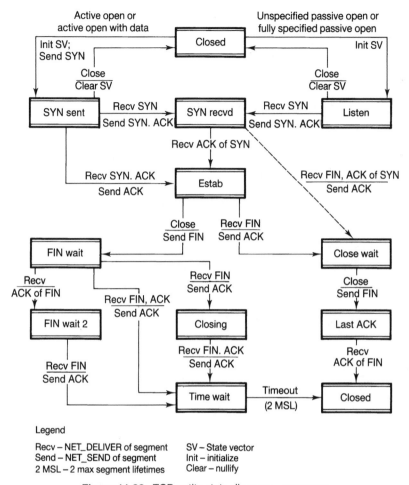

Figure 11.32. TCP entity state diagram—summary.

identify a host. The routing prefix is a tool for building routing tables. End stations make the prefix similar to the mask described in Chapter 10.

Existing protocols may use IPv6 addresses as well. However, they will have to change to understand 128-bit addressing as one might imagine.

11.4.6.2 IPv6 Header. Figure 11.33 illustrates the IPv6 header. Compare this to Figure 11.23, which shows the IPv4 header. All the fields have been changed except the Vers (version) field.

The following is a description of the IPv6 header fields:

- Version: (4 bits) Internet protocol version number = 6.
- Traffic class: (8 bits)
- Flow label: (20 bits)

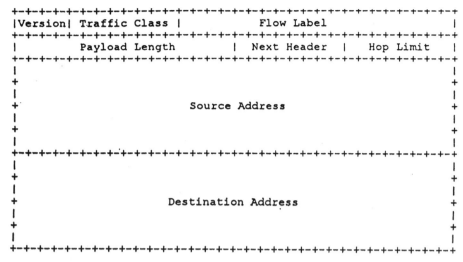

Figure 11.33. IPv6 header format. (From RFC-2460, Ref. 17.)

- Payload length: (16 bits) Unsigned integer. It is the length of the IPv6 payload, measured in octets. Note: Any extension headers present are considered part of the payload.
- Next header: (8 bits) This means it is an 8-bit selector. It identifies the type of header immediately following the IPv6 header. It uses the same values as the IPv4 protocol field (see RFC-1700).
- Hop limit: (8 bits) Unsigned integer. Decremented by 1 by each node that forwards the packet. The packet is discarded if the hop limit is decremented to zero.
- Source address: (128 bits) Address of the originator of the packet.
- Destination address: (128 bits) Address of the intended recipient of the packet. This may possibly not be the ultimate recipient if a routing header is present.

Discussion on Selected Fields

Traffic Class. The 8-bit Traffic Class field is available for use by originating nodes and/or forwarding routers to identify and distinguish between different classes or priorities of IPv6 packets. At the point in time at which specification was being written, there are a number of experiments underway in the use of the IPv4 Type of Service and/or Precedence bits to provide various forms of "Differentiated Service" for IP packets, other than through the use of explicit flow setup. The Traffic Class field in the IPv6 header is intended to allow similar functionality to be supported in IPv4. It is hoped by the IETF/Internet Society that those experiments will eventually lead to agreement on what sorts of traffic classifications are most useful for IP packets. Detailed definitions of the syntax and semantics of all or some of the IPv6 Traffic Class bits, whether experimental or intended for eventual standardization, are to be provided in separate documents.

The following general requirements apply to the Traffic Class field:

- The service interface to the IPv6 service within a node must provide a means for an upper-level protocol to supply the value of the Traffic Class bits in packets originated by that upper-layer protocol. The default value must be zero for all 8 bits.
- Nodes that support a specific (experimental or eventual standard) use of some or all of the Traffic Class bits are permitted to change the value of those bits in packets that they originate, forward or receive, as required for that specific use. Nodes should ignore and leave unchanged any bits of the Traffic Class field for which they do not support a specific use.
- An upper-layer protocol must not assume that the value of the Traffic Class bits in a received packet are the same as the value sent by the packet's source.

Flow Labels. The 20-bit Flow Label field in the IPv6 header may be used by a source to label sequences of packets for which it requests special handling by IPv6 routers, such as nondefault QoS or "real time" service. This aspect of IPv6 is still experimental and subject to change as the requirements for flow support in the Internet become clearer. Hosts or routers that do not support the functions of the Flow Label field are required to set the field to zero when originating a packet, pass the field on unchanged when forwarding a packet, and ignore the field when receiving a packet.

IPv6 Addressing. There are three types of addresses:

Unicast. An identifier for a single interface. A unique address delivered to a single notation.

Anycast. New for IP, an anycast address is an identifier for a set of interfaces (typically belonging to different nodes). This is similar to multicast, but a packet sent to an anycast address is delivered to one of the interfaces identified by that address (the "nearest" one, according to the routing protocols' measure of distance).

Multicast. An identifier for a set of interfaces (typically belonging to different nodes). A packet sent to a multicast address is delivered to all interfaces identified by that address.

The broadcast address is not defined in IPv6. It was superseded by the multicast address (Ref. 9).

IPv6 is based on RFC 2460 (Ref. 17).

11.5   USER DATAGRAM PROTOCOL (UDP)

UDP provides a procedure for application programs to send messages to other programs with a minimum of protocol mechanism. The protocol is transaction oriented and delivery and duplicate protection are not guaranteed. Applications requiring ordered reliable delivery of streams of data should use the transmission control protocol (see Section 11.4.5). UDP assumes that the internet protocol (IP—see Section 11.4) is used as the underlying protocol.

11.5.1   UDP Header Format and Fields

The UDP datagram header format is shown in Figure 11.34. The following is a description of the fields in Figure 11.34.

Source Port is an optional field, when meaningful, that indicates the port of the sending process. It may be assumed to be the port to which a reply should be addressed in the absence of any other information. If not used, a value of zero is inserted.

Destination Port has a meaning within the context of a particular IP destination address.

Length is the length in octets of this user datagram including this header and the data field. There is minimum value of length of eight (octets).

Checksum is the 16-bit one's complement of the one's complement sum of a pseudo-header of information from the IP header, the UDP header, and the data, padded with zero octets at the end (if necessary) to make a multiple of 2 octets.

The pseudo-header conceptually prefixed to the UDP header contains the source address, the destination address, the protocol, and the UDP length. It is shown in Figure 11.35. This information gives protection against misrouted datagrams. The checksum procedure is the same as used in TCP.

If the computed checksum is zero, it is transmitted as all ones (the equivalent in one's complement arithmetic). An all-zero transmitted check-

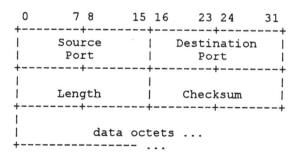

Figure 11.34. UDP datagram header format. (From RFC 768, Ref. 18.)

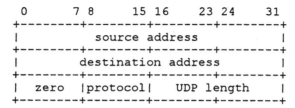

Figure 11.35. Format of pseudo-header. (Ref. 18)

sum value means that the transmitter generated no checksum (for debugging or for higher level protocols that don't care).

User Interface. A user interface allows the creation of new receive port. It also allows receive operations on the receive ports that return the data octets and an indication of source port and source address, and an operation that allows a datagram to be sent, specifying the data, source, and destination ports and addresses to be sent.

IP Interface. The UDP module must be able to determine the source and destination internet addresses and the protocol field from the internet header. One possible UDP–IP interface would return the whole internet datagram including all of the internet header in response to a receive operation. Such an interface would also allow the UDP to pass a full internet datagram complete with header to the IP to send. The IP would verify certain fields for consistency and compute the internet header checksum.

Protocol Application. The major uses of this protocol is the Internet Name Server and the Trivial File Transfer. This is protocol number 17 (21 octal) when used in the internet protocol. Other protocol numbers are listed in RFC 762.

Section 11.5 consists of abstracts from RFC 768.

11.6   THE CLNP PROTOCOL BASED ON ISO 8473 (Ref. 19)

CLNP (connectionless network protocol) was designed to be used in the context of the interworking protocol approach to the provision of the connectionless mode network service defined in ISO 8648. It is intended for use in the subnetwork independent convergence protocol (SNICP) role. CLNP operates to construct the OSI Network Service over a defined set of underlying services. CLNP was planned as a replacement of IPv4, however, IPv6 and other improvements resulted in the IT community remaining with IPv4 and IPv6. We have included CLNP, sometimes referred to as TUBA (TCP and

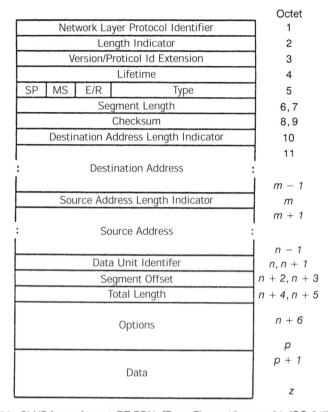

Octet

Network Layer Protocol Identifier	1
Length Indicator	2
Version/Proticol Id Extension	3
Lifetime	4
SP	MS	E/R	Type	5
Segment Length	6, 7
Checksum	8, 9
Destination Address Length Indicator	10
	11
Destination Address	$m - 1$
Source Address Length Indicator	m
	$m + 1$
Source Address	$n - 1$
Data Unit Identifer	$n, n + 1$
Segment Offset	$n + 2, n + 3$
Total Length	$n + 4, n + 5$
Options	$n + 6$
	p
Data	$p + 1$
	z

Figure 11.36. CLNP frame format, DT PDU. (From Figure 10, page 21, ISO-8473, Ref. 19.)

UDP over bigger addresses), because it is constantly referenced in the literature with some application in Europe.

11.6.1 CLNP Frame Formats

There are two different frame types used with CLNP. These are *Data* (*DT*) *PDU* and *Error Report PDU*. The formats for these frames are illustrated in Figures 11.36 and 11.37. Each consists of a fixed part and a variable part. The fixed part consists of those fields from the Network Layer Protocol Identifier through the Checksum.

11.6.2 Frame Field Description—FIXED PART

Network Layer Protocol Identifier (1 octet). The value of this field is set to binary 1000 0001 to identify this network layer protocol as ISO 8473. The value of this field is set to binary 0000 0000 to identify the inactive layer protocol subset.

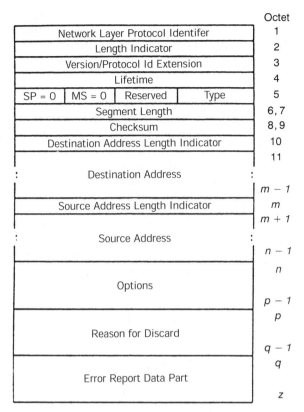

Figure 11.37. CLNP frame format, Error Report PDU. (From Figure 12, page 22, ISO-8473, Ref. 19.)

Length Indicator. The length is indicated by a binary number, with a maximum value of 254 (binary 1111 1110). The length indicated is the length in octets of the header. The value 255 (1111 1111) is reserved for possible future extensions.

Version / Protocol Identifier Extension (1 octet). The value of this field is binary 0000 0001, which identifies the standard Version 1 of ISO 8473.

PDU Lifetime (1 octet). The PDU Lifetime field is encoded as a binary number representing the remaining lifetime of the PDU, in units of 500 ms.

Flags

Segmentation Permitted (SP) (1 bit). The Segmentation Permitted flag indicates whether segmentation is permitted. Its value is determined by the originator of the PDU and cannot be changed by any other network entity for the lifetime of the Initial PDU and any Derived PDUs.

A value of 1 indicates that segmentation is permitted. A value of 0 indicates that segmentation is not permitted. When the value of 0 is selected, the segmentation part of the PDU header is not present, and the value of the Segment Length field gives the total length of the PDU.

More Segments (MS) (1 bit). The More Segments flag indicates whether or not the data part of this PDU contains (as its last octet) the last octet of the User Data in the NSDU (Network Service Data Unit). When the More Segments flag is set to 1, segmentation has occurred and the last octet of the NSDU is not contained in this PDU. The More Segments flag cannot be set to 1 if the Segmentation Permitted flag is not set to 1.

When the More Segments flag is set to 0, the last octet of the Data Part of the PDU is the last octet of the NSDU.

Error Report E/R (1 bit). When the Error Report flag is set to 1, the rules in Section 6.10 of Ref. 19 are used to determine whether to generate an Error Report PDU if it is necessary to discard this Data PDU.

When the Error Report flag is set to 0, discard of the Data PDU will not cause the generation of an Error Report PDU.

Type Code (5 bits). The Type Code identifies the type of PDU (Data of Error Report). If bits 5, 4, 3, 2, 1 are set to 111000, it is a Data PDU; if they are set to 00001, it is an Error Report PDU.

PDU Segment Length (2 octets). The Segment Length field specifies the entire length, in octets, of the PDU, including both header and data (if present). When the full protocol is employed and a PDU is not segmented, the value of this field is identical to the value of the Total Length field located in the Segmentation Part of the header.

When the nonsegmenting protocol subset is employed, no segmentation part is present in the header. In this case, the Segment Length field specifies the entire length of the Initial PDU, including both header and data (if present). The Segment Length field is not changed for the lifetime of the PDU.

PDU Checksum (2 octets). The checksum is computed on the entire PDU header. For a Data PDU, this includes the segmentation and option parts (if present). For an Error Report PDU, this includes the reason for discard field as well.

A checksum value of zero is reserved to indicate that the checksum is to be ignored. The operation of the PDU Header Error Detection function ensures that the value zero does not represent a valid checksum. A nonzero value indicates that the checksum will be processed.

Address Part

Destination and Source Address (Variable length). The Destination and Source Address used by this protocol are Network Service Access Point address as defined in ISO 8348/Add.2. The Destination and Source Addresses are of variable length. The Destination and Source Address fields are encoded as Network Protocol Address Information using the Preferred Binary Encoding defined in Section 8.3.1 of ISO 8348/Add.2.

The field consists of a subfield 1 octet long which is the Destination Address Length Indicator, then the Destination Address itself of variable length. Then there is the Source Address Length Indicator subfield that is 1 octet long. This is followed by the Source Address itself of variable length.

Segmentation Part. If the Segmentation Permitted Flag in the Fixed Part of the PDU Header is set to 1, the segmentation part of the header will be present. If the Segmentation Permitted flag is set to 0, the segmentation part is not present (the nonsegmenting protocol subset is in use).

The Segmentation part consists of three fields, each 2 octets long. These are Data Unit Identifier, Segment Offset, and Total Length. These three fields are described as follows.

Data Unit Identifier (2 octets). The Data Unit Identifier identifies an Initial PDU (and hence, its Derived PDUs) so that a segmented data unit may be correctly reassembled.

Segment Offset (2 octets). For each Derived PDU, the Segment Offset field specifies the relative position of the segment contained in the Data Part of the Derived PDU with respect to the start of the Data Part of the Initial PDU. The offset is measured in units of octets. The offset of the first segment (and hence the initial PDU) is 0, an unsegmented (Initial) PDU has a segment offset value of 0. The value of this field is a multiple of 8.

PDU Total Length (2 octets). The Total Length Field specifies the entire length of the Initial PDU in octets, including both the header and data. This field is not changed for the lifetime of the Initial PDU (and hence its Derived PDUs).

Options Part. If the options part is present, it may contain one or more parameters. The number of parameters that may be contained in the options part is constrained by the length of the options part, which is determined by the formula

Options part length = PDU header length

$-$ (length of fixed part + length of address part

+ length of segmentation part)

and by the length of the individual optional parameters.

TABLE 11.6   Encoding Option Parameters

Octets	Field
n	Parameter Code
$n + 1$	Parameter Length (e.g., m)
$n + 2$	
to	Parameter Value
$n + m + 1$	

Parameters defined in the options part may appear in any order. Duplication of options is not permitted. Receipt of a Protocol Data Unit with a duplicated option should be treated as a protocol error. The rules governing the treatment of protocol errors are described in Section 6.10 (Error Reporting Function) of the reference document.

The encoding parameters contained within the options part are listed in Table 11.6.

The *Parameter Code* field (1 octet) is encoded in binary and provides for a maximum of 255 different parameters. No parameter code uses bits 8 and 7 with the value 00, so the actual maximum number of parameters is lower. A parameter code of 255 (binary 1111 1111) is reserved for possible extensions.

The *Parameter Length* (1 octet) field indicates the length, in octets, of the *Parameter Value* field. The length is indicated by a positive binary number, m, with a theoretical maximum value of 254. The practical maximum value of m is lower. For example, in the case of a single parameter contained within the options part, 2 octets are required for the parameter code and the parameter length indicators. Thus, the value of m is limited to

$$m = 252 - (\text{length of fixed part} + \text{length of address part}$$

$$+ \text{length of segmentation part})$$

Accordingly, for each successive parameter the maximum value of m decreases.

The parameter field contains the value of the parameter identified in the parameter code field. The following parameters are permitted in the options part.

Parameters Permitted

Padding. The padding parameter is used to lengthen the PDU header to a convenient size as described in Section 6.12 of the reference document.

Parameter Code: 1100 1100
Parameter Length: variable
Parameter Value: any value is allowed

Security. This parameter allows a unique and unambiguous security level to be assigned to a PDU. (See Section 6.13 of the reference document.)

Parameter Code: 1100 0101
Parameter Length: variable
Parameter Value: The high-order 2 bits of the first octet specify the Security Format Code where:

Security Field Format	Type of Security Field
00	Reserved
01	Source Address Specific
10	Destination Address Specific
11	Globally Unique

The rest of the first octet is reserved and is set to 0. The remainder of the Parameter Value field specifies the security level.

Source Address Specific and the *Destination Address Specific* fields are coded 01 and 10, respectively, as indicated in the above table. The Security Format Code values of 01 and 10 indicate that the remaining octets of the parameter value field specify a security level which is unique and unambiguous in the context of the security classification system employed by the authority responsible for assigning the source NSAP (network service access point) Address and the destination NSAP Address, respectively.

Globally Unique is coded 11 and is similar to the above. However, in this case the security classification system is not specified.

Source Routing. Source routing, which is another parameter, specifies, either completely or partially, the route to be taken from Source Network Address to Destination Network Address.

Parameter Code: 1100 1000
Parameter Length: variable
Parameter Value: 2 octets of control information followed by a concatenation of network-entity title entries ordered from source to destination

The first octet of the parameter value is the type code, and has the following significance:

0000 0000 partial source routing
0000 0001 complete source routing
    (all other values are reserved)

The second octet indicates the octet offset of the next network-entity title entry to be processed in the list. It is relative to the start of the parameter, such that a value of 3 indicates that the next network-entity title entry begins immediately after this control octet. Successive octets are indicated by correspondingly larger values of this indicator.

The third octet begins the network-entity title list. The list consists of variable length network-entity title entries. The first octet of each entry gives the length of the network-entity title which comprises the remainder of the entry.

Recording of Route. The recording of route parameter identifies the intermediate systems traversed by the PDU.

Parameter Code: 1100 1011
Parameter Length: variable
Parameter Value: 2 octets of control information followed by a concatenation of network-entity title entries ordered from source to destination

The first octet of the parameter value is the type code and has the following significance:

0000 0000   Partial Recording of Route progress
0000 0001   Complete Recording of Route progress

The second octet identifies the first octet not currently used for a recorded network-entity title, and, therefore, also the end of the list. It is encoded relative to the start of the parameter value, such that a value of 3 indicates that no network-entity titles have yet been recorded. A value of all 1's is used to indicate that route recording has been terminated.

The third octet begins the network-entity title list. The list consists of variable length network-entity title entries. The first octet of each entry gives the length of the network-entity title comprising the remainder of the entry. Network-entity title entries are always added to the end of the list.

Other Parameters. There is a Quality of Service (QoS) Maintenance parameter which conveys information about quality of service requested by the originating Network-Service user. Another parameter is the Priority parameter that indicates the relative priority of the PDU. Intermediate systems that support this option make use of this information in routing and ordering PDUs for transmission.

11.6.3 Error Report PDU (ER)

The format of the Error Report PDU is shown in Figure 11.37. This format, similar to the Data PDU, contains a fixed part group of fields and a variable part. The only difference in the fixed parts is that the flag bits (SP, MS, and E/R) are always set to binary 0. The Destination Address specifies the network-entity title of the originator of the discarded PDU. The Source Address specifies the title of the intermediate-system or end-system network-entity initiating the Error Report PDU.

11.6.3.1 *Reason for Discard.* This parameter is valid only for the Error Report PDU.

Parameter Code: 1100 0001
Parameter Length: 2 octets
Parameter Value: type of error encoded in binary

The parameter values are listed in Table 11.7.

TABLE 11.7   Reason for Discard Parameter Values

Parameter Value	Class of Error	Meaning
0000 0000	General	Reason not specified
0001		Protocol Procedure Error
0010		Incorrect Checksum
0011		PDU Discarded due to Congestion
0100		Header Syntax Error (cannot be parsed)
0101		Segmentation needed but not permitted
0110		Incomplete PDU Received
0111		Duplicate Option
1000 0000	Address	Destination Address Unreachable
0001		Destination Address Unknown
1001 0000	Source	Unspecified Source Routing Error
0001	Routing	Syntax Error in Source Routing Field
0010		Unknown Address in Source Routing Field
0011		Path not Acceptable
1010 0000	Lifetime	Lifetime Expired while Data Unit in Transit
0001		Lifetime Expired during Reassembly
1011 0000	PDU	Unsupported Option not Specified
0001	Discarded	Unsupported Protocol Version
0010		Unsupported Security Option
0011		Unsupported Source Routing Option
0100		Unsupported Recording of Route Option
1100 0000	Reassambly	Reassembly Interference

Source: Table 7, page 23, ISO 8473, Ref. 19.

The first octet of the parameter value contains an error type code. If the error in the discarded Data PDU can be localized to a particular field, the number of the first octet of that field is stored in the second octet of the reason for discard parameter field; or if the error is a checksum error, then the value 0 is stored in the second octet of the reason for discard field.

Error Report Data Part. This field contains the entire header of the discarded Data PDU and may contain none, some, or all of the Data Part of the discarded PDU.

Section 11.6 has been abstracted from ISO 8473, 1988 (Ref. 19).

11.7   NETWORKING VIA VSATS

11.7.1   Rationale of VSAT Networks

There are several reasons for implementing a very-small-aperture terminal (VSAT) network. Much depends on the area of the world or even the country one is in. In industrial nations, VSAT networks are used as private data networks. The overwhelming application is to provide a star network connectivity as shown in Figure 11.38 where the hub is colocated with corporate headquarters and the VSAT terminals are in branch locations. These are generally data networks that feed data information to their headquarters

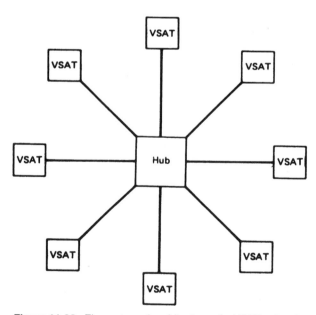

Figure 11.38. Elements and architecture of a VSAT network.

facility. In many cases such circuits are two-way. They are particularly attractive for large, dispersed industries and sales organizations. These networks completely bypass local telephone companies that otherwise might have provided such service.

In underdeveloped nations the rationale for VSAT networks is less economic and more service-oriented. In these areas the local telephone companies or administrations provide such poor service that owning one's own network becomes very attractive. Specialized data protocols developed to optimize usage are commonly used on hub-and-spoke configurations.

The hub-and-spoke architecture of Figure 11.38 is the most common topology. However, another topology is gaining popularity, especially in developing nations. This is a full mesh network where each and every VSAT terminal has connectivity to all other terminals in the network. There is no hub. Their antennas are larger and provide greater RF power to accommodate higher bit rates, often DS1 or E1 configurations (see Chapter 7). They provide both voice and data connectivity. We can expect to find the more common type of data protocols applied on these circuits.

11.7.2 Basic Description of a VSAT Network

The most common type of VSAT network has the hub-and-spoke configuration* as shown in Figure 11.38. There are three elements to the network: (1) the hub, which is a rather large facility, (2) the VSAT terminals, which are small, and (3) the satellite. The satellite will be in geostationary orbit. The round-trip delay, some 500 ms, may pose a problem.

The term VSAT contains the expression "very small aperture," meaning a terminal with a small antenna. Most satellite terminal antennas are "dishes." A dish means a parabolic antenna. VSAT antennas have diameters (of the parabolic dish) ranging from 0.8 m to 2.0 m. Hub stations have antenna diameters ranging from 5 m to 8 m. Aside from the hub, everything is small in a VSAT network. There seems to be no limit in the number of VSAT terminals a hub can service. One case we are familiar with has over 2000 accesses. Because the multiplier is 2000, to keep the system viable economically, the cost of a VSAT terminal must be reduced just as much as possible.

Of course, all terminals in the network must be within line-of-sight of the satellite being used. In fact, the terminals must have elevation angles greater than 5 degrees for C-band (i.e., 6/4 GHz) operation and greater than 10, or even 15, degrees of Ku-band (i.e., 14/12 GHz) operation. The definition of elevation angle is shown in Figure 11.39.

11.7.3 Disadvantages of a VSAT Network

There are essentially three drawbacks to the employment of VSAT networks. The first two deal with survivability. The problem is that there are two critical

---

*This is a star network, if you will.

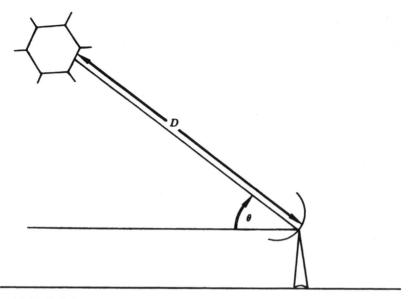

Figure 11.39. Definition of elevation angle for satellite communications. D, distance to the satellite; θ, elevation angle.

single-point failure locations. The first is the satellite. If we lose the satellite, the network goes down. The second is the hub. If we lose the hub for whatever reason, the network also goes down.

One alternative is to lease standby transponder space on another nearby satellite. This is fraught with danger. By nearby, we mean a satellite in a position such that antennas do not have to be slewed far to acquire the standby satellite. This reacquiring is not as easy as it seems. To keep costs low, we have no automatic pointing and tracking capability on the VSAT terminal antennas. Thus a well-trained technician must visit each site to go through the reacquisition exercise. Some terminals may be kept down for days because of this.

By regarding the hub as a single-point failure, we can build or lease a standby hub. Either way, it implies added cost.

The third drawback is that on conventional VSAT networks, for one VSAT to communicate with another requires routing through the hub involving a double hop (i.e., from VSAT1 up to satellite and down to hub, and from hub up to satellite and down to VSAT2). This is not done for voice operations, if that is what we want. It is not done because of excessive delay. If we use the proper protocols, the delay will have little effect on data connectivities. Seldom, however, is there a requirement for one VSAT to direct traffic to another. Nearly always the direction of traffic is from VSAT to hub, from hub to VSAT, or both. For instance, if it is a hotel chain installing a system, the hub will be located with the main reservation computer and service desk.

11.7.4 Important Design Variables

Probably the largest recurring cost of a VSAT network is space segment charges. By this we mean the cost of leasing transponder space from the satellite service provider.

Space segment charges are usually based on, or are a function of, transponder bandwidth required for the network. To reduce these recurring charges, we have to reduce the bandwidth as much as possible for a certain service quality.

How can we measure service quality? Several measures can come to mind besides BER. Time to carry out a transaction could well be one. However, for some data services, time to carry out a transaction may have secondary importance. Throughput, if the term is well-defined, is another measure of service quality.

Bit error rate (BER) behavior can be pretty well controlled except on VSAT networks using the Ku-band and higher frequencies. In frequency bands above 10 GHz, heavy rainstorms can severely degrade BER over the period of the storm. It causes a form of fading that is very similar to multipath fading except for some of the dispersive properties. There are three steps that can be taken to mitigate the effects of rainfall fading:

1. Accept the condition and let ARQ carry out its function. Throughput will degrade accordingly for the period of the rainstorm.
2. Overbuild the system as one would do to build up fade margin. This can be done by using larger antennas, increasing transmitter power and improving receiver noise performance. A subset of this is to make the circuit adaptive and only increase transmit power during the storm event.
3. Use channel coding with interleaving. This is a very effective measure if we choose the interleaving interval to be longer (statistically) than fade duration. Such a measure is fine for data circuits, but care must be taken on voice circuits because of the increase in delay caused by the interleaver.

Coding refers to channel coding (as opposed to source coding). With channel coding we systematically add redundancy to correct errors at the receiver. We described channel coding when we discussed FEC. They are the same. The problem with channel coding (or FEC) is that it corrects random errors and not burst errors. The fading due to rainfall gives burst errors. By using an interleaver, which pseudorandomizes the serial bit stream of data, we can correct for error bursts. This assumes that the duration of the block of data is longer than the duration of a fade (at least statistically). This "block" of data has nothing to do with blocks we have been working with in the past. This is a "chunk" of data more properly called a *span*. The interleaver stores the data for the span period and then pseudorandomly shuffles the data while it stores the next contiguous span. The reverse process is carried out at the distant end. Of course the receive interleaver must be in synchronism with the transmit interleaver. In such a way burst errors can be made to look like random errors. With this type of operation we can drastically reduce burst errors and pick up several decibels of coding gain as well.

11.7.4.1 *Definitions.* The term *inbound* means circuits and traffic from a VSAT terminal toward the hub. The converse of this is *outbound*, meaning circuits or traffic from the hub toward one or more related VSAT terminals.

11.7.4.2 *Optimizing Bandwidth Usage.* Here we are dealing with multiple access protocols. In Chapters 9 and 10 we dealt with multiple access protocols such as CSMA/CD, contention and polling, and token passing.

Conventional satellite systems use frequency division multiple access (FDMA) or time division multiple access (TDMA). FDMA is where a satellite transponder frequency slot is leased. These slots are typically less than 1 MHz up to the full bandwidth of a transponder (e.g., 36, 72 MHz). Here we have the bandwidth and have to pay for it whether we use it or not. TDMA is where we are assigned a time slot rather than a frequency slot. We *may* have to pay for it whether we use it or not. Demand assignment multiple access (DAMA) is where we are assigned a frequency slot for only the time we will use it. When we are finished, it is returned to the pool of free channels, ready for someone else to use.

Almost universally, outbound traffic is carried on a TDM link. A time division multiplex (TDM) link is one where frames, packets, or blocks are transmitted in one continuous serial bit stream and are received by all VSATs or all members of a particular VSAT family. Frame/block/packet message headers are examined by each receiving terminal, and only those frames with a terminal's address are delivered to the end-user. That same serial bit stream can also carry control information to control inbound traffic. One of the most common TDM bit rates is 56 kbps. If system traffic warrants, 128 and 384 kbps may be used. Some systems use 1.544 Mbps on the outbound link. Remember, as the bit rate increases, the antenna aperture will have to increase accordingly. If not, the hub may have to increase its radio-frequency power output assuming that the satellite transponder will act accordingly (i.e., it will also be able to increase downlink power).

For uplink traffic, a VSAT system may have a calling frequency and then a working frequency. When a VSAT terminal has traffic, it calls the hub and asks for a channel to pass its traffic downstream (to the hub). Often the calling channel is a common channel for all to use. To access the channel, some form of contention is used, such as pure ALOHA (similar to CSMA/CD, Chapter 10). Collisions are more probable here and their resolution more difficult because of the large delays involved.

Other approaches use a common channel for all inbound traffic and do not resort to a calling channel and working channel(s). Certainly as the user

(VSAT) population increases, the probability of collision increases. The tradeoff whether we approach the problem using a form of DAMA (i.e., calling and working channels) or contention scheme requires considerable study if we are to optimize bandwidth required.

Section 11.7 is based on material from Ref. 20.

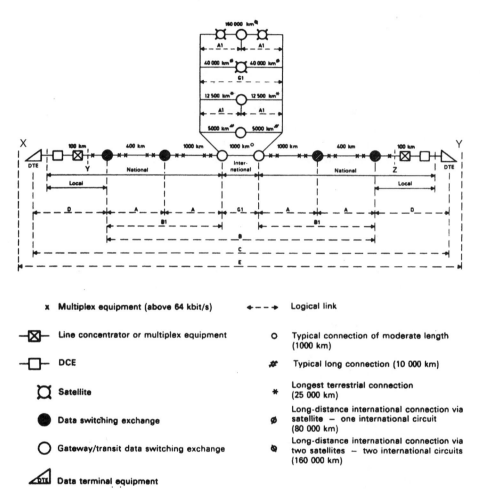

Figure 11.40. Hypothetical reference connections for public synchronous data networks. Link A is the data link between two adjacent data switching exchanges in a national network; link A1 is the data link between two adjacent gateway data switching exchanges in an international connection; link B is the data link between a source DSE and a destination DSE; link B1 is the data link between a local DSE and a gateway DSE; link G1 is the data link between a source gate DSE and a destination gateway DSE in an international connection; link C is the data link between source DTE and destination DTE; link D is the data link between source DTE and the source local DSE or the data link between destination DTE and destination local DSE; link E is the data link between communicating processes. (From Figure 1/X.92, page 296, CCITT Rec. X.92, Ref. 21.)

11.8 HYPOTHETICAL REFERENCE CONNECTIONS FOR PUBLIC SYNCHRONOUS DATA NETWORKS

Hypothetical reference circuits or connections (HRCs) are an aid to the network designer to apportion impairments to the various circuit segments. They are particularly useful on international connections and on long national connections involving several or many nodes. The HRCs presented here have been taken from CCITT Rec. X.92 (Ref. 21).

Figure 11.40 shows five hypothetical reference connections that are used to assess overall customer-to-customer performance objectives. These reference connections may be used for circuit-switched services, packet-switched services, and leased line services in public switched data networks (PSDNs).

In the figure, between points Y and Z, transmission takes place over 64-kbps digital paths, the standard digital network voice channel called DS0 in the United States and Canada and called E0 in Europe. Such paths may include sections using modems over analog facilities. It should be assumed that the signaling for the circuit-switched data call control follows the same

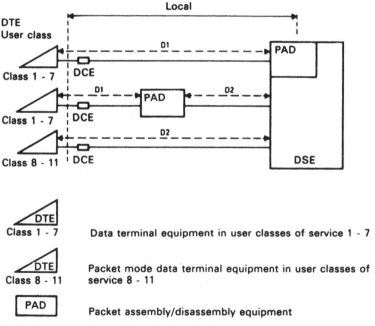

Figure 11.41. Variants to logical link D. Link D1 is the data link between data terminal equipment in user class of service 1–7 and PAD equipment; link D2 is the data link between data terminal equipment in user class of service 8–11 or PAD equipment and a local data switching exchange. *Note 1:* A user may see two different types of logical interfaces with the network (links D1 and D2). *Note 2:* Link D2 could provide an interface for a single-access terminal as well as for a multiple-access terminal. (From Figure 2/X.92, page 270, CCITT Rec. X.92, Ref. 21.)

route as the data connection. Logical links in the case of packet switching are indicated in Figure 11.40 by dashed lines.

Figure 11.41 illustrates the variants to logical link D of Figure 11.40 to allow incorporation of packet assembly/disassembly (PAD) facilities.

REFERENCES

1. Dixon R. Doll, *Data Communications: Facilities, Networks and Systems Design*, John Wiley & Sons, New York, 1978.

2. *Interface Between Data Terminal Equipment (DTE) and Data Circuit-Terminating Equipment (DCE) for Terminals Operating in the Packet Mode and Connected to Public Data Networks by Dedicated Circuit*, ITU-T Rec. X.25, ITU Geneva, October 1996.

3. *Interface Between Data Terminal Equipment (DTE) and Data Circuit-Terminating Equipment (DCE) for Synchronous Operation on Public Data Networks*, CCITT Rec. X.21, ITU Geneva, September 1992.

4. *Support of Packet Mode Terminal Equipment by an ISDN*, ITU-T Rec. X.31, ITU Geneva, March 1993.

5. Uyless Black, *TCP/IP and Related Protocols*, McGraw-Hill, New York, 1992.

6. *Packet-Switched Signaling System Between Public Networks Providing Data Transmission Services*, CCITT Rec. X.75, Fascicle VIII.3, CCITT Plenary Assembly, Melbourne, 1988.

7. *Internet Protocol*, RFC 791, DDN Network Information Center, SRI International, Menlo Park, CA, September 1981.

8. *A Standard for the Transmission of IP Datagrams Over IEEE 802 Networks*, RFC 1042, DDN Network Information Center, SRI International, Menlo Park, CA, 1988.

9. Mathew Naugle, *Illustrated TCP/IP*, John Wiley & Sons, New York, 1999.

10. *Assigned Numbers*, RFC 1060, DDN Network Information Center, SRI International, Menlo Park, CA, March 1990.

11. *Internet Protocol Transition Workbook*, SRI International, Menlo Park, CA, March 1982.

12. *Reverse Address Resolution Protocol*, RFC 903, DDN Network Information Center, SRI International, Menlo Park, CA, June 1984.

13. *Open Shortest Path First (OSPF)*, RFC 1131, DDN Network Information Center, SRI International, Menlo Park, CA, 1991.

14. *Internet Control Message Protocol*, RFC 792, DDN Network Information Center, SRI International, Menlo Park, CA, September 1981.

15. *Transmission Control Protocol*, RFC 793, DDN Network Information Center, SRI International, Menlo Park, CA, September 1981.

16. Military Standard, *Transmission Control Protocol*, MIL-STD-1778, U.S. Dept. of Defense, Washington DC, August 1983.

17. *Internet Protocol Version IPv6*, RFC-2460, from the Internet (Internet Society), December 1998.

18. *User Datagram Protocol (UDP)*, RFC-768, DDN Network Information Center, SRI International, Menlo Park, CA, September 1981.

19. *Information Processing Systems—Data Communications—Protocol for Providing Connectionless-Mode Network Service*, International Standards Organization, ISO-8473, 1st edition, Geneva, December 1988.

20. Roger L. Freeman, *Radio System Design for Telecommunications*, 2nd edition, John Wiley & Sons, New York, 1997.

21. *Hypothetical Reference Connections for Public Synchronous Data Networks*, CCITT Rec. X.92, Fascicle VIII.3, IXth Plenary Assembly, Melboune, 1988.

12

FRAME RELAY

12.1 HOW CAN NETWORKS BE SPEEDED UP?

It would seem from the meaning of "speeding up networks" that we were making bits travel faster down the pipe; that by some means we had broken the speed barrier by dramatically increasing the velocity of propagation. Of course this is eminently not true.

If somehow the bandwidth of the medium can be increased, the bit rate can be increased as well. Another alternative is to turn to a multilevel transmission scheme where we can squeeze more bits per hertz of bandwidth. Another approach is to work our way around the voice channel; turn to the digital voice channel, for example, which will provide 64 kbps. Still another commonly used practice is to use several digital voice channels in a grouping. For example, six of these will produce 384 kbps, the ISDN H0 channel.

Probably the greatest pressure to speed up the network came from LAN users who wished to extend LAN connectivity to distant destinations. Ostensibly this traffic, as described in Chapters 9 and 10, displays local data rates from 16 to 1000 Mbps. The initial resolution of this problem was to use X.25 wide area network connectivity (Chapter 11). It provided a robust service, but was slow and tedious. There is a better way.

What slows down X25 service? X.25 is feasible at 64 kbps, and it may be employed even at DS1/E1 rates. The intensive processing at every node (see Table 12.1), along with continual message exchange, reduces the progress of packets from node to originator and from node to destination. On many X.25 connectivities, multiple nodes are involved, slowing down service still further. One clue is that X.25 was designed for circuits with poor transmission performance degrading error rates. It worked well on circuits with bit error performance in the range of 1×10^{-4}. Meanwhile, the underlying digital networks of North America display BERs better than 5×10^{-10}. This begs the question of removing the responsibility of error recovery from the service provider. If statistically an error occurs in 1 bit in 2,000,000,000 bits, there is a strong argument for removing error recovery actions in their entirety.

TABLE 12.1 Functional Comparison of X.25 and Frame Relay

Function	X.25 in ISDN (X.31)	Frame Relay
Flag recognition / generation	X	X
Transparency	X	X
FCS checking / generation	X	X
Recognize invalid frames	X	X
Discard incorrect frames	X	X
Address translation	X	X
Fill interframe time	X	X
Manage V (S) state variable	X	
Manage V (R) state variable	X	
Buffer packets awaiting acknowledgment	X	
Manager timer T-1	X	
Acknowledge received I-frames	X	
Check received N (S) against V (R)	X	
Generation of rejection message	X	
Respond to poll / final bit	X	
Keep track of number of retransmissions	X	
Act upon reception of rejection message	X	
Respond to receiver not ready (RNR)	X	
Respond to receiver ready (RR)	X	
Multplexing of logical channels	X	
Management of D bit	X	
Management of M bit	X	
Management of W bit	X	
Management of P (S) packets sent	X	
Management of P (R) packets received	X	
Detection of out-of-sequence packets	X	
Management of network layer RR	X	
Management of network layer RNR	X	

In fact, with frame relay the following salient points emerge:

1. There is no process for error recovery by the frame relay service provider.
2. The service provider does not guarantee delivery of frames, nor are there any sort of acknowledgments provided.
3. Frame relay only uses the first two open system interconnection (OSI) layers (i.e., physical and data-link layer), thus removing layer 3 and its intensive processing requirements.
4. Frame overhead is kept to a minimum to reduce processing time and to increase useful throughput.
5. There is no control field and no sequence numbering.
6. Under certain circumstances frames may be discarded using the *leaky bucket concept*.

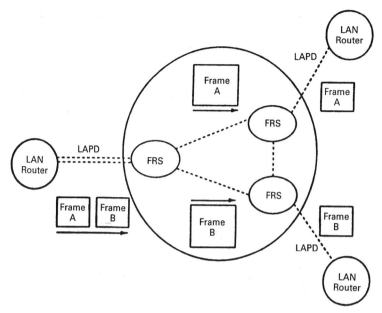

Figure 12.1. A typical frame relay network. FRS, frame relay switch.

7. Frames are discarded without notifying originator for such reasons as congestion and having encountered an error.
8. Frame relay operates on a statmultiplex concept.

In sum, the service that the network provides can be speeded up by increasing data rate, eliminating error recovery procedures, and reducing processing time. One source states that a frame relay frame takes some 20 ms to reach the distant end (statistically), where an X.25 packet takes in excess of 200 ms on terrestrial circuits inside CONUS over similar distances.

Another advantage of frame relay over a conventional static TDM connection is that it uses virtual connections. Data traffic is often bursty and normally would require much larger bandwidths to support the short data messages, and much of the time that bandwidth would remain idle. Virtual connections of frame relay only use the required bandwidth for the period of the burst or usage. This is one reason frame relay is used so widely to interconnect LANs over a WAN. Figure 12.1 shows a typical frame relay network.

12.1.1 The Genesis of Frame Relay

Frame relay derives from the Integrated Services Digital Network(s) (ISDN) link access procedure for the D-channel (LAPD), or D-channel layer 2. LAPD is discussed in Section 13.7. Its importance has taken on such a magnitude that the ITU-T Organization has formulated Rec. I.122, *Framework for Providing Additional Packet-Mode Bearer Services* (Ref. 1), and Rec. I.233, *ISDN Frame Mode Bearer Services* (Ref. 2). Even the term LAPD, although modified in certain cases, continues to be used for frame relay applications.

Frame relay has become an ANSI initiative. There is also the Frame Relay Forum, consisting of manufacturers and users of frame relay equipment, which many feel is leading this imaginative initiative. So when we discuss frame relay, we must consider what specifications a certain system is designed around:

- ANSI, based on ANSI specifications and their publication dates
- Frame Relay Forum with publication dates
- ITU-T organization and its most current recommendations

There are also equivalent ANSI specifications directly derived from ITU-T recommendations such as ANSI T1.617-1991 (Ref. 3). The term *core aspects* of ISDN LAPD, or *DL-CORE*, refers to a reduced subset of LAPD found in Annex A of ITU-T Rec. Q.922 (Ref. 4). The basic body of Q.922 presents CCITT/ITU-T specification for frame relay. This derivative is called LAPF rather than LAPD. For all intents and purposes, the material found in ANSI T1.618 (Ref. 5) is identical to that found in Annex A of Q.922.

To properly describe frame relay from our perspective, we briefly give an overview of the ANSI T1.618-1991 and T1.606-1990 (Ref. 6). This is followed by some fairly well identified variants.

12.2 INTRODUCTION TO FRAME RELAY

Frame relay may be considered a cost-effective outgrowth of ISDN meeting high-data-rate (e.g., 2 Mbps) and low-delay data communications requirements. Frame relay encapsulates data files. These may be considered "packets," although they are called frames. Thus frame relay is often compared to CCITT Rec. X.25 packet service. Frame relay was designed for current transmission capabilities of the network with its relatively wider bandwidths* and excellent error performance (e.g., BER better than 1×10^{-7}).

The incisive reader will note the use of the term *bandwidth*. It is used synonymously with bit rate. If we were to admit at first approximation 1 bit per hertz of bandwidth, such use is acceptable. We are mapping frame relay bits into bearer channel bits probably on a one-for-one basis. The bearer channel may be a DS0/E0 64-kbps channel, a 56-kbps channel of a DS1 configuration, or multiple DS0/E0 channels in increments of 64 kbps up to

---

*I would rather use the term *greater bit rate capacity*.

1.544/2.048 Mbps. We may also map the frame relay bits into a SONET or SDH configuration (Chapter 15). The final bearer channel may require more or less bandwidth than that indicated by the bit rate. This is particularly true for such bearer channels riding on radio systems, and to a lesser extent on a fiber-optic medium or other transmission media. The reader should be aware of certain carelessness of language used in industry publications.

Frame relay works well in the data rate range from 56 kbps up to 1.544/2.048 Mbps. It is being considered for the 45-Mbps DS3 rate for still additional *speed*.

ITU-T's use of the ISDN D channel for frame relay favors X.25-like switched virtual circuits (SVCs). However, ANSI recognized that the principal application of frame relay was interconnection of LANs, and not to replace X.25. Because of the high data rate of LANs (megabit range), dedicated connections are favored. ANSI thus focused on permanent virtual connections (PVCs). With PVCs, circuits are set up by the network,* not by the endpoints. This notably simplified the signaling protocol. Also, ANSI frame relay does not support voice or video.

As mentioned above, the ANSI frame relay derives from ISDN LAPD core functions. The core functions of the LAPD protocol that are used in frame relay (as defined here) are as follows:

- Frame delimiting, alignment, and transparency provided by the use of HDLC flags and zero-bit insertion/extraction[†]
- Frame multiplexing/demultiplexing using the address field
- Inspection of the frame to ensure that it consists of an integer number of octets prior to zero-bit insertion or following zero-bit extraction
- Inspection of the frame to ensure it is not too long or too short
- Detection of (but *not* recovery from) transmission errors
- Congestion control functions

In other words, ANSI has selected features from the LAPD structure/protocol, rejected others, and added some new features. For instance, the control field was removed, but certain control functions have been incorporated as single bits in the address field. These are the C/R bit (command/response), DE bit (discard eligibility), FECN bit, and BECN bit (forward explicit congestion notification and backward explicit congestion notification).

12.2.1 The Frame Structure

User traffic passed to a Frame Access Device (FRAD) is segmented into frames with a maximum length information field or with a default length of

---

*That is, set up by the switching nodes in the network.
[†]LAPD is a derivative of HDLC (see Chapter 3).

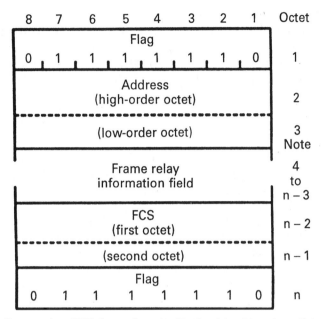

Figure 12.2. Frame relay ANSI frame format with 2-octet addressing. *Note:* The default address field length is 2 octets. By bilateral agreement, it may be extended to either 3 or 4 octets by using the *address field extension bit* **ea**. From ANSI T1.618-1991, Fig. 1, page 3, (Ref. 5.)

262 octets and a recommended length (ANSI) of at least 1600 octets when the application is LAN interconnectivity. The minimum information field is 1 octet.

Figure 12.2 shows the overall frame relay frame structure. As mentioned above, it uses HDLC opening and closing flags (01111110). The closing flag may also serve as the opening flag of the next contiguous frame; however, receivers must be able to accommodate reception of one or more consecutive flags on a bearer channel.

Address Field. The address field consists of at least 2 octets, but may be extended to 3 or 4 octets as shown in Figure 12.3. There is no control field as there is in HDLC, LAPD, or LAPB.

In its most reduced version, there are just 10 bits allocated to the address field in 2 octets of bit positions. (The remainder of the bits serve as control functions.) The 10 bits can support up to 1024 logical connections (i.e., $2^{10} = 1024$).

It should be noted that the number of addressable logical connections is multiplied because they can be reused at each nodal (switch) interface. That

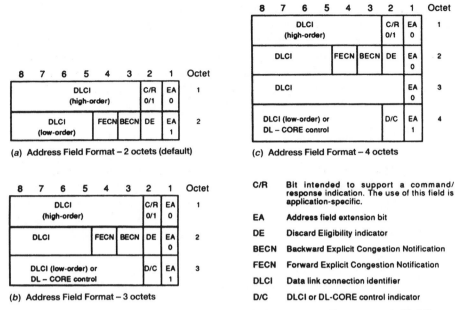

Figure 12.3. Address field formats. Source: ANSI T1.619-1991, Fig. 6, page 6, (Ref.5).

is, an address in the form of a data-link connection identifier (DLCI) has meaning only on one trunk between adjacent nodes. The switch (node) that receives a frame is free to change the DLCI before sending the frame onward over the next link. Thus, the limit of 1024 DLCIs applies to the link, not the network.

Information Field. Information field follows the address field and precedes the frame check sequence. The maximum size of the information field is an implementation parameter, and the default maximum is 262 octets. ANSI chose this default maximum to be compatible with LAPD on the ISDN D channel, which has a 2-octet control field and a 260-octet maximum information field. All other maximum values are negotiated between users and networks and between networks. The minimum information field size is 1 octet. The field must contain an integer number of octets; partial octets are not allowed. A maximum of 1600 octets is encouraged for applications such as LAN interconnects to minimize the need for segmentation and reassembly by user equipment.

A number of frame relay service providers have standardized their networks at considerably longer information fields, between 4000 and 5000 octets.

Transparency. As with HDLC, X.25, and LAPD, the transmitting data-link layer must examine the frame content between opening and closing flags and inserts a 0 bit after all sequences of five contiguous 1s (including the last 5 bits of the FCS) to ensure that a flag or an abort sequence is not simulated within the frame. At the other side of the link, a receiving data-link layer must examine the frame contents between the opening and closing flags and must discard any 0 bit that directly follows five contiguous 1s.

Frame Check Sequence. Frame check sequence (FCS) is based on the generator polynomial $X^{16} + X^{12} + X^5 + 1$. The CRC processing includes the content of the frame existing between, but not including, the final bit of the opening flag and the first bit of the FCS, excluding the bits inserted for transparency. The FCS, of course, is a 16-bit sequence. If there are no transmission errors (detected), the FCS at the receiver will have the sequence 00011101 00001111.

Order of Bit Transmission. The order of bit transmission is that the octets are transmitted in ascending numerical order and that inside an octet bit, 1 is the first bit transmitted.

Field Mapping Convention. When a field is contained within a single octet, the lowest bit number of the field represents the lowest-order value. When a field spans more than one octet, the order of bit values progressively decreases as the octet number increases within each octet. The lowest bit number associated with the field represents the lowest-order value.

An exception to the preceding field mapping convention is the data-link layer FCS field, which spans 2 octets. In this particular case, bit 1 of the first octet is the high-order bit and bit 8 of the second octet is the low-order bit.

Invalid Frames. An invalid frame is a frame that:

- is not properly bounded by two flags (e.g., a frame abort);
- has fewer than 3 octets between the address field and the closing flag;
- does not consist of an integral number of octets prior to zero bit insertion or following zero bit extraction;
- contains an FCS error;
- contains a single-octet address field; and
- contains a data-link connection identifier (DLCI) that is not supported by the receiver.

Invalid frames are discarded without notification to the sender, with no further action.

Frame Abort. Frame abort consists of seven or more contiguous 1 bits; upon receipt of an ABORT, the data-link layer ignores the frame currently being received.

12.2.2 Address Field Discussion

Figure 12.3 shows the ANSI-defined address field formats. Included in the field are the address field extension bits; a reserved bit to support a command/response (C/R) indication bit; forward and backward explicit congestion indicator (FECN and BECN) bits; discard eligibility indicator (DE); a data-link connection identification (DLCI) field; and, finally, a bit to indicate whether the final octet of a 3- or 4-octet address field is the low-order part of the DLCI or DL-CORE control information. The minimum and default length of the address field is 2 octets. However, the address field length may be extended to 3 or 4 octets. To support a larger DLCI address range, the 3- or 4-octet address fields may be supported at the user–network interface or network–network interface based on bilateral agreement.

12.2.2.1 Address Field Variables

Address Field Extension Bit (EA). The address field range is extended by reserving the first transmitted bit of the address field octets to indicate the final octet of the address field. If there is a 0 in this bit position, it indicates that another octet of the address field follows this one. If there is a 1 in the first bit position, it indicates that this octet is the final octet of the address field. As an example, for a 2-octet address field, bit 1 of the first octet is set to 0 and bit one of the second octet is set to 1.

It should be understood that a 2-octet address field is specified by ANSI. It is a user's option whether a 3- or 4-octet field is desired.

Command/Response Bit (C/R). The C/R bit is not used by the DL-CORE protocol, and the bit is conveyed transparently.

Forward Explicit Congestion Notification (FECN) Bit. This bit may be set by a congested network to notify the user that congestion avoidance procedures should be initiated, where applicable, for traffic in the direction of the frame carrying the FECN indication. This bit is set to 1 to indicate to the receiving end-system that the frames it receives have encountered congested resources. The bit may be used to adjust the rate of destination-controlled transmitters. While setting this bit by the network or user is optional, no network shall ever clear this bit (i.e., set to 0). Networks that do not provide FECN shall pass this bit unchanged.

Backward Explicit Congestion Notification (BECN) Bit. This bit may be set by a congested network to notify the user that congestion avoidance procedures should be initiated, where applicable, for traffic in the opposite direction of the frame carrying the BECN indicator. This bit is set to 1 to indicate to the receiving end-system that the frames it transmits may encounter congested resources. The bit may be used to adjust the rate of source-controlled transmitters.

While setting this bit by the network or user is optional according to the ANSI specification, no network shall ever clear (i.e., set to 0) this bit. Networks that do not provide BECN shall pass this bit unchanged.

Discard Eligibility Indicator (DE) Bit. This bit, if used, is set to 1 to indicate a request that a frame should be discarded in preference to other frames in a congestion situation. Setting this bit by the network or user is optional. No network shall ever clear (i.e., set to 0) this bit. Networks that do not provide DE capability shall pass this bit unchanged. Networks are not constrained to only discard frames with DE equal to 1 in the presence of congestion.

Data-Link Connection Identifier (DLCI). This is used to identify the logical connection, multiplexed within the physical channel, with which a frame is associated. All frames carried within a particular physical channel and having the same DLCI value are associated with the same logical connection.

The DLCI is an unstructured field. For 2-octet addresses, bit 5 of the second octet is the least significant bit. For 3- and 4-octet addresses, bit 3 of the last octet is the least significant bit. In all cases, bit 8 of the first octet is the most significant bit.

The structure of the DLCI field may be established by the network at the user–network interface subject to bilateral agreements.

In order to allow for compatibility of call control and layer management between B/H and D channels, the following ranges of DLCIs are reserved and preassigned (see Table 12.2). The DLCIs have local significance only.

DLCI on the D Channel. The six most significant bits (bits 8 to 3 of the first octet) correspond to the service access point identifier (SAPI) field in ANSI (standard) T1.602 (Ref. 7).

The DLCI subfield (bits 8 to 3 of the first octet) values that apply on a D channel are reserved for specific functions to ensure compatibility with operation of the D channel that may also use ANSI T1.602 protocols. A 2-octet address format for the DLCI is assumed when used on the D channel. *Note:* For frame relay in the D channel, only DLCI values in the range 512–991 (SAPI = 31–61) will be assigned.

Table 12.3 gives DLCI values for D channel.

DLCI or DL-CORE Control Indicator (D/C). The D/C indicates whether the remaining six usable bits of that octet are to be interpreted as the lower DLCI bits or as DL-CORE control bits. This bit is set to 0 to indicate that

TABLE 12.2 DLCI Values, Use of DLCIs

DLCI Range	Function
10 bits DLCIs (Note 1)	
0 (Note 2)	In channel signaling, if required
1–15	Reserved
16–511	Network option: on non-D channels, available for support of user information
512–991	Logical link identification for support of user information (Note 6)
992–1007	Layer 2 management of frame mode bearer service
1008–1022	Reserved
1023 (Note 2)	In channel layer 2 management, if required
16 bits DLCIs (Note 3)	
0 (Note 2)	In channel signaling, if required
1–1023	Reserved
1024–32,767	Network option: on non-D channels, available for support of user information
32,768–63,487	Logical link identification for support of user information (Note 6)
63,488–64,511	Layer 2 management of frame mode bearer service
64,512–65,534	Reserved
65,535 (Note 2)	In channel layer 2 management, if required
17 bits DLCIs (Note 4)	
0 (Note 2)	In channel signaling, if required
1–2047	Reserved
2048–65,535	Network option: on non-D channels, available for support of user information
65,536–126,975	Logical link identification for support of user information (Note 6)
126,976–129,023	Layer 2 management of frame mode bearer service
129,024–131,070	Reserved
131,071 (Note 2)	In channel layer 2 management, if required
23 bits DLCIs (Note 5)	
0 (Note 2)	In channel signaling, if required
1–131,071	Reserved
131,072–4,194,303	Network option: on non-D channels, available for support of user information

Note 1. These DLCIs apply when a 2-octet address field is used or when a 3-octet address field is used with D/C = 1.
Note 2. Only available within non-D channel.
Note 3. These DLCIs apply for non-D channels when a 3-octet address field is used with D/C = 0.
Note 4. These DLCIs apply for non-D channels when a 4-octet address field is used with D/C = 1.
Note 5. These DLCIs apply for non-D channels when a 4-octet address field is used with D/C = 0.
Note 6. The use of semipermanent frame mode connections may reduce the number of DLCIs available from this range.
Source: Table 1/Q.922, page 7, CCITT Rec. Q.922, ITU Geneva, 1992 (Ref. 4).

**TABLE 12.3 DLCI Values for D Channel
(2-Octet Address Format)**

DLCI Values	Function
512–991	Assigned using frame relay connection procedures

Source: Ref. 5.

the octet contains DLCI information. When this bit is set to 1, it indicates
that the octet contains DL-CORE control information. The D/C is limited
to use in the last octet of the 3- or 4-octet-type address field. The use of this
indication for DL-CORE control is reserved as there have not been any
additional control functions defined that need to be carried in the address
field. Thus this indicator has been added to provide possible future expansion of the protocol.

12.3 DL-CORE PARAMETERS (AS DEFINED BY ANSI)

DLCI VALUE parameter conveys the DLCI agreed to be used between core
entities in support of DL-CORE connection. Its syntax and usage are
described above.

DL-CORE connection endpoint identifier (CEI) uniquely defines the DL-
CORE connection.

Physical connection endpoint identifier (ph-CEI) uniquely identifies a physical connection to be used in support of a DL-CORE connection.

12.3.1 Procedures

For permanent frame relay bearer connections, information related to the
operation of the DL-CORE protocol in support of DL-CORE connection is
maintained by DL-CORE management. For demand frame relay bearer
connections, layer 3 establishes and releases DL-CORE connections on
behalf of the DL-CORE sublayer. Therefore, information related to the
operation of the DL-CORE protocol is maintained by coordination of layer 3
management and DL-CORE sublayer management through the operation of
the local system environment.

Connection Establishment. When it is necessary to notify the DL-CORE
sublayer entity (either because of establishment of a demand frame relay call,
because of notification of reestablishment of a permanent frame relay bearer
connection, or because of system initialization) that a DL-CORE connection
is to be established, the DL-CORE layer management entity signals an
MC-ASSIGN request primitive to the DL-CORE sublayer entity.

The DL-CORE sublayer entity establishes the necessary mapping between
supporting ph-connection, the core-CEI, and the DLCI. In addition, if it has

not already done so, it begins to transmit flags on the physical connection except on the D channel.

Connection Release. When it is necessary to notify the DL-CORE sublayer entity (either because of release of a demand frame relay call or because of notification of failure of a permanent frame relay bearer connection) that a DL-CORE connection is to be released, the DL-CORE layer management entity signals the MC-REMOVE request primitive to the DL-CORE sublayer entity.

12.4 TRAFFIC AND BILLING ON FRAME RELAY

Figure 12.4 shows a typical traffic profile on a conventional public telephone network, whereas Figure 12.5 shows a typical profile of bursty traffic over a frame relay network. Such bursty traffic is typical for a LAN. The primary employment of frame relay is to interconnect LANs at a distance.

Turning to Figure 12.5, the traffic is bursty. With conventional leased data circuits, we have to pay for the bandwidth* whether it is used or not. On the other hand, with frame relay, we only pay for the "time" used. Billing can be handled in one of three ways:

1. CIR (Committed Information Rate) is a data rate subscribed to by a user. This rate may be exceeded for short bursts during the peak period as shown in Figure 12.5.
2. We can pay a flat rate.
3. We can pay per packet (i.e., frame).

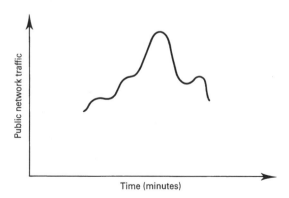

Figure 12.4. Typical traffic profile of a public switched telephone network. (From Ref. 8. Courtesy of Hewlett-Packard.)

*I prefer the use of the expression *bit rate capacity*.

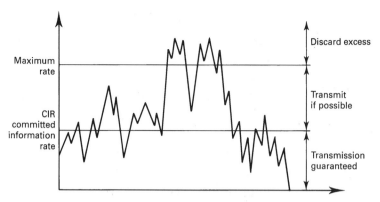

Figure 12.5. Typical bursty traffic of frame relay. Note the traffic levels indicated. (From Ref. 8. Courtesy of Hewlett-Packard.)

Note that on the right-hand side of Figure 12.5 there is the guaranteed transmission bit rate equivalent to the CIR. Depending on the traffic load and congestion, during short periods a user may exceed the CIR. However, there is a point where the network cannot sustain further increases in traffic without severe congestion resulting. Traffic above such levels is arbitrarily discarded by the network without informing the originator.

12.5   PVCs AND SVCs

Most frame relay service providers offer permanent virtual circuits (PVCs) to transport the frame relay traffic. Many frame relay standards describe PVCs as *dedicated* circuits whereas they often really mean *provisioned* circuits. "Provisioned" has a slightly different connotation than "dedicated." Provisioned means that a particular frame relay port accesses a preselected route, the end points being known. Once the circuit becomes "live" [i.e., the originating FRAD (frame relay access device) has frames to transmit], the contracted bit rate capacity is allocated on that route. With many service providers, there is an alternative backup route assigned in addition to the primary route. PVCs require *no* circuit setup signaling procedures, only verification that the route is "alive and healthy."

An SVC is a switched virtual circuit. In this case the circuit is not provisioned but set up by signaling procedures when there are frames to be delivered to a particular location. [Refer to ITU-T Rec. X.76, Amendment 1 (08/97) (Ref. 16)].

12.6   TWO TYPES OF INTERFACES: UNI AND NNI

A user–network interface (UNI) is that interface between a user FRAD and an FR switch. The interface between FR switches is called an NNI or

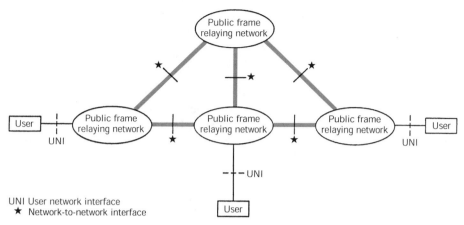

Figure 12.6. Illustration of interfaces: UNI and NNI. (From Figure 2/I.372, page 2, ITU-T Rec. I.372, Ref. 15).

network–network interface.* The UNI is described in ANSI T.618-1991 for OSI layer 2 (T.617), in ITU-T Rec. Q.933 for layer 3 SVC, in ITU-T Rec. Q.922 for layer 2, in X.36 for layer 2, and in X.36 Amd 1 for layers 2 and 3 for PVCs. The NNI is covered in ITU-T Rec. X.76 (SVC) and X.76 Amd 1. See Refs. 2–7, 10, and 14.

Specifically for PVCs, layer 3 circuit maintenance messages are employed, which are described in ANSI T.617A Annex D and ITU-T Q.933 Annex A (Ref. 11).

Figure 12.6 distinguishes UNIs and NNIs in frame relay networks. In the figure, an NNI or UNI can be either a PVC or SVC.

It is emphasized that all frame relay revenue bearing traffic is transported through layer 2 frames. A typical layer 2 frame is illustrated in Figure 12.3.

12.7 CONGESTION CONTROL: A DISCUSSION

Congestion in the user plane occurs when traffic arriving at a resource exceeds the network's capacity. It can also occur for other reasons such as equipment failure. Network congestion affects the throughput, delay, and frame loss experienced by the end user.

End-users should reduce their offered load in the face of network congestion. Reduction of offered load by an end-user may well result in an increase in the effective throughput available to the end-user during congestion.

Congestion avoidance procedures, including optional explicit congestion notification, are used at the onset of congestion to minimize its negative effects on the network and its users.

Explicit notification is a procedure used for congestion avoidance and is part of the data transfer phase. Users should react to explicit congestion

*An NNI is sometimes called a network–node interface.

notification (i.e., optional but highly desirable). Users that are not able to act on explicit congestion notification should have the capability to receive and ignore explicit notification generated by the networks.

Congestion recovery and the associated implicit congestion indication due to frame discard are used to prevent network collapse in the face of severe congestion. Implicit congestion detection involves certain events available to the protocols operating above the core function to detect frame loss (e.g., receipt of a REJECT frame, timer recovery). Upon detection of congestion, the user reduces the offered load to the network. Use of such reduction by users is optional.

12.7.1   Network Response to Congestion

Explicit congestion signals are sent in both forward (toward frame destination) and backward (toward frame source) directions. Forward explicit congestion notification is provided by using the FECN bit in the address field. Backward explicit congestion notification is provided by one of two methods. When timely reverse traffic is available, the BECN bit in an appropriate address field may be used. Otherwise, a single consolidated link layer management message may be generated by the network. The consolidated link layer management (CLLM) message travels on the U-plane physical path. The generation and transport of CLLM by the network are optional.

All networks shall transport the FECN and BECN bits without resetting.

12.7.2   User Response to Congestion

Reaction by the end-user to the receipt of explicit congestion notification is rate-based. Annex A to ANSI T1.618-1991 (Ref. 5) gives an example of user reaction to FECN and BECN.

End-User Equipment Employing Destination-Controlled Transmitters. End-user reaction to implicit congestion detection or explicit congestion notification (FECN indications), when supported, is based on the values of FECN indications that are received over a period of time. The method is consistent with commonly used destination-controlled protocol suites, such as OSI class 4 transport protocol operated over the OSI connectionless service.

End-User Equipment Employing Source-Controlled Transmitters. End-user reaction to implicit congestion notification (BECN indication), when supported, shall be immediate when a BECN indication or a CLLM is received. This method is consistent with implementation as a function of data-link-layer elements of procedure commonly used in source-controlled protocols such as CCITT Rec. Q.922 elements of procedure.

The following figure illustrates frame relay congestion control:

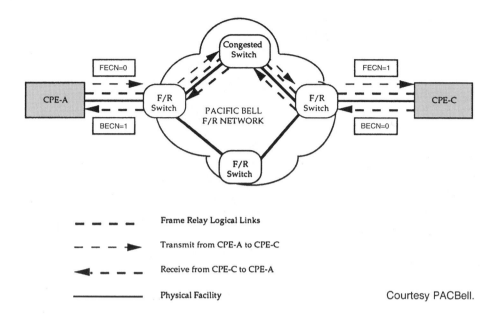

Frame Relay Logical Links

Transmit from CPE-A to CPE-C

Receive from CPE-C to CPE-A

Physical Facility   Courtesy PACBell.

12.7.3 Consolidated Link Layer Management (CLLM) Messages

The CLLM uses XID frames for the transport of functional information. We may remember in the discussion of HDLC (Section 3.5) that an XID frame was an *exchange identification frame*. It was used for reporting station identification. In frame relay, CLLM uses XID frames for network management as an alternative for congestion control. CLLM messages originate at network nodes to the frame relay interface usually housed in a router near or incorporated with the user equipment.

As we mentioned, BECN/FECN bits in frames must pass congested nodes in the forward or backward direction. Suppose for a given user no frames pass in either direction; thus that user has no knowledge of network congestion because at that moment the user is not transmitting or receiving frames. Frame relay standards do not permit a network to generate frames with the DLCI of the congested circuit. CLLM covers this contingency. It has a DLCI = 1023 reserved.

The use of CLLM is optional. If it is used, it may or may not operate in conjunction with BECN/FECN. The CLLM frame format dedicates 1 octet for the cause of congestion such as excessive traffic, equipment or facility failure, preemption, or maintenance action.

The CLLM message is based on the ISO 8885 definition of XID frames for transport of function information. The frame format of a CLLM message is illustrated in Figure 12.7. Each parameter is described in the CLLM using sequence-type-length value. The functional fields for CLLM are described

```
                    87654321
1             ┌──────────────┐  Address Octet 1
              │  11111010    │
2             │  11110001    │  Address Octet 2
3             │  10101111    │  XID Control Field
4             │  10000010    │  Format Identifier (130)
5             │  00001111    │  Group Identifier = 15
6             │              │  Group Length Octet 1
7             │              │  Group Length Octet 2
8             │  00000000    │  Parameter Identifier = 0
9             │  00000100    │  Parameter Length (4)
10            │  01101001    │  Parameter value = 105 (IA5 coded 1)
11            │  00110001    │  Parameter value = 49 (IA5 coded 1)
12            │  00110010    │  Parameter value = 50 (IA5 coded 2)
13            │  00110010    │  Parameter value = 50 (IA5 coded 2)
14            │  00000010    │  Parameter Identifier = 2 (cause id)
15            │  00000001    │  Parameter length = 1
16            │              │  Cause value
17            │  00000011    │  Parameter Identifier = 3 (DLCI Identifier)
18            │              │  Parameter Length
19            │      •       │  DLCI value Octet 1 (1st)
20            │      •       │  DLCI value Octet 2 (1st)
      •       │              │
      •       │              │
2n + 17       │              │  DLCI value Octet 1 (nth)
2n + 18       │              │  DLCI value Octet 2 (nth)
2n + 19       │              │  FCS octet 1
2n + 20       └──────────────┘  FCS octet 2
```

Figure 12.7. Frame format of a CLLM message, with 2-octet address field. IA5, International Alphabet No. 5 (Ref.12) (see Section 5.3). (From Figure C.1/X.36, page 33, ITU-T Rec. X.36, Ref. 10.)

below with reference to the figure. The CLLM message may be transmitted whenever a congestion control procedure is being performed as a result of network congestion, line or equipment failure, or the performance of maintenance functions. All fields are binary coded unless otherwise specified.

Field Descriptions, CLLM Message Format, Figure 12.7

Fields 1 and 2. Addressing, self-explanatory.

Field 3, XID Control Field. For a CLLM frame, coded 10101111 in binary.

Field 4, Format Identifier. Coded 130 decimal and 10000010 in binary.

Field 5, Group Identifier Field. Coded decimal 15 and 00001111 in binary.

Fields 6 and 7, Group Length Octets. This 16-bit field describes the "length," in octets, in the remainder of the group field. The maximum value of the group field is 256.

Field 8, Group Field Value Field. Consists of two or more parameter fields. The parameter set identification (with a parameter value of 0) identifies a set of private parameters within the group field in accordance with

TABLE 12.4   Codes for "Cause" in CLLM Message

Bits	Cause
87654321	
00000010	Network congestion due to excessive traffic—short term
00000011	Network congestion due to excessive traffic—long term
00000110	Facility or equipment failure—short term
00000111	Facility or equipment failure—long term
00001010	Maintenance action—short term
00001011	Maintenance action—long term
00010000	Unknown—short term
00010001	Unknown—long term
	All other values are reserved

Source: Table C.1/X.36, page 35, ITU-T Rec. X.36, ITU Geneva 4/95 (Ref. 10).

ISO 8885/DAD3 as an identifier to be determined. Other parameters appear in the following order: case identifier and DLCI identifier. Parameter Set Identification Field is always present; otherwise the CLLM message is ignored. There are three subfields under this heading. Octet 8 contains the parameter identification field for the first parameter and is set to 0 in accordance with ISO 8885/DAD3. Parameter 0 identifies the set of private parameters within this group.

Field 9, Parameter Set Identification Length Field. Contains the length parameter 0 and is set to binary value 100 (a decimal value of 4).

Fields 10–13, Parameter Field Value. Identify the usage of the XID frame private parameter group for (ITU-T Rec.) I.122 private parameters. Octet 10 contains IA5 (International Alphabet No. 5) value I (decimal 105). Octet 11 contains IA5 values 1 (decimal 49); and octets 12 and 13 each contains the IA5 value of 2 (decimal 50).

Fields 14–16, Cause Identifier Parameter Field. (See Table 12.4.) These fields are explained in the following text.

Field 14, Parameter Identifier Field. Is set to decimal 2 (binary 00000010)

Field 15, Parameter Length Field. It gives the length of the cause identifier and is set to binary 1 (i.e., 00000001)

Field 16, Cause Value. This octet gives the cause as determined by the congested network node whose layer management module originated the message. The value for the cause is to indicate the network status of the layer management entity issuing the message (e.g., congestion, failure, or maintenance operation). The binary values for cause are shown in Table 12.4.

Field 17 and Onwards, DLCI Identifier Parameter Field. This field is used to determine the DLCI corresponding to the cause listed in the CLLM described above. If the DLCI identifier is missing, then the frame is ignored. Octet 17 contains the parameter identifier field. When this field is set to decimal 3 (binary 00000011), the octets that follow this parameter contain the DLCI(s) of the frame relay connection(s) that are congested.

Field 18, Parameter Length. This contains the length of the DLCI(s) being reported, in octets. For example, if (n) DLCIs are being reported and they are of length 2 octets each, this will be 2 times (n) in octet size.

Field 19 Onward to FCS Field. This is the Parameter Value field. Here is a listing of the DLCI value(s) which identify the logical link(s) that have encountered a congested state. The DLCI field (in a frame relay frame) is 10 bits long and is contained in bits 8 to 3 of the first octet pair and bits 8 to 5 of the next octet pair. Bit 8 of the first octet is the most significant bit and bit 5 of the second octet is the least significant bit. Bits 2 to 1 in the first octet and 4 to 1 in the second octet are reserved.

Network Congestion, Use of the CLLM. When the network encounters resource congestion as a result of excessive DTE traffic levels and the level of traffic continues at these high levels, the network may be forced to discard traffic or bring the system to a halt for recovery. By transmitting a CLLM message to the DTE indicating the underlying cause of congestion, the network informs the DTE of the possibility of an action being taken. It should be noted that CLLM messages generated at times of congestion are intended only for providing notification in the opposite direction to the traffic congestion direction. CLLM messages may signal a congestion situation to the transmitting DTE when there is no traffic in the reverse direction.

12.7.4 Action of a Congested Node

When a node is congested, it has several alternatives it may use to mitigate or eliminate the problem. It may set the FECN and BECN bits to 1 in the address field and/or use the CLLM message. Of course, the purpose of explicit congestion notification is to either (a) inform the "edge" node at the network ingress of congestion so that the edge node can take the appropriate action to reduce the congestion, (b) notify the source that the negotiated throughput has been exceeded, or (c) do both.

One of the strengths of the CLLM is that it contains a list of DLCIs that correspond to the congested frame relay bearer connections. These DLCIs indicate not only the sources currently active causing the congestion but also those sources that are not active. The reason for the latter is to prevent those sources that are not active from becoming active and thus causing still further congestion. It may be necessary to send more than one CLLM message if all the affected DLCIs cannot fit into a single frame.

12.8 FLOW CONTROL AND POSSIBLE APPLICATIONS OF FECN AND BECN BITS

ANSI in T1.618-1991 Annex A (Ref. 5) presents some sample procedures and suggestions for the application of FECN and BECN information and user reaction to such congestion notifications.

12.8.1 FECN Usage

User Behavior on Receipt FECN Bit. ANSI recommends (Ref. 5) that users compare the number of frames in which the FECN bit is set with the number of frames in which the FECN bit is cleared over a measurement interval ▼. If the number of set FECN bits (a set bit in this context is a bit set to 1 indicating some form of congestion) is equal to or exceeds the number of clear FECN bits (bits set to 0) received during this period, then the user should reduce its throughput to $\frac{7}{8}$ (0.875) of its previous value. If the number of set FECN bits is less than the number of clear FECN bits, then the user may increase the information rate by $\frac{1}{16}$ of its throughput.

The suggested measurement interval ▼ is approximately equal to four times the end-to-end transit delay. However, other mechanisms not dependent on timers may be used by the terminal if the effect is similar.

A *slow-start* mechanism is recommended, so as to cause convergence toward equilibrium on the connection. ANSI suggests that the initial data rate should be set to the throughput value agreed at the time of call establishment, or less, in order to avoid an impulse load on the network at the time the user begins transmitting. If the connection has been idle for a long period of time (in the order of tens of seconds), the offered rate or less should be returned to the throughput.

Use of Windows as an Approximation to Rate-Based Control. For some implementations, it may be convenient to use a window-based mechanism as an approximation to rate-based control. Such implementations may or may not be able to measure their offered rate, or to relate to the throughput negotiated with the network during connection establishment. The actual offered rate is limited by end-to-end transit delay, access rate, window size, and frame size.

In the event that a windowed protocol is in use, then the user, in responding to the FECN bit, compares the number of frames received with the FECN bit set with the number of frames with the FECN bit clear. The measurement interval for this shall be twice the interval during which the number of frames equal to the current working window size is transmitted and acknowledged (i.e., two window turns). If the number of set FECN bits is equal to or exceeds the number of clear FECN bits received during this period, then the user will reduce its working window size variable to $\frac{7}{8}$ (0.875) of its previous value. However, it need never reduce its working window below the size of one frame. If the number of FECN bits is less than the number of clear FECN bits, then the user increases its window size variable by the size of one frame, not to exceed the maximum window size for that virtual circuit. After the adjustment is made, the set and clear FECN bit counters are reset to zero and the comparison begins again.

The working window is initialized at a small value, such as one frame, in order to avoid impulse load upon the network at the time a user begins transmitting. If the connection has been idle for a long period of time, in the order of tens of seconds, it may be appropriate to again reduce the window size to its initial value. There may be a maximum window size or information rate that a connection can accommodate, as limited by the end-system policies; the working window should not be adjusted beyond such a value. It should also be observed that this algorithm is relatively insensitive to the loss of acknowledgments carried in the user data when they are used to carry window adjustment information back to the source.

12.8.2 BECN Usage

BECN Congestion Control on a Step Count S. S, as defined in the ANSI specification Annex A, is a function of twice the end-to-end transit time, with the information rate in each direction multiplied by the ratio of the forward rate to the reverse data rate.

If a frame with a BECN bit is received and the user's offered rate is greater than the throughput, the user reduces its offered rate to the throughput rate agreed upon for the frame relay connection.

If S consecutive frames are received with the BECN bit set, the user reduces its rate to the next "step" rate below the current offered rate. Further rate reduction should not occur until S consecutive frames are received with the BECN bit set. The step rates are:

0.675 times throughput
0.5 times throughput
0.25 times throughput

Note: Networks should be engineered such that reduction below 0.5 times throughput never becomes necessary, and they may take alternative action (e.g., rerouting) to prevent such occurrence.

When the user has reduced its rate due to receipt of BECN, it may increase its rate by a factor of 0.125 times throughput after any $S/2$ consecutive frames are received with the BECN bit clear.

Network Procedures for Setting the BECN Bit. The network should, if possible, set the BECN bit before it becomes necessary to discard frames. The network should continue to set the BECN bit whenever it is in such condition, and it may elect to provide some hysteresis to prevent oscillation.

If the congestion condition deteriorates, the network should discard frames sent in excess of throughput at the access node, as well as discard those marked with the DE bit, in preference to other frames. At this stage, the network is in a moderately congested state, and BECN bits should continue to be set in frames that are not discarded.

If the congestion condition further deteriorates to the point that frames that are neither in excess of throughput nor are marked with the DE bit are being discarded, a severe congestion condition exists. The network should continue to use the BECN to encourage users to reduce their transmission rates and/or take further action such as clearing or rerouting calls to restore control.

12.9 POLICING A FRAME RELAY NETWORK

12.9.1 Introduction

Frame relay switches may carry out a policing function on accessing users. The result of such action is discarded frames. Similar policing actions are employed on ATM circuits (Chapter 16).

12.9.2 Definitions

Access Rate (AR). The data rate, expressed in bits per second, of the user access channel (D, B, or H channel). The rate at which users can offer data to the network is bounded by the access rate.

Excess Burst Size (B_e). The maximum amount of uncommitted data (in bits) that the network will attempt to deliver over the measurement interval (T). These data may or may not be contiguous (i.e., may appear in one frame or in several frames, possibly with interframe idle flags). B_e is negotiated at call establishment (for demand establishment of communication) or at service subscription time (for permanent establishment of communication). Excess burst data may be marked for discard eligibility (with the DE bit) by the network.

Measurement Interval (T). The time interval over which rates and burst sizes are measured. In general, the duration of T is proportional to the *burstiness* of the traffic. Except as noted below, T is computed as $T = B_c/\text{CIR}$ or $T = B_e/\text{AR}$.

Committed Information Rate (CIR). The rate, expressed in bits per second, at which the network agrees to transfer information under normal conditions. This rate is measured over the measurement interval T. CIR is negotiated at call establishment or service subscription time. Data marked *discard eligible* (DE) is not accounted for in CIR.

Committed Burst Size (B_c). The maximum amount of data (in bits) that a network agrees to transfer under normal conditions over a measurement interval T. These data may or may not be contiguous (i.e., they may appear in one frame or in several frames, possibly with interframe idle flags). B_c is negotiated at call establishment (for demand establishment of communication) or service subscription time (for permanent establishment of communication).

Fairness. An attempt by the network to maintain the negotiated quality of service for all users operating under normal conditions (i.e., within their CIR and B_c, without discard due to congestion). For example, the network may discard frames offered in excess of CIR, and it may reject any call attempts that would cause the network resources to be overcommitted.

Offered Load. The bits offered to the network by an end user, to be delivered to the selected destination. The information rate and burst length offered to the network could exceed the negotiated class of service parameters. The offered load consists of the user data portion of frames and therefore excludes flags, FCS, and inserted zeros.

12.9.3 Relationship Among Parameters

Figure 12.8 illustrates the relationship of access rate, excess burst, committed burst, committed information rate, discard eligibility indicator, and measurement interval parameters. The CIR, B_c, and B_e parameters are negotiated at call establishment time for demand establishment of communication or established by subscription for permanent establishment of communication. Access rate is established by subscription for permanent access connections or during demand access connection establishment. Each end-user and the network participate in the negotiation of these parameters to agreed-upon values.

These negotiated values are then used to determine the measurement interval parameter, T, and when the discard eligibility indicator (if used) is set. These parameters are also used to determine the maximum allowable end-user input levels. The relationship among parameters can be used at any instant of time, T_0, to measure the offered load over the interval $(T_0, T_c + T)$. Similarly, the offered load over any interval $(t - T, t)$ may be measured at any instant of time t, as long as the measurement function retains memory of user activity over the previous interval T. One way of doing this is by use of a "leaky bucket" algorithm described below.

The measurement interval is determined as shown in Table 12.5. The network and the end-users may control the operation of the discard eligibility indicator (DE) and the rate enforcer functions by adjusting at call setup the CIR, B_c and B_e parameters in relation to the access rate. If both the CIR

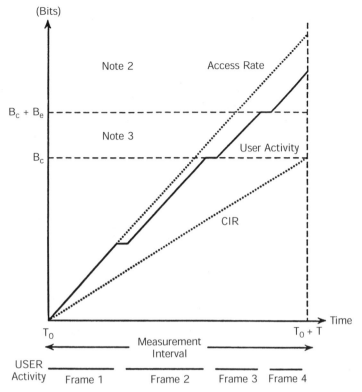

Figure 12.8. Illustration of relationships among parameters. *Note 1:* Number of frames and size of frames are for illustrative purposes only. *Note 2:* Frame may be discarded at ingress node. This is a region of rate enforcement. *Note 3:* Frame marked as discard eligible. (From Figure 7, page 9, ANSI T1.606a, Ref. 9.)

TABLE 12.5   Congestion Parameter State

CIR	B_c	B_e	Measurement Interval (T)
> 0	> 0	> 0	$T = B_c/\text{CIR}$
> 0	> 0	= 0	$T = B_c/\text{CIR}$
= 0	= 0	> 0	T is a network-dependent value

Source: Reference 9, paragraph 10.2.

and B_c parameters are not equal to zero, then $T = B_c/\text{CIR}$. In addition there are two special conditions:

1. When CIR = access rate, $B_c = 0$, and $B_e = 0$, both access rates must be equal (i.e., ingress = egress).
2. When CIR = 0 (B_c must = 0) and $B_e > 0$, then T is a *network-dependent* value.

Notes

1. The ingress and egress access rates do not have to be equal; however, when the ingress access rate is substantially higher than the egress access rate, continuous input of B_e frames at the ingress interface may lead to persistent congestion of the network buffers at the egress interface, and a substantial amount of the input B_e data will be discarded.

2. When a frame, entering the ingress node, consumes the remaining capacity of B_e or B_c, reducing it to zero, the action taken on that frame is network-dependent.

Figure 12.8 is a static illustration of the relationship among time, cumulative bits of user data, and rate. In this example, the user sends four frames during the measurement interval $(T_0 + T)$. The slope of the line marked "CIR" is B_c/T. Bits are received at the access rate (by the ingress node) of the access channel. Because the sum of the number of bits contained in frames 1 and 2 is not greater than B_c, the network does not mark these frames with the discard eligibility indicator (DE). The sum of the number of bits in frames 1, 2, and 3 is greater than B_c, but not greater than $B_c + B_e$; therefore frame 3 is marked discard eligible. Because the sum of the number of bits received by the network in frames 1, 2, 3, and 4 exceeds $B_c + B_e$, frame 4 is discarded at the ingress node. This figure does not address the case in which the end-user sets the DE bit.

This section is based on Ref. 9, paragraph 10.2.

12.10 QUALITY OF SERVICE PARAMETERS

The quality that frame-relaying service provides is characterized by the values of the following parameters. ANSI adds in Ref. 6 that the specific list of negotiable parameters is for further study.

- Throughput
- Transit delay
- Information integrity
- Residual error rate
- Delivered error(ed) frames
- Delivered duplicated frames
- Delivered out-of-sequence frames
- Lost frames
- Misdelivered frames
- Switched virtual call established delay

- Switched virtual call clearing delay
- Switched virtual call establishment failure
- Premature disconnect
- Switched virtual call clearing failure

12.11 NETWORK RESPONSIBILITIES

The frame are routed through the network on the basis of an attached label (i.e., the DLCI value in the frame). This label is a logical identifier with local significance. In the virtual call case, the value of the logical identifier and other associated parameters such as layer 1 channel delay may be requested and negotiated during call setup. Depending on the value of the parameters, the network may accept or reject the call. In the case of the permanent virtual circuit, the logical identifier and other associated parameters are defined by means of administrative procedures (e.g., at the time of subscription).

The user–network interface structure allows for the establishment of multiple virtual calls or permanent virtual circuits, or both, to many destinations over a single access channel.

Specifically, for each connection, the bearer service:

1. Provides bidirectional transfer of frames.
2. Preserves their order as given at one user–network interface if and when they are delivered at the other end. (*Note:* No sequence numbers are kept by the network. Networks are implemented in such a way that frame order is preserved.)
3. Detects transmission, format, and operational errors such as frames with an unknown label.
4. Transports the user data contents of a frame transparently. Only frame's address and FCS fields may be modified by network nodes.
5. Does not acknowledge frames.

At the user–network interface, the FRAD, as a minimum, has the following responsibilities:

1. Frame delimiting, alignment, and transparency provided by the use of HDLC flags and zero-bit insertion.
2. Virtual circuit multiplexing/demultiplexing using the address field of the frame.
3. Inspection of the frame to ensure that it consists of an integer number of octets prior to zero-bit insertion or following zero-bit extraction.
4. Inspection of the frame to ensure that it is not too short or too long.
5. Detection of transmission, format, and operational errors.

A frame received by a frame handler may be discarded if the frame:

1. Does not consist of an integer number of octets prior to zero-bit insertion or following zero-bit extraction.
2. Is too long or too short.
3. Has an FCS that is in error.

The network will discard a frame if it:

1. Has a DLCI value that is unknown.
2. Cannot be routed further due to internal network conditions.

A frame can be discarded for other reasons such as exceeding negotiated throughout.

12.12   FRAME RELAY SIGNALING FUNCTIONS

Frame relay SVC circuits require setup procedures. PVC circuits require periodic "health and welfare" checks. These functions are carried out by OSI layer 3 signaling actions. In the case of PVC, the signaling functions are reduced to a set of management processes. No setup or release procedures are necessary. On the other hand, SVC operation requires circuit setup and release processes and utilizes a somewhat extended group of management action messages.

12.12.1   PVC Management Procedures

These procedures are based on periodic transmission of a STATUS EN-QUIRY message by the DTE and a STATUS message by the DCE. These messages are principally used to make a provisioned circuit active; if it is not, then to determine the general "health" of the circuit. This includes link integrity verification of the DTE/DCE interface, notification to the DTE of the status of a PVC, notification of the addition of a PVC, and the detection by the DTE of the deletion of a PVC. These messages are transferred across the bearer channel using layer 2 unnumbered information (UI) frames.

12.12.1.1   Message Definition. Both messages are transferred on DLCI = 0. The FECN, BECN, and DE bits are not used and set to 0. The 3 octets following the address field have fixed values. Figure 12.9 illustrates the PVC management frame format.

Octet	8	7	6	5	4	3	2	1	
1	Flag								
2	0	0	0	0	0	0	0	0	Address field
3	0	0	0	0	0	0	0	1	DLCI = 0
4	0	0	0	0	0	0	1	1	UI P = 0
5	0	0	0	0	1	0	0	0	Protocol discriminator
6	0	0	0	0	0	0	0	0	Dummy call reference
	Message specific information element								
	FCS								
	Flag								

Figure 12.9. PVC management frame format with 2-octet addressing. (From Figure 7/X.36, page 14, ITU-T Rec. X.36, Ref. 10.)

These 3 octets are coded as follows:

- The first octet is the control field of a UI frame with P bit set to 0.
- The second octet is the protocol discriminator information element of the message.
- The third octet is the dummy call reference information element of the message.

The other information elements are reviewed below.

STATUS ENQUIRY Message. This message is sent to request the status of PVCs or to verify link integrity. Specific information elements for this message are described in Table 12.6, and they are in the order indicated in the table.

TABLE 12.6   Message-Specific Information Elements in STATUS ENQUIRY Message

Message type: STATUS ENQUIRY			
Significance: Local	Direction: Both		
Information Element	Direction	Type	Length[a]
Message type	Both	Mandatory	1
Report type	Both	Mandatory	3
Link integrity verification	Both	Mandatory	4

[a]Length in octets.
Source: Table 2/X.36, page 15, Ref. 10.

TABLE 12.7   **Message-Specific Information Elements in a STATUS Message**

Message type: STATUS
Significance: Local   Direction: Both

Information Element	Direction	Type	Length[a]
Message type	Both	Mandatory	1
Report type	Both	Mandatory	3
Link integrity verification	Both	Optional / Mandatory (Note)	4
PVC status	Both	Optional / Mandatory (Note)	5–7

Note: Optional or mandatory depending on the type of report.
[a]Length in octets.
Source: Table 3/X.36, page 15, Ref. 10.

STATUS Message. This message is sent in response to a STATUS EN-QUIRY message to indicate the status of PVCs or for a link integrity verification. Optionally, it may be sent at any time to indicate the status of a single PVC. Message-specific information elements for this message are given in Table 12.7, and they are given in the order indicated in the table. Moreover, the PVC status information element may occur several times.

Message Type Coding (Octet 7)

STATUS ENQUIRY message: 0111 0101
STATUS message: 0111 1101

Report Type Coding (Octets 8, 9, and 10)

Octet 8: 0101 0001
Octet 9: Length of report type contents
Octet 10: Type of report

   0000 0000   Full status (status of all PVCs on the DTE–DCE interface)
   0000 0001   Link integrity verification only
   0000 0010   Single PVC asynchronous status
   All other values are reserved.

The purpose of the Report Type information element is to indicate the type of enquiry requested when included in a STATUS ENQUIRY message or the contents of the STATUS message. The length of this information element is 3 octets.

Link Integrity Verification (Octets 11, 12, 13, and 14)

Octet 11:   0101 0011
Octet 12:   Length of link integrity verification contents = 2

Octet 13:   Send sequence number

Octet 14:   Receive sequence number

The purpose of the Link Integrity verification information element is to exchange sequence numbers between the DCE and DTE on a periodic basis. The length of the contents of the link integrity information element is binary encoded in octet 2. The send sequence number in octet 3 indicates the current send sequence number of the originator of the message, and the receive sequence number in octet 4 indicates the send sequence number received in the last received message.

PVC Status. The purpose of the PVC status information element is to indicate the status of existing PVCs at the interface. The information element can be repeated, as necessary, in a message to indicate the status of all PVCs on the DTE–DCE interface. The length of this information element depends on the length of the DLCIs being used on the DTE–DCE interface. The length of this information element is 5 octets when a default address format (2 octets) is used. The format of the PVC status information element is defined in Figure 12.10, where the default address format is used. Bit 6 of octet 3 is the most significant bit in the DLCI.

Bit 2 of the last octet of each PVC status information element is "Active bit," which is coded 1 to indicate the PVC is active and coded 0 to indicate the PVC is inactive. An active indication means that the PVC is available to be used for data transfer. An inactive indication means that the PVC is configured but is not available for data transfer.

Bit 3 of the last octet of each PVC information element is the "Delete bit," which is coded 1 to indicate that the PVC is deleted and coded 0 to indicate that the PVC is configured.

Bit 4 of the last octet for each PVC status information element is "New bit," which is coded 1 to indicate the PVC is newly configured and coded 0 to indicate that the PVC is already configured.

The PVC status information elements are arranged in the messages in ascending order of DLCIs; the PVC with the lowest DLCI is first, the second lowest is second, and so on. The maximum number of PVCs that can be indicated in a message is limited by the maximum frame size.

The "Delete bit" is only applicable for timely notification using the optional PVC asynchronous status report. When this bit is set to 1, the New and Active bits have no significance and are set to 0 upon transmission and not interpreted upon reception.

12.12.1.2 *Description of Operation: Health and Welfare, PVC Link.* The operation is based on periodic polling, as described below, to verify the integrity of the link and to report the status of PVCs.

Octet	8	7	6	5	4	3	2	1
1	0	1	0	1	0	1	1	1
2				Length of PVC status contents = 3				
3	0 ext.	0 spare	Data Link connection identifier (2nd most significant 4 bits)					
3a	1 ext.	0 spare		Data Link connection Identifier (Most significant 6 bits)			0	0
4	1 ext.	0	0 spare	0	New "N"	Delete "D"	Active "A"	0 reserved

Figure 12.10. PVC STATUS information element with 2-octet address. (From Figure 10-1/X.36, page. 18, ITU-T Rec. X.36, Ref. 10.)

464

Periodic Polling

General. The DTE initiates the polling as described below.

1. The DTE sends STATUS ENQUIRY message to the DCE and starts polling timer T391. When T391 expires, the DTE repeats the above action.

The STATUS ENQUIRY message typically requests a link integrity verification exchange only (report type 0000 0001). However, every N391 polling cycle, the DTE requests full status of all PVCs (report type 0000 0000).

2. The DCE responds to each STATUS ENQUIRY message with a STATUS message and starts or restarts the polling verification timer T392 used by the network to detect errors. The STATUS message sent in response to a STATUS ENQUIRY contains the link integrity verification and Report Type information elements. If the content of the Report Type information element specifies full status, then the STATUS message must contain one PVC status information element for each PVC configured on the DTE–DCE interface.
3. The DTE shall interpret the STATUS message depending upon the type of report contained in this STATUS message. The DCE may respond to any poll with a full status message in case of a PVC status change or to report the addition or deletion of PVC on the DTE–DCE interface. If it is a full status message, the DTE should update the status of each configured PVC.

Link Integrity Verification. The purpose of the link integrity verification information element is to allow the DTE and the DCE to determine the status of the signaling link (DLCI = 0). This is necessary since these procedures use the unnumbered information (UI) frame. Figure 12.11 shows the normal link integrity verification procedure.

The DTE and DCE maintain the following internal counters:

- The send sequence counter maintains the value of the send sequence number field of the last link integrity verification information element sent.
- The receive sequence counter maintains the value of the last received send sequence number field in the link integrity verification information element and maintains the value to be placed in the next transmitted received sequence number field.

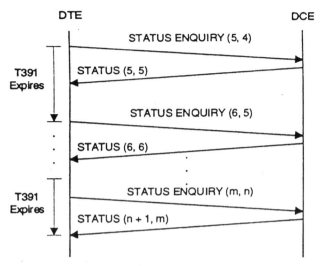

Figure 12.11. Link integrity verification. (From Figure 11/X.36, page 20, ITU-T Rec. X.36, Ref. 10.)

The following procedure is used:

1. Before any messages are exchanged, the DCE and DTE set the send sequence counter and receive sequence counter to zero.
2. Each time the DTE sends a STATUS ENQUIRY message, it increments the send sequence counter and places its value in the send sequence number field. It also places the current value of the receive sequence counter into the receive sequence number field of the link integrity verification information element. The DTE increments the send sequence counter using modulo 256. The value zero is skipped.
3. When the DCE receives a STATUS ENQUIRY from the DTE, the DCE checks the receive sequence number received from the DTE against its send sequence counter.

The received send sequence number is stored in the receive sequence counter. The DCE then increments its send sequence counter and places its current value in the send sequence number field and the value of the receive sequence number counter (the last received send sequence number) into the receive sequence number field of the outgoing link integrity verification information element. The DCE then transmits the completed STATUS message back to the DTE. The DCE increments the send sequence counter using modulo 256. The value zero is skipped.

4. When the DTE receives a STATUS from the DCE in response to a STATUS ENQUIRY, the DTE checks the receive sequence number received from the DCE against its send sequence counter. The received send sequence number is stored in the receive sequence counter.

Note: The value zero in the receive sequence number indicates that the field contents are undefined; this value is normally used after initialization. The value zero is not to be used in the send sequence number field so that the receive sequence number will never contain the value zero to differentiate the undefined condition from the normal modulo round-off.

Signaling the Activity Status of PVCs. In response to a STATUS ENQUIRY message sent by the DTE containing a Report Type information element sent to "full status," the DCE reports in a STATUS message to the DTE the activity status of each PVC configured on the DTE–DCE interface with PVC status information elements (one per PVC).

The Report Type information element in this STATUS message is set to "full status." Also in response to a STATUS ENQUIRY message set by the DTE containing a Report Type information element set to "link integrity verification only," the DCE may respond with a STATUS message containing a Report Type information element set to "full status" in case of a PVC status change. Each PVC status information element contains an Active bit indicating the activity status of that PVC.

The action that the DTE takes based on the value of the Active bit is independent of the action based on the New bit. The DTE could get a PVC status information element with the New bit set to 1 and the Active bit set to 0.

If the DTE receives a PVC status information element with the Active bit set to 0, the DTE shall stop transmitting frames on the PVC until it receives a PVC status information element for that PVC with the Active bit set to 1. When the Active bit is set to 1, the Delete bit must be set to 0 on transmission. The Delete bit is not interpreted in the full status reporting STATUS message. When the Delete bit set to 1 in the optional asynchronous status message, the active bit has no significance. Other action taken by the DTE is implementation-dependent.

Since there is a delay between the time the network makes a PVC active and the time the DCE transmits a PVC status information element notifying the DTE, there is a possibility of the DTE receiving frames on a PVC marked as inactive. The action the DTE takes on receipt of frames on an inactive PVC is implementation-dependent.

Since there is a delay between the time the network detects that a PVC has become inactive and the time the DCE transmits a PVC status information element notifying the DTE, there is a possibility of the DCE receiving frames on an inactive PVC. The action the DCE takes on receipt of frames for an inactive PVC is network-dependent and may include the dropping of frames on the inactive PVC.

The DCE will signal that a PVC is active if the following criteria are met:

- The PVC is configured and available for data transfer in the network from the local DCE to the remote DCE.

- There is no service affecting condition at both local and remote DTE–DCE interfaces.
- In the case bidirectional procedures are used at the remote DTE–DCE interface, the remote DTE indicates that the PVC is present and active.

Note that in the case bidirectional procedures are used at the local DTE–DCE interface, this indication is independent of the indication received from the local DTE.

This section is based on Ref. 10.

12.12.2 Signaling Required for SVCs

Switched virtual circuits (SVCs) at the UNI require call setup and release procedures. Certain link maintenance messages, similar to those discussed in Section 12.12.1, are also required for SVC operation. Table 12.8 lists call control messages employed in the various connection control procedures.

Note that the entries in Table 12.8 are presented in alphabetical order, not in chronological (operational) order. The reader's attention is called to the fact that STATUS ENQUIRY and STATUS messages that are described here differ from those given in Section 12.12.1. They are *not* interchangeable.

The most important message in the Call Establishment group is the SETUP message. This message is sent by the calling user to the network and by the network to the called user to initiate the actual call establishment procedure. Its 23 information elements in sequential order are shown in Table 12.9.

The first three information elements of all call control messages listed in Table 12.9 have some commonality. The first octet is the protocol discriminator. This simply tells the receiving device that this is a call control message. It distinguishes it from other message types. It is identical for all the call control messages listed in Table 12.8 and is coded 00001000. It should be noted that these call control messages are valid for UNI and are not necessarily valid for the NNI.

TABLE 12.8   Messages Required for Call Connection Control of SVCs

Message Type	Message Type
Call Establishment Messages	*Call Clearing Messages*
ALERTING	DISCONNECT
CALL PROCEEDING	RELEASE
CONNECT	RELEASE COMPLETE
CONNECT ACKNOWLEDGE	
PROGRESS	*Miscellaneous Messages*
SETUP	STATUS
	STATUS ENQUIRY

TABLE 12.9 SETUP Message Content

Message type: SETUP
Significance: Global   Direction: Both

Information Element	Direction	Type	Length
Protocol discriminator	Both	M	1
Call reference	Both	M	2–*
Message type	Both	M	1
Bearer capability	Both	M	4–5
Channel identification	Both	O (Note 1)	2–*
Data-link connection identifier	Both	O (Note 2)	2–6
Progress indicator	Both	O (Note 3)	2–4
Network-specific facilities	Both	O (Note 4)	2–*
Display	n → u	O (Note 5)	(Note 6)
End-to-end transit delay	Both	O (Note 7)	2–11
Packet layer binary parameters	Both	O (Note 8)	2–3
Link layer core parameters	Both	O (Note 9)	2–27
Link layer protocol parameters	Both	O (Note 10)	2–9
X.213 priority	Both	O (Note 11)	2–8
Calling party number	Both	O (Note 12)	2–*
Calling party subaddress	Both	O (Note 13)	2–23
Called party number	Both	O (Note 14)	2–*
Called party subaddress	Both	O (Note 15)	2–23
Transit network selection	u → n	O (Note 16)	2–*
Repeat indicator	Both	O (Note 17)	1
Low layer compatibility	Both	O (Note 18)	2–16
High layer compatibility	Both	O (Note 19)	2–4
User–user	Both	O (Note 20)	(Note 21)

*An asterisk in the length column indicates undefined length.

Note 1. Mandatory in the network-to-user direction. Included in the user-to-network direction when the user wants to indicate a channel. If not included, its absence is interpreted as "any channel acceptable." No channel negotiation is allowed in case A.

Note 2. Mandatory in the network-to-user direction. Included in the user-to-network direction when the user wants to indicate the DLCI value to be used for the frame mode call.

Note 3. Included in the event of interworking within a private network.

Note 4. Included by the calling user or the network to indicate network specific information (see Annex E/Q.931).

Note 5. Included if the network provides information that can be presented to the user.

Note 6. The minimum length is 2 octets; the maximum length is 82 octets.

Note 7. May be omitted in the user-to-network direction, if the calling user accepts default values for this quality of service parameter. Always included in the network-to-user direction to indicate cumulative end-to-end transit delay to the called user.

Note 8. Included in the user-to-network direction when the calling user wants to provide OSI network service requirements. Included in the network-to-user direction if the calling user included a packet layer binary parameter information element in the SETUP message.

Note 9. Included in the user-to-network direction when the calling user wants to indicate the proposed link layer core parameter values to the network. Always included in the network-to-user direction to indicate the proposed link layer core parameter values to the called user. If the link layer core parameter information element is not present in the user-to-network direction, the default values will be assumed, and the network will negotiate link layer core parameters with the called user based on the calling user default values.

Table notes continued on following page

TABLE 12.9 Notes (*Continued*)

Note 10. In case of Frame Relaying, it is included if the calling user wants to indicate link layer protocol parameters to the called user. It is carried transparently by the network. In case of Frame Switching, it is included in the user-to-network direction when the calling user wants to indicate the proposed link layer protocol parameter values to the network. Included in the network-to-user direction when the network wants to indicate the proposed link layer protocol parameter values to the called user. In this case these parameters have only local significance.

Note 11. Included in the user-to-network direction when the calling user wants to provide OSI network service requirements. Included in the network-to-user direction if the calling user included X.213 priority information element in the SETUP message.

Note 12. May be included by the calling user or the network to identify the calling user.

Note 13. Included in the user-to-network direction when the calling user wants to indicate the calling party subaddress. Included in the network-to-user direction if the calling user included a calling party subaddress information element in the SETUP message.

Note 14. The Called party number information element is included by the user to convey called party number information to the network. The Called party number information element is included by the network when called party number information is conveyed to the user.

Note 15. Included in the user-to-network direction when the calling user wants to indicate the called party subaddress. Included in the network-to-user direction if the calling user included a called party subaddress information element in the SETUP message.

Note 16. Included by the calling user to select a particular transit network (see Annex C/Q.931).

Note 17. The Repeat indicator information element is included immediately before the first Low layer compatibility information element when the low layer compatibility negotiation procedure is used (see Annex J/Q.931).

Note 18. Included in the user-to-network direction when the calling user wants to pass low layer compatibility information to the called user. Included in the network-to-user direction if the calling user included a Low layer compatibility information element in the SETUP message.

Note 19. Included in the user-to-network direction when the calling user wants to pass high layer compatibility information to the called user. Included in the network-to-user direction if the calling user included a High layer compatibility information element in the SETUP message.

Note 20. Included in the user-to-network direction when the calling user wants to pass user information to the called user. Included in the network-to-user direction if the calling user included User–user information element in the SETUP message.

Note 21. The minimum length is 2 octets; the standard default maximum length is 131 octets.

Source: Table 3-10/Q.933, pages 12 and 13, ITU-T Rec. Q.933, Ref. 11.

The second part of every message is the *call reference*. It has a minimum length of 2 octets. Its purpose is to identify the call or facility registration/cancellation at the local UNI to which the particular message applies. The call reference does not have end-to-end significance across the network.

Bits 8, 7, 6, and 5 of the first octet of the call reference are coded 0000; bits 4, 3, 2, and 1 provide the length of the call reference values. The second octet, bit 8, is a flag; the remainder of the octet and octets 3 onwards, if required, carry the actual call reference value.

The third information element is *message type*. Its intent is to identify the purpose of the message being sent. The coding of the message is shown in Table 12.10. Note that bit 8 (the first bit of an octet as shown) is coded binary 0 and is reserved for future use as a possible extension bit.

TABLE 12.10   Coding for Message Types

Bits 8765 4321	
0000 0000	Escape to nationally specified message types; see Note.
000- ----	*Call establishment messages:*
0 0001	ALERTING
0 0010	CALL PROCEEDING
0 0111	CONNECT
0 1111	CONNECT ACKNOWLEDGE
0 0011	PROGRESS
0 0101	SETUP
010- ----	*Call clearing messages:*
0 0101	DISCONNECT
0 1101	RELEASE
1 1010	RELEASE COMPLETE
011- ----	*Miscellaneous messages:*
0 0000	SEGMENT
1 1101	STATUS
1 0101	STATUS ENQUIRY

Note: When used, the message type is defined in the following octet (s), according to national specification. The extension mechanism (bit 8 of the message type) is independent of the escape mechanism for the message.

Source: Table 4-1/Q.933, page 15, Ref. 11.

Each information element has a unique *identifier coding* as illustrated in Table 12.11. This 1-octet element tells what the information element is going to do or what information it carries.

Due to space limitation, we cannot explore each call control message and information element. It will be useful, however, to explore several examples. The SETUP message was presented in Table 12.9. Another example is shown in Figure 12.12, which gives the link layer core parameters information element. The specific coding of the core parameters is given in Table 12.12, which consists of three parts. Figure 12.13 shows the format of the called party number information element. Table 12.13 gives the called party information element coding.

The information element "Link Layer Core Parameter," Figure 12.12 and the three subsequent tables deal with configuring an SVC. Among the configuration parameters are Outgoing and Incoming Maximum Frame Mode Information field size (octets 3a–d), Throughput (octets 4 and 5), and Committed and Excess Burst Size (octets 6–9). Some or all of these parameters can be included in the SLA (service level agreement) issued at contract signing or can be negotiated on-the-fly during SVC setup.

TABLE 12.11   Information Element Unique Identifier Coding

8765 4321		Max Length (Octets) (Note 1)
1:::: ----	*Single octet information elements:*	
000 ----	reserved	
001 ----	Shift (Note 2)	1
101 ----	Repeat indicator	1
0::: ::::	*Variable length information elements:*	
000 0000	Segmented message	
000 0100	Bearer capability	5
000 1000	Cause (Note 2)	32
001 0100	Call state	3
001 1000	Channel identification	(Note 4)
001 1001	Data-link connection identifier (Note 7)	6
001 1110	Progress indicator (Note 2)	4
010 0000	Network specific facilities (Note 2)	(Note 4)
010 1000	Display	82
100 0010	End-to-end transit delay	11
100 0100	Packet layer binary parameters (Note 7)	3
100 1000	Link layer core parameters (Note 7)	27
100 1001	Link layer protocol parameters (Note 7)	9
100 1100	Connected number	(Note 4)
100 1101	Connected subaddress	23
101 0000	X.213 priority (Note 7)	8
101 0001	Report type (Note 7)	3
101 0011	Link integrity verification (Note 7)	4
101 0111	PVC status (Notes 2 and 7)	5
110 1100	Calling party number	(Note 4)
110 1101	Calling party subaddress	23
111 0000	Called party number	(Note 4)
111 0001	Called party subaddress	23
111 1000	Transit network selection (Note 2)	(Note 4)
111 1100	Low layer compatibility (Note 6)	14
111 1101	High layer compatibility	4
111 1110	User–user	131
111 1111	Escape for extension (Note 3)	
	All other values are reserved (Note 5)	

Note 1. The length limits described for the variable length information elements below take into account only the present CCITT standardized coding values.

Note 2. This information element may be repeated.

Note 3. This escape mechanism is limited to code sets 5, 6, and 7. When the escape for extension is used, the information element identifier is contained in octet-group 3 and the content of the information element follows in the subsequent octets.

Note 4. The maximum length is network dependent.

Note 5. The reserved values with bits 5 to 8 coded "0000" are for future information elements for which comprehension by the receiver is required (see 5.8.7.1/Q.931).

Note 6. This information element may be repeated in conjunction with the Repeat indicator information element.

Note 7. Information elements defined in this Recommendation; not present in Recommendation Q.931.

Source: Table 4-2/Q.933, page 16, ITU-T Rec. Q.933, Ref. 11.

8	7	6	5	4	3	2	1	Octet
Link layer core parameters								
0	1	0	0	1	0	0	0	1 (Notes 1, 7)
Information element identifier								
Length of link layer core parameters contents								2
Maximum Frame Mode Information Field (FMIF) size								3
0 ext.	0	0	0	1	0	0	1	
0 ext.	Outgoing maximum FMIF size							3a
0/1 ext.	Outgoing maximum FMIF size (cont.)							3b
0 ext.	Incoming maximum FMIF size							3c*
1 ext.	Incoming maximum FMIF size (cont.)							3d*
Throughput								4*
0 ext.	0	0	0	1	0	1	0	(Note 2)
0 ext.	Outgoing magnitude			Outgoing multiplier				4a*
0/1 ext.	Outgoing multiplier (cont.)							4b*
0 ext.	Incoming magnitude			Incoming multiplier				4c*
1 ext.	Incoming multiplier (cont.)							4d*
Minimum acceptable throughput								5*
0 ext.	0	0	0	1	0	1	1	(Notes 3, 4)
0 ext.	Outgoing magnitude			Outgoing multiplier				5a* (Note 4)
0/1 ext.	Outgoing multiplier (cont.)							5b* (Note 4)
0 ext.	Incoming magnitude			Incoming multiplier				5c* (Note 4)
1 ext.	Incoming multiplier (cont.)							5d* (Note 4)
Committed burst size								6*
0 ext.	0	0	0	1	1	0	1	(Note 5)
0 ext.	Outgoing committed burst size value							6a*
0/1 ext.	Outgoing committed burst size value (cont.)							6b*
0 ext.	Incoming committed burst size value							6c*
1 ext.	Incoming committed burst size value (cont.)							6d*
Excess burst size								7*
0 ext.	0	0	0	1	1	1	0	(Note 6)
0 ext.	Outgoing excess burst size value							7a*
0/1 ext.	Outgoing excess burst size value (cont.)							7b*
0 ext.	Incoming excess burst size value							7c*
1 ext.	Incoming excess burst size value (cont.)							7d*
Committed burst size magnitude								
0 ext.	0	0	1	0	0	0	0	8* (Note 8)
1 ext.	Spare	Incoming Bc magnitude			Outgoing Bc magnitude			8a*
Excess burst size magnitude								
0 ext.	0	0	1	0	0	0	1	9* (Note 9)
1 ext.	Spare	Incoming Bc magnitude			Outgoing Bc magnitude			9a*

Figure 12.12. Format of the Link Layer Core Parameters information element. (From Figure 4-4/Q.933, page 21, ITU-T Rec. Q.933, Ref. 11.)

Note 1. All the parameters are optional and position independent. If certain parameters are not included, the network default value will be used. The term "outgoing" is defined in the direction from calling user to called user. The term "incoming" is defined in the direction from called user to calling user.

Note 2. When octet 4 is present, octet 4a and octet 4b shall also be present. Additionally, octet group 4c and 4d may be included.

Table notes continued on page 474.

Called party number, Figure 12.13, can be based on either ITU-T Rec. E.164 numbering or ITU-T Rec. X.121 numbering. Table 12.13 describes octet 3 (Figure 12.13), bits 4, 3, 2, and 1. If they are coded 0001, the E.164 numbering plan applies; if coded 0011, the X.121 numbering plan applies. Octet(s) 4 onwards gives the applicable number digits. IA5 refers to CCITT International Alphabet No. 5. Refer to CCITT Rec. T.50 (Ref. 13).

12.13 COMPATIBILITY ISSUES

Standardization agencies have not exactly standardized frame relay. Early on we identified three such agencies:

1. ITU
2. ANSI
3. The Frame Relay Forum

The Frame Relay Forum generally follows ITU-T practice. ANSI Frame Relay specifications parallel the ITU in many respects. We cite only a few of the differences. One is STATUS ENQUIRY and STATUS. Several information elements differ; even the number of information elements. Q.933 SVC STATUS ENQUIRY and STATUS for SVC differs from X.36 for PVC. Then there are proprietary frame relay systems. One such is AT & T's closed-loop system. All we need is one octet differing from one protocol system to another, and incompatibility results. Customer beware!

Notes to Figure 12.12 (*Continued*)

Note 3. When octet 5 is present, octet 5a and octet 5b shall also be present. Additionally, octet group 5c and 5d may be included.

Note 4. Included only in the SETUP message.

Note 5. When octet 6 is present, octet 6a and octet 6b shall also be present. Additionally, octet group 6c and 6d may be included.

Note 6. When octet 7 is present, octet 7a and octet 7b shall also be present. Additionally, octet group 7c and 7d may be included.

Note 7. "Throughput" and "Measurement interval (T)" are defined as "Committed Information Rate (CIR)" and "Committed rate measurement interval (T_c)" in Rec. I.370, respectively. See Section 12.9.2.

Note 8. Octet group 8 is used to indicate the committed burst size magnitude when the value cannot be coded in octet group 6. When the incoming committed burst size field is not included (in octet group 6), the incoming magnitude field has no significance.

Note 9. Octet group 9 is used to indicate the excess burst size magnitude when the value cannot be coded in octet group 7. When the incoming excess burst size field is not included (in octet group 6), the incoming magnitude field has no significance.

TABLE 12.12   Link Layer Core Parameters Information Element, Specific Coding

Maximum frame mode information field (FMIF) (octets 3, 3a, 3b, 3c, and 3d)

The frame mode information field size is the number of user data octets after the address field and before the FCS field in a frame mode frame. The count is done either before zero-bit insertion or following zero-bit extraction. If the frame mode information field is symmetrical, octets 3a and 3b indicate the size in both directions, and octets 3c and 3d are absent. The maximum size of the frame mode information field is a system parameter and is identified as N203.

Outgoing maximum FMIF size (octets 3a and 3b)

The outgoing maximum FMIF size is used to indicate the maximum number of end user data octets in a frame for the calling user to the called user direction. The size is in octets and is encoded in binary.

Incoming maximum FMIF size (octets 3c and 3d)

The incoming maximum FMIF size is used to indicate the maximum number of end user data octets in a frame for the called user to the calling user direction. The size is in octets and is encoded in binary.

Note 1. The maximum frame mode information field size allowed for the D-channel is 262 octets. For B- and H-channels a larger FMIF size is allowed (e.g., up to 4096 octets). The default value of the frame mode information field size is defined in Recommendation Q.922. The users may negotiate a maximum FMIF size that is lower than the maximum FMIF size that the network can deliver. The network is not required to enforce the negotiated FMIF value.

Throughput (octets 4, 4a, 4b, 4c, and 4d)

The purpose of the throughput field is to negotiate the throughput for the call. Throughput is the average number of "frame mode information" field bits transferred per second across a user-network interface in one direction, measured over an interval of duration "T."

This field, when present in the SETUP message, indicates requested throughput, which is the lesser of the throughput requested by the calling user and throughput available from the network(s), but is not less than the minimum acceptable throughput. When present in the CONNECT message, it indicates the agreed throughput, which is the throughput acceptable to the calling user, the called user, and the network(s).

If the throughput is asymmetrical (i.e., the values in the incoming and outgoing directions are different), octets 4a and 4b indicate throughput in the outgoing direction (from the calling user) and octets 4c and 4d indicate throughput in the incoming direction (to the calling user). If the throughput is symmetrical, octets 4a and 4b indicate throughput in both directions, and octets 4c and 4d are absent.

Throughput is expressed as an order of magnitude (in powers of 10) and an integer multiplier. For example, a rate of 192 kbps is expressed as 192×10^3.

Magnitude (octets 4a and 4c)

This field indicates the magnitude of the throughput. This is expressed as a power of 10.

Bits

7 6 5	
0 0 0	10^0
0 0 1	10^1
0 1 1	10^2
0 1 1	10^3
1 0 0	10^4
1 0 1	10^5
1 1 0	10^6

All other values are reserved.

Table continued on page 476

TABLE 12.12   (Contined)

Note 2. To ensure that various implementations will encode particular rates in a consistent fashion, the coding of the magnitude and multiplier shall be such that the multiplier is as small as possible; i.e. the multiplier shall not be evenly divisible by 10. For example, a rate of 192 kbps shall be expressed as 192×10^3, not as 1920×10^2.

Multipler (octets 4a, 4b, 4c, and 4d)

This field indicates, in binary, the value by which the magnitude shall be multiplied to obtain the throughput.

Minimum acceptable throughout

The purpose of the minimum acceptable throughput field is to negotiate the throughput for the call. Minimum acceptable throughput is the lowest throughput value that the calling user is willing to accept for the call. If the network or the called user is unable to support this throughout, the call shall be cleared.

This field, which is present only in the SETUP message, is carried unchanged through the network(s). Its value may not be greater than the requested throughput.

If the minimum acceptable throughput is asymmetrical (i.e., the values in the incoming and outgoing directions are different), octets 5a and 5b indicate minimum acceptable throughput in the outgoing direction (from the calling user) and octets 5c and 5d indicate minimum acceptable throughput in the incoming direction (to the calling user). If the minimum acceptable throughput is symmetrical, octets 5a and 5b indicate throughput in both directions, and octets 5c and 5d are absent.

Minimum acceptable throughput is expressed as an order of magnitude (in powers of 10) and an integer multiplier. For example, a rate of 192 kbps is expressed as 192×10^3.

Magnitude (octets 5a and 5c)

Same as octets 4a and 4c coding.

Multiplier (octets 5a, 5b, 5c, and 5d)

This field indicates, in binary, the value by which the magnitude shall be multiplied to obtain the minimum acceptable throughput.

Committed burst size

This field indicates the maximum amount of data (in bits) that the network agrees to transfer, under normal conditions, over a measurement interval (T). This data may or may not be interrupted (i.e., may appear in one frame or in several frames, possibly with interframe idle flags). T is calculated using the following combinations:

Throughput	Committed burst size (B_c) (Note 3)	Excess burst size (B_e)	Measurement interval (T)
> 0	> 0	> 0	$T = B_c/$Throughput
> 0	> 0	= 0	$T = B_c/$Throughput
= 0	= 0	> 0	Default (Note 4) $T = B_e/$Access rate

Note 3. The coding of this field is in octets. Therefore, the committed burst size is $8 \times$ the contents this field. If the committed burst size is symmetrical, octets 6a and 6b indicate the size in both directions, and octets 6c and 6d are absent.

TABLE 12.12 (*Continued*)

Note 4. However, the ingress and egress Access Rate (AR) do not have to be equal. However, when the ingress AR is substantially higher than the egress AR, continuous input of B_e frames at the ingress interface may lead to persistent congestion of the network buffers at the egress interface, and a substantial amount of the input B_e data may be discarded. Networks may define the interval T to be less than B_e/AR in which case the default is not used.

Outgoing committed burst size (octets 6a and 6b)
The outgoing committed burst size (in octets) is binary coded.

Incoming committed burst size (octets 6c and 6d)
The incoming committed burst size (in octets) is binary coded.

Excess burst size
This field indicates the maximum amount of uncommitted data (in bits) that the network will attempt to deliver over measurement interval (T). This data may appear in one frame or in several frames. If the data appears in several frames, these frames can be separated by interframe idle flags. Excess burst may be marked Discard Eligible (DE) by the network.
Note 5. The coding of this field is in octets. Therefore, the excess burst size is 8 × the contents of the field.
If the excess burst size is symmetrical, octets 7a and 7b indicate the size in both directions, and octets 7c and 7d are absent.

Outgoing excess burst size (octets 7a and 7b)
The outgoing excess burst size (in octets) is binary coded.

Incoming excess burst size (octets 7c and 7d)
The incoming excess burst size (in octets) is binary coded.

Committed burst size magnitude (octets 8 and 8a)
The purpose of the Committed burst size magnitude is to indicate the magnitude of the Committed burst size. It is expressed as a power of 10. It is multiplied by the Committed burst size value (octet group 6) to give the actual value of the Committed burst size. When the incoming Committed burst size field is not included (in octet group 6), the incoming magnitude has no significance.
The outgoing and incoming B_c magnitudes are coded as follows:

Bits

3/6	2/5	1/4	
0	0	0	10^0
0	0	1	10^1
0	1	0	10^2
0	1	1	10^3
1	0	0	10^4
1	0	1	10^5
1	1	0	10^6

All other values are reserved.

TABLE 12.12 (*Continued*)

Excess burst size magnitude (octets 9 and 9a)

The purpose of the Excess burst size magnitude is to indicate magnitude of the Excess burst size. It is expressed as a power of 10. It is multiplied by the Excess burst size value (octet group 7) to give the actual field of the Excess burst size. When the incoming Excess burst size value is not included (in octet group 7), the incoming magnitude has no significance.

The outgoing and incoming B_e magnitude are coded as a power of 10 as follows:

Bits

3/6	2/5	1/4	
0	0	0	10^0
0	0	1	10^1
0	1	0	10^2
0	1	1	10^3
1	0	0	10^4
1	0	1	10^5
1	1	0	10^6

All others values are reserved.

Source: Table 4-5/Q.933, ITU-T Rec. Q.933, Ref. 11.

Bits

8	7	6	5	4	3	2	1	Octets
			Called party number-					
0	1	1	1	0	0	0	0	1
			Information element identifier					
Length of called party number contents								2
1 Ext.	Type of number			Numbering plan identification				3
0	Number digits (IA5 characters) (Note)							4 etc.

NOTE – The number digits appear in multiple octet 4's in the same order in which they would be entered, that is, the number digit which would be entered first is located in the first octet 4.

Figure 12.13. Called party number information element. (From Figure 4-14/Q.931, page 75, ITU-T Rec. Q.931, ITU, Ref. 13.)

TABLE 12.13   Called Party Information Element Coding

Type of number (octet 3) (Note 1)

Bits

<u>7 6 5</u>

0 0 0   Unknown (Note 2)

0 0 1   International number (Note 3)

0 1 0   National number (Note 3)

0 1 1   Network specific number (Note 4)

1 0 0   Subscriber number (Note 3)

1 1 0   Abbreviated number (Note 5)

1 1 1   Reserved for extension

All other values are reserved.

Note 1. For the definition of international, national, and subscriber number, see ITU-T Recommendation 1.330.

Note 2. The type of number "unknown" is used when the user or the network has no knowledge of the type of number, (e.g., international number, national number, etc.). In this case the number digits field is organized according to the network dialing plan; for example, prefix or escape digits might be present.

Note 3. Prefix or escape digits shall not be included.

Note 4. The type of number "network specific number" is used to indicate administration/service number specific to the serving network (e.g., used to access an operator).

Note 5. The support of this code is network-dependent. The number provided in this information element presents a shorthand representation of the complete number in the specified numbering plan as supported by the network.

Numbering plan identification (octet 3)

Numbering plan (applies for type of number = 000, 001, 010, and 100)

Bits

<u>4 3 2 1</u>

0 0 0 0   Unknown (Note 6)

0 0 0 1   ISDN / telephony numbering plan (ITU-T Rec. E.164)

0 0 1 1   Data numbering plan (ITU-T Rec. X.121)

0 1 0 0   Telex numbering plan (ITU-T Rec. F.69)

1 0 0 0   National standard numbering plan

1 0 0 1   Private numbering plan

1 1 1 1   Reserved for extension

All other values are reserved.

Table continued on page 480

TABLE 12.13   (*Continued*)

Note 6. The numbering plan "unknown" is used when the user or network has no knowledge of the numbering plan. In this case the number digits field is organized according to the network dialing plan; for example, prefix or escape digits might be present.

Number digits (octets 4, etc.)

This field is coded with IA5 characters, according to the formats specified in the appropriate numbering/dialing plan.

Note that the E.164 and X.121 numbering plans may be found in the 3rd ed., *Reference Manual for Telecommunication Engineering*, 3rd edition, John Wiley & Sons, June 2001.
Source: Table 4-9, page 76, ITU-T Rec. Q.931, ITU Geneva, May 1998, Ref. 13.

REFERENCES

1. *Framework for Frame Mode Bearer Services*, ITU-T Rec. I.122, ITU, Geneva, March 1993.

2. *Frame Mode Bearer Services*, CCITT Rec. I.233, ITU Geneva, 1992.

3. *ISDN Signaling Specification for Frame Relay Bearer Service for Digital Subscriber Signaling System Number 1 (DSS1)*, ANSI T1.617-1991, ANSI, New York, 1991.

4. *ISDN Data Link Layer Service for Frame Mode Bearer Services*, CCITT Rec. Q.922, ITU Geneva, 1992.

5. *Integrated Services Digital Network (ISDN)—Core Aspects of Frame Protocol for Use with Frame Relay Bearer Service*, ANSI T1.618-1991, ANSI, New York, 1991.

6. *ISDN—Architectural Framework and Service Description for Frame Relaying Bearer Service*, ANST T1.606-1990, ANSI, New York, 1990.

7. *Telecommunications: Integrated Services Digital Network (ISDN)—Data Link Layer Signaling Specification for Application at the User–Network Interface*, ANSI T1.602-1989, ANSI, New York, 1989.

8. *Frame Relay & SMDS*, a seminar, Hewlett-Packard Co., Burlington, MA, October 1993.

9. *Integrated Services Digital Network (ISDN)—Architectural Framework and Service Description for Frame-Relaying Bearer Service (Congestion Management and Frame Size)*, ANSI T1.606a, ANSI, New York, 1992.

10. *Interface Between Data Terminal Equipment (DTE) and Data Circuit-Terminating Equipment (DCE) for Public Data Networks Providing Frame Relay Data Transmission Service by Dedicated Circuit*, ITU-T Rec. X.36, ITU Geneva, October 1995.

11. *Integrated Services Digital Network (ISDN) Digital Subscriber Signaling System No. 1 (DSS 1)—Signaling Specifications for Frame Mode Switched and Permanent Virtual Connection Control and Status Monitoring*, ITU-T Rec. Q.933, ITU Geneva, October 1995.

12. *International Reference Alphabet (IRA) (Formerly International Alphabet No. 5 or IA5)—Information Technology—7-bit Coded Character Set for Information Interchange*, CCITT Rec. T.50, ITU Geneva, September 1992.

13. *ISDN User–Network Interface Layer 3 Specification for Basic Call Control*, ITU-T Rec. Q.931, ITU Geneva, May 1998.

14. *Netwcrk-to-Network Interface Between Public Data Networks Providing the Frame Relay Data Transmission Service*, ITU-T Rec. X.76, ITU Geneva, April 1995.

15. *Frame Relay Bearer Service Network-to-Network Interface Requirements*, ITU-T Rec. I.372, ITU Geneva, March 1973.

16. *Switched Virtual Circuits*, ITU-T Rec. X.76, Amendment 1, ITU Geneva, August 1997.

13

INTEGRATED SERVICES
DIGITAL NETWORKS (ISDNs)

13.1 INTRODUCTION

Integrated services digital network (ISDN) might be defined as an extension
of the digital telephone network to the subscriber where both voice and data
are available in an integrated way. In theory, ISDN supports a wide variety of
telecommunication services such as:

- Voice
- Telex/Teletext
- Telemetry
- Data: circuit- and packet-switched
- Conference TV, slow scan TV
- Facsimile, including CCITT Group 4

ISDN is based on the public switched digital network, namely, the familiar
PCM network discussed in Chapter 7. In other words, the principal vehicle
for signal transport is the 64-kbps digital channel. When ISDN is implemented at a user premise, it assumes that access is available for 64-kbps
channels at the local serving exchange and that CCITT signaling system No. 7
is in place and operational for the intervening network end-to-end.

ISDN integrates services. It does so by providing a 64-kbps bearer channel
for all the services listed. In North America the digital channel may be a
56-kbps channel.

13.2 ISDN STRUCTURES

13.2.1 ISDN User Channels

At the ISDN user−network interface, the two most common structures for
user access transmission links are the B channel at 64-kbps transmission rate

and the D channel at 16 kbps. The B channel is the basic user channel and serves any one of the traffic types listed in Section 13.1.

The D channel is a 16-kbps channel primarily used for signaling. Secondarily, if capacity permits, it may be used for lower-speed data connectivity to the network.

Note: The D channel, when used for primary rate service, is a 64-kbps channel. In this case, it is used exclusively for signaling.

There are also H channels for higher transmission rates. The H0 channel is 384 kbps; there is an H1 group of transmission rates where H11 is 1536 kbps and H12 is 1920 kbps.

The H channels do not carry signaling. Their purpose is to serve the higher data rate user for such things as conference television, digitized program channels (for broadcasters), fast facsimile, and higher data rate service for computer data.

13.2.2 Basic and Primary Rate Interfaces

The *basic* interface structure is composed of two B channels and a D channel, commonly expressed as 2B + D. The D channel at this interface is 16 kbps. Often the literature uses the abbreviation *BRI* for basic rate interface.

Appendix I to CCITT Rec. I.412 (ITU-T I.412 in Ref. 1) states that the basic access may also be B + D or just D.

The *primary* rate B-channel interface structures are composed of n B channels and one D channel, where the D channel in this case is 64 kbps. There are two primary data rates:

1.544 Mbps = 23B + D (from the North American T1 configuration)

2.048 Mbps = 30B + D (from the European E1 configuration)

For the user–network access arrangement containing multiple interfaces, it is possible for the D channel in one structure not only to serve the signaling requirements of its own structure but also to serve another primary rate structure without an activated D channel. When a D channel is not activated, the designated time slot may or may not be used to provide an additional B channel, depending on the situation, such as 24B with 1.544 Mbps.

There are a number of H-channel interface structures where we would assume a primary rate configuration. H0, for example, is useful for video conferencing with its 384-kbps data rate (i.e., 6×64 kbps). Three H0 channels can be accommodated within the North American DS1 (T1) in time slots 1–6, 7–12, and 13–18. If a signaling channel is not required, then a fourth H0 channel can be handled in time slots 18–24.

At the E1 2.048-Mbps rate, five H0 channels can be accommodated and the preferred assignment is to use channels 1–6, 7–12, 13–19 (excluding channel 16), 20–25, and 26–31. Any time slot not used for H0 channels can be used for B channels. Of course, channel 16 in the E-1 configuration is the signaling channel.

There are two categories of H1 channels: H11 and H12. H11 is the 1.536-Mbps rate, and H12 is the 1920-kbps rate. This may be carried on the DS1 1.544-Mbps configuration, but a signaling channel, if required, would have to be carried on a separate DS1 configuration. For E-1, H1 would be carried in time slots 1–25 (excluding 16) of the 2.048-Mbps primary rate interface.

The industry often uses the acronym *PRI* to designate the primary rate interface.

13.3   USER ACCESS AND INTERFACE

13.3.1   General

The objective of the ISDN designers was to provide a telecommunication service that would be ubiquitous and universal. Whether this ambitious goal is being met can be argued, especially in light of the evolving Asynchronous Transfer Mode (ATM—Chapter 16) and the several varieties of Digital Subscriber Line (DSL).* This latter can provide downstream transmission rates as high as 6 Mbps on a subscriber pair. Other "last mile" broadband communication techniques also compete such as LMDS† and cable television (CATV). See Chapter 17.

ISDN was designed around the copper transmission medium (i.e., the twisted pair). Within 20 years much of the copper plant will be replaced by fiber optics (Chapter 15). Some futurists argue 10 years and others 30 years. Nevertheless, enterprise networks now demand transmission rates in excess of what ISDN has been designed for. In the business community, this demand, among other requirements, is for local area network (LAN) connectivity over wide area network (WAN) distances. Fiber can certainly do a better job than wire pair. Corporate intranets require megabit service rather than kilobit service offered by ISDN on copper pairs. ISDN is still very attractive for small businesses and residential service.

Figure 13.1 shows generic ISDN user connectivity to the network. We can select either the basic or primary rate service (e.g., 2B + D, 23B + D, or 30B + D) to connect to the ISDN network. The objectives of any digital interface design, and specifically those of ISDN access and interface, are:

1. Electrical and mechanical specification
2. Channel structure and access capabilities

---

*ISDN basic rate may also be considered a DSL (digital subscriber line) technique.
†LMDS stands for Local Multipoint Distribution System. This is a radio system that operates in the 28- to 30-GHz band.

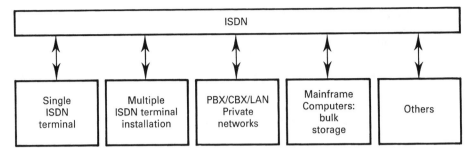

Figure 13.1. ISDN generic users.

3. User–network protocols
4. Maintenance and operation
5. Performance
6. Services

ISDN specifications may be found in the ITU-T I Recommendations and, for North American application, in relevant Telcordia and ANSI documents. In either case, expect these to cover the six items listed previously.

Figure 13.2 shows the conventional ISDN reference model. It delineates interface points for the user. In the figure, NT1 (network termination 1) provides the physical layer interface; it is essentially equivalent to open system interconnection (OSI) layer 1. The functions of the physical layer include:

- Transmission facility termination
- Layer 1 maintenance functions and performing monitoring
- Timing
- Power transfer
- Layer 1 multiplexing
- Interface termination, including multidrop termination employing layer 1 contention resolution

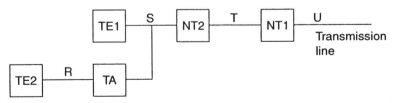

Figure 13.2. ISDN reference model.

Network termination 2 (NT2) can be broadly associated with OSI layers 1, 2, and 3. Among the examples of equipment that provide NT2 functions are user controllers, servers, LANs, and private automatic branch exchanges (PABXs). Among the NT2 functions are:

- Layers 1, 2, and 3 protocol processing
- Multiplexing (layers 2 and 3)
- Switching
- Concentration
- Interface termination and other layer 1 functions
- Maintenance functions

A distinction must be drawn here between North American and European practice. In Europe, telecommunication administrations have historically been, in general, national monopolies that are government controlled. In North America (i.e., United States and Canada) they are private enterprises, often very competitive. Thus in Europe, in many cases, the NT1 is the property of the telecommunication administration and the NT2 is the responsibility of the customer. Of course, in North America they both belong to the user, and there is a "U" interface between NT1 and ISDN switch. On the other hand, CCITT defines NT1 as part of the network even though it is on the user's premises.

It should be noted that there is an overall trend outside of North America to privatize telecommunications such as has happened in the United Kingdom and is scheduled to take place in other countries such as Germany, Mexico, and Venezuela.

TE1 in Figure 13.2 is the terminal equipment and has an interface that complies with ISDN terminal–network interface specifications at the S interface. We call this equipment *ISDN-compatible*. TE1 covers functions broadly belonging to OSI layer 1 and higher OSI layers. Among the equipment are digital telephones, computer workstations, and other devices in the user end-equipment category that are ISDN compatible.

TE2 in Figure 13.2 refers to equipment that does *not* meet the ISDN terminal–network interface at point S. TE2 adapts the equipment to meet the ISDN terminal–network interface. This process is assisted by the terminal adapter (TA).

Reference points T, S, and R are used to identify the interface available at those points. T and S are identical electrically, mechanically, and from the point of view of protocol. Point R relates to the TA interface, or, in essence, it is the interface of that nonstandard (i.e., non-ISDN) device. The U interface provides for transmission facility compatibility and is referenced in North American ISDN specifications.

We will return to user–network interfaces once the stage is set for ISDN protocols looking into the network from the user.

13.4 ISDN PROTOCOLS AND PROTOCOL ISSUES

When fully implemented, ISDN will provide both circuit and packet switching. For the circuit-switching case, now fairly broadly available in North America, the B channel is fully transparent to the network, permitting the user to utilize any protocol or bit sequence so long as there is end-to-end agreement on the protocol that will be used. Of course the protocol itself should be transparent to any bit sequence.

It is the D channel that carries the circuit-switching control function for its related B channels. Whether it is the 16-kbps D channel associated with BRI or the 64-kbps D channel associated with PRI, it is that channel which transports the signaling information from the user's ISDN terminal from NT to the first serving telephone exchange of the telephone company or administration. Here the D-channel signaling information is converted over to CCITT No. 7 signaling data employing ISUP (ISDN user part) of signaling system No. 7. Thus it is the D channel's responsibility for call establishment (setup), supervision, termination (takedown), and all other functions dealing with network access.

The B channel in the case of circuit switching is serviced by NT1 or NT2 using OSI layer 1 functions only. The D channel carries out OSI layers 1, 2, and 3 functions such that the B-channel protocol established by a family of ISDN end users will generally make layer 3 null in the B channel where the networking function is carried out by the associated D channel.

With packet switching, two possibilities emerge. The first basically relies on the B channel to carry out OSI layers 1, 2, and 3 functions at separate packet-switching facilities (PSFs). The D channel is used to set up the connection to the local switching exchange at each end of the connection. This type of packet-switched offering provides 64-kbps service. The second method utilizes the D channel exclusively for lower data rate switched service

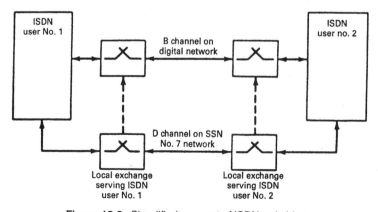

Figure 13.3. Simplified concept of ISDN switching.

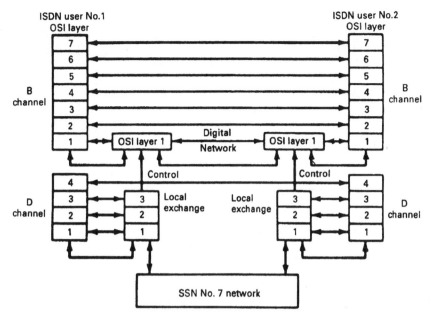

Figure 13.4. Detailed diagram of the ISDN circuit-switching concept.

where the local interface can act as a CCITT (ITU-T) X.25 data communication equipment (DCE) device.

Figure 13.3 is a simplified conceptual diagram of ISDN circuit switching. It shows the B channel riding on the public digital network and also shows the D channel, which is used for signaling. Of course, the D channel is a separate channel. It is converted to a CCITT No. 7 signaling structure and in this form may traverse several signal transfer points (STPs) and may be quasi-associated or disassociated from its companion B channel(s). Figure 13.4 is a detailed diagram of the same ISDN circuit-switching concept. The reader should note the following in the figure: (a) Only users at each end have a peer-to-peer relationship available for all seven OSI layers of the B channel. As the call is routed through the system, there is only layer 1 (physical layer) interaction at each switching node along the call route. (b) The D channel requires the first three OSI layers for call setup to the local switching center at each end of the circuit. (c) The D-channel signaling data are turned over to CCITT signaling system No. 7 (SS No. 7) at the near- and far-end local switching centers. (d) SS No. 7 also utilizes the first three OSI layers for circuit establishment, which requires the transfer of control information. In SS No. 7 terminology this is called the *message transfer part*. There is a fourth layer called the *user part* in SS No. 7. There are three user parts: telephone user part, data user part, and ISDN user part (ISUP) depending on whether the associated B channel is in telephone, data, or ISDN service for the user.

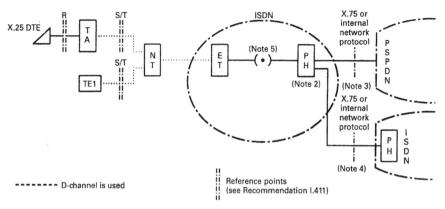

Figure 13.5. ISDN packet service for lower data rates using the D channel. TA, terminal adaptor; NT, network termination 2 and/or 1; ET, exchange termination; TE1, terminal equipment 1; PH, packet handling function. *Note 1:* This figure is only an example of many possible configurations and is included as an aid to the text describing the various interface functions. *Note 2:* In some implementations, the PH functions logically belonging to the ISDN may reside physically in a node of the PSPDN. The service provided is still the ISDN virtual circuit service. *Note 3:* See Recommendation X.325. *Note 4:* See Recommendation X.320. *Note 5:* This connection is either on demand or semipermanent, but has no relevance with the user \network procedures. Only internal procedures between the ET and the PH are required. (From Figure 2-3 / X.31, page 6, ITU-T Rec. X.31, Ref. 2.)

Figure 13.5 shows packet service for lower data rates where the D channel is involved.

13.5 ISDN NETWORKS

In this context, ISDN networking is seen as a group of access attributes connecting an ISDN user at either end to the local serving exchange (i.e., the local switch). This is shown in Figure 13.6, the basic architectural model of ISDN.

Figure 13.7 shows the ISDN reference configuration of public ISDN connection type, Figure 13.8 illustrates the access connection element model, Figure 13.9 shows the national tandem/transit connection element model, and Figure 13.10 shows the private ISDN access connection element.

The national transit network (CCITT terminology) is the public switched digital telephone network. That network, whether DS1–DS4-based or E1–E5-based,* provides the two necessary attributes for ISDN compatibility:

1. 64-kbps channelization
2. Separate channel signaling based on CCITT Signaling System No. 7

*See Chapter 7 for descriptions of DS1–DS4 and E1–E5 digital network configurations.

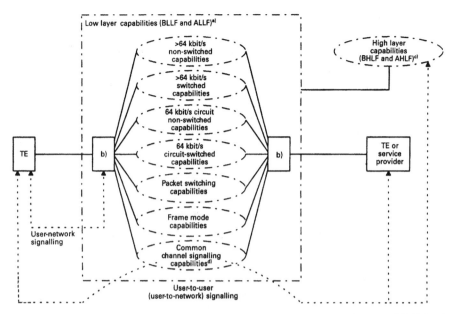

Figure 13.6. The basic architectural model of ISDN. TE, terminal equipment; BLLF, basic low-layer functions; ALLF, additional low-layer functions; BHLF, basic high-layer functions; AHLF, additional high-layer functions. *Note a:* In certain national situations, ALLF may also be implemented outside the ISDN, in special nodes or in certain categories of terminals. *Note b:* The ISDN local functional capabilities correspond to functions provided by a local exchange and possibly including other equipment, such as electronic cross-connect equipment, muldexes, and so on. *Note c:* These functions may either be implemented within ISDN or be provided by separate networks. Possible applications for basic high-layer functions and for additional high-layer functions are contained in Recommendation I.210. *Note d:* For signaling between international ISDNs, CCITT signaling system No. 7 shall be used. (From Figure 1/I.324, page 3, CCITT Rec. I.324, Ref. 3.)

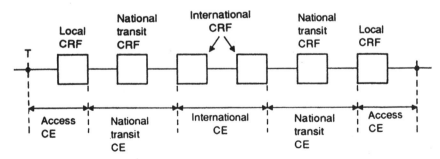

Figure 13.7. Reference configuration of public ISDN connection type. IRP, internal reference point; CRF, connection-related functions; CE, connection element. (From Figure 3/I.324, page 8, CCITT Rec. I.324, Ref. 3.)

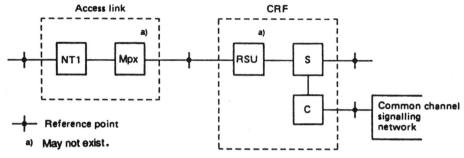

Figure 13.8. Access connection element model. NT1, network termination 1; S, 64-kbps circuit switch; C, signaling handling and exchange control functions; Mpx, (remote) multiplexer; RSU, remote switching unit and/or concentrator; CRF, connection-related function. (From Figure 4/I.324, page 9, CCITT Rec. I.324, Ref. 3.)

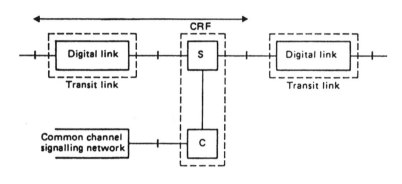

Figure 13.9. National tandem / transit connection element model. S, 64-kbps circuit switch; C, signaling handling and exchange control functions; CRF, connection-related function. (From Figure 5/I.324, page 10, CCITT Rec. I.324, Ref. 3.)

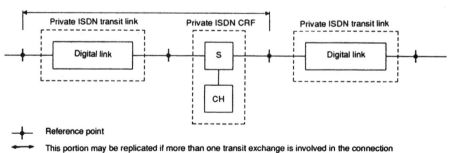

Figure 13.10. Private ISDN access connection element. S, 64-kbps circuit switch; CH, signaling handling and private ISDN exchange control function; CRF, connection-related function. (From Figure 8/I.324, page 11, CCITT Rec. I.324, Ref. 3.)

Connections from the user at the local connecting exchange interface include:

- Basic service (BRI) 2B + D = 192 kbps (CCITT) specified)
  = 160 kbps (North American ISDN 1)
  Both rates include overhead bits.
- Primary service 23B + D/30B + D = 1.544/2.048 Mbps

13.6 ISDN PROTOCOL STRUCTURES

13.6.1 ISDN and OSI

Figure 13.11 shows the ISDN relationship with OSI. (OSI was discussed in Chapter 2.) As is seen in the figure, ISDN concerns itself with only the first three OSI layers. OSI layers 4 to 7 are peer-to-peer connections and the end-user's responsibility. Remember that the B channel is concerned with OSI layer 1 only. We showed the one exception; that is, when the B channel is used for packet service, it will have those first three OSI layers to interface.

The D channel with its signaling and control functions is the exception to the previous statement. The D channel interfaces with CCITT Signaling System No. 7 at the first serving exchange. D channels handle three types of information: signaling (s), interactive data (p), and telemetry (t).

The layering of the D channel has followed the intent of the OSI reference model. The handling of the p and t data can be adapted to the OSI

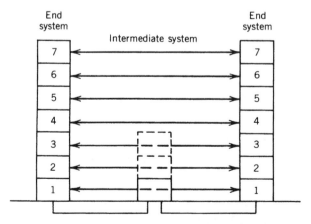

Figure 13.11. A generic communication context showing ISDN's relationship with the seven-layer OSI model. Note that the end-system protocol blocks may reside in the subscriber's TE or network exchanges or other ISDN equipment.

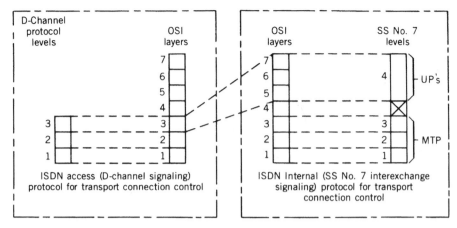

Figure 13.12. Correspondence among the ISDN D channel, CCITT SS No. 7, and the OSI model. MTP, message transfer part; UP, user part. (From Ref. 4. Copyright IEEE, New York, 1985.)

model; the s data, by its very nature, cannot.* Figure 13.12 shows the correspondence between D-channel signaling protocols, SS No. 7 levels, and the OSI seven-layer model.

13.6.2 Layer 1 Interface, Basic Rate (CCITT)

The S/T interface of the reference model, Figure 13.2 (or layer 1 physical interface), requires a balanced metallic transmission medium (i.e., copper pair) for each direction of transmission (four-wire) capable of supporting 192 kbps. Again, this is the NT interface of the ISDN reference model.

Layer 1 provides the following services to layer 2 for ISDN operation:

1. The transmission capability by means of an appropriately encoded bit stream of the B and D channels and any timing and synchronization functions that may be required.
2. The signaling capability and the necessary procedure to enable customer terminals and/or network terminating equipment to be deactivated when required and reactivated when required.
3. The signaling capability and necessary procedures to allow terminals to gain access to the common resource of the D channel in an orderly fashion while meeting the performance requirements of the D-channel signaling system.

*CCITT No. 7 signaling system, like any other signaling system, has the primary quality of service measure as "post-dial delay." This is principally the delay in call setup. To reduce the delay time as much as possible, it is incumbent upon system engineers to reduce processing time as much as possible. Thus SS No. 7 truncates OSI to 4 layers, because each additional layer implies more processing time, thus more post-dial delay.

TABLE 13.1   Primitives Associated with Layer 1

Generic	Specific Name		Parameter		Message Unit Content
	Request	Indication	Priority Indicator	Message Unit	
L1 ↔ L2					
PH-DATA	X (Note 1)	X	X (Note 2)	X	Layer 2 peer-to-peer message
PH-ACTIVATE	X	X	—	—	
PH-DEACTIVATE	—	X	—	—	
M ↔ L1					
MPH-ERROR	—	X	—	X	Type of error or recovery from a previously reported error
MPH-ACTIVATE	—	X	—	—	
MPH-DEACTIVATE	X	X	—	—	
MPH-INFORMATION	—	X	—	X	Connected/ disconnected

Note 1: PH-DATA request implies underlying negotiation between layer 1 and layer 2 for the acceptance of the data.
Note 2: Priority indication applies only to the request type.
Source: Table 1/I.430, ITU-T Rec. I.430, 11/95, page 2 (Ref. 6).

4. The signaling capability and procedures and necessary functions at layer 1 to enable maintenance functions to be performed.

5. An indication to higher layers of the status of layer 1.

13.6.2.1 *Primitives Between Layer 1 and Other Entities.* A *primitive* is data relating to the development or use of software that is used in developing measures or quantitative descriptions of software. Primitives are directly measurable or countable, or may be given a constant value or condition for a specific measure (IEEE definition, Ref. 5).

In this discussion, primitives represent in an abstract way the logical exchange of information and control between layer 1 and other entities or interfaces.

The primitives to be passed across the boundary between layers 1 and 2 or to the management entity are defined and summarized in Table 13.1. The parameter values associated with these primitives are also summarized in the table. For further information, the reader may consult ITU-T Rec. I.211 which describes the syntax and use of these primitives.

13.6.2.2 *Interface Functions.* The S and T functions for the BRI consist of three time division multiplexed bit streams: two 64-kbps B channels and one 16-kbps D channel for an aggregate bit rate of 192 kbps. Of this 192 kbps, the 2B + D configuration accounts for only 144 kbps. The remaining 48 kbps are overhead bits whose function is briefly described in the following text.

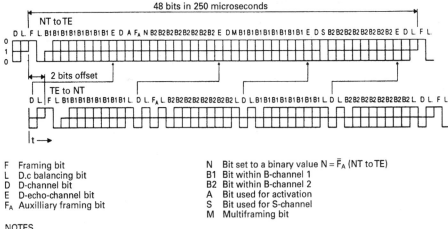

F   Framing bit
L   D.c balancing bit
D   D-channel bit
E   D-echo-channel bit
F_A Auxilliary framing bit

N   Bit set to a binary value N = $\bar{F}_A$ (NT to TE)
B1 Bit within B-channel 1
B2 Bit within B-channel 2
A   Bit used for activation
S   Bit used for S-channel
M   Multiframing bit

NOTES

1   Dots demarcate those parts of the frame that are independently d.c.-balanced
2   The F_A bit in the direction TE to NT is used as a Q bit in every fifth frame if the Q-channel capability is applied
3   The nominal 2-bit offset is as seen from the TE. The corresponding offset at the NT may be greater due to delay in the interface cable and varies by configuration

Figure 13.13. Frame structure at reference points S and T. This is the frame structure of the multiplexed bit stream for a 2B + D configuration, ITU-T interface, level 1. (From Figure 3/I.430, page 7, ITU-T Rec. I.430, Ref. 6.)

The functions covered at the interface include bit timing at 192 kbps to enable the TE and NT to recover information from the aggregate bit stream. This timing provides 8-kHz octet timing for the NT and TE to recover the time division multiplexed channels (i.e., 2B + D multiplexed). Other functions include D-channel access control, power feeding, deactivation, and activation.

Interchange circuits are required of which there is one in either direction of transmission (i.e., from the local switching point to and from the NT). They are used to transfer digital signals across the interface. All of the functions described above, except for power feeding, are carried out by means of a digitally multiplexed signal. The format of this signal is shown in Figure 13.13 and described in the next section.

It should be noted that NT1 in Europe is considered to be owned by the network, meaning the local telephone company or administration. In the United States, NT1 is owned by the customer. Transmission between the network and the customer is 2-wire. 2-wire to 4-wire conversion takes place in NT1. Echo suppressors at each end of the 2-wire transmission line allow two-way data to pass without harmful interference one to the other.

13.6.2.3 Frame Structure. In both directions, the bits are grouped into frames of 48 bits each. The frame structure is identical for all configurations, whether point-to-point or point-to-multipoint. However, the frame structures

TABLE 13.2A Notes on Bit Positions and Groups for a BRI Frame, Direction TE to NT

Bit Position	Group
1 and 2	Framing signal with balance bit
3 to 11	B1 channel (first octet) with balance bit
12 and 13	D-channel bit with balance bit
14 and 15	F_A auxiliary framing bit for Q bit with balance bit
16 to 24	B2 channel (first octet) with balance bit
25 and 26	D-channel bit with balance bit
27 to 35	B1 channel (second octet) with balance bit
36 and 37	D-channel bit with balance bit
38 to 46	B2 channel (second octet) with balance bit
47 and 48	D-channel bit with balance bit

Source: Table 2/I.430, page 7 (Ref. 6).

TABLE 13.2B Notes on Bit Positions and Groups for a BRI Frame, Direction NT to TE

Bit Position	Group
1 and 2	Framing signal with balance bit
3 to 10	B1 channel (first octet)
11	E-, D-echo channel bit
12	D-channel bit
13	Bit A used for activation
14	F_A auxiliary framing bit
15	N bit (coded as defined in 6.3 of I.430)
16 to 23	B2 channel (first octet)
24	E-, D-echo channel bit
25	D-channel bit
26	M, multiframing bit
27 to 34	B1 channel (second octet)
35	E-, D-echo channel bit
36	D-channel bit
37	S
38 to 45	B2 channel (second octet)
46	E-, D-echo channel bit
47	D-channel bit
48	Frame balance bit

Note: The use of the S bit is optional, and when not used it is set to binary ZERO.
Source: Table 3/I.430, page 8 (Ref. 6).

are different for each direction of transmission. These structures are shown in Figure 13.13, where the upper half of the figure shows the bit format direction NT to TE; the lower half, TE to NT. Table 13.2A gives explanatory notes for the direction TE to NT, and Table 13.2B, the direction NT to TE. Allow that NT refers to the network side of the circuit and TE refers to the terminal side. Note that each group is DC balanced by the last bit, the L bit.

13.6.2.4 *Line Code.* For both directions of transmission, pseudoternary coding is used with 100% pulse width, as shown in the following figure:

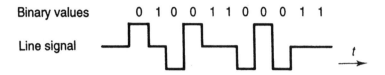

Coding is performed such that a binary 1 is represented by no line signal (0 volts), whereas a binary 0 is represented by a positive or negative pulse. The first binary signaling following the framing balance bit is the same polarity as the balance bit. Subsequent binary 0s alternate in polarity. A balance bit is a 0 if the number of 0s following the previous balance bit is odd. A balance bit is a binary 1 if the number of 0s following the previous balance bit is even. The balance bit tends to limit the buildup of a DC component on the line.

13.6.2.5 *Timing Considerations.* The NT derives its timing from the network clock. A TE synchronizes its bit, octet, and frame timing from the NT, which has derived its timing from the ISDN bit stream being received from the network. The NT uses this derived timing to synchronize its transmitter clock.

13.6.2.6 *BRI Differences in the United States.* Telcordia and ANSI prepared ISDN BRI standards fairly well modified from the CCITT I Recommendation counterparts. Telephone company administrations in the United States are at variance with those in most other countries. Furthermore, it was Telcordia's intention to produce equipment that was cost-effective and that would easily interface with existing the North American telephone plant.

One point, of course, is where the telephone company responsibilities end and customer responsibilities begin. This is called the "U" interface. The tendency toward a 2-wire interface, rather than 4-wire interface, is another point. Line waveform is yet another. Rather than a pseudoternary line waveform, the United States uses 2B1Q. The line bit rate is 160 kbps rather than that recommended by CCITT, namely, 192 kbps. More detailed explanation of North American practice follows. The 2B + D frames differ significantly from those recommended by CCITT.

DSL Bandwidth Allocation and Frame Structure. The *digital subscriber line* (DSL) modulation rate is 80 kilobauds, equivalent to 160 kbps. The effective 160-kbps signal is divided up into 12 kbps for synchronization words, 144 kbps for 2B + D of customer data, and 4 kbps of DSL overhead.

The synchronization technique used is based on transmission of nine (quaternary) symbols every 1.5 ms, followed by 216 bits of 2B + D data and 6 bits of overhead. The synchronization word provides a robust method of conveying line timing and establishes a 1.5-ms DSL "basic frame" for multiplexing subrate signals. Every eighth synchronization word is inverted (i.e., the 1s become 0s and the 0s become 1s) to provide a boundary for a 12-ms superframe composed of eight basic frames. This 12-ms interval defines an appropriate block of customer data for performance monitoring and permits a more efficient suballocation of the overhead bits among various operational functions.

The 48 overhead bits (sometimes referred to as *maintenance bits* or *M bits*) available per superframe are allocated into 24 bits for a 2-kbps *eoc*, 12 bits for a *crc* check covering the 1728 bits of customer data in a superframe, plus 8 particular overhead bits, and one *febe* bit for communicating block errors detected by the *crc* check to the far end. In determining relative positions of the various overhead bit functions, *febe* and *crc* bits were placed away from the start of the superframe to allow time for the *crc* calculation associated with the preceding superframe to be completed.

The *eoc* is the *embedded operations channel* and permits network operations systems to access essential operations functionality in the NT1. There are 12 cyclic redundancy check (*crc*) bits within each superframe; they cover the 1728 bits of 2B + D customer data plus the 8 "M4" bits, in the previous superframe. This provides a means of block error check. Two "*crc* bits" are assigned to each of the basic frames 3 through 8. This is illustrated in Figures 13.14 and 13.15.

The acronym *febe* stands for "far-end block error" bit which is assigned to each superframe. The *febe* bit indicates whether a block error was detected in the 2B + D customer data in a preceding superframe in the opposite direction of transmission. An outgoing *febe* bit with a value of 1 indicates that the *crc* of the preceding incoming superframe matched the value computed from the transmitted data; a value of 0 indicates a mismatch, which, in turn, indicates one or more bit errors.

2B + D Customer Data Bit Pattern. There are 216 2B + D bits placed in each 1.5-ms basic frame, for a customer data rate of 144 kbps. The bit pattern (before conversion to quaternary form and after reconversion to binary form) for the 2B + D data is

$$B_1 B_1 B_1 B_1 B_1 B_1 B_1 B_1 B_2 B_2 B_2 B_2 B_2 B_2 B_2 B_2 DD$$

where B_1 and B_2 are bits from the B_1 and B_2 channels, and D is a bit from the D channel. This 18-bit pattern is repeated 12 times per DSL basic frame.

The 2B1Q Waveform. It is convenient to express the 2B1Q waveform as $+3, +1, -1, -3$ because this indicates symmetry about zero, equal spacing

		FRAMING	2B+D	Overhead bits (M1–M6)					
	Quat positions	1–9	10–117	118s	118m	119s	119m	120s	120m
	Bit positions	1–18	19–234	235	236	237	238	239	240
Super frame #	Basic frame #	Sync word	2B+D	M1	M2	M3	M4	M5	M6
A	1	ISW	2B+D	eoc_{a1}	eoc_{a2}	eoc_{a3}	act	1	1
	2	SW	2B+D	eoc_{dm}	eoc_{i1}	eoc_{i2}	dea	1	febe
	3	SW	2B+D	eoc_{i3}	eoc_{i4}	eoc_{i5}	1	crc_1	crc_2
	4	SW	2B+D	eoc_{i6}	eoc_{i7}	eoc_{i8}	1	crc_3	crc_4
	5	SW	2B+D	eoc_{a1}	eoc_{a2}	eoc_{a3}	1	crc_5	crc_6
	6	SW	2B+D	eoc_{dm}	eoc_{i1}	eoc_{i2}	1	crc_7	crc_8
	7	SW	2B+D	eoc_{i3}	eoc_{i4}	eoc_{i5}	1	crc_9	crc_{10}
	8	SW	2B+D	eoc_{i6}	eoc_{i7}	eoc_{i8}	1	crc_{11}	crc_{12}
B, C, ...									

Figure 13.14. LT to NT1 2B1Q superframe technique and overhead bit assignments. NT1-to-LT superframe delay is offset by LT-to-NT1 superframe by 60 ± 2 quats (about 0.75 ms). All bits other than the Sync word are scrambled. 1, reserved bit for future standard (set = 1); ( )m and ( )s, "magnitude" bit and "sign" bit for given quat; act, activation bit; crc, cyclic redundancy check: covers 2B + D and M4; dea, deactivation bit; eoc, embedded operations channel; a, address bit; dm, data/message indicator; i, information (data/message); febe, far-end block error bit. (From Figure 3.1a, page 3-9, Ref. 7.)

between states, and convenient integer magnitudes. The block synchronization word (SW) contains nine quaternary elements repeated every 1.5 ms:

$$+3, +3, -3, -3, -3, +3, -3, +3, +3$$

Line Configurations. One basic access is provided by a single, 2-wire DSL interface. Operation on this 2-wire line is full-duplex. To avoid interface between transmitted and received signals on the same line, an echo canceler is used. Echo cancellation involves adaptively forming a replica of the echo signal arriving at the receiver from its local transmitter and subtracting it from the signal at the input of the receiver.

13.6.3 Layer 1 Interface, Primary Rate

This interface is applicable for the 1.544- or 2.048-Mbps data rates.

		FRAMING	2B+D	Overhead bits (M1–M6)					
	Quat positions	1–9	10–117	118s	118m	119s	119m	120s	120m
	Bit positions	1–18	19–234	235	236	237	238	239	240
Super frame #	Basic frame #	Sync word	2B+D	M1	M2	M3	M4	M5	M6
1	1	ISW	2B+D	eoc_{a1}	eoc_{a2}	eoc_{a3}	act	1	1/nib
	2	SW	2B+D	eoc_{dm}	eoc_{i1}	eoc_{i2}	ps_1	1	febe
	3	SW	2B+D	eoc_{i3}	eoc_{i4}	eoc_{i5}	ps_2	crc_1	crc_2
	4	SW	2B+D	eoc_{i6}	eoc_{i7}	eoc_{i8}	ntm	crc_3	crc_4
	5	SW	2B+D	eoc_{a1}	eoc_{a2}	eoc_{a3}	cso	crc_5	crc_6
	6	SW	2B+B	eoc_{dm}	eoc_{i1}	eoc_{i2}	1	crc_7	crc_8
	7	SW	2B+D	eoc_{i3}	eoc_{i4}	eoc_{i5}	1	crc_9	crc_{10}
	8	SW	2B+D	eoc_{i6}	eoc_{i7}	eoc_{i8}	1	crc_{11}	crc_{12}
2, 3, ...									

Figure 13.15. NT1 to LT 2B1Q superframe technique and overhead bit assignment. NT1-to-LT superframe delay is offset by LT-to-NT1 superframe 60 ± 2 quats (about 0.75 ms). All bits other than the Sync word are scrambled. 1, reserved bit for future standard (set = 1). ( )m and ( )s, "magnitude" bit and "sign" bit for given quat; act, activation bit; crc, cyclic redundancy check: covers 2B + D and M4; cso, cold start only bit; dea, deactivation bit; febe, far-end block error bit; eoc, embedded operations channel; a, address bit; dm, data / message indicator; i, information (data / message); ntm, NT1 in Test Mode bit; nib, network indicator bit from LULT and LUNT to LT; ps_1 and ps_2, power status bits (= 1 from NT1 to LT). (From Figure 3-1b, page 3-10, Ref. 7.)

13.6.3.1 Interface at 1.544 Mbps

Bit Rate and Synchronization

NETWORK CONNECTION CHARACTERISTICS. The network delivers (except as noted below) a signal synchronized from a clock having a minimum accuracy of 1×10^{-11} (stratum 1; see Ref. 8 for stratum definition). When synchronization by a stratum 1 clock has been interrupted, the signal delivered by the network to the interface will have a minimum accuracy of 4.6×10^{-6} (stratum 3).

While in normal operation, the TE1/TA/NT2 transmits a 1.544-Mbps signal having an accuracy equal to that of the received signal by locking the frequency of its transmitter signal to the long-term average of the incoming 1.544-Mbps signal or by providing equal signal frequency accuracy from another source. ITU-T Rec. I.431 (Ref. 9) advises against this latter alternative.

RECEIVER BIT STREAM SYNCHRONIZED TO A NETWORK CLOCK

Receiver Requirements. Receivers of signals across interface I_a operate with an average transmission rate in the range of 1.544 Mbps $\pm$ 4.6 ppm. However, operation with a received signal transmission rate in the range of 1.544 Mbps $\pm$ 32 ppm is required in any maintenance state controlled by signals/messages passed over the m bits and by an alarm indication signal (AIS). In normal operation the bit stream is synchronized to stratum 1.

Transmitter Requirements. The average transmission rate of signals transmitted across interface I_a by the associated equipment is the same as the average transmission rate of the received bit stream.

Note: The I_a and I_b interfaces are located at the input/output port of TE or NT.

TE1/TA OPERATING BEHIND AN NT2 THAT IS NOT SYNCHRONIZED TO A NETWORK CLOCK

Receiver Requirements. Receivers of signals across interface I_a operate with a transmission rate in the range of 1.544 $\pm$ 32 ppm.

Transmitter Requirements. The transmitted signal across interface I_a is synchronized to the received bit stream.

SPECIFICATION OF OUTPUT PORTS. The signal specification for output ports is summarized in Table 13.3.

TABLE 13.3 Digital Interface at 1.544 Mbps

Bit rate		1544 kbps
Pair(s) in each direction of transmission		One symmetrical pair
Code		B8ZS (Note 1)
Test load impedance		100 ohm resistive
Nominal pulse shape		See pulse mask (Note 2)[a]
Signal level	Power at 772 kHz	+12 dBm to +19 dBm
(Notes 2 and 3)	Power at 1544 kHz	At least 25 dB below the power at 772 kHz

Note 1: B8ZS is modified AMI code in which eight consecutive binary ZEROs are replaced with 000 + − 0 − + if the preceding pulse was positive (+) and with 000 − + 0 + − if the preceding pulse was negative (−).

Note 2: The pulse mask and power level requirements apply at the end of a pair having a loss at 772 kHz of 0 to 1.5 dB.

Note 3: The signal level is the power level measured in a 3-kHz bandwidth at the output port for an all binary ONEs pattern transmitted.

[a]Given in Ref. 17, Chapter 7, Figure 7.24.

Source: Table 4/I.431, page 13 (Ref. 9).

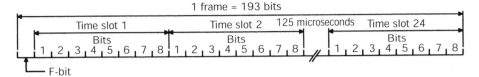

Figure 13.16. Frame structure of 1.544-Mbps interface.

TABLE 13.4 Multiframe Structure

Multiframe Frame Number	Multiframe Bit Number	F Bits Assignments		
		FAS	Control / Maintenance	CRC Bits
1	1	—	m	—
2	194	—	—	e_1
3	387	—	m	—
4	580	0	—	—
5	773	—	m	—
6	966	—	—	e_2
7	1159	—	m	—
8	1352	0	—	—
9	1545	—	m	—
10	1738	—	—	e_3
11	1931	—	m	—
12	2124	1	—	—
13	2317	—	m	—
14	2510	—	—	e_4
15	2703	—	m	—
16	2896	0	—	—
17	3089	—	m	—
18	3282	—	—	e_5
19	3475	—	m	—
20	3668	1	—	—
21	3861	—	m	—
22	4054	—	—	e_6
23	4247	—	m	—
24	4440	1	—	—

FAS, frame alignment signal
Source: Table 5/I.431, page 15 (Ref. 9).

Frame Structure. The frame structure is shown in Figure 13.16. Each time slot consists of consecutive bits, numbered 1 through 8. Each frame is 193 bits long and consists of an F bit (framing bit) followed by 24 consecutive time slots. The frame repetition rate is 8000 frames per second.

Table 13.4 shows the multiframe structure (called *extended superframe* in the United States) which is 24 frames long. It takes advantage of the more advanced search algorithms for frame alignment. There are 8000 F bits (frame alignment bits) transmitted per second (i.e., 8000 frames a second, 1 F bit per frame). In 24 frames, with these new strategies, only 6 bits are required for frame alignment as shown in Table 13.4. The remaining 18 bits are used as follows; There are 12 m bits used for control and maintenance, and the 6 e bits are used for CRC6 error checking.

Time Slot Assignment. Time slot 24 is assigned to the D channel, when this channel is present.

A channel occupies an integer number of time slots and the same time slot positions in every frame. A B channel may be assigned any time slot in the frame, an H0 channel may be assigned any six slots in a frame in numerical order (not necessarily consecutive), and an H11 channel may be assigned slots 1 to 24. The assignments may vary on a call-by-call basis.

CODES FOR IDLE CHANNELS, IDLE SLOTS, AND INTERFRAME TIME FILL. A pattern including at least three binary 1s in an octet is transmitted in every time slot that is not assigned a channel (e.g., time slots awaiting channel assignment on a per-call basis, residual slots on an interface that is not fully provisioned, etc.) and in every time slot of a channel that is not allocated to a call in both directions. Interframe (layer 2) time fill consists of contiguous HDLC flags* transmitted on the D channel when its layer 2 has no frames to send.

13.6.3.2 Interface at 2.048 Mbps

Frame Structure. There are 8 bits per time slot and 32 time slots per frame, numbered 0 through 31. The number of bits per frame is 256 (i.e., 32×8), and the frame repetition rate is 8000 frames per second. Time slot 0 provides frame alignment, and time slot 16 is assigned to the D channel when that channel is present. A channel occupies an integer number of time slots and the same time slot position in every frame. A B channel may be assigned any time slot in the frame; and an H0 channel may be assigned any six time slots, in numerical order, not necessarily consecutive. The assignment of time slots may vary on a call-by-call basis. An H12 channel is assigned time slots 1–15 and 17–31 in a frame. Time slots 1–31 provide bit-sequence-independent transmission.

Timing Considerations. The NT derives its timing from the network clock. The TE synchronizes its timing (bit, octet, framing) from the signal received from the NT and synchronizes accordingly the transmitted signal. In an unsynchronized condition—that is, when the access that normally provides network timing is unavailable—the frequency deviation of the free-running clock shall not exceed ± 50 ppm. A TE shall be able to detect and to interpret the input signal within a frequency range of ± 50 ppm.

---

*For information about HDLC flags, see Section 3.4.

Any TE which provides more than one interface is declared to be a multiple access TE and is capable of taking the synchronizing clock frequency from its internal clock generator from one or more than one access (or all access links) and synchronizing the transmitted signal at each interface accordingly.

Codes for Idle Channels and Idle Time Slots, Interframe Fill. A pattern including at least three binary 1s in an octet is transmitted in every time slot that is not assigned to a channel (e.g., time slots awaiting channel assignment on a per-call basis, residual slots on an interface is not fully provisioned, etc.), as well as in every time slot of a channel that is not allocated to a call in both directions.

Interframe (layer 2) time fill consists of contiguous HDLC flags (01111110) which are transmitted on the D channel when its layer 2 has no frames to send.

Frame alignment and CRC procedures can be found in ITU-T Rec. G. 706, paragraph 4 (Ref. 10).

13.7   OVERVIEW OF THE LAYER 2 INTERFACE: LINK ACCESS PROCEDURE FOR THE D CHANNEL (LAPD)

The link access procedure for the D channel (LAPD)* is used to convey information between layer 3 entities across the ISDN user–network interface using the D channel.

A *service access point* (SAP) is a point at which the data-link layer provides services to its next higher OSI layer or layer 3. Associated with each data-link layer (OSI layer 2) is one or more data-link connection endpoints (see Figure 13.17). A data-link connection endpoint is identified by a data-link connection endpoint identifier as seen from layer 3 and by a data-link connection identifier (DLCI) as seen from the data-link layer.

Cooperation between data-link-layer entities is governed by a specific protocol to the applicable layer. In order for information to be exchanged between two or more layer 3 entities, an association must be established between layer 3 entities in the data-link layer using a data-link-layer protocol. This association is provided by the data-link layer between two or more SAPs, as shown in Figure 13.18. Data-link message units are conveyed between data-link-layer entities by means of a physical connection. Layer 3 uses *service primitives* to request service from the data-link layer. A similar interaction takes place between layer 2 and layer 1.

---

*Note the similarity between LAPD described here and HDLC covered in Chapter 3. In fact, LAPD is a subset of HDLC.

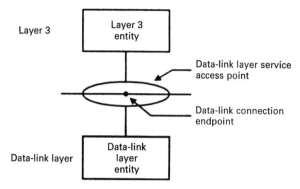

Figure 13.17. Entities, service access points (SAPs), and endpoints. (From Figure 2/Q.920, page 2, ITU-T Rec. Q.920, Ref. 11.)

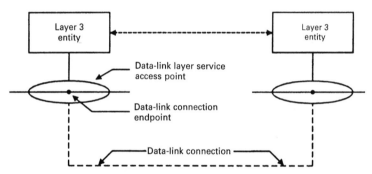

Figure 13.18. Data-link connections between two or more SAPs. (From Figure 3/Q.920, page 3, ITU-T Rec. Q.920, Ref. 11.)

Between the data-link layer and its adjacent layers there are four types of service primitives:

1. Request
2. Indication
3. Response
4. Confirm

These functions are shown diagrammatically in Figure 13.19.

The REQUEST primitive is used where a higher layer is requesting service from the next lower layer. The INDICATION primitive is used by a layer providing service to notify the next higher layer of activities related to the REQUEST primitive. The RESPONSE primitive is used by a layer to acknowledge receipt from a lower layer of the INDICATION primitive. The

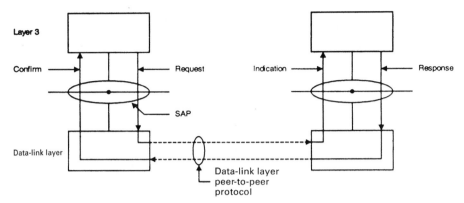

Figure 13.19. Functions of service primitives. *Note:* The same principle applies for data-link-layer–physical-layer interactions. (From Figure 4/Q.920, page 4, ITU-T Rec. Q.920, Ref. 11.)

CONFIRM primitive is used by the layer providing the requested service to confirm that the requested activity has been completed.

Figure 13.20 shows the data-link-layer reference model. All data-link-layer messages are transmitted in frames delimited by flags, where a flag is a unique sequence bit pattern. The frame structure as defined in ITU-T Rec. I.441 is briefly described in this section.

The LAPD includes functions for:

1. The provision of one or more data-link connections on a D channel. Discrimination between the data-link connections is by means of a DLCI contained in each frame.
2. Frame delimiting, alignment, and transparency, allowing recognition of a sequence of bits transmitted over a D channel as a frame.
3. Sequence, control, which maintains the sequential order of frames across a data-link connection.
4. Detection of transmission, format, and operational errors on a data link.
5. Recovery from detected transmission, format, and operational errors. Notification to the management entity of unrecovered errors.
6. Flow control.

There is unacknowledged and acknowledged operation. With unacknowledged operation, information is transmitted in unnumbered information (UI) frames. At the data-link layer the UI frames are unacknowledged. Transmission and format errors may be detected, but no recovery mechanism is defined. Flow control mechanisms are also not defined. With acknowledged

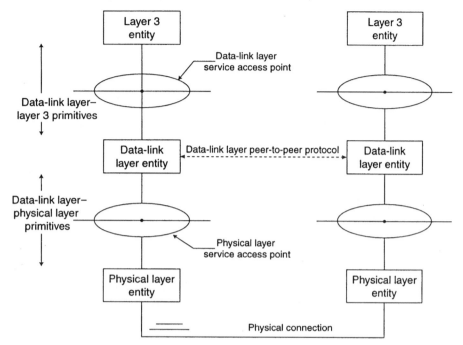

Figure 13.20. Data-link-layer reference model. (From Figure 5/Q.920, page 5, ITU-T Rec. Q.920, Ref. 11.)

operation, layer 3 information is transmitted in frames that are acknowledged at the data-link layer. Error recovery procedures based on retransmission of unacknowledged frames are specified. For errors that cannot be corrected by the data-link layer, a report to the management entity is made. Flow control procedures are also defined.

Unacknowledged operation is applicable for point-to-point and broadcast information transfer. However, acknowledged operation is applicable only for point-to-point information transfer.

Two forms of acknowledged information are defined:

1. Single-frame operation
2. Multiframe operation

For single-frame operation, layer 3 information is sent in sequenced information 0 (SI0) and sequenced information 1 (SI1) frames. No new frame is sent until an acknowledgment has been received for a previously sent frame. This means that only one unacknowledged frame may be outstanding at a time. With multiple-frame operation, layer 3 information is sent in numbered information (I) frames. A number of I frames may be outstanding at the same time. Multiple-frame operation is initiated by a multiple-frame establishment procedure using set asynchronous balanced mode/set asynchronous balanced mode extended (SABM/SABME) command.

13.7.1 Layer 2 Frame Structure for Peer-to-Peer Communication

There are two frame formats used on layer 2 frames:

1. Format A, for frames where there is no information field
2. Format B, for frames containing an information field

These two frame formats are shown in Figure 13.21.

The following discussion briefly describes the frame content (sequences and fields) for the LAPD (layer 2) frame.

Flag Sequence. All frames start and end with a flag sequence consisting of one 0 bit followed by six contiguous 1 bits and one 0 bit. These flags are called the *opening* and *closing* flags. If sequential frames are transmitted, the closing flag of one frame is the opening flag of the next frame.

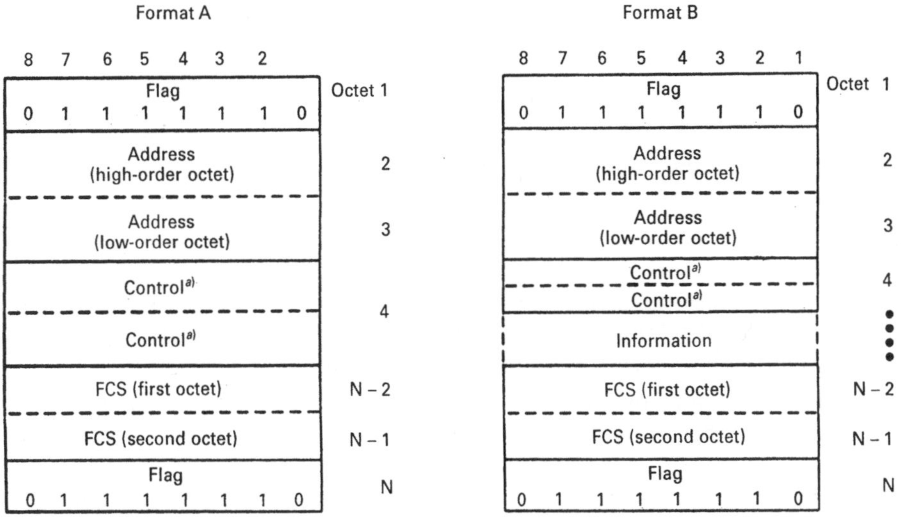

Format A		Format B	
8 7 6 5 4 3 2		8 7 6 5 4 3 2 1	
Flag 0 1 1 1 1 1 1 0	Octet 1	Flag 0 1 1 1 1 1 1 0	Octet 1
Address (high-order octet)	2	Address (high-order octet)	2
Address (low-order octet)	3	Address (low-order octet)	3
Control[a]		Control[a] / Control[a]	4
Control[a]	4	Information	• • •
FCS (first octet)	N – 2	FCS (first octet)	N – 2
FCS (second octet)	N – 1	FCS (second octet)	N – 1
Flag 0 1 1 1 1 1 1 0	N	Flag 0 1 1 1 1 1 1 0	N

[a] Unacknowledged operation – one octet

Multiple-frame operation – two octets for frames with sequence numbers;
 – one octet for frames without sequence numbers.

Figure 13.21. Frame formats for layer 2 frames. (From Figure 1/Q.921, page 20, ITU-T Rec. Q.921, Ref. 12.)

Address Field. As shown in Figure 13.21, the address field consists of 2 octets and identifies the intended receiver of a command frame and the transmitter of a response frame. A single-octet address field is used for LAPB operation.

Control Field. The control field consists of 1 or 2 octets. It identifies the type of frame, either command or response. It contains sequence numbers where applicable. Three types of control field formats are specified:

1. Numbered information transfer (I format)
2. Supervisory functions (S format)
3. Unnumbered information transfers and control functions (U format)

Information Field. The information field of a frame, when present, follows the control field and precedes the frame check sequence (FCS). The information field contains an integer number of octets:

- For SAPs supporting signaling, the default value is 128 octets.
- For SAPs supporting packet information, the default value is 260 octets.

Frame Check Sequence (FCS) Field. The FCS field is a 16-bit sequence and is the 1's complement of the modulo 2 sum of:

1. The remainder of X raised to the k power:

$$X^{15} + X^{14} \cdots X^1 + 1$$

   divided by the generating polynomial

$$X^{16} + X^{12} + X^5 + 1$$

   where k is the number of bits in the frame existing between but not including the final bit of the opening flag and the first bit of the FCS, excluding bits inserted for transparency, and
2. The remainder by modulo 2 division by the generating polynomial given above of the product of X^{16} by the content of the frame defined above.

Transparency, mentioned above, ensures that a flag or abort sequence is not imitated within a frame. On the transmit side the data-link layer examines the frame content between the opening and closing flag sequences and inserts a 0 bit after all sequences with five contiguous 1 bits (including the last 5 bits of the FCS). On the receive side the data-link layer examines the frame contents between the opening and closing flag sequences and discards any 0 bit that directly follows five contiguous 1 bits.

8	7	6	5	4	3	2	1	
SAPI						C/R	EA 0	Octet 2
TEI							EA 1	3

Figure 13.22. LAPD address field format. EA, address field extension bit; C/R, command/response field bit; SAPI, service access point identifier; TEI, terminal endpoint identifier. (From Figure 5/Q.921, page 23, ITU-T Rec. Q.920, Ref. 12.)

Address Field Format. The address field is shown in Figure 13.22. It contains address field extension bits (EA), command/response indication bit (C/R), a data-link-layer service access point identifier (SAPI) subfield, and a terminal endpoint identifier (TEI) subfield.

ADDRESS FIELD EXTENSION BIT (EA). The address field range is extended by reserving the first transmitted bit of the address field to indicate the final octet of the address field. The presence of a 1 in the first bit position of an address field octet signals that it is the final octet of the address field. The double-octet address field for LAPD operation has bit 1 of the first octet set to a 0 and bit 1 of the second octet set to a 1.

COMMAND/RESPONSE FIELD BIT (C/R). The C/R bit identifies a frame as either a command or a response. The user side sends commands with the C/R bit set to 0 and responses with the C/R bit set to 1. The network side does the opposite; that is, commands are sent with the C/R bit set to 1, and responses are sent with the C/R bit set to 0.

In keeping with HDLC rules,* commands use the address of the peer data-link entity while responses use the address of their own data-link-layer entity. In accordance with these rules, both peer entities on a point-to-point data-link connection use the same DLCI composed of a SAPI–TEI where SAPI and TEI conform to the definitions contained as follows.

Service Access Point Identifier (SAPI). The SAPI identifies a point at which data-link services are provided by a data-link-layer entity to a layer 3 or management entity. Consequently, the SAPI specifies a data-link-layer entity that should process a data-link-layer frame. The SAPI allows 64 SAPs to be specified, where bit 3 of the address field octet containing the SAPI is the least significant binary digit and bit 8 is the most significant. The SAPI values are allocated as shown in Table 13.5.

---

*LAPD, as we know, is a derivative of HDLC. See Chapter 3.

TABLE 13.5 Allocation of SAPI Values

SAPI Value	Related Layer 3 or Management Entity
0	Call control procedures
1	Reserved for packet mode communications using Q.931 call control procedures
16	Packet communication conforming to X.25 Level 3 procedures
63	Layer 2 management procedures
All Others	Reserved for future standardization

Terminal Endpoint Identifier (TEI). The TEI for a point-to-point link connection may be associated with a single terminal equipment (TE). A TE may contain one or more TEIs used for point-to-point data transfer. The TEI for a broadcast data-link connection is associated with all user-side data-link-layer entities containing the same SAPI. The TEI subfield allows 128 values where bit 2 of the address field octet containing the TEI is the least significant bit and bit 8 is the most significant bit. The following conventions apply for assignment of these values:

1. TEI for broadcast data-link connection. The TEI subfield bit pattern 111 1111 ($= 127$) is defined as the group TEI. The group TEI is assigned to the broadcast data-link connection associated with the addressed SAP.
2. TEI for point-to-point data-link connection. The remaining TEI values are used for point-to-point data-link connections associated with the addressed SAP. The range of TEI values are as follows:

Values 0–63:   Nonautomatic TEI assignment user equipment
Values 64–126: Automatic TEI assignment user equipment

Control Field Formats. Control field formats are shown in Table 13.6. The control field identifies the type of frame, either a command or a response. The control field contains sequence numbers, where applicable. Three types of control field formats are specified: numbered information transfer (I format), supervisory functions (S format), and unnumbered information transfers and control functions (U format).

Further discussion of control field formats may be found in Section 3.4.

LAPD Primitives. The following comments clarify the semantics and usage of primitives. Primitives consist of commands and their respective responses associated with the services requested of a lower layer. The general syntax of a primitive is:

XX-generic name-type: parameters

TABLE 13.6   Control Field Formats

Control Field Bits (Modulo 128)	8	7	6	5	4	3	2	1	
I format	N(S)							0	Octet 4
	N(R)							P	5
S format	X	X	X	X	S	S	0	1	Octet 4
	N(R)							P/F	5
U format	M	M	M	P/F	M	M̄	1	1	Octet 4

N(S) Transmitter send sequence number
N(R) Transmitter receive sequence number
S   Supervisory function bit

M   Modifier function bit
P/F   Poll bit when issued as a command, final bit when issued as a response
X   Reserved and set to 0

Source: Table 4/Q.921, page 26 (Ref.12).

TABLE 13.7   Primitives Associated with the Data-Link Layer

Generic Name	Type				Parameters		Message Unit Contents
	Request	Indication	Response	Confirm	Priority Indicator	Message Unit	
L3 ↔ L2							
DL-ESTABLISH	X	X	—	X	—	—	
DL-RELEASE	X	X	—	X	—	—	
DL-DATA	X	X	—	—	—	X	Layer 3 peer-to-peer message
DL-UNIT DATA	X	X	—	—	—	X	Layer 3 peer-to-peer message
M ↔ L2							
MDL-ASSIGN	X	X	—	—	—	X	TEI value, CES
MDL-REMOVE	X	—	—	—	—	X	TEI value, CES
MDL-ERROR	—	X	X	—	—	X	Reason for error message
MDL-UNIT DATA	X	X	—	—	—	X	Management function peer-to-peer message
MDL-XID	X	X	X	X	—	X	Connection management information
L2 ↔ L1							
PH-DATA	X	X	—	—	X	X	Data-link-layer peer-to-peer message
PH-ACTIVATE	X	X	—	—	—	—	
PH-DEACTIVATE	—	X	—	—	—	—	
M ↔ L1							
MPH-ACTIVATE	—	X	—	—	—	—	
MPH-DEACTIVATE	X	X	—	—	—	—	
MPH-INFORMATION	—	X	—	—	—	X	Connected/disconnected

L3 ↔ L2: Layer 3/data-link-layer boundary
L2 ↔ L1: Data-link-layer/physical-layer boundary
M ↔ L2: Management entity/data-link-layer boundary
M ↔ L1: Management entity/physical-layer boundary

Source: Table 6/Q.921, page 32 (Ref. 12).

where XX designates the layer providing the service. XX is DL for the data-link layer, PH for the physical layer, or MDL for the management entity to the data-link-layer interface. Table 13.7 gives the primitives associated with the data-link layer.

13.8 OVERVIEW OF LAYER 3

The layer 3 protocol of course deals with the D channel and its signaling capabilities. It provides the means to establish, maintain, and terminate network connections across an ISDN between communicating application entities. A more detailed description of the layer 3 protocol may be found in ITU-T Rec. Q.931, (Ref. 14).

Layer 3 utilizes functions and services provided by its data-link layer, as described in Section 13.7 under LAPD functions. These necessary layer 2 support functions are listed and briefly described as:

• Establishment of the data-link connection
• Error-protected transmission of data
• Notification of unrecoverable data-link errors
• Release of data-link connections
• Notification of data-link-layer failures
• Recovery from certain error conditions
• Indication of data-link-layer status

Layer 3 performs two basic categories of functions and services in the establishment of network connections. The first category directly controls the connection establishment. The second category includes those functions relating to the transport of messages in addition to the functions provided by the data-link layer. Among these additional functions are the provision of rerouting of signaling messages on an alternative D channel (where provided) in the event of D-channel failure. Other possible functions include multiplexing and message segmenting and blocking. The D-channel layer 3 protocol is designed to carry out establishment and control of circuit-switched and packet-switched connections. Also, services involving the use of connections of different types, according to user specifications, may be effected through "multimedia" call control procedures.

Functions performed by layer 3 include:

1. The processing of primitives for communicating with the data-link layer.
2. Generation and interpretation of layer 3 messages for peer-level communications.

3. Administration of timers and logical entities (e.g., call references) used in call control procedures.

4. Administration of access resources, including B channels and packet-layer logical channels (e.g., ITU-T X.25).

5. Checking to ensure that services provided are consistent with user requirements, such as compatibility, address, and service indicators.

The following functions may also be performed by layer 3:

1. *Routing and Relaying.* Network connections exist either between users and ISDN exchanges or between users. Network connections may involve intermediate systems which provide relays to other interconnecting subnetworks and which facilitate interworking with other networks. Routing functions determine an appropriate route between layer 3 addresses.

2. *Network Connection.* This function includes mechanisms for providing network connections making use of data-link connections provided by the data-link layer.

3. *Conveying User Information.* This function may be carried out with or without the establishment of a circuit-switched connection.

4. *Network Connection Multiplexing.* Layer 3 provides multiplexing of call control information for multiple calls onto a single data-link connection.

5. *Segmenting and Reassembly (SAR).* Layer 3 may segment and reassemble layer 3 messages to facilitate their transfer across a user–network interface.

6. *Error Detection.* Error detection functions are used to detect procedural errors in the layer 3 protocol. Error detection in layer 3 uses, among other information, error notification from the data-link layer.

7. *Error Recovery.* This includes mechanisms for recovering from detected errors.

8. *Sequencing.* This includes mechanisms for providing sequenced delivery of layer 3 information over a given network connection when requested. Under normal conditions layer 3 ensures the delivery of information in the sequence in which it is submitted by the user.

9. *Congestion Control and User Data Flow Control.* Layer 3 may indicate rejection or unsuccessful indication for connection establishment requests to control congestion within a network. Typical is the congestion control message to indicate the establishment or termination of flow control on the transmission of USER INFORMATION messages.

10. *Restart.* This function is used to return channels and interfaces to an idle condition to recover from certain abnormal conditions.

13.8.1 Layer 3 Specification

The layer 3 specification is contained in ITU-T Recs. Q.930/931 (Refs. 13 and 14). It includes both circuit-switched and packet-switched operation. There are 23 message types for circuit-mode connection control. These are shown in Table 13.8, and the content elements of each are given in over 50 tables in the specification. One typical table is Table 13.9. Additional tables are provided for packet switching.

The following are several explanatory notes for the tables found in ITU-T Rec. Q.931 and for Table 13.9. The letters "M" and "O" mean *mandatory* and *optional*, respectively. Letters "n" and "u" refer to *network* and *user*, respectively, and give the direction of traffic such as n → u (network to user) and u → n (user to network). An asterisk (∗) in the table means undefined maximum length.

TABLE 13.8 Messages for Circuit-Mode Connection Control

Call establishment messages:

ALERTING
CALL PROCEEDING
CONNECT
CONNECT ACKNOWLEDGE
PROGRESS
SETUP
SETUP ACKNOWLEDGE

Call information phase messages:

RESUME
RESUME ACKNOWLEDGE
RESUME REJECT
SUSPEND
SUSPEND ACKNOWLEDGE
SUSPEND REJECT
USER INFORMATION

Call clearing messages:

DISCONNECT
RELEASE
RELEASE COMPLETE

Miscellaneous messages:

CONGESTION CONTROL
FACILITY
INFORMATION
NOTIFY
STATUS
STATUS ENQUIRY

Source: Table 3-1/Q.931, page 16, Ref. 14.

TABLE 13.9   SETUP Message Content

Information Element	Direction	Type	Length
Protocol discriminator	Both	M	1
Call reference	Both	M	2–*
Message type	Both	M	1
Sending complete	Both	O (Note 1)	1
Repeat indicator	Both	O (Note 2)	1
Bearer capability	Both	M (Note 3)	4–13
Channel identification	Both	O (Note 4)	2–*
Facility	Both	O (Note 5)	2–*
Progress indicator	Both	O (Note 6)	2–4
Network specific facilities	Both	O (Note 7)	2–*
Display	n → u	O (Note 8)	Note 9
Keypad facility	u → n	O (Notes 10, 12)	2–34
Signal	n → u	O (Note 11)	2–3
Switchhook	u → n	O (Note 12)	2–3
Feature activation	u → n	O (Note 12)	2–4
Feature indication	n → u	O (Note 12)	2–5
Calling party number	Both	O (Note 13)	2–*
Calling party subaddress	Both	O (Note 14)	2–23
Called party number	Both	O (Note 15)	2–*
Called party subaddress	Both	O (Note 16)	2–23
Transit network selection	u → n	O (Note 17)	2–*
Low layer compatibility	Both	O (Note 18)	2–16
High layer compatibility	Both	O (Note 19)	2–4
User–user	Both	O (Note 20)	Note 21

Note 1: Included if the user or the network optionally indicates that all information necessary for call establishment is included in the SETUP message.

Note 2: The repeat indicator information element is included immediately before the first bearer capability information element when either the in-call modification procedure or the bearer capability negotiation procedure is used.

Note 3: May be repeated if the bearer capability negotiation procedure is used. For bearer capability negotiation, either two or three bearer capability information elements may be included in descending order of priority—that is, highest priority first.

Note 4: Mandatory in the network-to-user direction. Included in the user-to-network direction when the user wants to indicate a channel. If not included, its absence is interpreted as "any channel acceptable."

Note 5: May be included for functional operation of supplementary services.

Note 6: Included in the event of interworking or in connection with the provision of in-band information/patterns.

Note 7: Included by the calling user or the network to indicate network-specific facilities information.

Note 8: Included if the network provides information that can be presented to the user.

Note 9: The minimum length is 2 octets; the maximum length is network dependent and is either 34 or 82 octets.

Note 10: Either the called party number or the keypad facility information element is included by the user to convey called party number information to the network. The keypad facility information element may also be included by the user to convey other call establishment information to the network.

Note 11: Included if the network optionally provides additional information describing tones.

Note 12: As a network option, may be used for stimulus operation of supplementary services.

Note 13: May be included by the calling user or the network to identify the calling user.

Note 14: Included in the user-to-network direction when the calling user wants to indicate the calling party subaddress. Included in the network-to-user direction if the calling user included a calling party subaddress information element in the SETUP message.

Note 15: Either the called party number or the keypad facility information element is included by the user to convey called party number information to the network. The called party number information element is included by the network when called party number information is conveyed to the user.

Notes Continued

TABLE 13.9   Notes (*Continued*)

Note 16: Included in the user-to-network direction when the calling user wants to indicate the called party subaddress. Included in the network-to-user direction if the calling user included a called party subaddress information element in the SETUP message.

Note 17: Included by the calling user to select a particular transit network.

Note 18: Included in the user-to-network direction when the calling user wants to pass low-layer compatibility information to the called user. Included in the network-to-user direction if the calling user included a low-layer compatibility information element in the SETUP message.

Note 19: Included in the user-to-network direction when the calling user wants to pass high-layer compatibility information to the called user. Included in the network-to-user direction if the calling user included a high-layer compatibility information element in the SETUP message.

Note 20: Included in the user-to-network direction when the calling user wants to pass user information to the called user. Included in the network-to-user direction if the calling user included a user–user information element in the SETUP message.

Note 21: The minimum length is 2 octets; the standard default maximum length is 131 octets.

Source: Table 3–1/Q.931, page 30, Ref 14.

13.8.1.1 *General Message Format and Information Elements Coding.*

Within this protocol, every message consists of the following parts:

1. Protocol discriminator
2. Call reference
3. Message type
4. Other information elements, as required

Information elements 1, 2, and 3 are common to all messages and are always present, while information element 4 is specific to each message type. This organization is illustrated in the example shown in Figure 13.23.

The term "default" implies that the value defined should be used in the absence of any assignment, or in the negotiation of alternative values. When a field, such as the call reference value, extends over more than 1 octet, the order of bit values progressively decreases as the octet number increases. The least significant bit of the field is represented by the lowest-numbered bit of the highest-numbered octet field.

Protocol Discriminator. The purpose of the protocol discriminator is to distinguish messages for user–network call control from other messages (to be defined) within ITU-T Rec. Q.931. It also distinguishes messages of ITU-T Rec. Q.931 from those OSI network layer protocol units which are coded to other ITU-T Recommendations and other standards.

Call Reference. The purpose of the call reference is to identify the call or facility registration/cancellation request at the local user–network interface to which the particular message applies. The call reference does not have end-to-end significance across ISDNs.

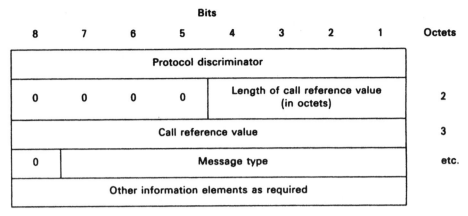

Bits

| 8 | 7 | 6 | 5 | 4 | 3 | 2 | 1 | Octets |

Figure 13.23. General message organization example. (From Figure 4-1/Q.931, page 67, ITU-T Rec. Q.931, ITU-T Rec. Q.931, Ref. 14.)

As a minimum, all networks and users must be able to support (a) a call reference value of 1 octet for a basic user–network interface and (b) a call reference value of 2 octets for a primary rate interface. The call reference information element includes the call reference value and the call reference flag.

Call reference values are assigned by the originating side of the interface for a call. These values are unique to the originating side within a particular D-channel layer 2 logical link connection. The call reference value is assigned at the beginning of a call and remains fixed for the lifetime of the call (except in the case of call suspension). After a call ends or after a successful suspension, the associated call reference value may be reassigned to a later call. Two identical call reference values on the same D-channel layer 2 logical link connection may be used when each value pertains to a call originated at opposite ends of the link.

The call reference flag can take the values of 0 or 1. The call reference flag is used to identify which end of the layer 2 logical link originated a call reference. The origination side always sets the call reference flag to "0." The destination always sets the call reference flag to "1."

Hence the call reference flag identifies who allocated the call reference value for this call, and the only purpose of the call reference flag is to resolve simultaneous attempts to allocate the same call reference value.

Message Type. The purpose of the message type is to identify the function of the message being sent. For instance, 00000101 indicates a call setup message.

Other Information Elements. Forty "other information elements" are listed, such as "sending complete," "congestion level," "call identity," "date/time," and "calling party number."

13.9 ISDN PACKET MODE REVIEW

13.9.1 Introduction

Two main services for packet-switched data transmission are defined for packet-mode terminals connected to the ISDN:

Case A: Access to a PSPDN (PSPDN services) (PSPDN = packet-switched public data network)

Case B: Use of an ISDN virtual circuit service

13.9.2 Case A: Configuration When Accessing PSPDN Services

This configuration is shown in Figure 13.24 and refers to the service of Case A, thus implying a transparent handling of packet calls through an ISDN. Only access via the B channels is possible. In this context, the only support that an ISDN gives to packet calls is a physical 64-kbps circuit-mode semiper-manent or demand transparent network connection type between the appro-priate PSPDN port and the X.25 DTE + TA or TE1 at the customer premises.

In the case of semipermanent access, the X.25 DTE + TA or TE1 is connected to the corresponding ISDN port at the PSPDN (AU). The TA, when present, performs only the necessary physical channel rate adaption

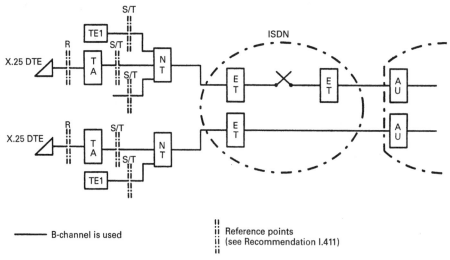

Figure 13.24. Case A: Configuration when accessing PSPDN services. AU, ISDN access unit ports; TA, terminal adaptor; NT, network termination 2 and / or 1; ET, exchange termination; TE1, terminal equipment 1. (From Figure 2-1/X.31, page 3, ITU-T Rec. X.31, ITU-T Rec. X.31, Ref. 2.)

between the user at the R reference point and the 64-kbps B-channel rate. D-channel layer 3 messages are not used in this case.

In the case of demand access to the PSPDN, which is shown in the upper portion of Figure 13.24, the X.25 DTE + TA or TE1 is connected to an ISDN port at the PSPDN (AU). The AU is also able to set up 64-kbps physical channels through the ISDN.

In this type of connection, an originating call will be set up over the B channel toward the PSPDN port using the ISDN signaling procedure prior to starting X.25 layer 2 and layer 3 functions. This is done by using either hot-line (e.g., direct call) or complete selection methods. Moreover, the TA, when present, performs user rate adaption to 64 kbps. Depending on the data rate adaption technique employed, a complementary function may be needed at the AU of the PSPDN.

In the complete selection case, two separate numbers are used for outgoing access to the PSPDN:

- The ISDN number of the access port of the PSPDN, given in the D-channel layer 3 SETUP message (Q.931)
- The address of the called DTE indicated in the X.25 call request packet

The corresponding service requested in the D-channel layer 3 SETUP message is ISDN circuit-mode bearer services.

For calls originated by the PSPDN, the same considerations as above apply. In fact, with reference to Figure 13.24, the ISDN port of the PSPDN includes both rate adaption (if required) and path setting-up functions. When needed, DTE identification may be provided to the PSPDN by using the call establishment signaling protocols in D-channel layer 3 (Q.931). Furthermore, DCE identification may be provided to the DTE, when needed, by using the same protocols.

For the demand access case, X.25 layer 2 and layer 3 operation in the B channel as well as service definitions are found in ITU-T Ref. X.32 (Ref. 15). Some PSPDNs may operate the additional DTE identification procedures defined in Rec. X.32 (Ref. 15) to supplement the ISDN-provided information in Case A.

13.9.3   Case B: Configuration for the ISDN Virtual Circuit Service

This configuration refers to the case where a packet handling (PH) function is provided within the ISDN. The configuration in Figure 13.25 relates to the case of X.25 link and packet procedures conveyed through the B channel. In this case, the packet call is routed, within an ISDN, to some PH function where the complete processing of the X.25 call can be carried out.

The PH function may be accessed in various ways depending on the related ISDN implementation alternatives. In any case, a B-channel connection set up to/from a PH port supports (a) the necessary processing for

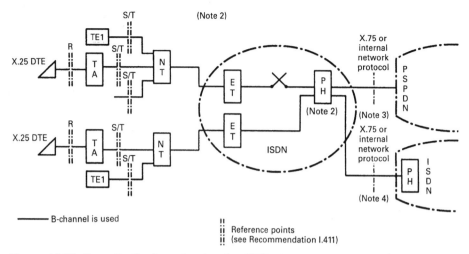

Figure 13.25. Case B: Configuration for the ISDN virtual circuit service (access via B channel). TA, terminal adaptor; NT, network termination 2 and/or 1; ET, exchange termination; TE1, terminal equipment 1; PH, packet handling function. *Note: 1:* This figure is only an example of many possible configurations and is included as an aid to the text describing the various interface functions. *Note 2:* In some implementations the PH functions logically belonging to the ISDN may reside physically in a node of the PSPDN. The service provided is still the ISDN virtual circuit service. *Note 3:* See Recommendation X.325. *Note 4:* See Recommendation X.320. (From Figure 2-2/X.31, page 5, ITU-T Rec. X.31, Ref. 2.)

B-channel packet calls, (b) standard X.25 functions for layer 2 and layer 3, and (c) possible path setting-up functions for layer 1 and possible rate adaption.

The configuration shown in Figure 13.5 (in Section 13.4) refers to the case of X.25 packet layer procedures conveyed through the D channel. In this case a number of DTEs can operate simultaneously through a D channel by using connection identifier discrimination at ISDN layer 2. The accessed port of PH is still able to support X.25 packet layer procedures.

It should be pointed out that the procedures for accessing a PSDTS (public switched data transmission service) through an ISDN user–network interface over a B or D channel are independent of where the service provider chooses to locate PH functions—that is, in either (a) a remote exchange or packet-switching module in an ISDN or, (b) the local exchange. However, the procedures for packet access through the B channel or the D channel are different.

In both cases of B- and D-channel accesses, in the service of Case B, the address of the called DTE is contained in the X.25 call request packet. The establishment of the physical connection from the TA/TE1 to the packet handling functions is done on the basis of the requested bearer service (ISDN virtual circuit service); therefore, the user does not provide any addressing information in the layer 3 procedures (D channel, Q.931).

13.9.4   Service Aspects

13.9.4.1   *Access to PSPDN Services—Case A*

Service Characteristics. The ISDN offers a 64-kbps circuit-switched or semipermanent transparent network connection type between the TA/TE1 and the PSPDN port (AU). In the switched access case the AU must be selected by the called address in the D-channel signaling protocol when the TA/TE1 sets up the circuit-switched connection to the AU. In the non-switched access case, layer 3 (D channel, Q.931) call control messages are not used.

Because the packet-switched service provider is a PSPDN, some DTEs are PSPDN terminals; they are handled by the PSPDN. Other DTEs may access the PSPDN without subscribing to the PSPDN permanently. In the first case, the same services as PSPDN services are maintained, including facilities, quality of service (QoS) characteristics, and DTE-DCE interfaces. In the case where a DTE is not subscribing to the PSPDN, it will be provided with a limited set of PSPDN facilities (see ITU-T Rec. X.32).

Every DTE will be associated with one or more ISDN (E.164) numbers. In addition, a DTE may be associated with one or more X.121 numbers assigned by the PSPDN(s) associated by the DTE. The method for X.25 packets to convey numbers from the ISDN numbering plan and the relationship with ITU-T Rec. X.121 are described in ITU-T Rec. E.166.

Basic Rules. Packet data communications, when using a switched B channel, will be established by separating the establishment phase of the B channel and the control phase of the X.25 virtual circuits using the X.25 protocol (link layer and packet layer). In general, ISDN has no knowledge of the customer terminal equipment or configuration. The incoming B-channel connection establishment will have to employ the D-channel signaling procedure in ITU-T Rec. Q.931.

13.9.4.2   *Access to the ISDN Virtual Circuit Service—Case B*

Service Characteristics. The virtual circuit service provided within the ISDN is aligned with what is described in ITU-T Series Recommendations (e.g., in terms of facilities, quality of service, etc.).

The service and facilities provided, as well as the quality of service characteristics, are those of the ISDN. Existing features of the X-Series Recommendations may be enhanced, and additional features may also be developed taking into account the new ISDN customer capabilities. A number from the ISDN numbering plan will be associated with one or more TA/TE1. The numbering plan is set out in ITU-T Rec. E.164 (Ref. 16).

User Access Capabilities. In this case both B and D channels can be used for accessing the ISDN virtual circuit service.

Access Through the B Channel—Basic Rules. Packet data communications, when using a switched B channel, is established by separating the establishment phase of the B channel and the control phase of the virtual circuits using the X.25 protocol (link layer and packet layer).

In general, an ISDN has no knowledge of the customer terminal equipment or configuration. In the demand access case, the incoming B-channel connection establishment uses the signaling procedures of clause 6 of Q.931 (Ref. 14) (D-channel layer 3).

Access Through the D Channel—Basic Rules. The following must be adhered to for TE access to the PSDTS, as defined in the X-Series Recommendations (particularly X.25).

A single SAPI = 16 LAPD link, as viewed by both the network and the user, must support multiplexing of logical channels at X.25 layer 3. Additionally, because the user may have a multipoint access, and because a single TA or TE1 is allowed to operate with more than one TE1 (terminal endpoint identifier), the network must support the presence of multiple SAPI = 16 LAPD logical links simultaneously operating at ISDN layer 2. This results in the requirement that the network be able to support simultaneous layer 2 and X.25 layer 3 multiplexing for D-channel packet mode connections.

All X.25 packets, including *call request* and *incoming call packets*, must be transported to and from the TE in numbered information frames (I-frames) in an SAPI = 16 LAPD link.

An *incoming call* packet is transmitted to a TE only after the public networks check at least the following:

- Compatibility of user facilities contained in the incoming call packet with the called subscriber profile when present
- Availability of the X.25 logical channel, either two-way or incoming, on which the incoming call packet is sent.

Section 13.9 is based on ITU-T Rec. X.31, Ref. 2.

REFERENCES

1. *ISDN User–Network Interfaces—Interface Structures and Access Capabilities*, CCITT Rec. I.412, Fascicle III.8, IXth Plenary Assembly, Melbourne, 1988.
2. *Support of Packet Mode Terminal Equipment by an ISDN*, ITU-T Rec. X.31, ITU Geneva, November 1995.
3. ISDN Network Architecture, CCITT Rec. I.324, ITU Geneva, 1991.
4. William Stallings, ed., *Tutorial: Integrated Services Digital Network (ISDN)*, IEEE Computer Society Press, Washington DC, 1985.
5. *IEEE Standard Dictionary of Electrical and Electronic Terms*, 6th edition, IEEE Press, New York, 1996.

6. *Basic User–Network Interface Layer 1 Specification*, ITU-T Rec. I.430, ITU Geneva, November 1995.

7. *ISDN Basic Access Transport System Requirements*, Technical Reference TR-TSY-000397, Issue 1, October 1988, Morristown, NJ.

8. Roger L. Freeman, *Telecommunication Transmission Handbook*, 4th edition, John Wiley & Sons, New York, 1998.

9. *Primary Rate User–Network Interface—Layer 1 Specification*, ITU-T Rec. I.431, ITU Geneva, March 1993.

10. *Frame Alignment and Cyclic Redundancy Check Procedures Relating to Basic Frame Structures Defined in Rec. G.704*, CCITT Rec. G.706, ITU Geneva, April 1991.

11. *Digital Subscriber Signaling System No. 1 (DSS1): ISDN User–Network Interface Data Link Layer—General Aspects*, ITU-T Rec. Q.920, ITU Geneva, March 1993.

12. *ISDN User–Network Interface—Data Link Layer Specification*, ITU-T Rec. Q.921, ITU Geneva, September 1997.

13. *ISDN User–Network Interface, Layer 3, General Aspects*, ITU-T Rec. Q.930, ITU Geneva, March 1993.

14. *ISDN User–Network Interface, Layer 3—For Basic Call Control*, ITU-T Rec. Q.931, ITU Geneva, May 1998.

15. *Interface Between Data Terminal Equipment (DTE) and Data Circuit Terminating Equipment (DCE) for Terminals Operating in the Packet Mode and Accessing the Packet-Switched Public Data Network through a Switched Telephone Network or an Integrated Services Digital Network or a Circuit-Switched Public Data Network*, ITU-T Rec. X.32, ITU Geneva, October 1996.

16. *The International Public Telecommunications Numbering Plan*, ITU-T Rec. E.164, ITU Geneva, May 1998.

17. Roger L. Freeman, *Reference Manual for Telecommunications Engineering*, 3rd edition, John Wiley & Sons, New York, to be published.

14

BUILDING AND CAMPUS WIRING
AND CABLING FOR DATA
COMMUNICATIONS

14.1 BACKGROUND AND OBJECTIVE

The enterprise network has come of age. Its performance is only as good as its underlying building wiring and cabling. The principal signal lead is the twisted pair and will remain so for quite some time. Longer runs are based on fiber-optic cable.

Ideally, a wiring and cabling plan for signal and ground leads and cabling should be incorporated in a building design layout prior to construction or renovation. Included in the plan is the strategic placement of wiring/equipment closets, signal entry locations, and convenient cross-connects (Ref. 1).

The principal objective of this chapter is to describe and suggest layouts of structured cabling systems for commercial buildings. Both single-tenant and multitenant structures are covered. Our concern, of course, is the cabling that will carry telecommunication signals with emphasis on data and enterprise networks.

Two subsidiary topics are also covered: (1) the effects of electromagnetic interference (EMI) and their mitigation and (2) grounding, bonding, and shielding. The chapter leans heavily on information provided by EIA/TIA-568A, *Commercial Building Telecommunications Cabling Standard* (Ref. 2).

Issues dealing with survivability and improved availability are flagged where appropriate.

The following is a list of common services and systems that should be considered in the design of a commercial building cabling system:

1. Voice communications and associated PABX
2. Data communications including LANs and WANs

3. Other building signaling systems

Common abbreviations: UTP, unshielded twisted pair; STP, shielded twisted pair.

14.2 MAJOR ELEMENTS OF TELECOMMUNICATION BUILDING LAYOUT

Figure 14.1 gives an overview of a campus infrastructure consisting of four buildings that are interconnected, showing the cabling system indicating its major components. We discuss, in separate sections, the several system architectural components that include horizontal cabling, backbone cabling, work areas (previously called work stations), telecommunication closets, equipment rooms, and entrance facilities.

14.3 HORIZONTAL CABLING

Horizontal cabling connects work areas to the horizontal cross-connect(s) in a telecommunication closet (TC). Consider that a work area has a telephone on an employee's desk connected by four-pair UTP and a workstation PC connected (separately) to cross-connects in the telecommunication closet by two-pair STP or four pair UTP via wall connectors in each case. Of course, cross-connects allow access to circuits for testing and ease of rearrangements. EIA/TIA recommends that distances from wall connectors to a TC should be no greater than 90 m or 295 ft.

In metallic cable installations, the proximity of horizontal cabling to electrical facilities that generate a high level of electromagnetic interference (EMI) should be taken into account. Motors and transformers, copiers, and PCs are typical examples of these sources.

The topology of horizontal cabling is of the star type. Each work area is served by a telecommunication closet on the same floor and should be connected to a horizontal cross-connect installed in that closet. Bridged taps and splices should not be permitted as part of the copper horizontal cabling.

It is recommended that a length limitation be enforced of no more than 6 m (20 ft) for cross-connect jumpers and patch cords in cross-connection facilities.

Three types of transmission media are recommended for use in the horizontal cabling system. These are:

(a) Four-pair 100-Ω UTP cables
(b) Two-pair 150-Ω STP-A cables
(c) Two-fiber, 62.5/125-μm optical fiber cable

The use of 50-Ω cable should be deprecated.

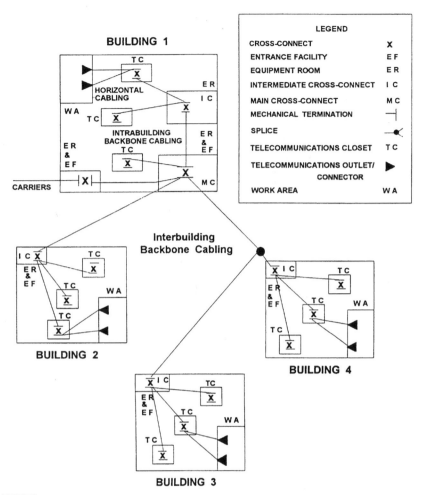

Figure 14.1. Campus wiring/cabling, interbuilding backbone cabling. (From Figure 1-1, EIA/TIA-568A, Ref. 2.)

NOTES:

1. This figure is not meant as an all-inclusive representation of the telecommunications cabling system but only as a typical example.

2. All cross-connects located in TCs in this figure are horizontal cross-connects (HCs).

14.3.1 Selection of Media

An individual work area should be provided two telecommunication outlet/connectors. One outlet/connector is associated with voice telephony, and the other is associated with data connectivity. The voice telephony circuit should be supported by a four-pair 100-Ω UTP cable, category 3 or higher. The several categories of UTP cable are described in Section 14.7.1. The data circuit should be supported by one of the following:

- Four-pair 100-Ω UTP cable, category 5
- Two-pair 150-Ω STP-A cable
- Two-fiber, 62.5/125-μm optical fiber cable

14.4 BACKBONE CABLING

Backbone cabling provides interconnection among telecommunication closets, equipment rooms, and entrance facilities in the basic telecommunication cabling structure shown in Figure 14.1. Backbone cabling also includes intermediate and main cross-connects, mechanical terminations and patch cords or jumpers used for backbone-to-backbone cross-connection. The cabling among buildings in a campus environment is also included in backbone cabling.

14.4.1 Topology of Backbone Cabling

As shown in Figure 14.2, backbone cabling is based on a hierarchical star topology. Here each horizontal cross-connect in a telecommunication closet is cabled to a main cross-connect or an intermediate cross-connect and is then cabled to a main cross-connect (except for certain cabling between telecommunication closets). The hierarchy consists of no more than two levels of cross-connects. From the horizontal cross-connect, no more than one cross-connect should be passed through to reach the main cross-connect.

Therefore, interconnects between any two horizontal cross-connects should pass through three or fewer cross-connects. Only a single cross-connect should be passed through to reach the main cross-connect. This concept is illustrated in Figure 14.2.

In certain situations it may be convenient to cable directly between telecommunication closets. This may hold for bus or ring type LAN installations. Such cabling is in addition to the connections for the basic star topology described in the preceeding text.

14.4.2 Selecting Cable Media

Transmission cable media should be selected from among the following:

(a) 100-Ω UTP cable
(b) 150-Ω STP-A cable
(c) 62.5/125-μm optical fiber cable
(d) single-mode optical fiber cable

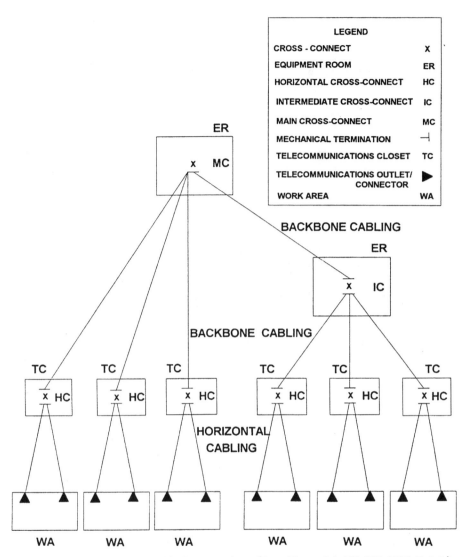

Figure 14.2. Backbone hierarchical star topology. (From Figure 5-1, EIA/TIA-568A, Ref. 2.)

EIA/TIA-568A sees future use of coaxial cable as deprecated. (There may be a notable exception if CATV broadband last-mile communication capability is desired.) Table 14.1 gives recommended cable types versus spectral bandwidth versus maximum run distance.

The 90-m distances in Table 14.1 assume that 5 m (16 ft) is needed at each end for jumpers connecting to the backbone.

Figure 5.2 of EIA/TIA-568-A for voice telephony and lower-speed data suggests greater distances than those shown in Table 14.1—for example,

TABLE 14.1   Intra- and Interbuilding Distances

Cable Type	Spectral Bandwidth	Maximum Distance
Category 3 UTP multipair	5–16 MHz	90 m (295 ft)
Category 4 UTP multipair	10–20 MHz	90 m (295 ft)
Category 5 UTP multipair	20–100 MHz	90 m (295 ft)
150-Ω STP-A	20–300 MHz	90 m (295 ft)

between the horizontal cross-connect and intermediate cross-connect 500-m (1640-ft) maximum with UTP, STP-A, or fiber-optic cable; and the total distance from horizontal cross-connect to main cross-connect, no more than 800 m (2624 ft).

14.5   TELECOMMUNICATION CLOSETS

The principal function of a telecommunication closet is to terminate horizontal cable distribution. It also is the location for the termination of backbone cable. The cross-connection of horizontal and backbone cable terminations using jumpers or patch cords allows flexible connectivity when extending various services to telecommunication outlet/connectors. Connecting hardware, jumpers, and patch cords used for this purpose are collectively referred to as *horizontal cross-connect*.

A telecommunication closet may also contain the intermediate cross-connect or the main cross-connect for different portions of the backbone cabling system. Sometimes backbone-to-backbone cross-connections in the telecommunications closet are used to tie different telecommunication closets together in a ring, bus, or tree configuration as discussed above.

A telecommunication closet also provides a controlled environment to house telecommunication equipment, connecting hardware, and splice closures serving a portion of the building.

14.5.1   Cabling Practices

Certain precautions should be observed such as the elimination of cable stress caused by cable tension. Cables should not be routed in tightly cinched bundles. Appropriate cable routing and dressing fixtures should be used for effective organization and management of the different types of cables in telecommunication closets. Horizontal and backbone building cables should be terminated on connecting hardware that meets the requirements of EIA/TIA-568-A. These cable terminations should not be used to administer cabling system moves, adds, and changes. All connections between horizontal and backbone cables should be made through a horizontal cross-connect.

Equipment cables that consolidate several ports on a single connector should be terminated on dedicated connecting hardware. Equipment cables that extend a single port appearance may be either permanently terminated or interconnected directly to horizontal or backbone terminations. Direct interconnections reduce the number of connections in a link but may also reduce flexibility.

Equipment rooms are considered distinct from telecommunication closets because of the nature and complexity of the equipment they contain. Any or all of the functions of a telecommunication closet may alternatively be provided by an equipment room. Such a room provides a controlled environment to house telecommunication equipment, connecting hardware, splice closures, grounding and bonding facilities, and protection apparatus where applicable. From a cabling perspective, an equipment room contains either the main cross-connect or the intermediate cross-connect used in the backbone cabling hierarchy.

14.6 ENTRANCE FACILITIES

An entrance facility serves as the entry points for cables, connecting hardware, protection devices, and other equipment needed to connect the outside plant facilities to the premises cabling. Such components may be used for public network services, private network customer premises services, or both. The demarcation point between regulated carriers/service providers and customer premise cabling may be part of the entrance facility. The entrance facilities include connections between cabling used in the outside environment and cabling authorized for in-building distribution. The connections may be accomplished via splices or other means.

14.7 100-Ω UNSHIELDED TWISTED-PAIR (UTP) CABLING SYSTEMS

14.7.1 UTP Category Definitions

The recognized categories of UTP cabling are as follows:

(a) *Category 3.* This designation applies to 100-Ω UTP cables and associated connecting hardware whose transmission characteristics are specified up to 16 MHz.

(b) *Category 4.* This designation applies to 100-Ω UTP cables and associated connecting hardware whose transmission characteristics are specified up to 20 MHz.

(c) *Category 5.* This designation applies to 100-Ω UTP cables and associated connecting hardware whose transmission characteristics are specified up to 100 MHz.

14.7.2   Horizontal UTP Cable

The transmission characteristics discussed here apply to cables consisting of four UTP of 24 AWG thermoplastic insulated solid conductors and enclosed by a thermoplastic jacket. Four-pair, 22 AWG cables that meet or exceed these requirements may also be used. The diameter of the insulated conductor should be 1.22-mm (0.048-in.) maximum. The pair twist length normally are selected by the cable manufacturer to comply with the crosstalk requirements of EIA/TIA-568-A.

The color codes of this cable are as shown in Table 14.2.

The diameter of the completed cable should be less than 6.35 mm (0.25 in.).

14.7.2.1 *Specific Transmission Characteristics.* See Table 14.3.

Attenuation. See Table 14.5.

Near End Crosstalk (NEXT) Loss. See Table 14.6.

Propagation Delay. At 10 MHz, not to exceed 5.7 ns/m.

TABLE 14.2   Color Codes for Horizontal 100-Ω UTP Cable

Conductor Identification	Color Code	Abbreviation
Pair 1	White-blue, Note 1; Blue, Note 2	(W-BL) (BL)
Pair 2	White-orange, Note 1; Orange, Note 2	(W-O) (O)
Pair 3	White-green, Note 1; Green, Note 2	(W-G) (G)
Pair 4	White-brown, Note 1; Brown, Note 2	(W-BR) (BR)

Note 1. The wire insulation is white, and a colored marking is added for identification. For cables with tightly twisted pairs [all less than 38.1 mm (1.5 in.) per twist], the mate conductor may serve as the marking for the white conductor.
Note 2. A white marking is optional.

TABLE 14.3   Transmission Characteristics

Parameter	Value
DC resistance	9.38 Ω per 100 m (328 ft)
DC resistance unbalance	No more than 5% between any two conductors
Mutual capacitance at 1 kHz	Not to exceed 6.6 nF/100 m (328 ft) for Category 3; 5.6 nF/100 m for Category 4 and Category 5
Capacitance unbalance: Pair to ground	At 1 kHz any pair, not to exceed 330 pF/100 m (328 ft)
Characteristic impedance from 1 MHz to highest reference frequency	100 Ω ± 15%
Structural return loss (SRL)	See Table 14.4

Source: Based on Ref. 2.

TABLE 14.4   Horizontal UTP Structural Return Loss (Worst Pair)

Frequency (MHz)	Category 3 (dB)	Category 4 (dB)	Category 5 (dB)
1–10	12	21	23
10–16	$12 - 10\log(f/10)$	$21 - 10\log(f/10)$	23
16–20	—	$21 - 10\log(f/10)$	23
20–100	—	—	$23 - 10\log(f/20)$

Source: From Table 10-2, Ref. 2.

TABLE 14.5   Horizontal UTP Cable Attenuation (per 100 m [328 ft] at 20°C)

Frequency (MHz)	Category 3 (dB)	Category 4 (dB)	Category 5 (dB)
0.064	0.9	0.8	0.8
0.256	1.3	1.1	1.1
0.512	1.8	1.5	1.5
0.772	2.2	1.9	1.8
1.0	2.6	2.2	2.0
4.0	5.6	4.3	4.1
8.0	8.5	6.2	5.8
10.0	9.7	6.9	6.5
16.0	13.1	8.9	8.2
20.0	—	10.0	9.3
25.0	—	—	10.4
31.25	—	—	11.7
62.5	—	—	17.0
100.0	—	—	22.0

Note: The attenuation of some category 3 UTP cables, such as those with PVC insulation, exhibits a significant temperature dependence. A temperature coefficient of attenuation of 1.5% per °C is not uncommon for such cables. In particular installations where the cable will be subjected to higher temperatures, a less-temperature-dependent cable may be required.
Source: Table 10-4, Ref. 2.

14.8   BACKBONE UTP CABLING

Backbone UTP cabling consists of cables in pair sizes greater than four pairs made up of 24 AWG thermoplastic insulated copper conductors that are formed into one or more units of unshielded twisted pairs. The units are assembled into binder groups of 25 pairs or part thereof following standard industry color code. It should be noted that multipair 22 AWG cables that meet the transmission requirements listed below are also acceptable for backbone UTP application. The diameter over the insulation will be 1.22 mm (0.048 in.) maximum.

14.8.1   Transmission Performance Requirements

See Tables 14.3, 14.4, 14.5, and 14.6.

TABLE 14.6   Horizontal UTP Cable NEXT Loss (Worst Pair Combination) [100 m (328 ft)]

Frequency (MHz)	Category 3 (dB)	Category 4 (dB)	Category 5 (dB)
0.150	53	68	74
0.772	43	58	64
1.0	41	56	62
4.0	32	47	53
8.0	27	42	48
10.0	26	41	47
16.0	23	38	44
20.0	—	36	42
25.0	—	—	41
31.25	—	—	39
62.5	—	—	35
100.0	—	—	32

Note: 0.150 MHz is for reference purposes only.
Source: From Table 10.5, Ref. 2.

Note on Crosstalk Interference. In multipair cable, a given pair receives crosstalk interference from other energized pairs sharing the same sheath. The total crosstalk energy that a pair receives is specified as the power sum crosstalk. The power sum crosstalk for uncorrelated disturbing pairs can be calculated from the individual pair-to-pair crosstalk measurements at a given frequency. Generally, the power sum crosstalk energy is dominated by the couplings between pairs in close proximity and is relatively unaffected by pairs in separate binder groups. Therefore, it is desirable to separate services with different signal levels or services that are susceptible to impulse noise into separate binder groups.

The propagation delay of any pair at 10 MHz should not exceed 5.7 ns/m.

14.8.2   Connecting Hardware for UTP Cable

Under this heading we include telecommunication outlet/connectors, patch panels, transition connectors, and cross-connect blocks. It should be noted that requirements for connector categories 3, 4, and 5 are not sufficient in themselves to ensure required cabling system performance. Link performance also depends on cable characteristics (including cross-connect jumpers and patch cords), the total number of connections, and the care with which they are installed and maintained.

It is desirable that hardware used to terminate UTP cables be of the insulation displacement contact (IDC) type.

Connection hardware for the 100-Ω UTP cabling system is installed at the following:

(a) Main cross-connect
(b) Intermediate cross-connect
(c) Horizontal cross-connect
(d) Horizontal cabling transition points
(e) Telecommunication outlet/connectors

Typical cross-connect facilities consist of cross-connect jumpers or patch cords and terminal blocks or patch panels that are connected directly to horizontal or backbone cabling.

Attenuation. Worst-case attenuation of any pair within a connector should not exceed the values listed in Table 14.7 at each specified frequency for a given performance category.

NEXT Loss. Table 14.8 shows the NEXT loss for the connecting hardware for the three categories of interest.

Return Loss. Connector return loss is a measure of the degree of impedance matching between the UTP cable and connector and is derived from swept frequency voltage measurements on short lengths of 100-Ω twisted-pair test leads before and after inserting the connector under test.

Because the return loss characteristics of category 3 connecting hardware are not considered to have a significant effect on the link performance of category 3 UTP cabling, return loss requirements are not specified for category 3 connectors.

**TABLE 14.7   Attenuation of Connecting Hardware Used for
100-Ω UTP Cable**

Frequency (MHz)	Category 3 (dB)	Category 4 (dB)	Category 5 (dB)
1.0	0.4	0.1	0.1
4.0	0.4	0.1	0.1
8.0	0.4	0.1	0.1
10.0	0.4	0.1	0.1
16.0	0.4	0.2	0.2
20.0	—	0.2	0.2
25.0	—	—	0.2
31.25	—	—	0.2
62.5	—	—	0.3
100.0	—	—	0.4

Source: Table 10-8, Ref. 2.

TABLE 14.8   UTP Connecting Hardware NEXT Loss

Frequency (MHz)	Category 3 (dB)	Category 4 (dB)	Category 5 (dB)
1.0	58	65	65
4.0	46	58	65
8.0	40	52	62
10.0	38	50	60
16.0	34	46	56
20.0	—	44	54
25.0	—	—	52
31.25	—	—	50
62.5	—	—	44
100.0	—	—	40

Source: Table 10-9, Ref. 2.

DC Resistance. The DC resistance between the input and output connections of the connecting hardware (not including the cable stub, if any) used for 100-Ω cabling should not exceed 0.3-Ω.

14.9   150-Ω SHIELDED TWISTED-PAIR CABLING SYSTEMS

This section covers the requirements for 150-Ω STP-A inside cable for use in the horizontal cabling system. The cable is restricted to two-pair size consisting of 22 AWG thermoplastic insulated solid conductors enclosed by a shield and an overall thermoplastic jacket. The pair twist lengths are selected by the manufacturer to ensure compliance with the crosstalk requirements of EIA/TIA-568-A. The pair 1 color code is red-green, and for pair 2 it is orange-black.

14.9.1   Transmission Performance Requirements

DC Resistance. The resistance of any conductor should not exceed 5.71 Ω/100 m (328 ft).

DC Resistance Unbalance. The resistance unbalance between the two conductors of any pair should not exceed 4%.

Capacitance Unbalance: Pair to Ground. The capacitance unbalance to ground at 1 kHz of any pair should not exceed 100 pF/100 m (328 ft).

Balanced Mode Attenuation. The cable balanced mode attenuation is given in Table 14.9.

Common Mode Attenuation. The common mode attenuation of any pair, at or corrected to a temperature of 25 $\pm$ 3°C and measured in accordance with

TABLE 14.9   Cable Balanced Mode Attenuation
(Horizontal STP-A Cable)

Frequency (MHz)	Maximum Attenuation [dB/100 m (328 ft)]
0.0096	0.30
0.0384	0.50
4.0	2.2
8.0	3.1
10	3.6
16	4.4
20	4.9
25	6.2
31.25[a]	6.9
62.5	9.8
100	12.3
300	21.4

[a]Value not clear in reference
Source: Table 11-2, Ref. 2.

ASTM D 4566, shall not exceed:

$$95.0\sqrt{\frac{f}{50}} \text{ dB/km}, \quad \text{for all frequencies from 50 MHz to 600 MHz,}$$

where f = frequency in MHz. The attenuation measurements from 50 MHz to 600 MHz shall be performed on cable lengths of 100 m (328 ft) to 305 m (1000 ft).

NEXT Loss. The NEXT loss between the two pairs within a cable should not exceed:

- $+$ 58.0 dB at 9.6 kHz terminated in 270 Ω
- $+$58.0 dB at 38.4 kHz terminated in 185 Ω
- $+$58.0 dB for all frequencies between 0.1 MHz and 5 MHz terminated in 150 Ω and
- $+$58.0 $-$ 15 $\log(f/5)$ dB for all frequencies from 5 MHz to 300 MHz terminated in 150 Ω where f is expressed in MHz.

Propagation Delay. The propagation delay of any pair at 10 MHz should not exceed 5.7 ns/m.

14.10   OPTICAL FIBER CABLING SYSTEMS

62.5/125-μm optical fiber cable is recommended for horizontal cable, and both 62.5/125-μm and single-mode optical fiber cables may be used for backbone cables.

For the 62.5/125-μm cable, at 850-nm wavelength, budget 3.75 dB/km loss and a minimum information capacity of 160 MHz-km; at 1300 nm, budget 1.5 dB/km loss and a minimum information transmission capacity of 500 MHz-km.

14.10.1   Backbone Optical Fiber Cable

The optical fiber construction consists of 62.5/125-μm optical fibers or single-mode optical fibers, or both, typically formed into groups of 6 or 12 fibers each. These groups and individual fibers should be identifiable in accordance with EIA/TIA-598. These groups are assembled to form a single compact core, and the core is covered by a protective sheath. The sheath consists of an overall jacket and may contain an underlying metallic shield and one or more layers of dielectric material applied over the core.

Customer premises optical fiber backbone cabling has been and continues to be primarily 62.5/125-μm multimode fiber based because this fiber can use LED transmitters. With the rapidly growing bit rate requirements, consideration should be given to installing single-mode optical fiber in addition to the multimode fiber. Single-mode fiber systems inherently have higher bit rate capacities and longer distance capabilities than the conventional multimode optical fiber.

14.11   GROUNDING AND BONDING INFRASTRUCTURE FOR COMMERCIAL BUILDINGS

14.11.1   Rationale

A well-designed and installed grounding and bonding infrastructure in a commercial building in support of telecommunications can pay off in two ways:

- It can be a major step in assuring electrical efficiency.
- It can negate or mitigate the effects of EMI on the telecommunication system as a whole.

These goals can be accomplished by providing a ground reference for telecommunication systems within the telecommunication entrance facility, telecommunication closet(s), and equipment room(s). To further these objectives, the bonding and connecting of pathways, cable shields, conductors, and hardware at telecommunication closets, equipment rooms, and entrance facilities will be required.

Figure 14.3 shows a typical layout for bonding and grounding of fairly large, multistory commercial buildings with multiple backbones. The telecommunications grounding and bonding infrastructure originates with a connection to the service equipment (power) ground and extends throughout the

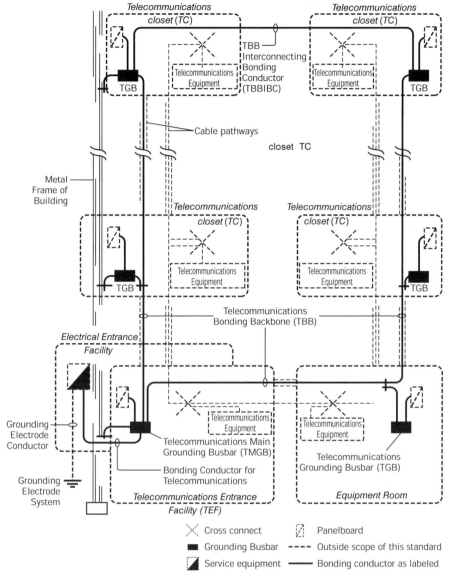

Figure 14.3. Bonding and grounding layout for a large commercial building. (From Figure 2.1-1, EIA/TIA-607, Ref. 3.)

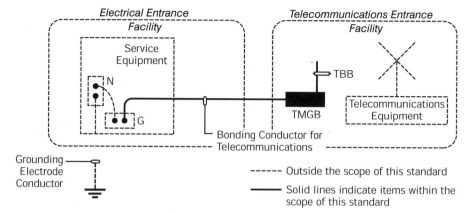

Figure 14.4. Connectivity to the service equipment (power) ground. (From Figure 5.2-1, EIA/TIA-607, Ref. 3.)

building. It comprises five major components:

(a) Bonding conductor for telecommunications
(b) Telecommunications main grounding busbar (TMGB)
(c) Telecommunications bonding backbone (TBB)
(d) Telecommunications grounding busbar (TGB)
(e) Telecommunications bonding backbone interconnecting bonding conductor (TBBIBC)

All bonding conductors should be insulated and of copper. Bonding conductors should not be placed in ferrous metallic conduit. Bonding conductors should be marked green in color.

The bonding conductor for telecommunications bonds the TMGB to the service equipment (power) ground as shown in Figure 14.4. As a minimum, the bonding conductor should be the same size as the TBB.

14.11.2   The Telecommunications Bonding Backbone (TBB)

A TBB is a conductor that interconnects all TGBs with the TMGB. A TBB's basic function is to reduce or equalize potential differences between telecommunication systems bonded to it. A TBB is not intended to serve as the only conductor providing a ground fault return path.

A TBB(s) originates at the TMGB, extends throughout the building using the telecommunication backbone pathways, and connects to the TGB(s) in all telecommunication closets and equipment rooms. The interior water piping system of the building should not be used as a TBB, nor should the metallic cable shield.

Bonding and Sizing the TBB. A TBB is an insulated copper conductor. Its minimum conductor size is No. 6 AWG; a larger diameter is preferable, even 3/0 AWG.

Whenever two or more vertical TBBs are used within a multistory building, the TBBs are bonded together with a TBB interconnecting bonding conductor (TBBIBC) at the top floor and at a minimum of every third floor in between. A TBB may be spliced, provided that all applicable requirements of Section 5 of EIA/TIA-607 are met. The TBB is connected to the TMGB as specified in Section 5.4.7.1 of EIA/TIA-607.

The TBB conductors are installed without splices, where practicable. If splices are necessary, there should be a minimum number which are accessible and located in telecommunication spaces. Joined segments of a TBB should be connected using irreversible compression-type connectors, exothermic welding, or the equivalent.

14.11.3   The Telecommunication Main Grounding Busbar (TMGB)

The TMGB serves as a dedicated extension of the building grounding electrode system for the telecommunications infrastructure. The TMGB also serves as the central attachment point for telecommunication bonding backbones (TBB) and equipment, and it is located such that it is accessible to telecommunication personnel. Extensions of the TMGB (i.e., other busbars) are telecommunication grounding busbars (TGB). Typically, there should be a single TMGB per building.

The ideal location of the TMGB is in the telecommunication entrance room or space. However, the TMGB should be located such as to minimize the length of the bonding conductor for telecommunication.

TMGB Description. The TMGB is a predrilled copper busbar with standard NEMA bolt hole sizing and spacing for the type of connectors to be used. It should be sized in accordance with the immediate requirements of the application and with consideration of future growth. Its minimum dimensions are 6 mm thick × 100 mm wide and variable in length. For reduced contact resistance, the busbar should be electrotin-plated.

Connections to the TMGB. The connections to the bonding conductor for telecommunications and the TBBs to the TMGB should use listed 2-hole compression connectors, exothermic-type welded connections, or the equivalent. The connection of conductors for bonding telecommunication equipment to the TMGB may use one-hole lugs or equivalent; however, 2-hole compression connectors are preferable.

All metallic raceways for telecommunication cabling located within the same room or space as the TMBG should be bonded to the TMGB.

A practical location for the TMGB is to the side of the panelboard (where provided). The vertical location of the TMGB should take into consideration whether the grounding and bonding conductors are routed in an access floor or overhead cable tray.

14.11.4 The Telecommunications Grounding Busbar (TGB)

The TGB is the common central point of grounding connection for telecommunication systems and equipment in the location served by that telecommunication closet or equipment.

Description of the TGB. The TGB consists of a predrilled copper busbar provided with standard NEMA bolt hole sizing and spacing for the type of connectors to be used. It will have the minimum dimensions of 6 mm thick × 50 mm wide and vary in length to meet the applications requirement with consideration for future growth.

The TGB should be electrotin-plated for reduced contact resistance. If not plated, the busbar should be cleaned prior to fastening the conductors to the busbar.

Bonds to the TGB. TBBs and other TGBs within the same space are bonded to the TGB by means of at least a No. 6 AWG conductor, preferably with a 3/0 AWG conductor. The bonding conductor between a TBB and a TGB should be continuous and routed in the shortest possible straight-line path.

Where a panelboard for telecommunications is located within the same room or space as the TGB, the panelboard's ACEG bus (when equipped) or the enclosure should be bonded to the TGB. The TGB should be as close to the panelboard as practicable and should be installed to maintain clearances required by applicable electrical codes.

Where a panelboard for telecommunications is not located within the same room or space as the TGB, consideration should be given to bonding the panelboard's ACEG bus (when equipped) or the enclosure to the TGB. The TGB should be bonded to the TBBIBC as required. All metallic raceways for telecommunication cabling located within the same room or space as the TGB should be bonded to the TGB.

Connections of TBBs to the TGB should use listed 2-hole compression connectors.

A practical location for the TGB is to the side of the panelboard (where provided). The vertical location of the TGB should take into consideration whether the grounding and bonding conductors are routed in an access floor or overhead cable tray.

14.11.5 Bonding to the Metal Frame of a Building

All bonding conductors and connectors for bonding the metal frame of a building will be listed for the purpose intended and approved by an NRTL. In buildings where metal frames (structural steel) are effectively grounded, each TGB should be bonded to the metal frame within the room using a No. 6 AWG conductor. Where the metal frame is external to the room and readily accessible, the metal frame should be bonded to the TGB and TMGB separately with a No. 6 AWG conductor.

When practical because of shorter distances and other considerations, and where horizontal steel members are permanently bonded to vertical column members, TGBs may be bonded to these horizontal column members in lieu of vertical column members.

14.11.6 Telecommunications Entrance Facility (TEF)

The TEF includes the entrance point where telecommunications services enter. There is generally a room or other space devoted to this. It is also where the joining of inter- and intrabuilding backbone facilities takes place and where the proper grounding and bonding of these facilities is accomplished. The TEF may also contain antenna entrances and electronic equipment serving telecommunication functions.

The TEF is the desirable location for the TMGB. The TMGB may serve as the TGB for collocated equipment in the TEF as appropriate.

14.11.6.1 Placement of the TMGB. Placement of the TMBG should be such that it results in the straightest route considering the total length of the bonding conductor from the telecommunication primary protectors to the TMGB. This bonding conductor is intended to conduct lightning and ac fault currents from the telecommunication primary protectors. At least a 300-mm (1-ft) separation should be maintained between this insulated conductor and any dc power cables, switchboard cable, or high-frequency cables, even when isolated by metallic conduit or EMT (electrical metallic tubing).

The TMGB is the common point in the TEF to which grounding connections for that room are made. It should be as close to the panelboard for telecommunications as practicable and should be installed to maintain clearances required by applicable electrical codes. When a panelboard for telecommunications is not installed in the TEF, the TMGB should be located near the backbone cabling and associated terminations. In addition, the TMGB should be placed such that the bonding conductor for telecommunications is as short and as straight as possible.

When the TEF pathway incorporates an isolation gap, the pathway on the building side of the gap should be bonded to the TMGB. In buildings where the backbone cable incorporates a shield or metallic member, this shield or metallic member should be bonded to the TMGB/TGB.

The TMGB is intended to be the location for connecting grounding bars of telecommunication equipment located in the TEF. This typically may be multiplex equipment (e.g., DS1 channel banks) or optical fiber termination equipment.

14.11.7   Telecommunication Closets and Equipment Rooms

Each telecommunication closet and equipment room should contain a TGB. It should be located inside the closet/room and be insulated from its support. A 50-mm (2-in.) separation is recommended. Multiple TGBs may be installed within the same closet to minimize conductor lengths and terminating space. In all cases, multiple TGBs within a closet should be bonded together with at least a No. 6 AWG conductor.

Where a panelboard for telecommunications is not installed in the telecommunications closet, the TGB should be placed near the backbone cabling and associated terminations. In addition, the TGB should be located so that the grounding conductors are as short and straight as possible.

14.12   CUSTOMER-OWNED OUTSIDE PLANT (OSP) INFRASTRUCTURE

14.12.1   OSP Cabling Infrastructure Defined

The outside plant cabling infrastructure provides interconnections between building entrance facilities, structures on a campus or other multibuilding environment, or telecommunication pedestals or cabinets. The customer-owned OSP consists of backbone cables, splices, terminations, and patchcords or jumpers used for backbone-to-backbone interconnection.

14.12.2   OSP Topology

A star configuration is the recommended topology for customer-owned OSP cabling infrastructure. Figure 14.5 illustrates a campus with a star backbone topology. In the example, building "A" is the center or hub of the star with backbone cables extending to other campus buildings. ("B, C, D, E, F") and an outdoor telecommunication pedestal ("G"). The example also shows an optical fiber backbone cable passing from building "A" to building "F" through an intermediate building ("E").

In the example, Figure 14.5, building "A" provides a point of service for a microwave communication link to a second campus. The backbone cables can be used for distributing these applications from "A" to all or to selected buildings. Of course, "A" was selected only as an example.

Campus telecommunication applications require use of both intrabuilding and interbuilding backbone cabling. Figure 14.6 shows the relationship between the campus star backbone and the intrabuilding backbones of building "E." This illustrates the intrabuilding cabling topology from an individual work area through the intrabuilding backbone cabling to the campus backbone main interconnect facility in building "A."

A simple star topology is not always feasible, particularly on very large campuses. The distances between buildings may exceed the maximum allow-

CAMPUS OUTSIDE CABLE PLANT LOGICAL DIAGRAM

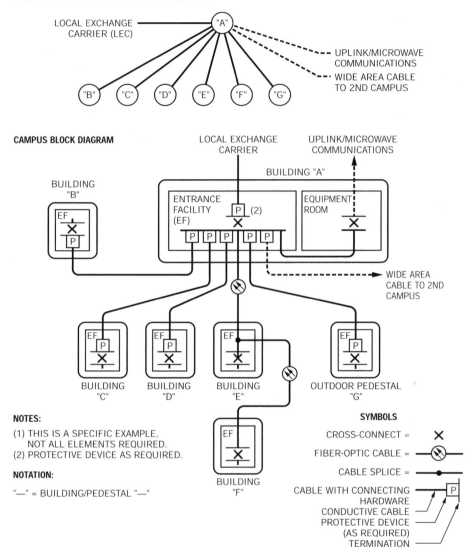

Figure 14.5. An example of a campus star topology showing outside plant installation. (From Figure 3, TIA/EIA-758, Ref. 4.)

able cable lengths. This problem can often be solved by turning to a hierarchical star topology. Each campus segment may connect to a hub location that would support the area as a simple star topology. These hub locations may be connected with other topologies to support equipment and technologies normally used for wide area applications. Here SONET over cable or LOS microwave may answer the interconnectivity issues.

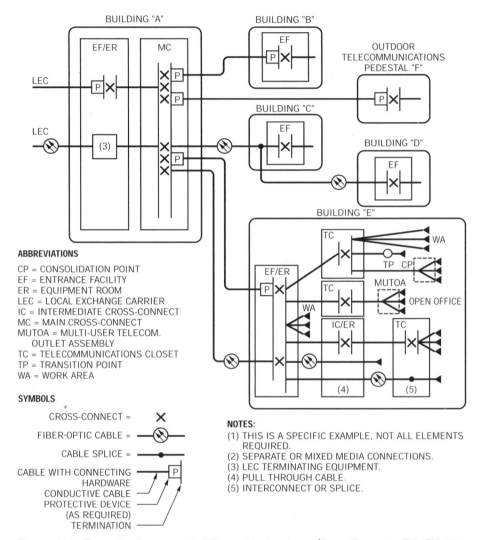

Figure 14.6. Example of campus/building cable topology. (From Figure 4, EIA/TIA-758, Ref. 4.)

14.12.3 Recognized Cable Media and Media Selection

Customer-owned OSP cabling must support a wide range of services and site sizes. Therefore, more than one transmission medium is recognized. This standard specifies recognized transmission media that may be used individually or in combination. The recognized media include:

- 50/125-μm optical fiber cable
- 62.5/125-μm optical fiber cable

- Single-mode optical fiber cable
- 100-Ω twisted-pair cable
- 75-Ω coaxial

The specific performance characteristics for recognized cables, associated connecting hardware, cross-connect jumpers, and patch cords are specified in this text.

Customer-owned OSP cabling specified by EIA/TIA-758 (Ref. 4) is applicable to a wide range of user requirements. Media choices must be made depending upon the characteristics of the applications, and distance. Where a single cable type may not satisfy all user requirements, it will be necessary to use more than one media type in the OSP cabling. Where possible, the different media should use the same physical pathway architecture and space for connecting hardware. In making this choice, factors to be considered include:

- Flexibility with respect to supported services
- Required useful life of backbone cabling
- Site size and user population

REFERENCES

1. Fred J. McClimans, *Communications Wiring and Interconnection*, McGraw-Hill, New York, 1992.
2. *Commercial Building Telecommunications Cabling Standard*, ANSI/EIA/TIA-568A, Telecommunication Industry Association, Washington DC, October 1995.
3. *Commercial Building Grounding and Bonding Requirements for Telecommunications*, ANSI/EIA/TIA-607, Telecommunication Industry Association, Washington DC, August 1994.
4. *Customer-Owned Outside Plant Telecommunications Cabling Standard*, ANSI/EIA/TIA-758, Telecommunication Industry Association, Washington DC, April 1999.

15

BROADBAND DATA
TRANSPORT TECHNIQUES

15.1 CHAPTER OBJECTIVE

Data is one of several binary media that is transmitted electrically. Voice and video can also be represented in a binary format for electrical transmission. The demand for bandwidth* to accommodate these media with ever-increasing transmission rates needed to be satisfied. The general theme is transmission over long distances. Long-distance may be just across the street or around the world.

The chapter covers three methods of data transport:

* Fiber optics, with particular emphasis here
* Line-of-sight microwave
* Satellite communications

The question arises: "How *broad* is broad?"

Line-of-sight (LOS) microwave and satellite communications can provide in the order of 500 MHz, the bandwidth being limited by international treaty. Fiber-optics transmission can provide 75,000 gigahertz or more, being limited by the laws of physics. Present technology permits 10-Gbps transmission on one wavelength on a single fiber with 40-Gbps operation on the drawing boards. Through shrewd bit-packing techniques, LOS radio can achieve 655 Mbps on a single radio frequency (RF) carrier; satellite communications has the capability to achieve the same bit rate.

It also became apparent that the DS1/E1 digital hierarchy described in Chapter 7 had transmission rate limitations and lacked flexibility. We can define *format* as a way of organizing the bits in a serial bit stream so that the resulting representation will be meaningful to a digital receiver.

---

*A better expression might be *bit rate capacity*.

To overcome the bit rate limitations of the DS1/E1 family of the *plesiochronous digital hierarchy* (PDH), two new format structures were developed:

- *Synchronous optical network* (SONET)
- *Synchronous digital hierarchy* (SDH)

The former was developed and is widely deployed in North America, whereas the latter is a European development and is deployed to some lesser degree in Europe and in those countries of European hegemony.

SONET was developed in the United States, and its standards are issued by Telcordia and ANSI. SDH is of European origin and its standards are issued by the ITU-T Organization and by ETSI (European Telecommunication Standardization Institute). The second part of this chapter covers these advanced digital multiplexing schemes.

We hasten to add that there is an implication in terminology that these two formats are only compatible with optical fiber transmission. This statement is patently untrue. Given the necessary operational bandwidth, SONET and SDH operate with LOS microwave and satellite communications just as well as with optical fiber. With a 40-MHz bandwidth assignment, the 622-Mbps rates may be transported. This rate is denominated STS-12 by SONET and STM-4 by SDH. Even higher rates can be accommodated in the higher frequency bands such as 28 and 40 GHz. However, the length of these links is severely limited by rainfall considerations. Typical link lengths may be in the range of 3–5 km depending on the desired time availability (Ref. 1).

Some fiber-optics applications implement links from less than 1 m to 10 m or more. We should expect to find these extremely short links inside digital switches, for example. On the other side of the coin, nearly all transoceanic communication is done by optical fiber today. It is the medium of choice.

Such transport systems as discussed in this chapter carry immense amounts of data. The data are in the formats discussed in previous chapters such as IP and frame relay. They may also have an ATM format, which is covered in Chapter 16.

15.2   INTRODUCTION TO FIBER-OPTICS TRANSMISSION

15.2.1   What Is So Good About Fiber-Optic Transmission?

An *optical fiber* is a circular waveguide of tiny dimensions, about the diameter of a human hair. We generally think of optical fiber being made of glass. Most fiber *is* made of glass; however, other materials such as plastic may also be used. There are also new advances in halides and bromides. It is expected that once these materials appear on the scene in the form of optical waveguide, we should expect to see dramatic improvements in the performance of transmitting data. For example, early tests have shown reductions in loss per unit length of at least one order of magnitude, possibly two (Ref. 2).

I tell my students in seminars that optical fiber has nearly an infinite bandwidth. If we consider the operational band between 1550 and 1635 nm, the available bandwidth is around 10 THz (teraHz) or 10,000 GHz. The usable radio-frequency spectrum is from about 10 kHz to 100 GHz, or, rounded off, 100 GHz of bandwidth for the total spectrum. So we can fit about 100 total RF spectrums in the 1550 nm band. That's pretty close to infinity in my book.

Another point is the optical fiber needs no equalization as coaxial cable does.

When we discuss light, we will be using wavelength rather than frequency to describe a certain emission location in the frequency spectrum. Chronologically, as fiber development proceeded, we started with the shorter wavelengths of around 820 nm (0.820 μm) and then progressively moved to longer wavelengths of around 1330 nm (1.330 μm) and 1550 nm (1.550 μm). We can convert frequency (F in Hz) to wavelength (λ in meters), and vice versa, with the following formula:

$$F\lambda = 3 \times 10^8 \text{ m/s}$$

Remember, F is measured in Hz and λ is measured in meters. For example, 1550 nm = 1550×10^{-9} m. If you wish to convert this to frequency, which you may be a bit more familiar with, use this formula. For example, using 1550-nm wavelength:

$$F_{\text{Hz}} = 3 \times 10^8 / 1550 \times 10^{-9}$$

$$= 1.93548 \times 10^{14} \text{ Hz} \quad \text{or} \quad 193.548 \times 10^{12} \text{ Hz} = 193.548 \text{ THz}$$

15.2.2 Advantages of Optical Fiber

Light and Small. Fiber cable is very light and small in size. Over 200 fibers can be placed in 0.5-in. cable. Compare this to copper pair or coaxial tube cables. Another example is the wiring in a large U.S. aircraft carrier. It is estimated that if all signal leads were changed over from copper to fiber, a 200-ton saving of topside weight would accrue.

Repeater Spacing. One of the greatest impairments to digital transmission is *jitter*. Jitter accumulation is a function of the number of regenerative repeaters there are in tandem on a certain connectivity. Fiber-optic systems require about 1/100th the number of repeaters compared to comparable coaxial cable systems.

Electromagnetic Compatibility (EMC). Fiber-optic systems do not radiate RF energy and in themselves are immune from radiation from other sources. For this reason, power companies run aerial fiber cable to support their telecommunication systems on the same tower line as their high-tension transmission lines. Fiber is immune to the rampant electrical noise of high-tension lines.

Crosstalk. On metallic cable systems, crosstalk can be a limiting impairment. There is no possibility of crosstalk on fiber-optic cable systems because there is no optical coupling from one fiber to another.

Temperature and Humidity. Fiber-optic systems are relatively unaffected by temperature and humidity extremes, unlike their metallic counterparts. This may not be the case for active light components, which, however, can be kept in a protected environment such as in repeater casings.

Cost. The cost of fiber-optic cable and related electronics has been eroding considerably. In many circumstances, fiber-optic cable is much more economic than equivalent coaxial cable, especially considering related electronics (e.g., there are fewer repeaters per unit distance for fiber cable).

Frequency Allocation and Licensing. Unlike microwave and satellite radio systems, fiber-optic cable systems (and metallic cable systems) require no frequency allocations, nor must they go through extensive and expensive licensing procedures.

15.2.3 Overview of an Optical Fiber Link

As illustrated in Figure 15.1, a fiber-optic link consists of a light source (i.e., transmitter), a fiber-optic cable with connectors and/or splices, and a detector (i.e., receiver). Long fiber-optic links require intermediate repeaters that

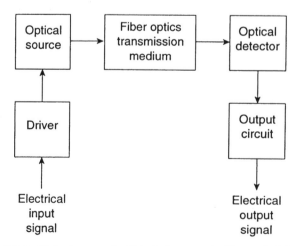

Figure 15.1. A conceptual block diagram of a fiber-optic link.

are regenerative (in the case of digital transmission). The distance between repeaters is much greater than with coaxial cable systems, and very much greater than when dealing with wire-pair systems. The distance is a function of source and detector parameters, the bit rate and BER desired, and, of course, fiber loss. That distance is greatly extended with the use of erbium-doped amplifiers.

There are two basic impairments to fiber-optic transmission: fiber loss and dispersion. Standard glass fiber experiences losses as low as 0.25 dB per kilometer at the longer wavelengths. When monomode fiber is used, we expected that dispersion could be neglected. This is true for the hundreds of Mbps transmission range. However, at the higher bit rates (e.g., > 1 Gbps), chromatic dispersion becomes a factor.

15.2.4 Optical Fiber Transmission

15.2.4.1 Optical Fiber Composition. As shown in Figure 15.2, an optical fiber is composed of a central cylindrical core of higher refractive index (n_1) and a concentric cladding of lower refractive index (n_2). The most common fibers for communication links have glass or fused silica core and cladding, although plastic-clad glass or all-plastic construction is used on applications which allow less bandwidth and/or more attenuation. Glass fiber is usually covered by an outer protective plastic coating which has no effect on light propagation. The coating protects the glass fiber from mechanical damage and from strength and transmission degradation due to exposure to air. It protects the fiber and gives it pulling strength.

Optical Fiber as a Waveguide. Fibers act as tubular mirrors, with the cladding–core interface confining light to the fiber core. Light coupled into an optical fiber is prevented from escaping by inward reflection within the core.

Snell's Law. Snell's law deals with the change of direction of light rays as they pass from a medium of one refractive index to another. Understanding of two terms is important to the discussion of refraction. The *normal* is the

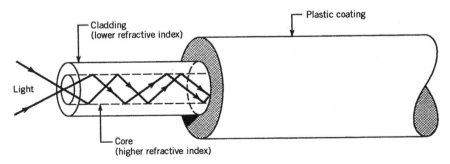

Figure 15.2. Basic composition of an optical fiber.

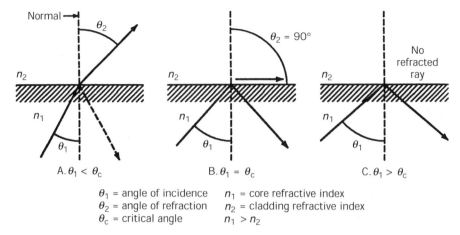

Figure 15.3. Reflection, refraction, and critical angle. Note that the critical angle is where the light ray just grazes the surface as shown in part B.

imaginary line perpendicular to the interface of the two materials; in the case of fiber, it is perpendicular to the interface between the core and cladding. The *angle of incidence* is the angle between the normal and the incident ray. Referring to Figure 15.3 for the definition of angles θ_1 and θ_2, Snell's law states

$$n_1 \sin \theta_1 = n_2 \sin \theta_2$$

where n_1 and n_2 represent media of higher and lower indices of refraction, respectively.

The refracted ray entering the lower index area is bent away from the normal, as shown in Figure 15.3A. At the *critical angle*, the refracted ray is bent $90°$ from the normal and just grazes the interface ($\theta_2 = 90°$), as illustrated in Figure 15.3B. The critical angle of incidence can be described by

$$\sin \theta_c = n_2/n_1, \qquad n_1 > n_2$$

When the angle of incidence exceeds the critical angle, the light ray is totally reflected at the interface, a characteristic called *total internal reflection* (see Figures 15.3C and 15.4). At angles less than the critical angle, most of the energy of the light ray escapes, as shown in Figure 15.3A. The discussion that follows concerns only rays passing through the axis of the fiber (meridional rays). Skew rays propagate without passing through the axis of the fiber and are not dealt with here, due to the mathematical complexity of their behavior. For most purposes, it is sufficient to assume that skew rays will behave in much the same way as meridional rays (Ref. 2).

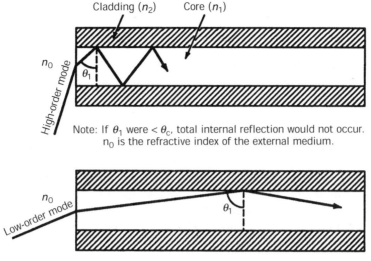

Note: If θ_1 were $< \theta_c$, total internal reflection would not occur.
n_0 is the refractive index of the external medium.

Figure 15.4. Total internal reflection within a fiber.

Numerical Aperture (NA). The angle of incidence for total reflection depends on the relative values of the indices of refraction of the core and the cladding. Total internal reflection takes place only for rays striking the interface at angles greater than or equal to the critical angle. This gives rise to the *acceptance cone* shown in Figure 15.5. It is described by the acceptance angle θ_A, which is the maximum angle with respect to the fiber axis at which an entering ray will experience total internal reflection. The sine of the acceptance cone half-angle is called the *numerical aperture* (NA) and is a measure of the light-gathering ability of the fiber. NA is directly related to the refractive index of the core and cladding. Applying Snell's law, and assuming air (refractive index of 1) as the external medium, the following equation can be derived, showing that NA is a function of the refractive

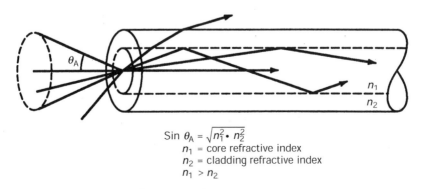

$$\text{Sin } \theta_A = \sqrt{n_1^2 \cdot n_2^2}$$
n_1 = core refractive index
n_2 = cladding refractive index
$n_1 > n_2$

Figure 15.5. Acceptance cone.

indices of the core and cladding in a step-index fiber:

$$\text{NA} = \sin \theta_A = \left(n_1^2 - n_2^2\right)^{1/2}$$

Note: Rays of light are shown entering the center of the fiber in Figure 15.5. Similar light cones obeying the same analysis are incident at an infinite number of points on the core–air interface. This does not affect the calculation of NA (Ref. 3).

15.2.4.2 Distortion and Dispersion

Distortion and Dispersion Differentiated. Distortion is an inherent property of all optical fibers. It causes pulse broadening in digital transmission as shown in Figure 15.6. When pulses broaden to the point where they begin to merge, intersymbol interference (isi) begins to occur. The upper limit of bandwidth-distance factor is a function of distortion.

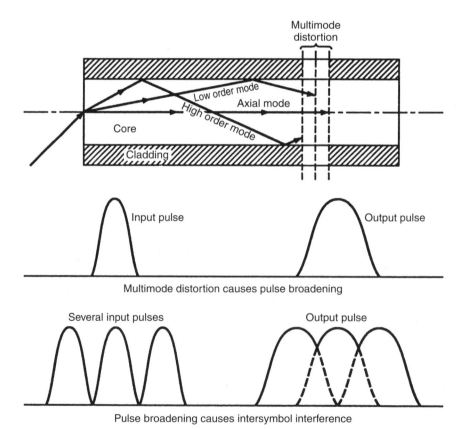

Figure 15.6. Multimode distortion, pulse broadening, and intersymbol interference.

Why is there this broadening? As we can see in the upper drawing of Figure 15.6, light launched into multimode fiber contains several transmission modes: low order and high order. We send a pulse down this "pipe," and it contains signal energy of the low-order mode and signal energy in the high-order mode. As can be seen in the figure, the low-order mode has fewer reflections as it travels down the pipe; thus to reach the distant receiver this energy arrives first because it travels less distance than the energy in the high-order mode. Such indeed is what causes pulse broadening. When there are several pulses, then, broadened energy of the first pulse spills into the second pulse position. This confuses the receiver's decision circuit: Was it a 1 or a 0 that was received?

Distortion can be the determining factor limiting link length in very high bit rate systems. The higher the bit rate on a system, the more dispersion and distortion have to be controlled. Distortion is usually measured in picoseconds per kilometer.

DISTORTION VERSUS DISPERSION. We tend to use the terms distortion and dispersion synonomously. Reference 3 (MIL-HDBK-415) states that this can cause a great deal of confusion. The IEEE dictionary calls distortion "temporal" and dispersion "spectral" (Ref. 4).

DISCUSSION OF DISPERSION AND DISTORTION

Multimode Distortion. In multimode fibers, light can travel different paths within the fiber core. These paths, or propagation modes, vary in length. Because of this variance, light will exit a fiber over a time interval slightly longer than that over which it entered. This phenomenon, known as *multimode distortion*, causes an undesirable broadening of the signal waveform as it progresses along with fiber as shown in Figure 15.6. Multimode distortion is most prominent in multimode step-index fibers, where it is the major form of distortion. It is less evident in graded-index fibers and nonexistent in single-mode fibers.

Dispersion. Within a given fiber, different wavelengths of light have different propagation characteristics. This variance with wavelength is called *dispersion* and, like multimode distortion, can cause pulse broadening. Unlike multimode distortion, dispersion affects all types of fiber, even single-mode. Three types of dispersion fall within the scope of this chapter: material dispersion, waveguide dispersion, and chromatic dispersion.

Material Dispersion. The speed of light through a fiber varies with refractive index, which itself is wavelength-dependent. Light having a wavelength at which the refractive index is lower will travel faster than light at a wavelength at which the refractive index is higher. Even the light source with the narrowest spectral width emits over a finite band of wavelengths. As illustrated in Figure 15.7, choice of either a source with narrow spectral width or

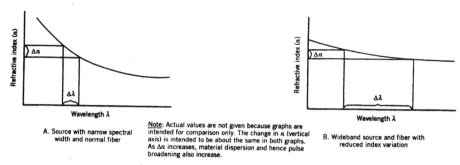

Figure 15.7. Effect of source spectral width and fiber index variations on material dispersion.

of a fiber material with reduced variation of index with wavelength will lessen the effects of material dispersion. Material dispersion is a major contributor to total dispersion on fiber-optic systems using light-emitting diodes (LEDs) for the light source (transmitter) because of the wide spectral width LEDs display.

Waveguide Dispersion. Waveguide dispersion is attributed to the dependence of phase and group velocities on the geometric characteristics of the waveguide. Because light wave peaks occur at different locations in the waveguide with changing frequency, intensity differences occur when all reflections are added together. This causes broadening, but waveguide dispersion usually has much less effect than material dispersion and can be ignored in most cases.

Chromatic Dispersion. Chromatic dispersion affects only fiber-optic links carrying very high rate digital systems (e.g., > 1 Gbps). Such systems will use laser diode sources with monomode fiber. A laser diode has a narrow emission line width, perhaps several nanometers wide. With distributed feedback lasers, this line width can be narrowed to 0.5-nm widths and less. Even this narrow width carries many emission wavelengths. Some of these wavelengths travel slightly faster than others. As the bit rate increases, the pulse width decreases, as one might expect. We are interested in half-pulse width where the detector will make a 1 or 0 decision. The durations of these half-pulse widths start approaching the time difference of arrival of signal energy giving rise to chromatic dispersion. Again it is the problem of the slower signal energy spilling into the subsequent bit position confusing the receiver.

Zero Dispersion Near 1300 Nanometers. In fused-silica single-mode fibers, there is sometimes an operating wavelength at which material dispersion and waveguide dispersion are equal in amplitude but opposite in sign. This results in total dispersion approaching zero. The exact frequency at which this occurs is a function of core diameter and refractive index profile. The range extends from about 1300 nm for a 10-μm core to about 1600 nm for a 4-μm core. Choice of materials and core diameter make it possible to bring the wavelengths of minimum dispersion and minimum attenuation to coincide.

15.2.4.3   *Attenuation.* Attenuation (absorption and scattering) of light in a fiber is caused by material variations, microbends induced by mechanical stress, and surface imperfections at the core–cladding interface.

Absorption and Scattering. Absorption, whereby light is converted to heat in the fiber core, and scattering, whereby energy is ejected from the core into surrounding areas, have the same result: a decrease in the energy reaching the detector. Absorption is caused by impurities and hydroxyl (OH^-) ions present even in the purest glass fibers. Broad infrared and ultraviolet absorption regions contribute additional absorption. Finally, Rayleigh scattering due to variations in density and composition in the core material causes further attenuation. The overall loss resulting from all of these factors is least in wavelength windows around 1300 and 1500 nm. Conveniently, dispersion in single-mode fibers can be minimized at these wavelengths as explained above. Figure 15.8 shows optical fiber attenuation versus wavelength.

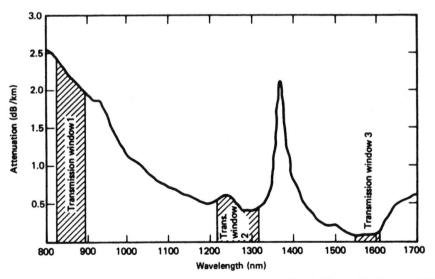

Figure 15.8. Optical fiber attenuation versus wavelength. (From Figure 11.3, page 723, Ref. 2.)

15.2.5   Types of Optical Fiber

One method of classifying optical fibers is by their refractive index profile and the number of modes propagated. Classifying them in this manner derives three types of fiber:

1. Multimode, step-index
2. Multimode, graded-index
3. Single-mode fiber

15.2.5.1   Multimode Fibers. The core diameter of multimode fibers allows several modes of light to propagate, producing multimode distortion. Core sizes of glass multimode fibers range from 50 to 200 μm or even larger.

Multimode Step-Index Fiber. Conceptually, the simplest fiber is the step-index fiber, with a core of high refractive index and a concentric lower index cladding, with a sharp interface between the two as shown in Figure 15.9A. Step-index fiber produces high multimode distortion.

Multimode Graded-Index Fiber. In a graded-index fiber, refractive index decreases continuously with radial distance from the center of the core, giving the fiber a nearly parabolic index profile. This is illustrated in Figure 15.9B. Light propagation occurs through refraction, with light rays turned back toward the core axis less sharply than the reflection angles in a step-index fiber. Light rays travel a wavelike course down the fiber.

Graded-index fiber displays less multimode distortion than does step-index fiber. Light rays travel a longer path further from the axis, but they travel faster due to decreasing index of refraction further from the axis. A specialized type of graded-index fiber, called a depressed cladding or W-index fiber for its characteristic index profile, exhibits lower attenuation than does conventional graded-index fiber. This fiber has a core and thin cladding surrounded by an external higher index region. Such a unique composition prevents leaky modes from the cladding from reentering the core.

Multimode fiber is tending to become obsolete in favor of single-mode fiber.

15.2.5.2   Single-Mode Fiber. The core of a fiber can be reduced in size (typically 2–10 microns) where only the axial mode can propagate. This is called single-mode fiber, and its propagation characteristic is shown in Figure 15.9C. Single-mode fiber, sometimes called monomode fiber, eliminates multimode distortion because only one transverse mode can propagate, giving these fibers the broadest bandwidth. A type of material dispersion, however, does affect single-mode fibers, particularly at very high data rates, with pulse components of different wavelengths taking different times to traverse the fiber. This is referred to as *chromatic dispersion*. (See Section 15.2.4.2.)

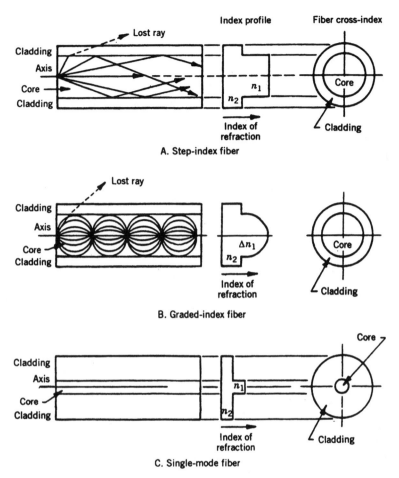

Figure 15.9. Index profiles and modes of propagation.

15.2.6 Fiber-Optic Cable

The delicate optical fibers are cabled for protection against mechanical and environmental abuse during installation and operation. Fibers are cabled using many of the same materials used to cable metallic media, but employing different cabling techniques. Unlike metallic conductors, optical fibers do not contribute to the strength of the cable. In fact, the fiber itself must be decoupled from tensile, thermal, and vibrational loads.

15.2.6.1 Cable Composition. Strengthened fiber-optic cables are used in all outdoor and most indoor applications. Unstrengthened cables are used in low-cost multiple-line links in controlled environments, such as within a rack. The following is a list of typical components of fiber-optic cable. It should be pointed out that not every cable will contain each constituent, and configurations will vary.

1. Coated optical fiber with plastic or silicon elastomer coating to protect the fiber from scratches and microbending.
2. Single-fiber buffer tubes or multifiber ribbons to cushion fibers and isolate them from longitudinal stresses during installation and from microbending during operation. Buffer tubes are not to be confused with the protective plastic or silicon fiber coating, also called a *fiber jacket*.
3. Central and distributed strength members composed of either metallic or dielectric materials to ensure adequate tensile strength during installation and to resist kinking and crushing of the cable.
4. Polyester binder tape to hold buffer tubes in place.
5. Moisture-barrier compounds.
6. Outer and inner nonconducting jackets composed of low-density polyethylene resistant to abrasion, chemicals, fungus, moisture, and decomposition by ultraviolet radiation. High-density polyethylene is used in armored cable.
7. Corrugated steel tape for protection from rodents.
8. Messenger cable for support of some aerial cables.

15.2.7 Fiber-Optic Transmitters

A conventional fiber-optic transmitter consists of an electrical signal driver, a light source, and a pigtail that connects the light source to a fiber in the optical fiber cable, as shown in Figure 15.10. A transmitter in a fiber-optic communication system converts electrical energy into optical energy and launches the converted signal into the optical fiber transmission medium. Conventional fiber-optic light sources are either light-emitting diodes (LEDs) or the higher-performance laser diodes (LDs). Of course, the operating wavelength of the transmitter must be compatible with its companion receiver.

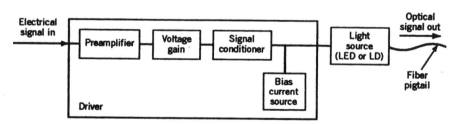

Figure 15.10. Block diagram of a typical fiber-optic transmitter.

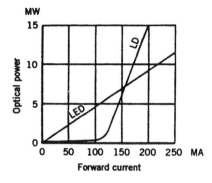

Figure 15.11. Optical power versus drive current for typical light sources.

15.2.7.1 *Light-Emitting Diodes.*

An LED is the preferred light source when less stringent system requirements permit its use. An LED source tends to have a longer life, is generally less expensive, and requires lower drive current than the higher-performance laser diode (LD). An LED is stable over a long lifetime. LEDs are suitable for analog applications because of their linearity; optical power output is a nearly linear function of drive current, as shown in Figure 15.11. However, because of an LED's broad spectral width of its emission and because the emitted area of light is larger than the core, the LED suffers from low coupling efficiency. LEDs have a broad spectral linewidth, as shown in Figure 15.12, resulting in increased material dispersion. Spectral linewidth is the wavelength interval between half-power (3 dB) points. In addition, modulation rates are limited. Table 15.1 compares LEDs and LDs.

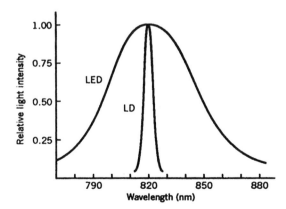

Figure 15.12. Typical spectral linewidths for LED and LD. Note that the peak intensities have been normalized to the same value. The actual peak intensity of an LD is much greater than that of an LED.

TABLE 15.1 Comparison of Typical LEDs and LDs

Characteristic	LED	LD
Coupled power[a] (μW) (max) (into 50-μm core)	> 50	3000
Radiant power (mW) (max)	20	20
Spectral linewidth (nm)	30 to 90	< 1 to 5
Wavelength (nm) (max)	1550	1550
Drive current (mA)	10 to 200	10 to 200
Modulation bandwidth (GHz) (max)	1	10
Life (hours) (estimated, not guaranteed)	10^5 to 10^6	10^4 to 10^5
Cost	Lower	Higher

[a]Power remaining after coupling losses have been taken into account.
Source: Reference 1.

15.2.7.2 Laser Diodes.

Laser diodes (LDs) are used for light sources when a fiber-optic link requires a higher-performance transmitter. Because of fast risetime, LDs are capable of higher modulation rates. They have narrower spectral linewidths than do LEDs, as shown in Figure 15.12. As a result, an LD produces less material dispersion within the fiber than does an LED. However, LDs require an auxiliary high-voltage power supply for external biasing to achieve the lasing effect. LDs have a shorter life, are temperature sensitive, and are less stable than LEDs. They are best suited for digital transmission. In links where the LD light source overloads the detector, an attenuator (neutral density filter) can be placed between the light source and the fiber to lessen the light coupled into the fiber, yet allow the light source to operate within optimum range.

15.2.8 Receivers

A fiber-optic receiver consists of the fiber pigtail or coupler, light detector, opto-electronic circuitry, and preamplifier as shown in Figure 15.13. Positive-intrinsic-negative (PIN) photodiodes and avalanche photodiodes (APDs) are the two common types of light detectors used in fiber-optic communication systems. The receiver converts received optical power into an electrical signal.

15.2.8.1 PIN Photodiodes.

Positive-intrinsic-negative (PIN) photodiodes have a longer life and are less subject to thermal instability than APDs. Peak responsivity is in the 800- to 900-nm range and 1300- to 1600-nm range. They have a wide dynamic range and good linearity. PINs have no internal gain, and therefore they exhibit less sensitivity than do APDs. PIN diodes introduce shot noise from photocurrent and leakage current noise. PIN diodes and APDs are compared in Table 15.2. PINs have a large dynamic range and

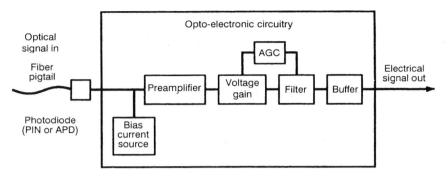

Figure 15.13. Simplified block diagram of a fiber-optic receiver.

TABLE 15.2 Comparison of Typical PIN Photodiodes and APDs

Characteristic	PIN Photodiode	APD
Sensitivity (dBm for BER or SNR)	−30 to −45	−40 to −50
Spectral response range (nm)	200 to 1700	200 to 1700
Noise, equivalent power (W/$\sqrt{Hz}$)	1×10^{-10} to 27×10^{-14}	1×10^{-14}
Cost	Lower	Higher

Source: Reference 1.

have bandwidths exceeding 2 GHz. There is quantum noise associated with the photodetection process. Thermal noise in the electronic preamplifier is the controlling noise of the receiver output.

15.2.8.2 Avalanche Photodiodes. An avalanche photodiode (APD) has internal gain, thereby providing a sensitivity approximately 5–20 dB greater than the PIN while generating some internal noise. APDs require an external bias voltage. Peak responsivity is in the 800- to 900-nm range and in the 1300- to 1600-nm range. APDs introduce quantum noise (i.e., shot noise from multiplied photocurrent) and surface leakage noise (i.e., shot noise from nonmultiplied leakage current). They require temperature compensation to prevent damage and maintain gain. They are suited to long-haul systems and for high data rates.

15.2.8.3 Noise. Noise appears at the detector and is due to quantum or photon noise arising in the optical beam itself and thermal noise arising in the resistances associated with the front-end amplifier. Quantum noise or photon noise is more evident in highly sensitive detectors such as APDs, whereas thermal noise is more evident with PIN diodes. An important aspect of the spectral density of thermal noise is that it is quite constant with frequencies up to near 10^{12} Hz. Above this, thermal noise density begins to decline and drops off rapidly above 10^{13} Hz, so that, although thermal noise is common at microwave frequencies, it is essentially absent at optical frequencies (i.e., above 10^{14} Hz). On the other hand, quantum noise becomes evident in the 10^{12}- to 10^{13}-Hz range right where thermal noise is dropping off, and it increases directly with frequency. Quantum noise becomes altogether dominant at optical frequencies and sometimes is referred to as *blue noise* because of its spectral density characteristics. With current optical systems, the conversion to electronic signals before detection is not carried out, and thus the detector front-end amplifier will exhibit both types of noise. This will change when heterodyne systems begin to be implemented.

15.2.9   Repeaters

Repeaters reshape, retime, and regenerate digital signals by converting optical signals to electrical signals and back to optical signals again. (Signal conditioning occurs in the electrical domain. Current technology does not permit such signal conditioning in the optical domain). A repeater may be viewed as a receiver back-to-back with a transmitter, with necessary control and signal conditioning equipment. Some repeaters also contain diagnostic equipment to transmit alarm status, monitor bit error rate (BER), and provide orderwire functions. To allow for various combinations of emitters and detectors, some repeaters use plug-in module technology. Repeaters are usually powered by a serving switch or have an auxiliary power feed. Other repeaters are powered by a dedicated power supply using local prime power along with backup battery.

Repeater spacing varies from 15 to over 150 km and depends on the type of fiber, type of transmitter and detector used, and the bit rate being transmitted. Figure 15.14 is a functional block diagram of a repeater.

15.2.10   Fiber-Optic Amplifiers

The most common type of amplifier found in fiber-optic systems today is the erbium-doped fiber amplifier (EDFA). These amplifiers operate in the 1550-nm band, which is propitious because it is the band of lowest loss. The amplifier's bandwidth is on the order of 100s to 1000s of GHz range, which can support most WDM (wavelength division multiplex) operation. Typical

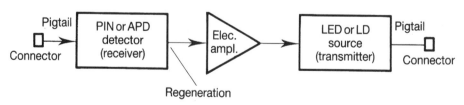

Figure 15.14. A functional block diagram of a fiber-optic cable repeater.

amplifier gains are from 30 to 50 dB, and power output gains are from $+10$ to $+20$ dBm.

EDFA amplification is based on a pumping principle, where the most common pumping wavelengths are 1480 or 980 nm. The pump output of from 50 to 100 mW is coupled into a length of fiber from 25 to 100 m in length, where the longer length displays the greatest gain. Over 100 m, the gain starts to tail off. For a more detailed description of EDFAs, consult Refs. 1 and 2.

EDFAs are usually installed at a light source output and at the light detector input. In such a situation with two amplifiers, the gain accrued, assuming 40 dB per amplifier, is an 80-dB net gain. If we allowed just 0.35 dB/km including splice loss on a link without amplifiers, and now installed two amplifiers as we have indicated with a net gain of 80 dB, we can add to the distance between repeaters some 80/0.35 km or an additional 228 km. This provides many benefits: less active devices with improved reliability (i.e., less probability of failure), fewer repeaters thus reduced jitter, some improvement in error performance, and so on. Amplifiers make DWDM (dense wavelength division multiplex) feasible on fiber-optic systems. The amplifiers compensate for the rather large insertion losses of DWDM systems.

15.2.11 The Joining of Fibers: Splices and Connectors

15.2.11.1 Splices. Splices must provide precise alignment of fiber ends when completed. Alignment of the prepared ends may be achieved through the use of an alignment tube, with the fiber ends sealed with epoxy once alignment has been achieved. More positive alignment may be achieved with a V-groove guide or a square tube, so that fibers introduced at a slight angle to the axis of the tube or groove are forced to line up face-to-face. Epoxy is then applied. Sometimes an index-matching fluid is added to minimize Fresnel reflection and scattering loss. An alternate method is the use of fusion splicing. Here the fiber ends are manipulated under a microscope or by monitoring optical throughputs, and the ends are then fused by an electric arc or other source of heat. Fusion splice losses of 0.15–0.25 dB per splice are common.

15.2.11.2 Connectors. Connectors require efficient signal transfer across the joined fiber ends, and they are designed to be mated and demated many times. Insertion losses of connectors vary from as low a 0.5 dB to as high as 1.5 dB. Connectors are either lens- or butt-coupled. Some butt-coupled connectors use index-matching fluid to reduce loss, whereas others rely on optical contact (Ref. 1).

15.2.12 Modulation and Coding

Digital rather than analog modulation is better suited to most fiber-optic transmission systems because of the relatively nonlinear characteristics of fiber-optic components. Currently, both analog and digital modulation are accomplished by intensity modulation of the light source. The most common application for fiber-optic transmission is telecommunications using PCM (see Chapter 8), or on LANs (such as FDDI covered in Section 10.1.4) using other digital modulation schemes. Analog transmission is widely used in the cable TV (CATV) arena.

15.2.12.1 Digital Signals

Polarity. This section covers bipolar signals only (which some call just *polar*). Bipolar signals have two nonzero polarities and are usually symmetrical about the zero axis.

Coding. Digital signals are also classified according to whether the line signal state returns to or maintains the zero level during each bit interval, or uses polarity changes to signify a designated logic element. This is referred to as *coding*. Signal codes commonly used in fiber-optic communication systems are illustrated in Figure 15.15.

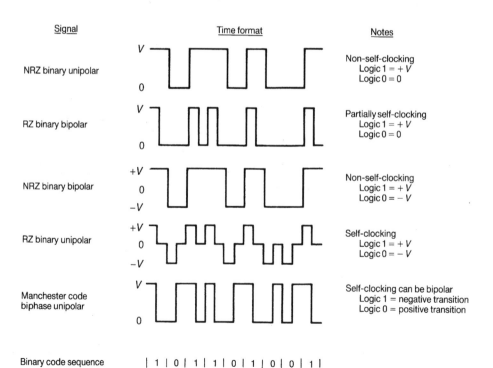

Figure 15.15. Digital coding (line formats). (From Ref. 2.)

15.2.12.2 *Optical Line Coding and Transmission Rate*

Coding. RZ (return-to-zero) bipolar and bipolar AMI (alternate mark inversion) are ternary, three-state electrical signals. The optical regime in digital applications, however, supports only binary, two-state signals. Therefore, the ternary signals must be converted to binary for applications to optical transmitters. This is done by assigning a two-digit binary code (i.e., 00, 01, 10) to each of the three states. NRZ (non-return-to-zero) polar, biphase level, and conditioned diphase are binary signals and can be applied directly to the optical transmitter.

Rate. Each data unit interval in ternary coding can contain a maximum of two states. Because each state is represented by two digits after conversion to optical line code, the optical line rate is four times the electrical code bit rate. Biphase level and conditioned diphase coding depend on mid-interval transitions. Consequently, the optical line rate is twice the electrical code bit rate. NRZ is the only code of those treated here for which the optical line rate is the same as the electrical bit rate (Ref. 3).

15.3 HIGHER-ORDER DIGITAL MULTIPLEXING FORMATS: SONET AND SDH

15.3.1 Synchronous Optical Network (SONET)

15.3.1.1 Introduction. SONET provides a digital multiplex format for the wide bandwidth of optical fiber cable. Prior to the implementation of SONET, multiplex formats had capacities in the range of 100–500 Mbps. SONET also incorporates some innovations such as "slipless" operation and the ability to accommodate digital broadcast-quality TV. SONET also supports B-ISDN.

BellCore describes SONET as a transporter of standard (digital) hierarchical signals (DS1, DS1C, DS2, and DS3). It can also transport E-1 hierarchical signals and ATM. See Figures 15.31 and 15.32 at the end of this section for system block diagrams. Just because SONET has the word "optical" in its name does not mean that its use is exclusively on optical networks. If the transmission medium has the bandwidth (and other parameters such as group delay meeting transmission requirements), it most certainly can be used to carry SONET line signals. Millimeter-wave radio is an example.

15.3.1.2 SONET Rates and Formats. SONET has its own digital hierarchy supporting various digital rates for user transport. The SONET hierarchy is built upon a basic signal of 51.840 Mbps and a byte interleaved multiplex scheme that results in a family of digital rates and formats defined at a rate of N times 51.840 Mbps. The basic signal has a portion of its capacity devoted to overhead, and the remaining portion carries payload.

TABLE 15.3   Current Allowable OC-*N* Line Rates

OC Level	Line Rate (Mbps)
OC-1	51.840
OC-3	155.520
OC-9	466.560
OC-12	622.080
OC-18	933.120
OC-24	1244.160
OC-36	1866.240
OC-48	2488.320
OC-192	9953.28

SONET's basic modular signal is called STS-1 with its 51.840-Mbps data rate. STS-1's optical counterpart is called OC-1 (OC stands for optical carrier). OC-1 is directly converted from STS-1 after frame synchronous scrambling.

Higher-level SONET signals are obtained by synchronously multiplexing lower-level signals. Higher-level signals are integer multiples of the basic STS-1 rate and are denoted STS-*N* and OC-*N*, where, again, *N* is an integer. The current values of *N* are 1, 3, 9, 12, 24, 36, 48, and 192 (Ref. 5). These data rates are shown in Table 15.3.

15.3.1.3 *STS-1 Frame Structure.* The STS-1 frame is shown in Figure 15.16. It consists of 90 columns and 9 rows of 8-bit bytes for a total of 810 bytes. With a frame length of 125 microseconds (derived from 8000 frames per second), STS-1 has a bit rate of 51.840 Mbps. The order of transmission

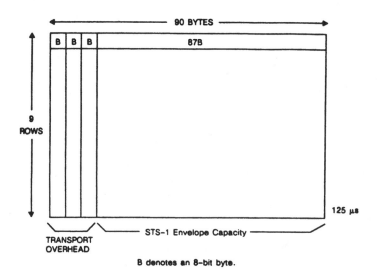

Figure 15.16. STS-1 frame structure.

of bytes is row-by-row, from left to right. The most significant bit is transmitted first, and the bits are numbered 1–8 (Ref. 6).

As shown in Figure 15.16, the first three columns are the transport overhead, containing overhead bytes of sections and line layers. Of the 27 bytes assigned, 9 are assigned to line overhead and 18 to section overhead. The remaining capacity, consisting of 87 columns, constitutes the STS-1 envelope capacity.

Figures 15.17, 15.18, and 15.19 show the STS-1 synchronous payload envelope (SPE). It consists of 87 columns and 9 rows of bytes for a total of 783 bytes. Column 1 is called the STS path overhead (POH) and contains 9 bytes with the remaining 774 bytes available for payload. The path overhead provides the facilities, such as alarm and performance monitoring, required to support and maintain the transportation of the SPE between end locations (known as *path terminations*) where the SPE is either assembled or disassembled. The STS-1 SPE may begin anywhere in the STS envelope capacity. Typically, it begins in one frame and ends in the next or can be wholly contained in one frame. A key innovation in SONET is the payload pointer. The STS-1 payload pointer is contained in the transport overhead. It designates the location of the byte where the STS-1 SPE begins.

SPE Assembly Process. The concept of a tributary signal such as a DS3 signal being assembled into an SPE, to be transported end-to-end across a

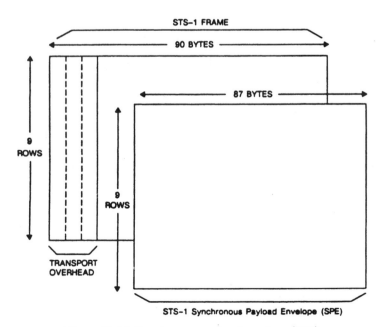

Figure 15.17. Synchronous payload envelope (SPE).

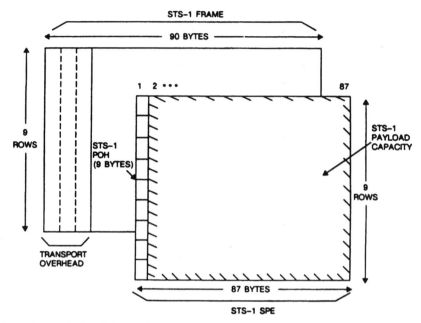

Figure 15.18. STS-1 SPE with STS-1 POH and STS-1 payload capacity. Based on Ref. 6.

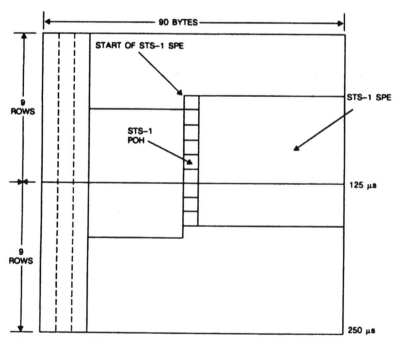

Figure 15.19. STS-1 SPE in the interior of an STS-1 frame. Based on Ref. 6.

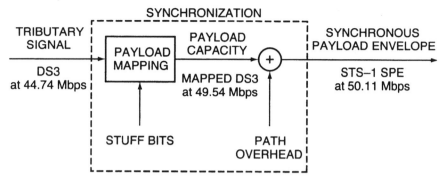

Figure 15.20. SPE assembly process. (From Ref. 7. Courtesy of Hewlett-Packard Company.)

synchronous network, is fundamental to SONET operation. The process of assembling the tributary signal into an SPE is referred to as *payload mapping.*

To provide uniformity across all SONET transport capabilities, the payload capacity provided for each individual tributary signal is always slightly greater than that required by the tributary signal. Thus, the essence of the mapping process is to synchronize the tributary signal with the payload capacity provided for transport. This is achieved by adding extra stuffing bits (also called *justification bits*) to the signal bit stream as part of the mapping process. Thus, for example, a DS3 tributary signal at a nominal rate of 44 Mbps needs to be synchronized with a payload capacity of 49.54 Mbps provided by the STS-1 SPE. Addition of the POH completes the assembly of the STS-1 SPE and increases the bit rate of the composite signal to 50.11 Mbps. This process is shown in Figure 15.20 (Ref. 7).

SPE Disassembly Process. At the point of exit from the synchronous network, the tributary signal that has been transported over the network needs to be recovered from the SPE which provided the transportation facilities. This process of disassembling the tributary signal from the SPE is referred to as *payload demapping.*

The SPE comprises the POH, the tributary signal, and additional stuffing bits that have been added in order to synchronize the data transmission rate of the tributary signal to the payload capacity available for transportation. In essence the demapping process is to desynchronize the tributary signal from the composite SPE signal and to reproduce this tributary signal, as nearly as possible, in its original form. For example, as STS-1 SPE carrying a mapped DS3 payload arrives at the tributary disassembly location with a signal rate of 50.11 Mbps. Stripping the POH and the stuffing bits from the SPE results in a discontinuous signal representing the transported DS3 signal with an average data rate of 44.74 Mbps. These timing discontinuities are reduced by

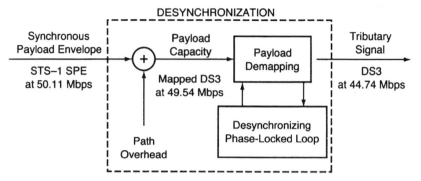

Figure 15.21. SPE disassembly process. (From Ref. 7. Courtesy of Hewlett-Packard Company.)

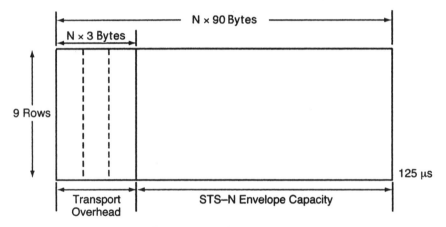

Figure 15.22. STS-*N* frame.

means of a desynchronizing phase-locked loop (PLL) in order to produce a continuous DS3 signal at the same average data rate. Figure 15.21 shows this process.

15.3.1.4 STS-N Frame Structure. Figure 15.22 shows the STS-*N* frame structure. It is made up of *N* STS-1 byte interleaved signals. Frame alignment of the transport overhead channels of individual STS-1 signals is carried out prior to byte interleaving. Because of the unique payload pointer, associated STS SPEs do not require alignment.

SONET provides super rate services that require multiples of the STS-1 rate. Typical of such a rate is the ISDN H4 channel. For these services, the super rates are mapped into an STS-*N*c SPE and transported as a concatenated STS-*N*c whose constituent STS-1s are kept together. The STS-*N*c is multiplexed, switched, and transported through the network as a single entity. A concatenation indicator, which is used to show that the STS-Nc is to be kept together, is contained in the STS-1 payload pointer. Figure 15.23 shows the STS-3c transport overhead assignments, and Figure 15.24 shows the STS-Nc SPE.

In the STS-Ne SPE, only one set of STS POH is required. The STS-Ne SPE is carried within the STS-Ne in such a way that the STS POH always appears in the first of the N-STS-1s that make up the STS-Ne (Ref. 10).

15.3.1.5   *The Structure of Virtual Tributary.* The sub-STS-1 payloads are transported and switched with the virtual tributary (VT) structure. There are four sizes of VTs (VTx): VT1.5 (1.728 Mbps), VT2 (2.304 Mbps), VT3 (3.456 Mbps), and VT6 (6.912 Mbps). These VT sizes are shown in Figure 15.25. In the 9-row structure of the STS-1 SPE, these VTs occupy 3 columns, 4 columns, 6 columns, and 12 columns, respectively.

The VT-structured STS-1 SPE is divided into seven VT groups which efficiently accommodate mixes of VTs. Each VT group occupies 12 columns of the 9-row structure and may contain 4, 3, 2, or 1 VT(s). The size (x) of the VT determines the number of VTs contained in that particular VT group. VT groups can carry 4, 3, 2, or 1 VT accommodating 4 VT1.5s, 3 VT2s,

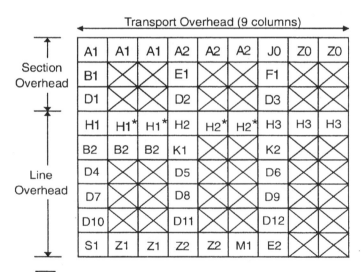

: Undefined Overhead byte (All-zeros pattern as an objective)

: Concatenation Indication

H1* = 1001XX11

H2* = 11111111

Figure 15.23. Transport overhead assignment, OC-3 carrying an STS-3c SPE. (From Telcordia GR-253-CORE, Issue 2, Revision 2, Figure 3-8, page 3-9. Copyright Telcordia, Jan. 1999.)

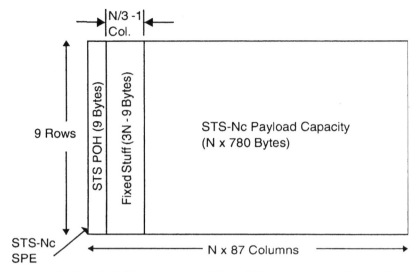

Figure 15.24. STS-*N*c SPE. (From Telcordia GT-253-CORE, Issue 2, Revision 2, Figure 3-7, page 3-8. Copyright Telcordia, Jan. 1999.)

2 VT3s, or 1 VT6, respectively. However, only one size is allowed in a particular VT group, but a different size is allowed for each VT group in an STS-1.

Floating and locked are the two possible modes of operation of the VT structure. The locked mode minimizes interface complexity in distributed 64-kbps switching. The floating mode minimizes delay for distributed VT switching.

Floating VT Mode. Four consecutive 125-μs frames of the STS-1 SPE are organized into a 500-μs superframe in the floating VT mode. The phase of the superframe is indicated by the multiframe indicator byte (H4) in the STS POH. This defines a 500-μs structure for each of the VTs, which is called the *VT superframe.* The VT payload pointer and the VT SPE are contained in the VT superframe. Four bytes of the superframe are for VT pointer use (V1, V2, V3, and V4). The remaining bytes define the VT envelope capacity, which is different for each size, as shown in Figure 15.26.

V1 is the first byte in the superframe V2 through V4 appear as the first byte in the succeeding VT frames, regardless of size.

The VT payload pointer provides a flexible and dynamic tool for alignment of the VT SPE with the VT envelope capacity, independent of other VT SPEs. Each VT SPE contains one byte (V5) of VT POH, and the remaining bytes constitute the VT payload capacity, which is different for each VT size.

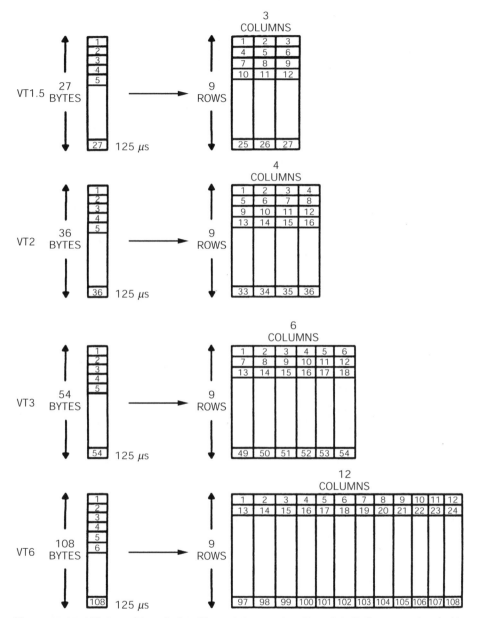

Figure 15.25. VT sizes. (From Ref. 6. Figure 3-9, page 3-4. Copyright Bellcore, reprinted with permission.)

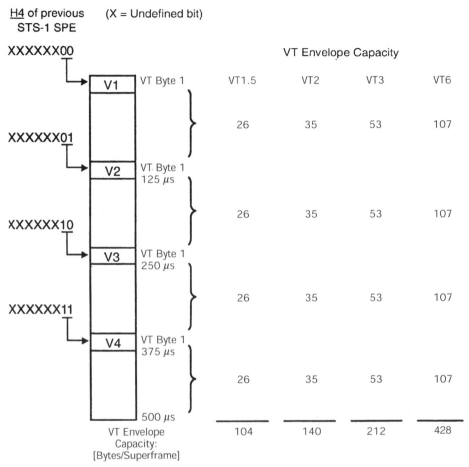

Figure 15.26. VT superframe and envelope capacity of a VT superframe. Designations: V1, VT Pointer 1, V2, VT Pointer 2, V3, VT Pointer 3 (Action), V4, VT Reserved. (Adapted from Ref. 6.)

Locked VT Mode. Here the VT structure contains synchronous payloads that are "locked" to the STS-1 SPE. There are no pointers to process in this mode because the tributary information is fixed and immediately identifiable with respect to the STS-1 pointer.

Because the locked VTs remain fixed with respect to the STS-1 SPE, the V1 through V4 pointer bytes and the 500-μs VT superframe are not used. The multiframe indicator byte (H4) in this mode defines the 3-millisecond superframe for DS0 transmission. Also in the VT locked mode there is no VT POH byte. The two bytes of the 125-μs locked VT frame that correspond to the VT pointer and V5 bytes are reserved. In each 125-μs period, the

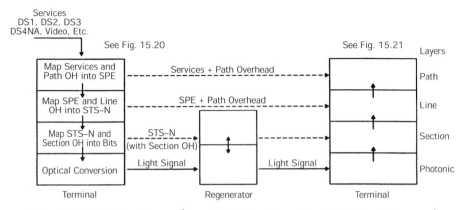

Figure 15.27. Interface layers. (From Ref. 7. Courtesy of Hewlett-Packard Company.)

remaining $N - 2$ bytes of information capacity is equivalent in the two modes. A mix of the two modes within an STS-1 SPE is not permitted.

15.3.1.6 SONET Layers and Transport Functions.
There are four SONET layers: physical, section, line, and path. The layers have a hierarchical relationship and are considered from the top down. Figure 15.27 may be used in showing how each layer requires the services of all lower-level layers to perform its function.

Bellcore (Ref. 6) gives the following example of this interdependability. Suppose that two path layer processes are exchanging DS3s. The DS3 signal and the STS path overhead (POH) are mapped into an STS-1 SPE, which is then given to the line layer. The line layer multiplexes several inputs from the path layer (frame and frequency aligning each one) and adds line overhead (e.g., required for protection switching). Finally the section layer provides framing and scrambling before optical transmission by the physical layer. Figures 15.28 and 15.29 show section, line, and path definitions and will help to place these terms in perspective.

Interface Layers

Physical Layer. The physical layer transports the bits as optical or electrical pulses across the physical medium. There is no overhead associated with the physical layer. The principal function of this layer is to convert STS signals and make them suitable for transmission on the selected medium, which may be optical, metallic, or radio. The physical layer deals with such things as pulse shape, power levels, and line code.

Section Layer. This layer prepares STS signals for delivery to the physical layer. It is concerned with framing, scrambling, section error monitoring, and

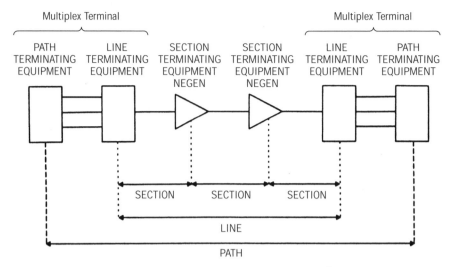

Figure 15.28. Simplified diagram defining section, line, and path. (From Ref. 7. Courtesy of Hewlett-Packard Company.)

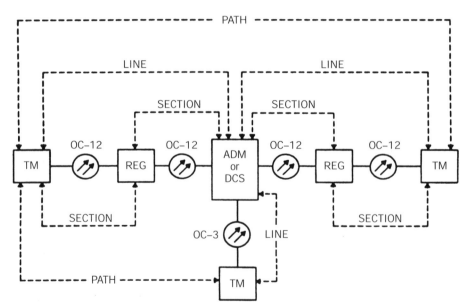

Figure 15.29. More detailed section, line, and path definitions. DCS, digital cross-connect system; ADM, add-drop multiplex; REG, regenerator; TM, terminal multiplex. (From Figure 2-2, page 2-2, Ref. 6. Copyright Telcordia, reprinted with permission.)

the communication of section level overhead, such as a local orderwire. The section terminating equipment (STE) interprets or creates the overhead associated with the section layer.

Line Layer. This layer provides synchronization and multiplexing functions for the path layer. The overhead associated with these functions includes that required for maintenance and protection and is inserted in the line overhead channels. The line terminating equipment (LTE) interprets and modifies or creates the line overhead. The LTE carries out section layer functions and therefore the LTE is also an STE.

Path Layer. This layer deals with the transport of network services between SONET terminal multiplexing equipment. Such services include DS1, DS3, and DS4NAs.

The path layer maps these services into the format required by the line layer and communicates end-to-end with the POH. The path terminating equipment interprets and modifies or creates the overhead defined for this layer.

Layer Interaction. Figure 15.27 shows the interactions of the interface layers. Similar to OSI (open system interconnection), each layer communicates horizontally to peer equipment in that layer. It also processes certain information and passes it vertically to the next layer.

The path layer transmits horizontally to its peer entities the services and the path layer overhead. The path layer maps the services and POH into SPEs that it passes vertically to the line layer.

The line layer transmits SPEs with line layer overhead to its peer entities. At this time the SPEs are synchronized and multiplexed. Then the STS-*N* signal is passed to the section layer.

The section layer communicates with its peer entities, transmitting to them STS-*N* signals and section layer overhead (e.g., local orderwire). This layer maps STS-*N* and section overhead into bits that are handed over to the path layer, which transmits optical or electrical pulses to its peer entities.

It should be noted that access to all four layers is not required of every SONET NE.* As an example, an OC-*N* regenerator uses only the first two layers (physical and section layers). Also, an NE that merely routes SPEs and does not accept any new inputs from the path layer only uses the first three layers. However, in this case these NEs may only monitor the overhead of layers that they do not terminate (Ref. 6).

15.3.2 Synchronous Digital Hierarchy

Synchronous digital hierarchy (SDH) resembles SONET in many respects. The SDH design is behind SONET in maturity. It represents the European

---

*NE stands for network element.

TABLE 15.4   SDH Bit Rates

SDH Level	Hierarchical Bit Rate (kbps)
1	155,520
4	622,080
16	2,488,320
64	9,953,280

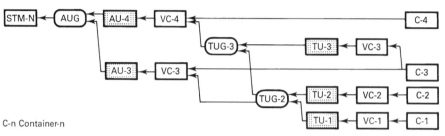

C-n Container-n

Figure 15.30. Generalized SDH multiplexing structure. C-*n* stands for container-*n*. (From Figure 2-1/G.708, page 3, ITU-T Rec. G. 708, Ref. 9.)

hegemony much like the E-1 hierarchy does. It is generally understood to be more flexible than SONET and was designed more with the PTT (government-owned telecommunications administrations) in mind. The SONET design, on the other hand, has more flavors of the private network.

15.3.2.1   SDH Standard Bit Rates. The standard operational SDH bit rates are shown in Table 15.4. ITU-T Rec. G.707 (Ref. 8) states "that the first level of the digital hierarchy shall be 155,520 kbps and that higher synchronous digital hierarchy bit rates shall be obtained as integer multiples of the first level bit rate."

15.3.2.2   Interface and Frame Structure of SDH. Figure 15.30 illustrates the relationship between various multiplexing elements that are given below and shows possible multiplexing structures. Figures 15.31, 15.32, and 15.33 show specific derived multiplexing methods.

Definitions

Synchronous Transport Module (STM). An STM is the information structure used to support section layer connections in the SDH. It is analogous to STS in the SONET regime. STM consists of information payload and section overhead (SOH) information fields organized in a block frame structure which repeats every 125 μs. The information is suitably conditioned for serial transmission on selected media at a rate which is

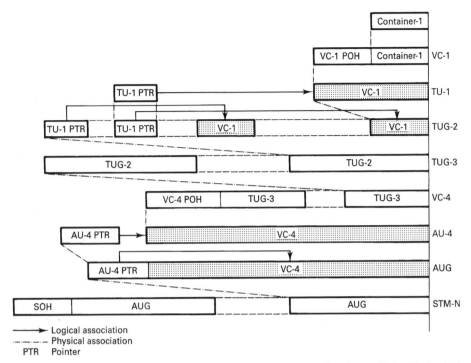

Figure 15.31. Multiplexing method directly from container 1 using AU-4. *Note:* Unshaded areas are phase-aligned. Phase alignment between the unshaded and shaded areas is defined by the pointer (PTR) and is indicated by the arrow. (From Figure 6-2/G.707, page 7, ITU-T Rec. G.707, Ref. 8.)

synchronized to the network. A basic STM (STM-1) is defined at 155,520 kbps. Higher-capacity STMs are formed at rates equivalent to N times multiples of this basic rate. STM capacities for $N = 4$ and $N = 16$ are defined, and higher values are under consideration by ITU-T.

An STM comprises a single administrative unit group (AUG) together with the SOH. STM-N contains N AUGs together with SOH.

Container, C-n ($n = 1$ to $n = 4$). This element is a defined unit of payload capacity which is dimensioned to carry any of the levels currently defined in Section 15.3.2.1 and may also provide capacity for transport of broadband signals which are not yet defined by CCITT (ITU-T organization).

Virtual Container-n (VC-n). A virtual container is the information structure used to support path layer connection in the SDH. It consists of information payload and POH information fields organized in a block frame which repeats every 125 or 500 μs. Alignment information to identify VC-n frame start is provided by the server network layer. Two types of virtual container have been identified:

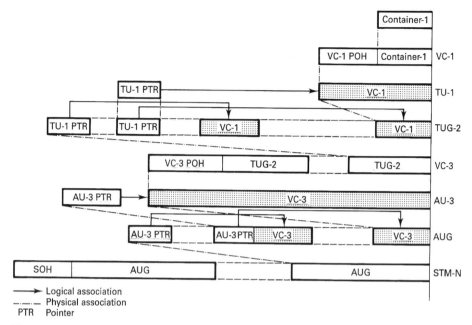

Figure 15.32. SDH multiplexing method directly from container 1 using AU-3. *Note:* Unshaded areas are phase-aligned. Phase alignment between the unshaded and shaded areas is defined by the pointer (PTR) and is indicated by the arrow. (From Figure 6-3/G.707, page 8, ITU-T Rec. G.707, Ref. 8.)

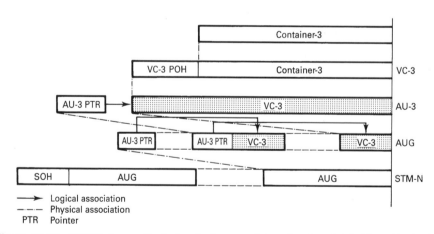

Figure 15.33. Multiplexing method directly from container 3 using AU-3. *Note:* Unshaded areas are phase-aligned. Phase alignment between the unshaded and shaded areas is defined by the pointer (PTR) and is indicated by the arrow. (From Figure 6-4/G.707, page 9, ITU-T Rec. G.707, Ref. 8.)

Lower-order virtual container-n, to VC-n ($n = 1, 2$). This element comprises a single C-n ($n = 1, 2$), plus the basic virtual container POH appropriate to that level.

Higher-order virtual container-n, to VC-n ($n = 3, 4$). This element comprises a single C-n ($n = 3, 4$), an assembly of tributary unit groups (TUG-2s), or an assembly of TU-3s, together with virtual container POH appropriate to that level.

Administrative Unit-n (AU-n). An administrative unit is the information structure which provides adaptation between the higher-order path layer and the multiplex section. It consists of an information payload (the higher-order virtual container) and an administrative unit pointer which indicates the offset of the payload frame start relative to the multiplex section frame start. Two administrative units are defined. The AU-4 consists of a VC-4 plus an administrative unit pointer which indicates the phase alignment of the VC-4 with respect to the STM-N frame. The AU-3 consists of a VC-3 plus an administrative unit pointer which indicates the phase alignment of the VC-3 with respect to the STM-N frame. In each case the administrative unit pointer location is fixed with respect to the STM-N frame (Ref. 9).

One or more administrative units occupying fixed, defined positions in an STM payload are termed an *administrative unit group* (AUG). An AUG consists of a homogeneous assembly of AU-3s or an AU-4.

Tributary Unit-n, TU-n. A tributary unit is an information structure which provides adaptation between the lower-order path layer and the higher-order path layer. It consists of an information payload (the lower-order virtual container) and a tributary unit pointer which indicates the offset of the payload frame start relative to the higher-order virtual container frame start.

The TU-n ($n = 1, 2, 3$) consists of a VC-n together with a tributary unit pointer.

One or more tributary units occupying fixed, defined positions in a higher-order VC-n payload are termed a *tributary unit group* (TUG). TUGs are defined in such a way that mixed-capacity payloads made up of different-sized tributary units can be constructed to increase flexibility of the transport network.

A TUG-2 consists of a homogeneous assembly of identical TU-1s or a TU-2. A TUG-3 consists of a homogeneous assembly of TUG-2s or a TU-3.

Container-n ($n = 1-4$). A container is the information structure which forms the network synchronous information payload for a virtual container. For each of the defined virtual containers there is a corresponding container. Adaptation functions have been defined for many common network rates into a limited number of standard containers. These include standard E-1/DS-1 rates defined in ITU-T Rec. G.702.

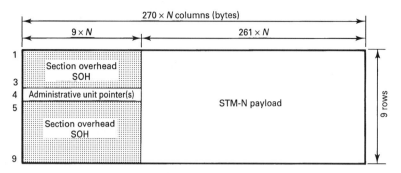

Figure 15.34. STM-*N* frame structure. (From Figure 6-6/G.707, page 11, ITU-T Rec. G.707, Ref. 8.)

15.3.2.3 *Frame Structure.*

The basic frame structure, STM-*N*, is shown in Figure 15.34. The three main areas of the STN-*N* frame are section overhead, AU pointers, and information payload.

Section overhead is shown in rows 1–3 and 5–9 of columns 1–9 × *N* of the STM-*N* in Figure 15.34.

Administrative Unit (AU) Pointers. For columns 1–9 × *N* in Figure 15.34, row 4 is available for AU pointers. The positions of the pointers of the AUs for different organizations of the STM-1 payload are shown in Table 15.5. See ITU-T Rec. G.709 for application of pointers and their detailed specifications.

The following (taken from Ref. 8) summarizes the rules for interpreting the AU-*n* pointers:

1. During normal operation, the pointer locates the start of the VC-*n* within the AU-*n* frame.

2. Any variation from the current pointer value is ignored unless a consistent new value is received three times consecutively or it is preceded by one of the rules 3, 4, or 5. Any consistent new value received three times consecutively overrides (i.e., takes priority over) rules 3 or 4.

3. If the majority of the I bits of the pointer word are inverted, a positive justification operation is indicated. Subsequent pointer values are incremented by one.

4. If the majority of the D bits of the pointer word are inverted, a negative justification operation is indicated. Subsequent pointer values are decremented by one.

5. If the NDF (new data flag) is set to "1001," then the coincident pointer value replaces the current one at the offset indicated by the new pointer value unless the receiver is in a state that corresponds to a loss of pointer.

TABLE 15.5 AU-*n* / TU-3 Pointer (H1, H2, H3) Coding

H1	H2	H3
1 2 3 4 5 6 7 8	9 10 11 12 13 14 15 16	
N N N N S S I D	I D I D I D I D	

|←——— 10-bit pointer value ———→| ↑ Negative justification opportunity   ↑ Positive justification opportunity

I Increment bit
D Decrement bit
N New data flag bit

New data flag
—Enabled "1001"
—Disabled "0110"

Pointer value (bits 7—16)
—Normal range
AU-4, AU-3: 0-782 decimal
TU-3:   0-764 decimal

Negative justification
—Invert 5 D bits
—Accept majority vote

Concatenation indication

—1001SS1111111111
(S bits are unspecified)

SS values	AU-*n* / TU-*n* type
10	AU-4, AU-3, TU-3

Null pointer indication (NPI)

—1001SS1111100000
(S bits are unspecified)

Positive justification
—Invert 5 I bits
—Accept majority vote

[a]NPI value applies only to TU-3 pointers. The pointer is set to all "1"s when an AIS occurs.
Source: Figure 8-3/G.707, page 36, Ref. 8.

Administrative Units in the STM-N. The STM-*N* payload can support *N* AUGs where each AUG may consist of either one AU-4 or three AU-3s.

The VC-*n* associated with each AU-*n* does not have a fixed phase with respect to the STM-*N* frame. The location of the first byte of the VC-*n* is indicated by the AU-*n* pointer. The AU-*n* pointer is in a fixed location in the STM-*N* frame. This is shown in Figures 15.35 and 15.36.

The AU-4 may be used to carry, via the VC-4, a number of TU-*n*s ($n = 1, 2, 3$) forming a two-stage multiplex. An example of this arrangement is illustrated in Figures 15.35a and 15.36a. The VC-*n* associated with each TU does not have a fixed-phase relationship with respect to the start of the VC-4. The TU-*n* pointer is in a fixed location in the VC-4, and the location of the first byte of the VC-*n* is indicated by the TU-*n* pointer.

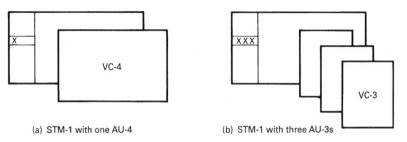

(a) STM-1 with one AU-4   (b) STM-1 with three AU-3s

Figure 15.35. Administrative units in the STM-1 frame, X denotes AU-*n* pointer; AU-*n* denotes AU-*n* pointer + VC-*n*. (From Figure 6-7/G.707, page 12, ITU-T Rec. G.707, Ref. 8.)

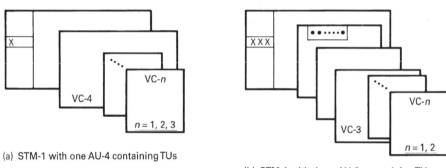

(a) STM-1 with one AU-4 containing TUs

(b) STM-1 with three AU-3s containing TUs

Figure 15.36. Two-stage multiplex. X denotes AU-*n* pointer; ● denotes TU-*n* pointer; AU-*n* denotes AU-*n* pointer + VC-*n*; TU-*n* denotes TU-*n* pointer + VC-*n*. (From Figure 6-8/G.707, page 12, ITU-T Rec. G.707, Ref. 8.)

The AU-3 may be used to carry, via the VC-3, a number of TU-*ns* ($n = 1, 2$) forming a two-stage multiplex. An example of this arrangement is illustrated in Figures 15.35b and 15.36b. The VC-*n* associated with each TU-*n* does not have a fixed-phase relationship with respect to the start of the VC-3. The TU-*n* pointer is in a fixed location in the VC-3, and the location of the first byte of the VC-*n* is indicated by the TU-*n* pointer.

15.3.2.4 *Interconnection of STM-1s.* The SDH has been designed to be universal, allowing transport of a layer variety of signals such as those specified in ITU-T Rec. G.702, including North American 1.544-Mbps and European 2.048-Mbps regimes. However, different structures can be used for the transport of virtual containers. The following interconnection rules

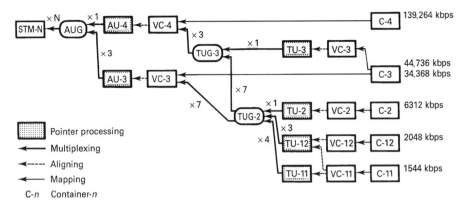

Figure 15.37. SDH multiplexing structure. *Note:* G.702 tributaries associated with containers C-x are shown. Other signals (e.g., ATM) can also be accommodated. (From Figure 6-1/G.707, page 6, ITU-T Rec. G.707, Ref. 8.)

are used:

1. The rule for interconnecting two AUGs based upon two different types of administrative unit, namely AU-4 and AU-3, is to use the AU-4 structure. Therefore, the AUG based upon AU-3 is demultiplexed to the TUG-2 or VC-3 level according to the type of the payload, and then it is remultiplexed within an AUG via the TUG-3/VC-4/AU-4 route.
2. The rule for interconnecting VC-11s transported via different types of tributary unit, namely TU-11 and TU-12, is to use the TU-11 structure. VC-11, TU-11, and TU-12 are described in ITU-T Rec. G.709.

15.3.2.5 Basic SDH Multiplexing Structure. The SDH multiplexing structure is shown in Figure 15.37.

15.4 LINE-OF-SIGHT MICROWAVE AND MILLIMETER-WAVE TRANSMISSION SYSTEMS

15.4.1 Broadband Radio Systems

Radio systems operating in frequency ranges of 3.7 GHz and above generally are assigned some 500 MHz of bandwidth. There are several reasons why one user cannot use the entire 500-MHz "pipe." In most circumstances, full-duplex operation is required. Leaving aside interference considerations, this drops the usage "pipe" width in half to 250 MHz (i.e., 250 MHz in each direction). Further, there is a high probability that there will not be uniform behavior across bandwidths as wide as 250 MHz. In addition, such frequency bands must be shared with other nearby users. Thus usable bandwidths are reduced to 30 or 40 MHz per radio-frequency carrier for line-of-sight microwave and 100 MHz or less for satellite radio systems. Systems operating above 28 GHz have bandwidths of 2.5 GHz, with some bandwidths in the mid-millimeter range over 5 GHz wide. The problem with these higher-frequency systems is cost. There is available technology, but production runs are low because of low demand.

Why use radio at all when fiber-optic systems have almost unlimited coherent bandwidths? Again it is cost. If it is cheaper, with equal perfor- mance and equally dependable, then use it. This proves out very often for line-of-sight microwave radio systems in heavily urbanized areas and in areas of difficult terrain. Typically, the area for the first application would be New York City, and that for the second would be the jungled mountains of Venezuela.

Satellite radio systems can have much broader applications, especially for private networks. Again the driving factor is cost, performance, and depend- ability. In this regard, space segment charges become a cost factor; further- more, the space segment can be a point failure if network availability/survi- vability is a primary consideration, and it may be a drawback.

15.4.2 An Overview of Line-of-Sight Microwave

15.4.2.1 *Line of Sight.* A microwave terminal consists of a radio transmit- ter, a receiver, and an antenna subsystem. The antenna is mounted on a tower and is connected to the transmitter and receiver by means of a transmission line, usually waveguide. Economic factors limit tower height to some 100-m (300-ft) maximum.

This leads to the so-called line-of-sight (LOS) concept. The maximum distance from one microwave terminal to the next is limited by LOS, which is a function of tower height. Radio waves at such high frequencies do not bend around obstacles in the conventional sense; they are either diffracted or blocked entirely. There is, however, some bending which we will explain.

The term "line of sight" gives the connotation that microwave radio rays behave exactly like light waves. One difference is that radiowaves at these microwave frequencies are somewhat bent because of variations in the refractive index of the atmosphere through which they progress. In about 80% of circumstances we can get some 15% more range between microwave terminals than we would get using light as the transmission medium.

A microwave radio ray beam is either diffracted or blocked by an obstacle as we mentioned above. The most common obstacle is the horizon. If there are no other obstacles, which microwave engineers call "smooth earth," then the distance to the *radio* horizon can be calculated for a given antenna tower height (h) by the following formulas:

$$d_{mi} = \sqrt{(2h)}$$

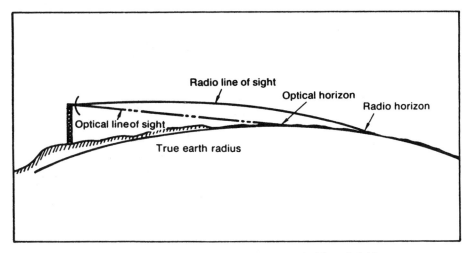

Figure 15.38. Radio line of sight versus optical line of sight.

where d is in statute miles and h, the antenna (tower) height, is measured in feet, and

$$d_{km} = 2.9\sqrt{(2h')}$$

where d is in kilometers and h' is measured in meters (Ref. 2).

If a tower is 100 m high, the radio distance to the horizon is 41 km. This concept is shown in Figure 15.38. If we build a second tower directly in line also 100 m high, the link would be 2×41 km long (i.e., 82 km*), there would be smooth earth, and there would be no other intervening obstacles with standard refraction.

Of course, we can achieve much greater distances if we take advantage of natural and man-made high spots. Use tops of mountains, ridges, cliffs, and tops of high buildings, or lease space on high towers (typically TV towers). Such techniques can notably extend LOS distances.

15.4.2.2 Typical Line-of-Sight Microwave Installation. Figure 15.39 is a simplified conceptual block diagram of a microwave terminal. The transmitter emits an analog carrier wave that, in our case, is modulated digitally. We might transmit an OC-3 configuration (SONET) or STM-1 bit rate of 155 Mbps. How can we get 155 Mbps in a 30- or 40-MHz bandwidth? The trick is similar to those we described in Section 6.3.2, where we squeezed much greater data rates into a voice channel than it appeared it could accommodate. In essence we converted to a higher-level digital signal using what is called *M*-ary modulation techniques often of a hybrid nature. Typical of

*This calculation is only good to a first approximation.

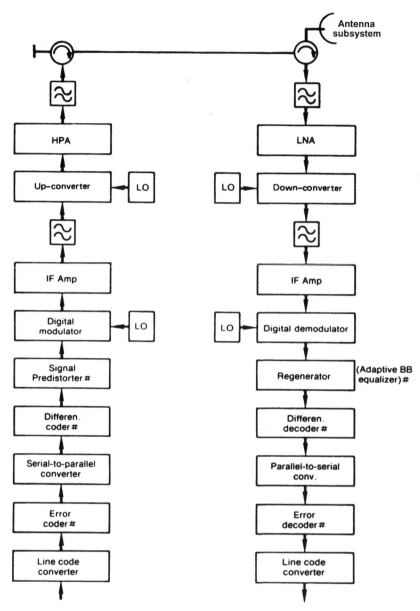

Figure 15.39. A functional block diagram of a digital LOS microwave terminal. Typical PCM line codes: AMI, polar, BNZS, HDB3/CMI. BB, baseband; #, optional feature. (From Ref. 2.)

these was the quadrature amplitude modulation (QAM), which is hybrid in that it combined both amplitude and phase modulation.

Because digital transmission is notoriously wasteful of RF bandwidth, regulatory authorities such as the U.S. Federal Communications Commission (FCC) require a certain minimum *bit packing*. In the case of the FCC, at least 4.5 bits per hertz of bandwidth are required. 64-QAM provides 6 bits/Hz theoretically and about 4.5 bits/Hz practically. Thus 30 MHz can accommodate 30 MHz × 4.5 bps, or 135 Mbps.* For STM-1/STS-3 rate, probably 128-QAM might be more appropriate.

Microwave transmitters generally have just 1-watt (0-dBW) outputs, and 10-watt traveling wave tube (TWT) or solid-state amplifiers (SSA) are available off-the-shelf if required. Often, in the case of millimeter wave installations, transmitter output may be measured in milliwatts.

Nearly all microwave/millimeter wave links suffer from fading. Simplistically, this is the variation in level of the received signal from its free space value. Fading can cause intersymbol interference or insufficient signal-to-noise ratio, both affecting error performance. It can even cause loss of frame alignment. We mitigate effects of fading by overbuilding the link. Here we use larger antennas than required for an unfaded link; or we can increase power output of the transmitter by using a 10-watt HPA; or we can improve receiver noise performance by using low-noise amplifiers (LNAs). There are still other techniques available to the link design engineer such as coding gain devices and diversity.

Because of user congestion in the popular RF band from 1 to 10 GHz, more and more new installations are being forced to use frequencies above 10 GHz. Frequency bands above this 10-GHz arbitrary borderline suffer from excess attenuation (over the free space value) due to heavy (downpour) rainfall. This will impact the time availability of a link. *Time availability* is the decimal value or percentage of time that a particular link meets performance objectives. We can use similar techniques overbuilding a link. Usually this is sufficient to meet an objectives criterion. In practice, in either case, we generally recommend that antennas no greater than 3.5 m (12 ft) be used. The reason, of course, is economic—not so much in antenna cost as in the cost of mechanically stiffening towers to meet twist and sway maximums.† Large parabolic dish antennas act like a big sail surface on top of a tower, and wind places twisting torques on the antenna and tower such that severe fading or circuit outages result.

It is again emphasized that as time progresses, there will be more and more pressure to use the higher frequencies (i.e., > 23 GHz) because of the ever relentless demand for frequencies and greater bandwidth allocation.

---

*Without filter roll-off properties such as the raised cosine filter.
†Also, ray beamwidths get progressively smaller as antenna gains increase, further magnifying the twist and sway situation; it gets easier and easier for the transmit ray beam to fall out of the capture range of the receive antenna, and vice versa.

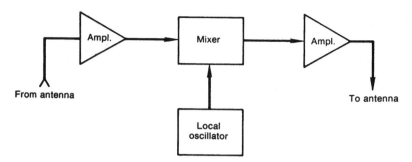

Figure 15.40. A simplified diagram of a satellite communication payload.

There are several high-absorption bands due to the gaseous content of the atmosphere. There is one particularly severe absorption band between 58 and 62 GHz. If we stay away from these absorption bands, then we can either live with the rainfall problem or overbuild a link to mitigate. In this regard, channel coding and interleaving can be very effective, if properly designed. Thus these "higher" frequencies will be particularly attractive even for STS-3/STM-1 and higher SONET/SDH configurations.

15.4.3 Satellite Communications

15.4.3.1 Concept and Applications. Satellite communications is an extension of LOS microwave. In fact, in nearly every case it uses the same frequency bands as LOS microwave. In other words, both services share the same bands, which can be a drawback. Again we deal with 500-MHz bandwidths. This means that the total allocation for the service in a particular band is 500 MHz. One exception to the 500-MHz rule is the 30/20-GHz band which allocates 2.5 GHz of bandwidth.

The nominal 500 MHz of bandwidth is usually broken down into segments of 36 MHz; however, in some cases the segments are 72 MHz and in a few cases they are 100 MHz or slightly greater. A satellite is an RF repeater as shown in Figure 15.40. The RF repeating device is called a *transponder*, and there is a transponder tuned to the center frequency of each frequency segment and with sufficient amplitude response (frequency response) to cover the 36-MHz (or 72-MHz) bandwidth. A typical satellite might have 24 such transponders.

To increase satellite capacity, frequency reuse is implemented. Satellite capacity can be doubled or tripled by this technique, reusing the same frequencies. If there is doubling of capacity by frequency reuse, there is a doubling of the number of transponders. Spot beam antennas with low sidelobes plus good polarization isolation permit frequency reuse, cutting interference from one system into the other down to a reasonable level.

15.4.3.2   Uplinks and Downlinks. An uplink carries the digital/data signal from an earth station to the satellite. In the satellite it is converted in frequency and returned to a distant earth station on a downlink. With only few exceptions, the downlink is always the *weak link*. There are several reasons why, and these reasons give us insight into satellite communications.

On an uplink there are few limitations. We can easily spew all the RF power necessary up into space, provided it is confined to reach only the satellite of interest. There is no limitation on prime power generation and little chance of causing interference, provided that earth station sidelobes are held in check and the angular spread of the ray beam is contained. RF power not used by the satellite in its receiving system spreads out into space, which harms no one.

Quite different conditions hold for the downlink. The downlink power must be limited for two important reasons. First, the satellite spews RF power onto the earth's surface in a band shared by LOS microwave. If the power level is too great, harmful interference will occur on LOS links encompassed by the satellite ray beam. Thus downlink power must be limited to limit interference on LOS microwave service. Second, downlink power is also very expensive for three reasons: (i) Prime power on the satellite is generated by solar cells; that is, more power requires more solar cells. (ii) If more power is required, larger batteries are needed for darkness periods and solar eclipse. (iii) If more power is needed, larger components (such as power supplies) are required on the satellite, and thus more weight is implied. The cost of launching a satellite is really a function of its weight (Ref. 2).

Figure 15.41 illustrates a satellite connectivity with uplinks and downlinks.

15.4.3.3   Satellite Orbits. Satellites are in elliptical orbits. As we remember, a circle is a special form of an ellipse, and most satellites used for commercial communications in the western world are in circular orbits 35,786 km (approximately 22,254 miles) above the equator. These orbits are called

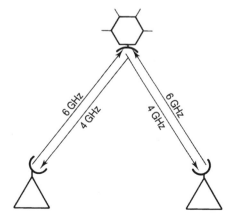

Figure 15.41. A satellite connectivity showing uplinks and downlinks.

geostationary where the satellite appears fixed over a point on the earth's surface. The primary advantage of such an orbit is that earth station antenna tracking in many cases is not required. The major drawback is the resulting propagation delay up to the satellite and down again, some 125 ms each way. Secondarily is the increased free space loss incurred by the large distance.

Another class of satellites used for telecommunication are the low earth orbit (LEO) satellites such as the Iridium system operated by Motorola. Here the altitude of the satellite is around 500 km, vastly reducing the delay and reducing free space loss accordingly. Such satellites ordinarily would require tracking from earth stations; however, quasi-omnidirectional antennas are used on the satellites as well as on the earth terminals. The primary market for Iridium is worldwide cellular and personal communication services (PCS). Another is the TELEDESIC system with hundreds of LEO satellites to be used for the specialized fixed-service data transport market.

15.4.3.4 Digital Satellite System Operation and Access. There are three generic access techniques available for satellite communications. By access we mean how we can use these satellite resources. Such generic techniques are:

- Frequency division multiple access (FDMA)
- Time division multiple access (TDMA)
- Code division multiple access (CDMA)

FDMA. An uplink user is assigned a frequency segment inside a transponder bandwidth. Other users are assigned other segments. There are guard bands between contiguous segments. These user frequency slots or segments are transmitted uplink to the satellite where there is a frequency translation and the segments are passed onwards on the downlink for reception at distant earth stations. At a distant earth station, only those segments are removed that are destined for it. The FDMA concept is shown in Figure 15.42a. Each frequency slot or segment may carry a frequency division multiplex configuration for an analog network, a digital configuration for digital network interconnectivity, or a pure data configuration for private data network connectivity. A segment, for instance, might be 10 MHz wide and be used for LAN interconnectivity. The modulation in this case might be quadrature phase-shift keying (QPSK), easily accommodating a 10-Mbps data signal.

TDMA. This access technique operates in the time domain where each user is assigned a time slot, rather than a frequency slot as in FDMA. Each user, in his/her assigned time slot, uses the entire transponder bandwidth with a data burst for the time slot duration. A user bursts and waits for his/her next turn, bursts and waits. Thus TDMA only operates in the digital domain where traffic storage is practical to permit the "burst and wait" operation. TDMA time slots can be varied in their duration in accordance with traffic demand. Figure 15.42b is a conceptual diagram of TDMA.

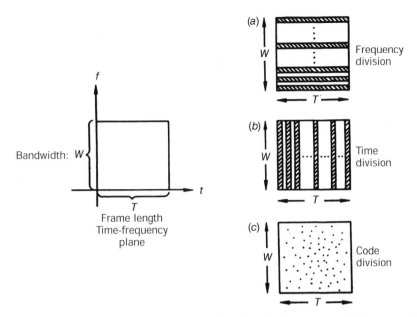

Figure 15.42. Conceptual drawings of (*a*) FDMA, (*b*) TDMA, and (*c*) CDMA in the time-frequency plane. (From Ref. 1.)

CDMA. Code division multiple access uses one of two forms of spread spectrum wave form: frequency hop (FH) or direct sequence (DS). Direct sequence is also called *pseudo-noise*. In either case, users can be stacked one on the other using the same center frequency. For FH each user has a different hopping pattern; for DS each user has a different sequence determined by a *key variable*. It can be argued that spread spectrum systems allow more users per unit bandwidth than does TDMA or FDMA. Figure 15.42c is a conceptual diagram of CDMA. Spread spectrum systems are ideally suited for digital data transmission.

VSAT Networks. There is one very special and particularly attractive application of satellite communications for the data end user for a wide area network (WAN). This is the very-small-aperture terminal (VSAT) approach. VSAT networks were discussed in Chapter 11, Section 11.7.

REFERENCES

1. Roger L. Freeman, *Reference Manual for Telecommunications Engineering*, 3rd edition, John Wiley & Sons, New York, 2001.
2. Roger L. Freeman, *Telecommunications Transmission Handbook*, 4th edition, John Wiley & Sons, New York, 1998.
3. *Design Handbook for Fiber Optic Communications Systems*, MIL-HDBK-415, U.S. Dept. of Defense, Washington DC, 1985.

4. *The IEEE Standard Dictionary of Electrical and Electronics Terms*, 6th edition, IEEE, New York, 1996.

5. *Synchronous Optical Network (SONET) Common Generic Criteria*, GR-253-CORE, Issue 1, Telcordia/Bellcore, Morristown, NJ, 1994.

6. *Synchronous Optical Network (SONET) Transport Systems: Common Generic Criteria*, TR-TSY-GR-253-CORE, Issue 2, Revision 2, Telcordia/Bellcore, Morristown, NJ, Jan. 1999.

7. *Introduction to SONET*, an H-P seminar, Hewlett-Packard Company, Burlington, MA, October 1993.

8. *Synchronous Digital Hierarchy Bit Rates*, ITU-T Rec. G.707, ITU Geneva, March 1996.

9. *Network Node Interface for the Synchronous Digital Hierarchy*, ITU-T Rec. G.708, ITU Geneva, March 1993.

10. Curtis A. Siller and Mansoor Shafi, *SONET/SDH A Sourcebook of Synchronous Networking*, IEEE Press, New York, 1996.

16

BROADBAND ISDN (B-ISDN)
AND THE ASYNCHRONOUS
TRANSFER MODE (ATM)

16.1 WHERE ARE WE GOING?

Frame relay (Chapter 12) was the beginning of the march toward an optimized* format for multimedia transmission (voice, data, video, facsimile).
There were new and revolutionary concepts in frame relay. There was a
trend toward simplicity where the header was considerably shortened. A
header is pure overhead, so it was cut back as much as practically possible.
A header also implies processing. By reducing the processing, delivery time
could be speeded up. Frame relay is a low-latency system.

In the effort to speed up delivery, operation was unacknowledged (at least
at the frame relay level); there was no operational error correction scheme. It
was unnecessary because it was assumed that the underlying transport system
had excellent error performance (better than 1×10^{-9}). There was error
detection for each frame, and a frame found in error was thrown away. Now
that is something that we never did for those of us steeped in old-time data
communication. It is assumed that the higher OSI (open system interconnection) layers would request repeats of the few frames missing (i.e., thrown
away).

Frame relay also moved into the flow control arena with the BECN and
FECN bits and the CLLM. The method of handling flow control has a lot to
do with its effectiveness in this case. It also uses a discard eligibility (DE) bit
that set a type of priority to a frame. If the DE bit were set, the frame would
be among the first to be discarded in a time of congestion.

The distributed queue dual bus (DQDB) developed by the IEEE 802.6
committee provides a simple and unique access scheme. Even more impor-

---

Compromise might be a more appropriate word.

tant, its data transport format is based on the *cell* that the IEEE calls a *slot*. The DQDB slot or cell has a format very similar to that of the ATM cell, which we discuss at length in this chapter. It even has the same number of octets (i.e., 53), 48 of which is the payload capacity. This is identical to ATM. Also enter a comparatively new concept of the header check sequence (HCS) for detecting errors in the header. The DQDB has no error detection for the body or info portion of the slot or cell. There is a powerful CRC32 in the trailer of the IMPDU (initial MAC protocol data unit), an upper layer of DQDB. It also employs the BOM, COM, and EOM (beginning of message, continuation of message, and end of message, respectively) as well as the message identifier (MID) in the segmentation and reassembly (SAR) process. These fields are particularly helpful in message reassembly.

Switched multimegabit data service (SMDS) came on the scene using DQDB as its access protocol. DQDB uses a 5-octet ACF/header and a 48-octet payload with no trailer, which is very similar to the ATM cell. SMDS has a 44-octet payload, a 7-octet header, and a 2-octet trailer, for a total of 53 octets in what we will call a cell. Many of the strategies of DQDB followed on down to SMDS. However, SMDS has a credit manager at the switching system (SS) which polices users. This policing concept came from frame relay. It is carried onward into ATM.

We consider ATM to be a prime candidate to transport data. Data traffic is bursty, and the short cells, which are the basic transport format for ATM, can handle short bursts very efficiently. IP and frame relay, because of their comparatively long frames, are less amenable to short burst traffic. It is most attractive to the user when comparatively high speed circuits are available such as DS3 or E3 (see Chapter 7). Service can also be provided down to DS1 and E1 circuits. Reference 1 states that ATM tends to be expensive.

Broadband-ISDN (B-ISDN) provides the framework for the transport of ATM cells. Its nomenclature is similar to that of ISDN (Chapter 13). Broadband is not a very well-defined term. In this context it is any digital channel with a data rate greater than DS0 (56 or 64 kbps) or E0 (64 kbps). Certainly the ISDN H0 channel falls in this category with its 384-kbps date rate. We like to think of B-ISDN being that transport framework starting at the ITU primary rate of 1.544 and 2.048 Mbps.

16.2   INTRODUCTION TO ATM

ATM is based on a 53-octet cell. A typical cell has a 5-octet header and a 48-octet information field. All traffic is transported in the information field of that cell, whether voice, data, or image.

As we discussed, ATM is an outgrowth of the several data transmission format systems covered previously. Admittedly, some may argue this point. ATM cells can be mixed and matched as shown in Figure 16.1, where a cell carrying a data payload may be contiguous with another cell carrying voice. ATM cells can be transported on SONET, SDH (Chapter 15), E1/DS1

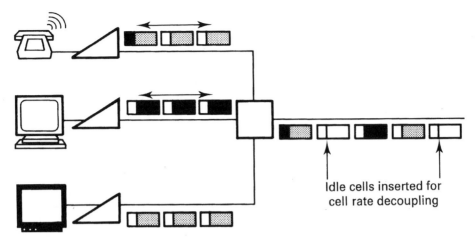

Figure 16.1. ATM links simultaneously carry a mix of voice, data, and image information.

(Chapter 7), and other popular digital formats. Cells can also be transported contiguously without an underlying digital network format.

Philosophically, voice and data must be handled differently regarding time sensitivity. Voice cannot wait for long processing and ARQ delays. Most types of data can. So ATM must distinguish the type of service such as constant bit rate (CBR) and variable bit rate (VBR) services. Voice service is typical constant bit rate or CBR service.

Signaling is another area of major philosophical difference. In data communications, "signaling" is carried out within the header of a data frame (or packet). As a minimum the signaling will have the destination address and quite often the source address as well. And this signaling information will be repeated over and over again on a long data file that is heavily segmented. On a voice circuit, a connectivity is set up, and the destination address, and possibly the source address, are sent just once during call setup. There is also some form of circuit supervision to keep the circuit operational throughout the duration of a telephone call. ATM is a compromise, stealing a little from each of these separate worlds.

Like voice telephony, ATM is fundamentally a connection-oriented telecommunication system. Here we mean that a connection must be established between two stations before data can be transferred between them. An ATM connection specifies the transmission path, allowing ATM cells to self-route through an ATM network. Being connection-oriented also allows ATM to specify a guaranteed quality of service (QoS) for each connection.

By contrast, most LAN protocols are connectionless. This means that LAN nodes simply transmit traffic when they need to, without first establishing a specific connection or route with the destination node.

In that ATM uses a connection-oriented protocol, bandwidth is allocated only when the originating end user requests a connection. This allows ATM to efficiently support a network's aggregate demand by allocating bit rate capacity on demand based on immediate user need. Indeed it is this concept which lies in the heart of the word *asynchronous*. An analogy would help. New York City is connected to Washington, DC, with a pair of railroad tracks for passenger trains headed south and another pair of tracks for passenger trains headed north. On those two pairs of tracks we would like to accommodate everybody we can when they would like to ride. The optimum for reaching this goal is to have a continuous train of coupled passenger cars. As the train enters Union Station, it disgorges its passengers and connects around directly for the northward run to Pennsylvania Station. Passenger cars are identical in size, and each has the same number of identical seats.

Of course at 2 A.M. the train will have very few passengers and many empty seats. Probably from 7 to 9 A.M. the train will be full, no standees allowed, so we will have to hold potential riders in the waiting room. They will ride later; those few who try to be standees will be bumped. Others might seek alternate transportation to Washington, DC.

Here we see that the railroad tracks are the transmission medium. Each passenger car is a SONET/SDH frame. The seats in each car are our ATM cells. Each seat can handle a person no bigger than 53 units. Because of critical weight distribution, if a person is not 53 units in size/weight, we will place some bricks on the seat to bring the size/weight to 53 units exactly. Those bricks are removed at the destination. All kinds of people ride the train because America is culturally diverse, analogous to the fact that ATM handles all forms of traffic. The empty seats represent idle or unassigned cells. The header information is analogous to the passengers' tickets. Keep in mind that the train can only fill to its maximum capacity of seats. We can imagine the SONET/SDH frame as being full of cells in the payload, some cells busy and some idle/unassigned. At the peak traffic period, all cells will be busy, and some traffic (passengers) may have to be turned away.

We can go even further with this analogy. Both Washington, DC, and New York City attract large groups of tourists, and other groups travel to business meetings or conventions. A tour group has a chief tour guide in the lead seat (cell) and an assistant guide in the last seat (cell). There may be so many in the group that they extend into a second car or may just intermingle with other passengers on the train. The tour guide and assistant tour guide keep an exact count of people on the tour. The lead guide wears a badge that says BOM, all tour members wear badges that say COM, and the assistant tour guide wears a badge that says EOM.

Each group has a unique MID (message ID). We also see that service is connection-oriented (Washington, DC, to New York).

Asynchronous means that we can keep filling the seats on the train until we reach its maximum capacity. If we look up the word, it means *nonperiodic*, whereas the familiar E-1/T-1 are periodic (i.e., synchronous). One point that seems to get lost in the literature is that the train has a maximum capacity. Thus the concept "bandwidth on demand" is that we can use the "bandwidth" until we fill to rated capacity. Again we have the unfortunate use of the word *bandwidth*, because our capacity will be measured in octets, not hertz. For example, SONET's STS-1 has a payload capacity of 87×9 octets (see Chapter 15), not 87×9 Hz.

16.3 USER–NETWORK INTERFACE (UNI), CONFIGURATION, AND ARCHITECTURE

ATM is the underlying message format of B-ISDN. At times in this section, we use the terms ATM and B-ISDN interchangeably. Figures 16.2 and 16.3 interrelate the two. Figure 16.2 relates B-ISDN access reference configuration with the ATM user–network interface (UNI). Figure 16.3 is the traditional ITU-T Rec. I.121 (Ref. 2) B-ISDN protocol reference model showing the extra layer necessary for switched service. There is absolutely no relationship between this model and the OSI reference model.

The ATM Forum (Ref. 3) provides the following definitions applicable to Figure 16.3:

U-Plane. The user plane provides for the transfer of user application information. It contains physical layer, ATM layer, and multiple ATM adaptation layers required for different service users such as constant bit rate service (CBR) and variable bit rate service (VBR).

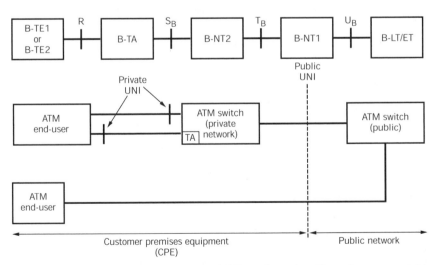

Figure 16.2. B-ISDN user–network interface (UNI) configuration. The reference model is at the top of the drawing. Note its similarity to the ISDN model, Figure 13.2. The "R" reference point is where nonstandard ISDN equipment is made B-ISDN-compatible via B-TA. In this case, the TA functionality is limited to physical layer conversion.

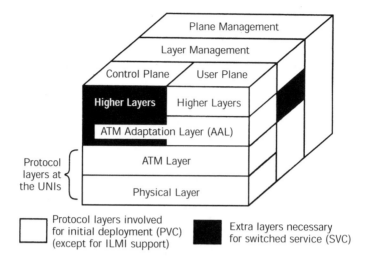

Figure 16.3. B-ISDN protocol reference model.

C-Plane. The control plane protocols deal with call establishment and call release and other connection control functions necessary for providing switched services. The C-plane structure shares the physical and ATM layers with the U-plane, as shown in Figure 16.3. It also includes ATM adaptation layer (AAL) procedures and higher-layer signaling protocols.

M-Plane. The management plane provides management functions and the capability to exchange information between the U-plane and the C-plane. The M-plane contains two sections: layer management and plane management. Layer management performs layer-specific management functions, while the plane management performs management and coordination functions related to the complete system.

We return to Figure 16.3 and B-ISDN/ATM layering and layer descriptions in Section 16.6.

16.4   THE ATM CELL: KEY TO OPERATION

16.4.1   ATM Cell Structure

The ATM cell consists of 53 octets; 5 of these make up the header, and 48 octets carry the payload or "info" portion of the cell. The basic structure is shown in Figure 16.4, where contiguous cells flow on a circuit from left to right. Figure 16.5 shows the detailed structure of the cell header at the UNI.

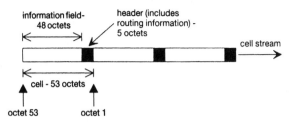

Figure 16.4. Basic ATM cell structure is shown where a stream of contiguous cells flow from left to right.

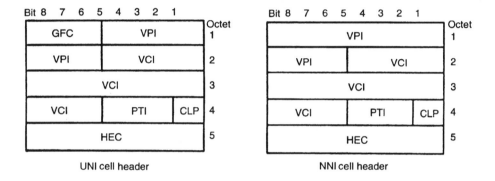

Figure 16.5. Header structure at the UNI (left); header structure at the NNI (right). GFC, generic flow control; VPI, virtual path identifier; VCI, virtual channel identifier; PTI, payload type; CLP, cell loss priority, HEC, header error control.

Now let us examine each of the bit fields that make up the ATM cell header.

Generic Flow Control (GFC). These first 4 bits are either part of the VPI, if the cells is traveling between two switches, or make up a reserved field if the cell is traveling between an end-node and a switch. In this case it has local significance only and can be used to provide standardized local flow control function on the customer site.

VPI / VCI Fields (Routing Field). Twenty-four bits are available for routing a cell from a customer site. There are 8 bits for virtual path identifier (VPI) and 16 bits for virtual channel identifier (VCI). Preassigned combinations of VPI and VCI values are given in Table 16.1. The VCI value of zero is not available for user virtual channel identification. The bits within the VPI and

TABLE 16.1 Combinations of Preassigned VP, VCI, PT, and CLP Values at the UNI

Use	VPI	VCI (Note 8)	PTI	CLP
Unassigned cell	00000000	00000000 00000000	Any value	0
Invalid	Any VPI value other than 0	00000000 00000000	Any value	B
Metasignaling (see Rec. I.311)	XXXXXXXX (Note 1)	00000000 00000001 (Note 5)	0AA	C
General broadcast signaling (see Rec. I.311)	XXXXXXXX (Note 1)	00000000 00000010 (Note 5)	0AA	C
Point-to-point signaling (see Rec. I.311)	XXXXXXXX (Note 1)	00000000 00000101 (Note 5)	0AA	C
Segment OAM F4 flow cell (see Rec. I.610)	Any VPI value	00000000 00000011 (Note 4)	0A0 (Note 11)	A
End-to-end OAM F4 flow cell (see Rec. I.610)	Any VPI value	00000000 00000100 (Note 4)	0A0 (Note 11)	A
VP resource management cell (see Rec. I.371)	Any VPI value	00000000 00000110 (Note 10)	110 (Note 9)	A
Reserved for future VP functions (Note 6)	Any VPI value	00000000 00000111 (Note 10)	0AA (Note 11)	A
Reserved for future functions (Note 7)	Any VPI value	00000000 000SSSSS (Notes 2 and 10)	0AA	A
Reserved for future functions (Note 7)	Any VPI value	00000000 000TTTTT (Note 3)	0AA	A
Segment OAM F5 flow cell (see Rec. I.610)	Any VPI value	Any VCI value other than 00000000 00000000, 00000000 00000011, 00000000 00000100, 00000000 00000110, or 00000000 00000111	100	A
End-to-end OAM F5 flow cell (see Rec. I.610)	Any VPI value	Any VCI value other than 00000000 00000000, 00000000 00000011, 00000000 00000100, 00000000 00000110, or 00000000 00000111	101	A
VC Resource management cell (see Rec. I.371)	Any VPI value	Any VCI value other than 00000000 00000000, 00000000 00000011, 00000000 00000100, 00000000 00000110, or 00000000 00000111	110	A
Reserved for future VC functions	Any VPI value	Any VCI value other than 00000000 00000000, 00000000 00000011, 00000000 00000100, 00000000 00000110, or 00000000 00000111	111	A

The GFC field is available for use with all of these combinations.
A Indicates that the bit may be 0 or 1 and is available for use by the appropriate ATM layer function.
B Indicates the bit is a "don't care" bit
C Indicates the originating entity shall set the CLP bit to 0. The value may be changed by the network.

Tables notes continued on page 609

TABLE 16.1 (*Continued*)

Note 1: XXXXXXXX: Any VPI value. For VPI value equal to 0, the specific VCI value specified is reserved for user signaling with the local exchange. For VPI values other than 0, the specified VCI value is reserved for signaling with other signaling entities (e.g., other users or remote networks).
Note 2: SSSSS: Any value from 01000 to 01111.
Note 3: TTTTT: Any value from 10000 to 11111.
Note 4: Transparency is not guaranteed for the OAM F4 flows in a user-to-user VP.
Note 5: The VCI values are preassigned in every VPC at the UNI. The usage of these values depends on the actual signaling configurations. (See Rec. I.311.)
Note 6: VCI value is reserved to provide the same function for VPs as PTI 111 is reserved to provide for VCs.
Note 7: These VCI values are reserved for future standardization for specific functions.
Note 8: Cells with VCI values 1, 2, 5, 16 through 31, and greater than 31 are monitored by the VP OAM function. Cells with other VCI values are not monitored by the VP OAM function. Cells with other VCI values are not monitored by the VP OAM function. (See Rec. I.610 and Section 16.13.) Whether a cell with a particular VCI value is conveyed transparently between the endpoints of the VPC is described in 3.1.4.1e/I.150.
Note 9: This specifies the allowed coding of the PTI field on transmission. This VCI value shall only be used for the stated functions regardless of the coding of the PTI field. It is an implementation option on how to process errored cells received with VCI = 6 and PTI not equal to 110. In particular, such cells may be processed as VP RM cells.
Note 10: Transparency of these VCI values is not guaranteed; that is, cells with these VCI values may be extracted or inserted at midpoints of a VP. The specific situations under which this may occur are for further study. In the absence of this further study these VCIs shall be transparently transported in a VP.
Note 11: This specifies the allowed coding of the PTI field on transmission. These VCI values shall only be used for the stated functions regardless of the coding of the PTI field. On reception, the PTI field is not used for the purpose of identifying the cell type. For example, a cell with VC = 4 will be treated as an End-to-end F4 OAM cell regardless of the coding of the PTI field.
Source: Table 4/I.361, page 8, ITU-T Rec. I.361 (11/95), Ref. 4.

VCI fields used for routing are allocated using the following rules:

- The allocated bits of the VPI field are contiguous.
- The allocated bits of the VPI field are the least significant bits of the VPI field, beginning at bit 5 of octet 2.
- The allocated bits of the VCI field are contiguous.
- The allocated bits of the VCI field are the least significant bits of the VCI field, beginning at bit 5 of octet 4.

Comment: Consider a very high-speed backbone carrying millions of messages. The split between VPI and VCI saves routers in the backbone from requiring that their call-mapping database keep track of millions of individual calls. In this situation, the backbone routers use only the VPI portion of the call identifier. Because thousands of VCs might be going on the same VP, switches inside can treat all the VCs for that VP as a unit.

Outside the backbone, the switches treat the VPI/VCI together as one combined field. The term *VP-switching* refers to switches that are looking at only the VPI portion of the field. *VC-switching* refers to switches that are looking at the entire field.

Payload Type (PT) Field. The first bit = 0 indicates data rather than ATM control information. For data, the middle bit indicates congestion experienced, and the last bit is used by AAL5 (see Section 16.6.2.2) to indicate last cell of a packet. Payload Type Identifier (PTI) values are given in Table 16.2. The main purpose of the PTI is to discriminate between user cells (i.e., cells carrying user information) and nonuser cells. The first four code groups in the table (000-011) are used to indicate user cells. Within these four, 2, and 3 (010 and 011) are used to indicate congestion experienced. The fifth and sixth code groups (100 and 101) are used for VCC-level management functions.

Any congested network element, upon receiving a user data cell, may modify the PTI as follows: Cells received with PTI = 000 or PTI = 010 are transmitted with PTI = 010. Cells received with PTI = 001 or PTI = 011 are transmitted with PTI = 011. Noncongested network elements should not change the PTI.

Cell Loss Priority (CLP) Field. Depending on network conditions, cells where the CLP is set (i.e., CLP value is 1) are subject to discard prior to cells where the CLP is not set (i.e., CLP value is 0). The concept here is identical with that of frame relay and the DE bit. ATM switches may tag CLP = 0 cells detected by the usage parameter control (UPC) to be in violation of the traffic contract by changing the CLP bit from 0 to 1.

Header Error Control (HEC) Field. The HEC is an 8-bit field, and it covers the entire cell header. The code used for this function is capable of either

TABLE 16.2   PTI Coding

	PTI Coding	Interpretation
Bits	4 3 2	
	0 0 0	User data cell, congestion not experienced. ATM-user-to-ATM-user indication = 0
	0 0 1	User data cell, congestion not experienced. ATM-user-to-ATM-user indication = 1
	0 1 0	User data cell, congestion experienced. ATM-user-to-ATM-user indication = 0
	0 1 1	User data cell, congestion experienced. ATM-user-to-ATM-user indication = 1
	1 0 0	OAM F5 segment associated cell
	1 0 1	OAM F5 end-to-end associated cell
	1 1 0	Resource management cell
	1 1 1	Reserved for future functions

Source: Table appearing on page 6 in Section 2.2.4, ITU-T Rec. I.361 (11/95), Ref. 4.

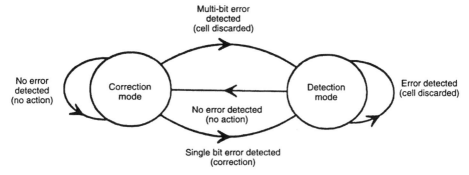

Figure 16.6. HEC receiver modes of operation.

single-bit error correction or multiple-bit error detection. Briefly, the transmitting side computes the HEC value. The receiver has two modes of operation, as shown in Figure 16.6. In the default mode there is the capability of single-bit error correction. Each cell header is examined; if an error is detected, one of two actions takes place. The action taken depends on the state of the receiver. In the *correction mode*, only single-bit errors can be corrected and the receiver switches to the *detection mode*. In the "detection mode," all cells with detected header errors are discarded. When a header is examined and found not to be in error, the receiver switches to the "correction mode." The term *no action* in Figure 16.6 means that no correction is performed and no cell is discarded.

Figure 16.7 is a flow chart showing the consequences of errors in the ATM cell header. The error protection function provided by the HEC provides for (a) recovery from single-bit errors and (b) a low probability of delivery of cells with errored headers under bursty error conditions. ITU-T Rec. I.432.1 (Ref. 5) states that error characteristics of fiber-optic transmission systems appear to be a mix of single-bit errors and relatively large burst errors. Thus, for some transmission systems the error correction capability might not be invoked.

16.4.2 Header Error Control Sequence Generation

The transmitter calculates the HEC value across the entire ATM cell header and inserts the result in the appropriate header field.

The notation used to describe header error control is based on the property of cyclic codes. For example, code vectors such as 1000000100001 (it has 13 elements or bits, count them) can be represented by a polynomial $P(x) = X^{12} + X^5 + 1$. The elements of an n-element code word are thus the coefficients of a polynomial of the order $n - 1$. In this application, these coefficients can have the value 0 or 1 and the polynomial operations are performed using modulo 2 operations. The polynomial representing the

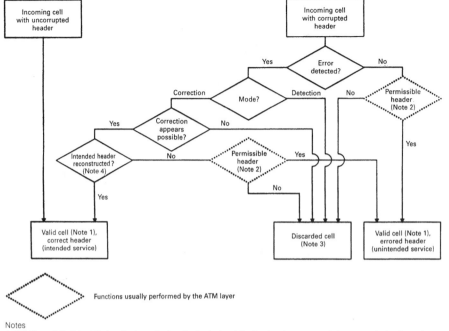

Figure 16.7. Consequences of errors in an ATM cell header. (From Figure 4/I.432.1, page 4, ITU-T Rec. I.432.1, Ref. 5.)

content of a header excluding the HEC field is generated using the first bit of the header as the coefficient of the highest-order term.

The HEC field is an 8-bit sequence. It is the remainder of the division (modulo 2) by the generator polynomial $X^8 + X^2 + X + 1$ of the product X^8 multiplied by the content of the header excluding the HEC field.

At the transmitter, the initial content of the register of the device computing the remainder of the division is preset to all 0s and is then modified by division of the header excluding the HEC field by the generator polynomial (as described above); the resulting remainder is transmitted as the 8-bit HEC.

To improve the cell delineation performance in the case of bit-slips, the following is recommended:

- The check bits calculated by the use of the check polynomial are added (modulo 2) to an 8-bit pattern before being inserted in the last octet of the header.
- The recommended pattern is "01010101" (the left bit is the most significant bit).

TABLE 16.3   Header Pattern for Idle Cell Identification

	Octet 1	Octet 2	Octet 3	Octet 4	Octet 5
Header pattern	00000000	00000000	00000000	00000001	HEC = Valid code 01010010

Note 1: The content of the information field is "01101010" repeated 48 times.
Note 2: There is no significance to any of these individual header fields from the point of view of the ATM layer, as idle cells are not passed to the ATM layer.
Source: Table 3/I.432.1, page 11, ITU-T Rec. I.432.1 (08/98), Ref. 5.

- The receiver must subtract (equal to add modulo 2) the same pattern from the 8 HEC bits before calculating the syndrome of the header.

This operation in no way affects the error detection/correction capabilities of the HEC.

As an example, if the first 4 octets of the header were all zeros, the generated header before scrambling would be "00000000 00000000 00000000 00000000 01010101." The starting value for the polynomial check is 0s (binary).

16.4.3   Idle Cells

Idle cells cause no action at a receiving node except for cell delineation including HEC verification. Idle cells are inserted and discarded for cell rate decoupling. Idle cells are identified by the standardized pattern for the cell header shown in Table 16.3.

16.5   CELL DELINEATION AND SCRAMBLING

16.5.1   Delineation and Scrambling Objectives

Cell delineation allows identification of cell boundaries. In other parts of this book we simply called this *alignment*. The cell HEC field achieves cell delineation. Keep in mind that the ATM signal must be self-supporting in that it has to be transparently transported through every network interface without any constraints from the transmission systems used.

Scrambling is used to improve security and robustness of the HEC cell delineation mechanism discussed below. In addition, it helps the randomizing of data in the information field for possible improvement in transmission performance.

Any scrambling operation must not alter the ATM header structure, header error control, or cell delineation algorithm.

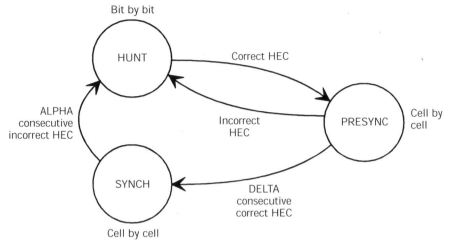

Figure 16.8. Cell delineation state diagram. *Note:* The "correct HEC" means that the header has no bit error (syndrome is zero) and has not been corrected. (From Figure 5/I.432.1, page 6, ITU-T Rec. I.432.1, Ref. 5.)

16.5.2 Cell Delineation Algorithm

Cell delineation is performed by using the correlation between the header bits to be protected (32 bits or 4 octets) and the HEC octet, which are the relevant control bits (8 bits) introduced in the header using a shortened cyclic code with the generating polynomial $X^8 + X^2 + X + 1$.

Figure 16.8 shows the state diagram of HEC cell delineation method. A discussion of the figure follows:

1. In the HUNT state, the delineation process is performed by checking bit by bit for the correct HEC (i.e., syndrome equals zero) for the assumed header field. For the cell-based* physical layer, prior to scrambler synchronization, only the last 6 bits of the HEC are used for cell delineation checking. For the SDH-based interface, all 8 bits are used for acquiring cell delineation. Once such an agreement is found, it is assumed that one header has been found, and the method enters the PRESYNC state. When octet boundaries are available within the receiving physical layer prior to cell delineation as with the SDH-based interface, the cell delineation process may be performed octet by octet.

2. In the PRESYNC state, the delineation process is performed by checking cell by cell for the correct HEC. The process repeats until the correct HEC has been confirmed *Delta* times consecutively. If an incorrect HEC is found, the process returns to the HUNT state.

---

*Only cell-based and SDH-based interfaces are covered by current ITU-T recommendations. Besides these, we cover cells riding on other transport means at the end of this chapter.

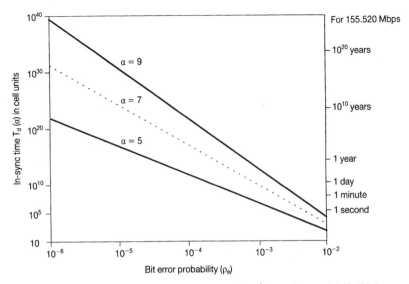

Figure 16.9. In-sync time versus bit error probability. (From Figure I.1/I.432.1, page 13, ITU-T Rec. I.432.1, Ref. 5.)

3. In the SYNCH state, the cell delineation will be assumed to be lost if an incorrect HEC is obtained *Alpha* times consecutively.

The parameters *Alpha* and *Delta* are chosen to make the cell delineation process as robust and secure as possible while satisfying QoS requirements. Robustness depends on *Alpha* when it is against false misalignments due to bit errors. And robustness depends on *Delta* when it is against false delineation in the resynchronization process.

For the SDH-based physical layer, values of *Alpha* = 7 and *Delta* = 6 are suggested by the ITU-T organization (Rec. I.432, Ref. 5); and for cell-based physical layer, values of *Alpha* = 7 and *Delta* = 8. Figures 16.9 and 16.10 give performance information of the cell delineation algorithm in the presence of random bit errors, for various values of *Alpha* and *Delta*.

16.6   ATM LAYERING AND B-ISDN

The B-ISDN reference model is given in Figure 16.3, and its several planes are described therein. We reiterate that B-ISDN layering has no relationship whatsoever with the OSI reference model.

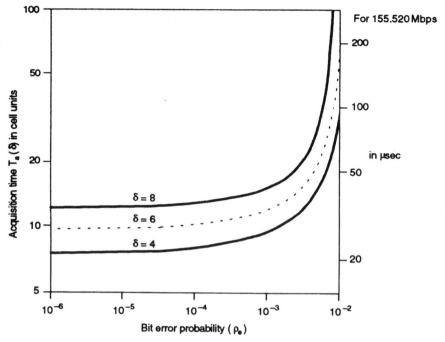

Figure 16.10. Acquisition time versus bit error probability. (From Figure I.2/I.432.1, page 14, ITU-T Rec. I.432.1, Ref. 5.)

16.6.1 Functions of Individual ATM/B-ISDN Layers

Figure 16.11 illustrates B-ISDN/ATM layering and sublayering of the protocol reference model. It identifies the functions of the physical layer, the ATM layer and the AAL and related sublayers.

16.6.1.1 Physical Layer. The physical layer consists of two sublayers. The physical medium (PM) sublayer includes only PM-dependent functions. The transmission convergence (TC) sublayer performs all functions required to transform a flow of cells into a flow of data units (i.e., bits) which can be transmitted and received over a physical medium. The service data unit (SDU) crossing the boundary between the ATM layer and the physical layer is a flow of valid cells. The ATM layer is unique (meaning independent of the underlying physical layer). The data flow inserted in the transmission system payload is PM-independent and self-supported. The physical layer merges the ATM cell flow with the appropriate information for cell delineation, according to the cell delineation mechanism described above, and carries the operations and maintenance (OAM) information relating to this cell flow.

	Higher layer functions	Higher layers	
	Convergence	CS	AAL
	Segmentation and reassembly	SAR	
	Generic flow control Cell header generation/extraction Cell VPI/VCI translation Cell multiplex and demultiplex	ATM	
Layer management	Cell rate decoupling HEC header sequence generation/verification Cell delineation Transmission frame adaption Transmission frame generation/recovery	TC	Physical layer
	Bit timing Physical medium	PM	

Figure 16.11. B-ISDN/ATM functional layering. CS, convergence sublayer; PM, physical medium; SAR, segmentation and reassembly (sublayer); TC, transmission convergence.

The physical medium sublayer provides bit transmission capability including bit transfer and bit alignment as well as line coding and electrical–optical transformation. Of course, the principal function is the generation and reception of waveforms suitable for the medium, the insertion, and extraction of bit timing information and line coding where required. The primitives identified at the border between the PM and TC sublayers are a continuous flow of logical bits or symbols with this associated timing information.

Transmission Convergence Sublayer Functions. Among the important functions of this sublayer are the generation and recovery of the transmission frame. Another function is transmission frame adaptation which includes the actions necessary to structure the cell flow according to the payload structure of the transmission frame (transmit direction) and to extract this cell flow out of the transmission frame (receive direction). The transmission frame may be a cell equivalent (i.e., no external envelope is added to the cell flow), an SDH/SONET envelope, an E-1/T-1 envelope, and so on.

In the transmit direction, the HEC sequence is calculated and inserted in the header. In the receive direction, we include cell header verification. Here cell headers are checked for errors and, if possible, header errors are corrected. Cells are discarded where it is determined that headers are errored and are not correctable.

Another transmission convergence function is cell rate decoupling. This involves the insertion and removal of idle cells in order to adapt the rate of valid ATM cells to the payload capacity of the transmission system. In other words, cells must be generated to exactly fill the payload of SDH/SONET (for example) whether the cells are idle or busy.

Section 16.12 gives several examples of transporting cells using the convergence sublayer.

16.6.1.2 The ATM Layer. Table 16.4 shows the ATM layer functions supported at the UNI (U-plane). The ATM layer is completely independent of the physical medium. One important function of this layer is *encapsulation*. This includes cell header generation and extraction. In the transmit direction, the cell header generation function receives a cell information field from a higher layer and generates an appropriate ATM cell header except for the HEC sequence. This function can also include the translation from a service access point (SAP) identifier to a virtual path (VP) and virtual circuit (VC) identifier.

In the receive direction, the cell header extraction function removes the ATM cell header and passes the cell information field to a higher layer. As in the transmit direction, this function can also include a translation of a VP and VC identifier into an SAP identifier.

In the case of the NNI, the generic flow control (GFC) is applied at the ATM layer. The flow control information is carried in assigned and unassigned cells. Cells carrying this information are generated in the ATM layer.

TABLE 16.4   ATM Layer Functions Supported at the UNI

Functions	Parameters
Multiplexing among different ATM connections	VPI / VCI
Cell rate decoupling (unassigned cells)	Preassigned header field values
Cell discrimination based on predefined header field values	Preassigned header field values
Payload type discrimination	PT field
Loss priority indication and selective cell discarding	CLP field, network congestion state
Traffic shaping	Traffic descriptor

In a switch, the ATM layer determines where the incoming cells should be forwarded to, resets the corresponding connection identifiers for the next link, and forwards the cell. The ATM layer also handles traffic management functions and buffers incoming and outgoing cells. It indicates to the next higher layer (the AAL) whether or not there is congestion during transmission. The ATM layer monitors both transmission rates and conformance to the service contract; this is called *traffic shaping* and *traffic policing*.

Cell Rate Decoupling. Cell rate decoupling includes insertion and suppression of idle cells in order to adapt the rate of valid ATM cells to the payload capacity of the transmission system (Ref. 6, ITU-T Rec. I.321). According to the ITU, this function is carried out in the TC (transmission convergence) portion of the physical layer. The ATM Forum provides the following statement: The cell rate decoupling function at the sending entity adds unassigned cells to the assigned stream (i.e., cells with valid payloads) to be transmitted, transforming a noncontinuous stream of assigned cells into a continuous stream of assigned and unassigned cells. At the receiving entity the opposite operation is performed for both unassigned and invalid cells. The rate at which the unassigned cells are inserted/extracted depends on the bit rate (and rate variation) of the assigned cell and/or the physical layer transmission rate. The ATM Forum carries out the function in the ATM layer and uses different terminology (i.e., "assigned and unassigned" cells), whereas ITU-T carries out the function in the TC sublayer and uses the term "idle cells." Note the possible incompatibility here. Also note the "unassigned cell" coding in Table 16.1, and note "idle cell" coding in Table 16.3.

Physical layers that have synchronous cell time slots generally require cell rate decoupling (typically SONET/SDH, DS3/E3, etc.), whereas physical layers that have asynchronous cell time slots do not require this function because no continuous flow of cells need to be provided.

Cell Discrimination Based on Predefined Header Field Values. Certain predefined header field values at the UNI are given in Table 16.1. The functions of several of these are described as follows:

Meta-Signaling cells are used by the meta-signaling protocol for establishing and releasing signaling virtual channel connections. For virtual channels allocated permanently (PVC), meta-signaling is not used.

General Broadcast signaling cells are used by the ATM network to broadcast signaling information independent of service profiles.

F4 OAM cells. The virtual path connection (VPC) operation flow (F4 flow) is carried via specially designated OAM cells. F4 flow OAM cells have the same VPI value as the user-data cells transported by the VPC but are identified by two unique preassigned virtual channels within this VPC. At the UNI, the virtual channel identified by a VCI value of 3 is used for VP-level management functions between ATM nodes on both sides of the UNI (i.e., single VP link segment), while the virtual channel identified by a VCI value of 4 can be used for VP-level end-to-end (user-to-user) management functions.

B-ISDN operation and maintenance (OAM) is a responsibility of the physical and ATM layers. It is discussed in Section 16.13.

16.6.2   The ATM Adaptation Layer (AAL)

The basic purpose of the AAL is to isolate the higher layers (see Figure 16.3) from the specific characteristics of the ATM layer by mapping the higher-layer protocol data units (PDUs) into the information field of the ATM cell and vice versa.

16.6.2.1   Sublayering of the AAL. To support services above the AAL, a number of independent functions are required of the AAL. These functions are organized in two logical sublayers: the convergence sublayer (CS) and the segmentation and reassembly sublayer (SAR). The principal functions of these sublayers are:

SAR. Here the functions are (a) segmentation of higher-layer information into a size suitable for the information field at an ATM cell and (b) reassembly of the contents of ATM cell information fields into higher-layer information.

CS. Here the prime function is to provide the AAL service at the AAL-SAP. This sublayer is service-dependent.

16.6.2.2   Service Classification for the AAL. The service classification is based on the following parameters:

- Timing relation between source and destination (this refers to urgency of traffic): required or not required
- Bit rate: constant or variable
- Connection mode: connection-oriented or connectionless

When we combine these parameters, four service classes (A, B, C, and D) emerge as shown in Table 16.5. There are five different AAL types or categories. The simplest is AAL-0. It just transmits cells down a pipe. That pipe is commonly a fiber-optic link. Ideally, we would like the bit rate to be some multiple of 53×8 or 424 bits. For example, 424 Mbps would handle 1 million cells per second.

TABLE 16.5 Service Classification for AAL

Parameters	Class A	Class B	Class C	Class D
Timing compensation	Required	Required	Not required	Not required
Bit rate	Constant	Variable	Variable	Variable
Connection mode	Connection-oriented	Connection-oriented	Connection-oriented	Connectionless
Example	Circuit emulation, voice, video	Variable bit rate video or audio	Connection-oriented data transfer	Connectionless data transfer
AAL Type	Type 1	Type 2	Type 3, Type 5	Type 4

AAL-1. AAL-1 is used to provide transport for synchronous bit streams. Its primary application is to adapt ATM cell transmission to circuits representative of E1/T1 and SDH/SONET. Typically, AAL-1 is for voice communications (POTS—plain old telephone service). AAL-1 checks that mis-sequencing of information does not occur by verifying a 3-bit sequence counter and allows for regeneration of the original clock timing of the data received at the far end of the link. The SAR-PDU format of AAL-1 is shown in Figure 16.12. The 4-bit sequence number (SN) is broken down into a 1-bit convergence sublayer indicator (CSI) and sequence count. The SNP (sequence number protection) contains a 3-bit CRC and a parity bit. Clock recovery is via a synchronous residual time stamp (SRTS) and common network clock by means of a 4-bit residual time stamp extracted from CSI of

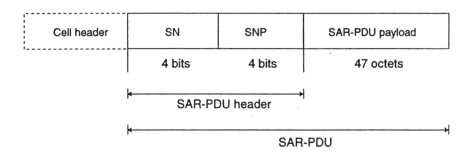

SN   Sequence number (4 bits); to detect lost or misinserted cells. A specific value of the sequence number may indicate a special purpose, e.g. the existence of convergence sublayer functions. The exact counting scheme is for further study.

SNP   Sequence number protection (4 bits). The SNP field may provide error detection and correction capabilities. The polynomial to be used is for further study.

Figure 16.12. SAR-PDU format for AAL-1. SN, sequence number; SNP, sequence number protection.

cells with odd sequence numbers. The residual time stamp is transmitted over 8 cells. Alarm indication in this adaptation layer is via a check of the one's density. When the one's density of the received cell stream becomes significantly different from the density used for the particular PCM line coding scheme in use, it is determined that the system has lost signal and alarm notifications are given (Ref. 7).

AAL-2. AAL-2 handles the VBR scenario such as MPEG* video. Functions performed by AAL-2 include:

- Segmentation and reassembly (SAR) of user information
- Handling of cell delay variation
- Handling of lost and misinserted cells
- Source clock frequency recovery at the receiver
- Recovery of the source data structure at the receiver
- Monitoring of AAL-PCI[†]
- Handling of AAL-PCI bit errors
- Monitoring of user information field for bit errors and possible corrective action

AAL-2 is still in ITU-T definitive stages. However, an example of an SAR-PDU format for AAL-2 is given in Figure 16.13. The following informa-

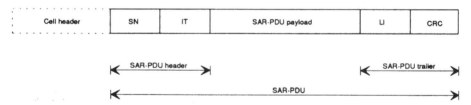

Figure 16.13. Example of an SAR-PDU format for AAL-2. The need of each of these fields, the position of these fields, and their size require further study by the ITU-T Organization. SN is the sequence number, used to detect lost or misinserted cells. A specific value of the sequence number may indicate a specified purpose. IT is information type, used to indicate beginning of message (BOM), continuation of message (CM), and end of message (EOM), provides timing information, and also is a component of the audio or video signal; LI is length indicator, checks to indicate that the number of octets of the CS-PDU are included in the SAR-PDU payload field. CRC is cyclic redundancy check, used to correct up to two correlated bit errors. (From Figure 1/I.363, page 4, ITU-T Rec. I.363, Ref. 8; see also Ref. 9.)

*MPEG is a set of digital video compression schemes. MPEG stands for Motion Picture Experts Group.
[†]PCI stands for protocol control information.

tion is passed from the user plane to the management plane:

- Errors in the transmission of user information
- Loss of timing/synchronization
- Lost or misinserted cells
- Cells with errored AAL-PCI
- Buffer underflow or overflow

AAL-2 Convergence Sublayer (CS). This sublayer performs the following functions:

1. Clock recovery for variable bit rate audio and video services is performed by means of the insertion of a time stamp or real-time synchronization word in the CS-PDU.
2. Sequence number processing is performed to detect the loss or misinsertion of ATM-SDUs. The handling of lost and misinserted ATM-SDUs is also performed in this sublayer.
3. For audio and video services, forward error correction may also be performed.

The following is a comment by Radia Perlman (taken out of context, Ref. 10, page 174): "Recently a new AAL was proposed for multiplexing several voice channels into a single cell. Instead of being called AAL-6, it's being called AAL-2 and is under development."

AAL-3/4. Initially, in CCITT Rec. I.363 (1991, Ref. 8), there were two separate AALs, one for connection-oriented variable bit rate data services (AAL-3) and one for connectionless service (AAL-4). As the specifications evolved, the same procedures turned out to be necessary for both of these services, and the specifications were merged to become the AAL-3/4 standard. It is used for ATM transport of SMDS, CBDS (connectionless broadband data services, an ETSI initiative), and frame relay.

AAL-3/4 has been designed to take variable-length frames/packets (maximum 65,535 octets) and segment them into cells. The segmentation is done in a way that protects the transmitted data from corruption if cells are lost or mis-sequenced.

Radia Perlman again comments (taken out of context, Ref. 10): "AAL-3/4 attempts to provide service that allows you to send a reasonable sized datagram. It was done in a complex and inefficient way, and people stepped in and designed AAL-5, which is better in all ways, and AAL-3/4 will eventually go away."

Modes of Service. AAL-3/4 offers two different modes of service: Message mode and streaming mode.

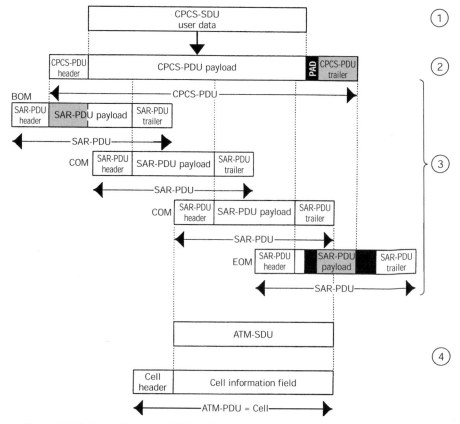

Figure 16.14. Formation of an ATM cell from a user frame, showing terminology used.

With the message mode, a CPCS-PDU is passed across the CPCS inter-face in its entirety in exactly one CPCS-IDU (IDU, interface data unit). In other words, it handles a complete user frame as one piece. In the case of the streaming mode, parts of a user frame can be sent at different points in time, or a long frame may be sent on to a receiver before the entire frame has been received by the transmitter (Ref. 11).

Figure 16.14 depicts the steps required to accept a variable length user data frame or packet and to develop ATM cells for transmission. These steps are numbered 1 through 4:

1. Variable length packets, typically from frame relay, are padded to an integral word length and encapsulated with a header and a trailer to form what is called the *convergence sublayer PDU* (CPCS).
2. Padding is added to make sure that fields align themselves to 32-bit boundaries, allowing the efficient implementation of the operations in

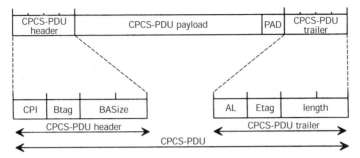

Figure 16.15. Common part convergence sublayer (CPCS) PDU format.

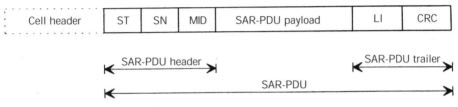

Figure 16.16. SAR-PDU format for AAL-3/4.

hardware at lower layers. The CPCS adds a 4-octet header and a 4-octet trailer to form the CPCS PDU. The CPCS header and trailer contain identifiers (Btag and Etag) and length indicators (BAsize and Length) that help the receiving end allocate appropriate buffer space, reassemble the CS data unit, and do basic error checks (see Figure 16.15).

3. The CPCS-PDU is now broken down into 44-octet segments by the SAR sublayer, which, in its turn, adds a 2-octet header and a 2-octet trailer to each of the segments, forming what is referred to as an SAR Protocol Data Unit (SAR-PDU) (see Figure 16.16). This SAR-PDU has a total of 48 octets including its headers and trailers.

4. These 48 octets now become the information field of a standard ATM cell.

DISCUSSION OF FIGURES 16.15 AND 16.16

Discussion of Figure 16.15. The user data (Figure 16.14) is encapsulated with a 4-octet header and a 4-octet trailer. The header consists of the following fields:

Common Part Indicator (CPI field, 1 octet). This field is used to interpret subsequent fields for the CPCS functions in the header and trailer. The counting units for the values specified in the BAsize and Length fields may be indicated; other users are under study by the ITU-T Organization.

Beginning Tag (Btag) Field (1 octet). This field allows the association of the CPCS-PDU header and trailer. The sender inserts the same value in the Btag and the Etag in the trailer for a given CPCS-PDU and changes the values of each successive CPCS-PDU. The receiver checks the value of the Btag in the header with the value of the Etag in the trailer. It does not check the sequence of the Btag/Etags in successive CPCS-PDUs.

As an example, a suitable mechanism is as follows: The sender increments the value placed in the Btag and Etag field for each successive CPCS-PDU sent over a given MID value. Btag values are cycled up to modulo 256.

Buffer Allocation Size (BAsize) Indication Field (2 octets). This field indicates to the receiving peer entity the maximum buffering requirements to receive the CPCS-SDU. In message mode, the BAsize value is encoded equal to the CPCS-PDU payload length. In streaming mode, the BAsize value is encoded equal to or greater than the CPCS-PDU payload length. The buffer allocation size is binary encoded as number of counting units. The size of the counting units is identified by the CPI field.

Padding (PAD) Field (0 to 3 octets). Between the end of the CPCS-PDU payload and the 32-bit aligned CPCS-PDU trailer, there will be from 0 to 3 unused octets. These unused octets are called the Padding (PAD) field. They are strictly used as filler octets and do not convey any information. It may be set to "0," and its value is ignored at the receiving end. This padding field complements the CPCS-PDU payload to an integral multiple of 4 octets.

Alignment (AL) Field (1 octet). The function of the alignment field is to achieve 32-bit alignment in the CPCS-PDU trailer. The alignment field complements the CPCS-PDU trailer to 32 bits. This unused octet is strictly used as a filler octet and does not convey any information. The alignment field is set to zero.

End Tag (Etag) Field (1 octet). For a given CPCS-PDU, the sender inserts the same value in this field as was inserted in the Btag field in the header to allow the association of the trailer with its header.

Length Field (2 octets). The length field is used to encode the length of the CPCS-PDU payload field. This field is also used by the receiver to detect the loss or gain of information. The length is binary encoded as number of counting units. The size of the counting units is identified by the CPI field.

Note: The length of the CPCS-PDU payload is limited to the maximum value of the length field multiplied by the value of the counting unit.

Discussion of Figure 16.16

Segment Type (ST) Field (2 bits). The segment-type indication identifies an SAR-PDU as containing a Beginning of Message (BOM) coded 10, a Continuation of Message (COM) coded 00, an End of Message (EOM) coded 01, or a Single Segment Message (SSM) coded 11.

Sequence Number (SN) Field (4 bits). The 4 bits allocated to this field allow the stream of SAR-PDUs of a CPCS-PDU to be numbered modulo 16.

Each SAR-PDU belonging to an SAR-PDU (and hence associated with a given MID value) has its sequence number incremented by one relative to its previous sequence number. The receiver checks the sequence of the sequence number field of SAR-PDUs derived from one SAR-SDU and does not check the sequence number of the sequence number field of the SAR-PDUs derived from successive SAR-SDUs. As the receiver does not check the sequence number continuity between SAR-SDUs, the sender may set the sequence number field to any value from 0 to 15 at the beginning of each SAR-SDU.

Multiplexing Identification (MID) Field (10 bits). This field is used for multiplexing. If no multiplexing is used, this field is set to zero.

In connection-oriented applications, it may be used to multiplex multiple SAR connections on a single ATM connection.

In connectionless and connection-oriented applications, all SAR-PDUs of an SAR-SDU will have the same MID field value. The MID field is used to identify SAR-PDUs belonging to a particular SAR-SDU. The MID field assists in the interleaving of SAR-PDUs from different SAR-SDUs and reassembly of these SAR-SDUs.

SAR-PDU Payload Field (44 octets). The SAR-SDU information is left justified within the SAR-PDU payload field. The remaining octets of the SAR-PDU payload field may be set to 0 and are ignored at the receiving end.

Length Indication (LI) Field (6 bits). The length indication field is binary encoded with the number of octets of SAR-SDU information that are included in the SAR-PDU payload field. The permissible values of the LI field, depending on the coding of the ST field, are BOM, COM 44, EOM $4 \ldots 44$, 63, and SSM $8 \ldots 44$. The 63 value for an EOM is used in the Abort-SAR-PDU.

CRC Field (10 bits). The CRC-10 is used to detect bit errors in the SAR-PDU. It is the remainder of the division (modulo 2) by the generator polynomial of the product of x^{10} and the content of the SAR-PDU, including the SAR-PDU header, SAR-PDU payload, and the LI field of the SAR-PDU trailer. Each bit of the concatenated fields mentioned above are considered coefficients (modulo 2) of a polynomial of degree 373. The CRC-10 generator polynomial is

$$G(x) = 1 + x + x^4 + x^5 + x^9 + x^{10}$$

The result of the CRC calculation is placed with the least significant bit right-justified in the CRC field.

AAL-5. This type of AAL was designed specifically to carry data traffic typically found in today's LANs. AAL-5 evolved after AAL-3/4, which was found to be too complex and inefficient for LAN traffic. Thus AAL-5 got the name "SEAL" for simple and efficient AAL layer. Only a small amount of overhead is added to the CPCS-PDU, and no extra overhead is added when the AAL-5 segments them into SAR-PDUs. There is no AAL-level cell multiplexing. In AAL-5 all cells belonging to an AAL-5 CPCS-PDU are sent sequentially.

As with AAL-3/4, AAL-5 provides both message mode service and streaming mode service. Both modes offer nonassured operation peer-to-peer. The following points should be considered:

- Integral CPCS-SPU may be delivered, lost or corrupted.
- Lost and corrupted CPCS-SDUs are not corrected by retransmission. There is a corrupted data delivery option.
- Flow control may be provided as an option.

The CPCS has the following service characteristics:

- Nonassured data transfer of user data frames with any length measured in octets from 1 to 65,535 octets
- The CPCS connection is established by management or by the control plane
- Error detection and optional indication (bit error and cell loss or gain)
- CPCS-SDU sequence integrity on each CPCS connection

CPCS-PDU Structure and Coding. The CPCS-PDU format is illustrated in Figure 16.17. The CPCS functions require an 8-octet CPCS-PDU trailer. The trailer is always located in the last 8 octets of the last SAR-PDU of the CPCS-PDU. Therefore, a padding field provides for a 48-octet alignment of the CPCS-PDU. The trailer together with the passing field and the CPCS-PDU payload comprise the CPCS-PDU. The coding of the CPCS-PDU is discussed in the following text. (Refer to Figure 16.17.)

CPCS-PDU Payload. The payload is used to carry to CPCS-SDU. The field is octet-aligned and can range from 1 to 65,535 octets in length.

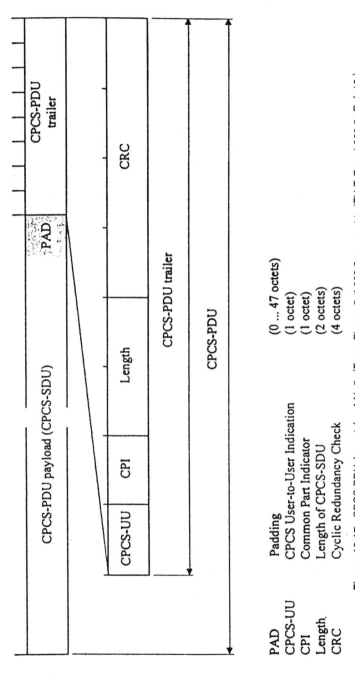

Figure 16.17. CPCS-PDU format for AAL-5. (From Figure 5/I.363.5, page 11, ITU-T Rec. I.363.5, Ref. 13.)

PAD	Padding	(0 ... 47 octets)
CPCS-UU	CPCS User-to-User Indication	(1 octet)
CPI	Common Part Indicator	(1 octet)
Length	Length of CPCS-SDU	(2 octets)
CRC	Cyclic Redundancy Check	(4 octets)

Padding (PAD) Field. Between the end of the CPCS-PDU payload and the CPCS-PDU trailer, there will be from 0 to 47 unused octets. These unused octets are called the padding (PAD) field. They are strictly used as filler octets and do not convey any information. Any coding is acceptable. This padding field complements the CPCS-PDU (including CPCS-PDU payload, padding field, and CPCS-PDU trailer) to an integral multiple of 48 octets.

CPCS User-to-User Indication (CPCS-UU) Field. The CPCS-UU field is used to transparently transfer CPCS user-to-user information.

Common Part Indicator (CPI) Field. One of the functions of the CPI field is to align the CPCS-PDU trailer to 64 bits. Other functions are for further study by the ITU-T Organization. When only the 64-bit alignment function is used, this field is coded as zero.

Length Field. The length field is used to encode the length of the CPCS-PDU payload field. The length field value is also used by the receiver to detect the loss or gain of information. The length is binary encoded as number of octets. A length field coded as zero is used for the abort function.

CRC Field. The CRC-32 is used to detect bit errors in the CPCS-PDU. The CRC field is filled with the value of a CRC calculation which is performed over the entire contents of the CPCS-PDU, including the CPCS-PDU payload, the PAD field, and the first 4 octets of the CPCS-PDU trailer. The CRC field shall contain the 1 complement of the sum (modulo 2) of:

1. The remainder of x^k $(x^{31} + x^{30} + \cdots + x + 1)$ divided (modulo 2) by the generator polynomial, where k is the number of bits of the information over which the CRC is calculated and

2. The remainder of the division (modulo 2) by the generator polynomial of the product of x^{32} by the information over which the CRC is calculated.

The CRC-32 generator polynomial is

$$G(x) = x^{32} + x^{26} + x^{23} + x^{22} + x^{16} + x^{12} + x^{11} + x^{10} + x^8$$

$$+ x^7 + x^5 + x^4 + x^2 + x + 1$$

The result of the CRC calculation is placed with the least significant bit right justified in the CRC field.

As a typical implementation at the transmitter, the initial content of the register of the device computing the remainder of the division is preset to all

1s and is then modified by division by the generator polynomial on the information over which the CRC is to be calculated; the ones complement of the resulting remainder is put into the CRC field.

As a typical implementation at the receiver, the initial content of the register of the device computing the remainder of the division is preset to all 1s. The final remainder, after multiplication by x^{32} and then division (modulo 2) by the generator polynomial of the serial incoming CPCS-PDU, will be (in the absence of errors)

$$C(x) = x^{31} + x^{30} + x^{26} + x^{25} + x^{24} + x^{18} + x^{15} + x^{14} + x^{12} + x^{11}$$

$$+ x^{10} + x^8 + x^6 + x^5 + x^4 + x^3 + x + 1$$

SAR-PDU Structure and Coding. The SAR sublayer utilizes the ATM-User-to-ATM-User indication (AUU) parameter of the ATM layer primitives (the relationship between the AUU parameter and the ATM layer PTI encoding is defined in Section 2.2.4 of ITU-T Rec. I.361) to indicate that an SAR-PDU contains the end of an SAR-SDU. An SAR-PDU where the value of the AUU parameter is 1 indicates the end of an SAR-SDU; the value of 0 indicates the beginning or continuation of an SAR-PDU. The structure of the SAR-PDU is shown in Figure 16.18.

Figure 16.19 illustrates the formation (segmentation) of AAL-5 ATM cells from CPCS user data.

The following are Radia Perlman's comments on AAL-5: "So indeed AAL5 is superior in every way to AAL3/4. Because it supports variable bit rate, it can also do everything anyone would have wanted to do with AAL2." So almost certainly the only AALs that will survive are AAL1 and AAL5.

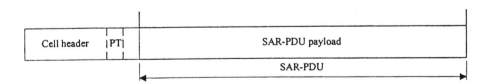

PT   Payload Type

NOTE – The Payload Type field belongs to the ATM header. It conveys the value of the AUU parameter end-to-end.

Figure 16.18. SAR-PDU format for AAL-5. (From Figure 4/I.363.5, page 9, ITU-T Rec. I.363.5, Ref. 13.)

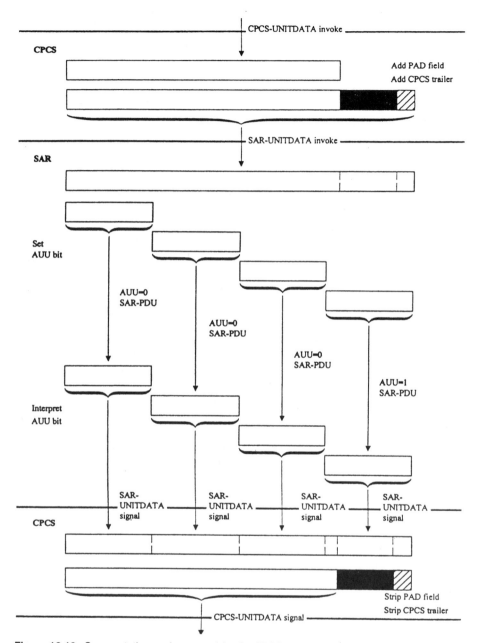

Figure 16.19. Segmentation and reassembly of a CPCS user PDU. (From Figure B.3/I.363.5, page 20, ITU-T Rec. I.363.5, Ref. 13.)

16.7 SERVICES: CONNECTION-ORIENTED AND CONNECTIONLESS

The issues such as routing decisions and architectures have a major impact on connection-oriented services, where B-ISDN/ATM end nodes have to maintain or get access to lookup tables which translate destination addresses into circuit paths. These circuit path lookup tables that differ at every node must be maintained in a quasi-real-time fashion. This will have to be done by some kind of routing protocol.

One way to resolve this problem is to make it an internal network problem and use a connectionless service as described in ITU-T Rec. I.364 (Ref. 14). We must keep in mind that ATM is basically a connection-oriented service. Here we are going to adapt it to provide a connectionless service.

16.7.1 Functional Architecture

The provision of connectionless data service in the B-ISDN is carried out by means of ATM switches and connectionless service functions (CLSF). ATM switches support the transport of connectionless data units in the B-ISDN between specific functional groups where the CLSF handles the connectionless protocol and provides for the adaptation of the connectionless data units into ATM cells to be transferred in a correction-oriented environment. As shown in Figure 16.20, CLSF functional groups may be located outside the B-ISDN, in a private connectionless network, or in a specialized service provider or inside the B-ISDN itself.

The ATM switching is performed by the ATM nodes (ATM switch/cross-connect) that are a functional part of the ATM transport network. The CLSF functional group terminates the B-ISDN connectionless protocol and includes functions for the adaptation of the connectionless protocol to the intrinsically connection-oriented ATM layer protocol. These latter functions are performed by the ATM adaptation layer Type 3/4 (AAL-3/4), while the CLSF group terminations are carried out by the services layer above the AAL called the connectionless network access protocol (CLNAP). The CL protocol includes functions such as routing, addressing, and QoS selection. In order to perform the routing of CL data units, the CLSF has to interact with the control/management planes of the underlying ATM network.

The general protocol structure for the provision of connectionless data service is shown in Figure 16.21. Figure 16.22 illustrates the protocol architecture for supporting connectionless layer service. The CLNAP layer uses the Type 3/4 AAL unassured service and includes the necessary functionality to provide the connectionless layer service. The connectionless layer service provides for transparent transfer of variable size data units from a source to one or more destinations in a manner such that lost or corrupted data units are not retransmitted. The transfer is performed using a connectionless technique, including embedding destination and source addresses into each data unit.

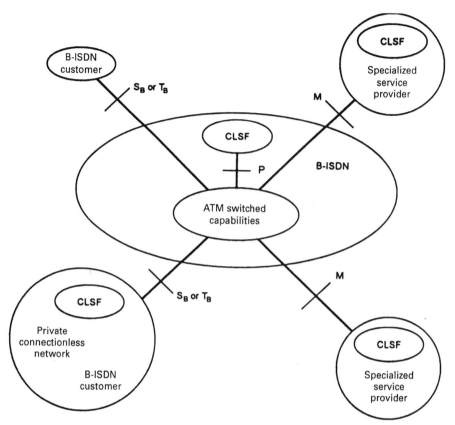

Figure 16.20. Reference configuration for connectionless data service in the B-ISDN. CLSF, connectionless service functions. P, M, S, and T are reference points. (From Figure 1 / I.364, page 2, ITU-T Rec. I.364, Ref. 14.)

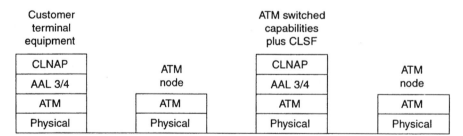

Figure 16.21. General protocol structure for connectionless data service in the B-ISDN. CLNAP, connectionless network access protocol; CLSF; connectionless service functions.

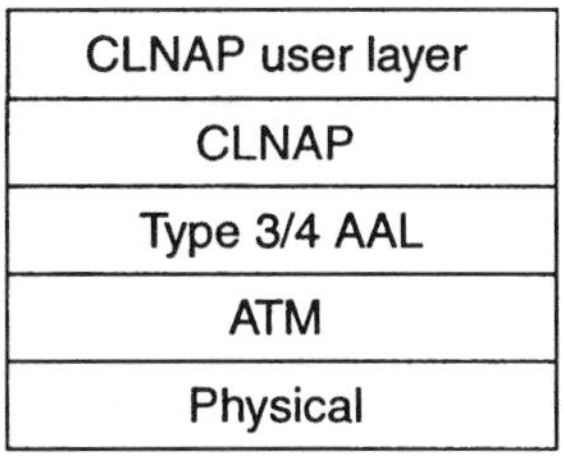

| CLNAP user layer |
| CLNAP |
| Type 3/4 AAL |
| ATM |
| Physical |

Figure 16.22. Protocol architecture for supporting connectionless service.

16.7.2   CLNAP Protocol Data Unit (PDU) Structure and Encoding

Figure 16.23 shows detailed structure of the CLNAP-PDU, which contains the following fields:

Destination Address and Source Address. These 8-octet fields each contain a 4-bit *address type* subfield, followed by the 60-bit *address* subfield. The address type subfield indicates whether the address subfield contains a publicly administered 60-bit individual address or a publicly administered 60-bit group address. The address subfield indicates to which CLNAP entity(ies) the CLNAP-PDU is destined; and in the case of the source address, it indicates the CLNAP entity that sourced the CLNAP-PDU. The encoding of the address type and address subfields are shown in Figures 16.24 and 16.25. The address is structured in accordance with ITU-T Rec. E.164 (Ref. 15).

Higher-Layer Protocol Identifier (HLPI). This 6-bit field is used to identify the CLNAP user layer entity which the CLNAP-SDU is to be passed to at the destination node. It is transparently carried end-to-end by the network.

PAD Length. This 2-bit field gives the length of the PAD field (0–3 octets). The number of PAD octets is such that the total length of the user-information field and the PAD field together is an integral multiple of 4 octets (32 bits).

CRC Indication Bit (CIB). This 1-bit field indicates the presence (if CIB = 1) or absence (if CIB = 0) of a 32-bit CRC field.

Header Extension Length (HEL). This 3-bit field can take on any value from 0 to 5 and indicates the number of 32-bit words in the header extension field.

Reserved. This 16-bit field is reserved for future use. Its default value is 0.

Header Extension. This variable-length field can range from 0 to 20 octets. Its length is indicated by the value of the header extension length field (see above). In the case where the HEL is not equal to zero, all unused octets in the header extension are set to zero. The information carried in

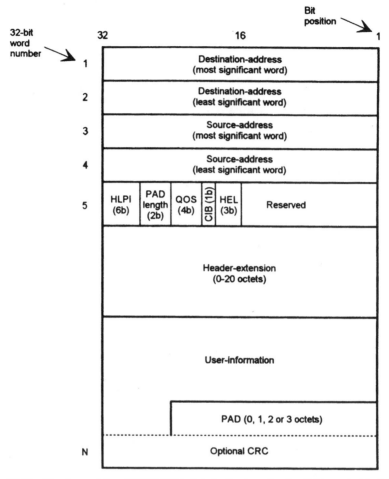

Figure 16.23. Structure of the CLNAP-PDU. (*nb* is the length of a field, *n*, in bits.) (From Figure 5/I.364, page 7, ITU-T Rec. I.364, Ref. 14.)

Address type	Meaning
1100	60-bit publicly administered individual address
1110	60-bit publicly administered group address

Figure 16.24. Destination address field.

Address type	Meaning
1100	60-bit publicly administered individual address

Figure 16.25. Source address field.

the header extension is structured into information entities. An information entity (element) consists of (in this order) element length, element type, and element payload, described as follows:

Element Length. This is a 1-octet field and contains the combined lengths of the element length, element type, and element payload in octets.

Element Type. This is also a 1-octet field and contains a binary coded value which indicates the type of information found in the element payload field.

Element Payload. This is a variable-length field and contains the information indicated by the element type field.

User Information. This is a variable-length field up to 9188 octets and is used to carry the CLNAP-SDU.

PAD. This field is 0, 1, 2, or 3 octets in length and is coded as all zeros. Within each CLNAP-PDU the length of this field is selected such that the length of the resulting CLNAP-PDU is aligned on a 32-bit boundary.

CRC. This optional 32-bit field may be present or absent as indicated by the CIB field. The field contains the result of a standard CRC-32 calculation performed over the CLNAP-PDU with the "reserved" field always treated as if it were coded as all zeros.

16.8 ASPECTS OF A B-ISDN/ATM NETWORK

16.8.1 ATM Routing and Switching

An ATM transmission path supports VPs (virtual paths), and inside VPs are VCs (virtual channels) as shown in Figure 16.26.

As we discussed in Section 16.4.1, each ATM cell contains a label in its header to explicitly identify the VC to which the cell belongs. This label consists of two parts: a virtual channel identifier (VCI) and a virtual path identifier (VPI). See Figure 16.5.

16.8.1.1 Virtual Channel Level. *Virtual channel* (VC) is a generic term used to describe a unidirectional communication capability for the transport of ATM cells. A VCI identifies a particular VC link for a given virtual path

Figure 16.26. Relationship between the VC, VP, and the transmission path.

connection (VPC). A specific value of VCI is assigned each time a VC is switched in the network. A VC link is a unidirectional capability for the transport of ATM cells between two consecutive ATM entities where the VCI value is translated. A VC link is originated or terminated by the assignment or removal of the VCI value.

Routing functions of virtual channels are done at a VC switch/cross-connect.* The routing involves translation of the VCI values of the incoming VC links into the VCI values of the outgoing VC links.

Virtual channel links are concatenated to form a virtual channel connection (VCC). A VCC extends between two VCC endpoints or, in the case of point-to-multipoint arrangements, more than two VCC endpoints. A VCC endpoint is the point where the cell information field is exchanged between the ATM layer and the user of the ATM layer service.

At the VC level, VCCs are provided for the purpose of user–user, user–network, or network–network information transfer. Cell sequence integrity is preserved by the ATM layer for cells belonging to the same VCC.

16.8.1.2 *Virtual Path Level.* *Virtual path* (VP) is a generic term for a bundle of virtual channel links; all the links in a bundle have the same endpoints.

A VPI identifies a group of VC links, at a given reference point, that share the same VPC. A specific value of VPI is assigned each time a VP is switched in the network. A VP link is a unidirectional capability for the transport of ATM cells between two consecutive ATM entities where the VPI value is translated. A VP link is originated or terminated by the assignment or removal of the VPI value.

Routing functions for VPs are performed at a VP switch/cross-connect. This routing involves translation of the VPI values of the incoming VP links into the VPI values of the outgoing VP links. VP links are concatenated to form a VPC. A VPC extends between two VPC endpoints; however, in the case of point-to-multipoint arrangements, there are more than two VPC endpoints. A VPC endpoint is the point where the VCIs are originated, translated, or terminated. At the VP level, VPCs are provided for the purpose of user–user, user–network, and network–network information transfer.

When VPCs are switched, the VPC supporting the incoming VC links are terminated first and a new outgoing VPC is then created. Cell sequence integrity is preserved by the ATM layer for cells belonging to the same VPC. Thus cell sequence integrity is preserved for each VC link within a VPC.

---

*A VC cross-connect is a network element which connects VC links. It terminates VPCs and translates VCI values and is directed by management plane functions, not by control plane functions.

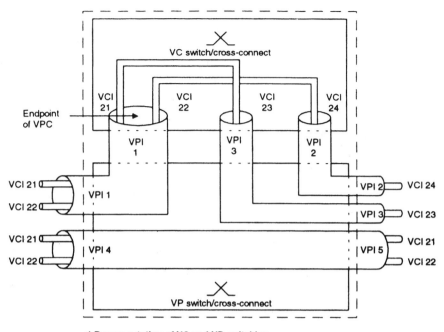

a) Representation of VC and VP switching

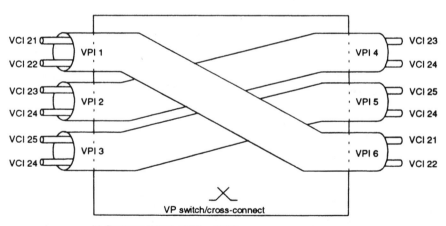

b) Representation of VP switching

Figure 16.27. Representation of the VP and VC switching hierarchy. (From Figure 4/I.311, page 5, ITU-T Rec. I.311, Ref. 16.)

Figure 16.27 is a representation of the VP and VC switching hierarchy where the physical layer is the lowest layer composed of, from bottom up, a regenerator section level, digital section level, and transmission path level. The ATM layer resides just above the physical layer and is composed of the VP level and, just above that, the VC level.

16.9 SIGNALING REQUIREMENTS

16.9.1 Setup and Release of VCCs

The setup and release of VCCs at the UNI can be performed in various ways:

- Without using signaling procedures. Circuits are set up at subscription with permanent or semipermanent connections.
- By using metasignaling procedures where a special VCC is used to establish or release a VCC used for signaling. Metasignaling is a simple protocol used to establish and remove signaling channels. All information interchanges in metasignaling are carried out via single cell messages.
- By employing user-to-network signaling procedures such as a signaling VCC to establish or release a VCC used for end-to-end connectivity.
- By employing user-to-user signaling procedures such as a signaling VCC to establish or release a VCC within a preestablished VPC between two UNIs.

16.9.2 Signaling Virtual Channels

16.9.2.1 Requirements for Signaling Virtual Channels. For a point-to-point signaling configuration, the requirements for signaling virtual channels are as follows:

1. One virtual channel connection in each direction is allocated to each signaling entity. The same VPI/VCI value is used in both directions. A standardized VCI value is used for point-to-point signaling virtual channel (SVC).
2. In general, a signaling entity can control, by means of associated point-to-point SVCs, user-VCs belonging to any of the VPs terminated in the same network element.
3. As a network option, the user-VCs controlled by a signaling entity can be constrained such that each controlled user-VC is in either upstream or downstream VPs containing the point-to-point SVCs of the signaling entity.

For point-to-multipoint signaling configurations, the requirements for signaling VCs are as follows:

1. *Point-to-Point Signaling Virtual Channel.* For point-to-point signaling, one VC connection in each direction is allocated to each signaling entity. The same VPI/VCI value is used in both directions.
2. *General Broadcast Signaling Virtual Channel.* The general broadcast signaling virtual channel (GBSVC) may be used for call offering in all cases. In cases where the "point" does not implement service profiles or where "the multipoints" do not support service profile identification, the GBSVC is used for call offering. The specific VCI value for general broadcast signaling is reserved per VP at the UNI. Only when metasignaling is used in a VP is the GBSVC activated in the VP.
3. *Selective Broadcast Signaling Virtual Channels.* Instead of the GBSVC, a VC connection for selective broadcast signaling (SBS) can be used for call offering, in cases where a specific service profile is used. No other uses for SBSVCs are foreseen.

16.9.3   Metasignaling

16.9.3.1   Metasignaling Functions at the User Access. In order to establish, check, and release point-to-point and selective broadcast signaling virtual channel connections, metasignaling procedures are provided. For each direction, metasignaling is carried in a permanent virtual channel connection having a standardized VCI value. This channel is called the metasignaling virtual channel. The metasignaling protocol is terminated in the ATM layer management entity.

The metasignaling function will be required to:

• Manage the allocation of capacity to signaling channels
• Establish, release, and check the status of signaling channels
• Provide a means to associate a signaling endpoint with a service profile if service profiles are supported
• Provide the means to distinguish between simultaneous requests

It may be necessary to support metasignaling on any VP. Metasignaling can only control signaling VCs within its VP.

16.9.3.2   Relationship Between Metasignaling and the User Access Signaling Configuration. A point-to-multipoint signaling configuration exists when the network supports more than one signaling entity at the user side. In this configuration, terminals must use the metasignaling protocol to request allocation of their individual point-to-point signaling virtual channels.

A point-to-point signaling configuration exists when the network supports only one signaling entity at the user side. When this configuration is known, terminals can use the specific VCI value reserved for the point-to-point signaling virtual channel. In this case no broadcast signaling virtual channel will be provided.

In a user-to-user signaling configuration, the metasignaling protocol can optionally be used over a user-to-user VPC to manage a user-to-user signaling virtual channel. It is recommended (but it is a user choice) to use the standardized VCI value for the user-to-user metasignaling channel. In this case, metasignaling shall not have an impact on the network.

The metasignaling protocol can be used to manage signaling virtual channel between a user and another network over the same user access. In this case, VPIs other than VPI $= 0$ are used and the VCI value is the standardized one.

Normally, the VPI value equal to zero is used to manage signaling virtual channels to the local exchange. For the case where communication of a user to an alternative local exchange over the same user access is required, another VPI value different from zero will be used.

16.9.3.3 *Metasignaling Functions in the Network.* metasignaling is not used to assign signaling channels between two network signaling endpoints. Therefore, every VP within the network has one VCI value reserved for point-to-point signaling and activated in case signaling is used on this VP.

16.9.3.4 *Metasignaling Requirements*

(a) *Scope of Metasignaling.* A metasignaling virtual channel is able to manage signaling virtual channels only within its own VP pair. In VPI $= 0$, the metasignaling virtual channel is always present and has a standardized VCI value.

(b) *Initiation of Metasignaling for SVCI Assignment.* Metasignaling VC may be activated at VP establishment. Other possibilities are for further study. The signaling virtual channel (SVC) should be assigned and removed when necessary.

(c) *Metasignaling VCI and VP.* A specific VCI value for metasignaling is reserved per VP at the UNI. For a VP with point-to-multipoint signaling configuration, metasignaling is required and the metasignaling VC within this VP will be activated. For a VP in point-to-point signaling configuration, the use of metasignaling is for further study.

(d) *SVC Bandwidth.* The user should have the possibility to negotiate the bandwidth parameter value. The bandwidth parameter values are for further study.

(e) *Metasignaling-Virtual Channel (MSVC) Bandwidth.* MSVC has default bandwidth value. The bandwidth can be changed by mutual agreement between network operator and user. The default value is for further study.

16.9.4 Practical Signaling Considerations

There are several network interfaces, each with its own signaling variant. The following interfaces must be taken into account:

- User–network interface (UNI)
  - Interface with the public ATM network
  - Interface with a private network
- Network-to-network interface (NNI) (e.g., interface between ATM switches) *Note:* This interface is often referred to as the B-ICI or broadband intercarrier interface. It loosely resembles a Signaling System No. 7 (SS7) interface, and the signaling protocols derive from ISDN SS7 procedures.

Great care must be taken when specifying interfaces. There are two recognized standards-making bodies:

- ITU
- ATM Forum

There are various incompatibilities.

When we start to dig into practical signaling procedures, we will ask: Haven't we been over this ground before? The answer is a resounding yes. In particular, in Chapter 12 dealing with frame relay, many of the protocol elements are nearly identical; certainly sets of procedures are identical, such as setting up PVCs and SVCs.

The ATM signaling protocol at the UNI is derived from the narrowband ISDN call control procedures described in ITU-T Rec. Q.931. The applicable protocol for ATM is Q.2931, which defines Digital Subscriber Signaling System No. 2 (DSS2) (Ref. 17). The equivalent ATM Forum Specification is UNI 3.0. Subsequently, the ATM Forum published version 3.1 to fix the disparities there were with Q.2931. However, they are still not identical.

The P-NNI (private network–network interface) is exclusively defined by the ATM Forum because it is out of scope for the ITU. It is called the Interim Interswitch Signaling Protocol or IISP (Ref. 18). It uses UNI 3.1 with some minor procedural differences. There is no routing protocol specified. Instead, routing information is configured manually and statically at each private network station.

At the B-ICI signaling interface, ITU-T Recs. Q.2761 and Q.2764 (Refs. 19 and 20) refer. The ITU protocol at this interface is commonly known as "BISUP," where the name is taken from SS7 ISDN User Part or ISUP. The ATM Forum has defined an equivalent specification called the B-ICI Specification Version 2.0 (Ref. 21).

16.9.4.1 *ATM Addressing.* Similar to frame relay signaling, there are two recognized address formats: NSAP and E.164. We are familiar with E.164 from frame relay; this is the recommended international telephone address-

ing scheme. An address consists of two parts: network prefix (13 bytes) and the user part (7 bytes). The addressing formats are shown and compared in Figure 16.28. Some of the standards involved use other terminology. The first part is called the international domain identifier (IDI), and the second part is the domain-specific part (DSP).

The AFI is the first octet (digit) and is common to all three. AFI stands for Authority and Format Identifier and indicates which address format is being used. The DCC (data country code) specifies the country to which the address is registered as defined in ISO 3166.

Within each one of these domains, there is the domain-specific part or DSP (i.e., the remainder of the address). The remaining formats are identical for both NSAP address types whether DCC or ICD. The DSP format identifier (DFI) specifies the meaning of the remainder of the address. The Administrative Authority field identifies the organization that is responsible for administering the remainder of the address. This may be a carrier (U.S. terminology), private network, or manufacturer. The remainder of the DSP is identical for all domains.

The routing domain (RD) identifies a unique domain within the IDI format. This is further subdivided using a 2-octet AREA field. The end system identifier (ESI) and SELector (SEL) portions of the DSP are identical for all IDI (Ref. ISO 10589). The ESI can be globally unique such as an IEEE 802 MAC address (48 bits). The SELector (SEL) field is not used for routing but may be used by end-users or end systems (ES).

The E.164 numbering plan is in BCD code and is padded with zeros on the left-hand side to the standard 15 digits to correspond to the standard's

Figure 16.28. ATM addressing plans, comparison. AFI, authority and format identifier; DCC, data country code; IDI, international domain identifier; DFI, DSP format identifier; RD, routing domain; ESI, end system identifier; SEL, SELector; DSP, domain-specific part. derived from ATM Forum, version 3.0 specification (Ref. 22.)

requirements. This is the international ISDN number. This is broken down to a country code (CC) of one to three digits. The remainder of the address is a nationally significant number (NSN). The NSN breaks down to a national destination code (NDC) and a subscriber number (SN).

Addresses are registered throughout an ATM network using the interim local management interface (ILMI). It is used to convey configuration and management information across the UNI over a dedicated virtual channel (VPI = 0, VCI = 16). This protocol uses the Simple Network Management Protocol (SNMP) with a management information base (MIB) describing the ATM UNI. (For SNMP, see Chapter 18.) The ILMI procedures are used to register end-station ATM addresses. Each end system (e.g., ATM network interface card, ATM-LAN bridge, and so forth) is expected to know its own ESI value.

16.9.4.2 Circuit Setup Using ATM Signaling. The procedures described below are very similar to those discussed in Section 12.12.2. Figure 16.29 shows a typical exchange of messages for call establishment. The discussion below is based on ITU-T Q.2931 (Ref. 12) and (Ref. 13).

Connection Establishment. A call is initiated at the end-user side by sending a SETUP message across the UNI. A SETUP message has 24 information

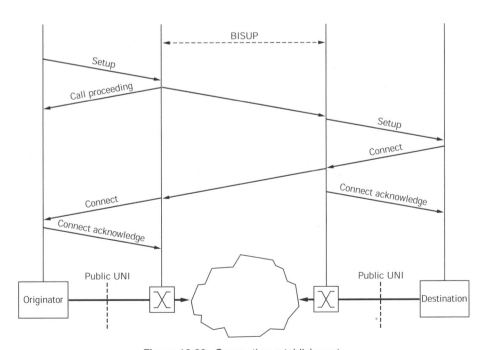

Figure 16.29. Connection establishment.

elements. Several of the key information elements are briefly described as follows:

- AAL parameters: AAL type, CBR bit rate, CPCS-SDU size, source clock frequency
- Broadband bearer capability: traffic type, timing requirements, CBR, or VBR
- Called party number: addressing information described above. NSAP or E.164
- Calling party number using ATM address of user originating the call
- End-to-end transit delay: indicates nominal maximum delay acceptable
- Quality of service (QoS): indicates QoS class backwards and forwards
- Transit network selection: indicates the requested public carrier network for call (e.g., Sprint, MCI, BT, AT & T, etc.).

When the SETUP message is received by the network, it is acknowledged with a CALL PROCEEDING message. This message contains the connection identifier employed to notify the user which VPI/VCI will be used for the connection. The SETUP message continues to pass through the network, eventually arriving at the destination terminal with all the information elements (IEs) the originating user specified, plus the connection identifier. Keep in mind that the network always chooses the VPI/VCI values for the connection.

Once the called user accepts the call, it replies sending a CONNECT message. It follows back across the network from UNI, thence from switch to switch through a UNI to the originator. Note in Figure 16.29 that there is also a CONNECT ACKNOWLEDGE exchange. To identify this call from other calls that may be in progress, a unique call reference value is assigned.

There is equivalent NNI signaling between switches on the routing of the call. When signaling messages cross the public UNI, the network switch continues the call's progress by sending the appropriate NNI messages. The protocol involved here is known as the BISUP. Breaking out the acronym, we have B for broadband (as in B-ISDN) and ISUP, which is an SS7 term for ISDN User Part. BISUP messages have a different format than equivalent UNI signaling messages, although with similar objectives. BISUP messages have parameters whereas UNI messages have information elements. Routing through the network is determined by the SETUP message with two of its IEs (information elements): transit network selection and called party number. The originating switch then sends an IAM* (initial address message) which sets up the path for traffic.

---

*Note the similarity to Signaling System No. 7.

16.10   QUALITY OF SERVICE (QoS)

16.10.1   ATM Service Quality Review

A basic performance measure for any digital data communication system is bit error rate (BER). Well-designed fiber-optic links predominate in the national network all the way down to the local serving exchange. We may expect BERs from such links between 1×10^{-10} and 1×10^{-12}; with end-to-end performance better than 1×10^{-9}. Thus other performance issues may dominate the scene. These may be called ATM-unique QoS items, namely:

- Cell transfer delay
- Cell delay variation
- Cell loss ratio
- Mean cell transfer delay
- Cell errored ratio
- Severely error cell block ratio
- Cell misinsertion rate

16.10.2   Definitions

Cell Event

1. A "cell exit event" occurs when the first bit of an ATM cell has completed transmission out of an end-user device to a private ATM network element across the private UNI measurement point, or out of a private ATM network element to a public ATM network element across the public UNI measurement point, or out of an end-user device to a public ATM network across the public UNI measurement point.

2. A "cell entry event" occurs when the last bit of an ATM cell has completed transmission into an end-user device from a private ATM network element across the private UNI measurement point, or into a private ATM network from a public ATM network element across the public UNI measurement point, or into an end-user device from a public ATM network element across the public UNI measurement point.

ATM Cell Transfer Outcome. The following are possible cell transfer outcomes between measurement points for transmitted cells (ITU-T definitions):

1. *Successful Cell Transfer Outcome.* The cell is received corresponding to the transmitted cell with a specified time $T_{\max}$. The binary content of the received cell conforms exactly to the corresponding cell payload,

and the cell is received with a valid header field after header error control procedures are completed.

2. *Errored Cell Outcome.* The cell is received corresponding to the transmitted cell within a specified time T_{max}. The binary content of the received cell payload differs from that of the corresponding transmitted cell or the cell is received with an invalid header field after the header error control procedures are completed.

3. *Lost Cell Outcome.* No cell is received corresponding to the transmitted cell within a specified time T_{max} (examples: "never showed up" or "late").

4. *Misinserted Cell Outcome.* This involves a received cell for which there is no corresponding transmitted cell.

5. *Severely Errored Cell Block Outcome.* This is when M or more lost cell outcomes, misinserted cell outcomes, or errored cell outcomes are observed in a receiver cell block of N cells transmitted consecutively on a given connection.

16.10.3 Cell Transfer Delay

Cell transfer delay is defined as the elapsed time between a cell exit event at measurement point 1 (e.g., at the source UNI) and the corresponding cell entry event at measurement point 2 (e.g., the destination UNI) for a particular connection. The cell transfer delay between two measurement points is the sum of the total inter-ATM node transmission delay and the total ATM node processing delay between MP_1 and MP_2.

In addition to the normal delay that one would expect for a cell to traverse a network, extra delay is added in the ATM network at each ATM switch. One cause of this delay is asynchronous digital multiplexing. Where this method is employed, two cells directed toward the same output port of an ATM switch or cross-connect can result in contention.

One or more cells are held in a buffer until the contention is resolved. Thus the second cell suffers additional delay. Delay of a cell depends on the traffic intensity within a switch which influences the probability of contention.

The asynchronous path of each ATM cell also contributes to cell delay. Cells can be delayed one or many cell periods, depending on traffic intensity, switch sizing, and the transmission path taken through the network.

16.10.4 Cell Delay Variation

ATM traffic, by definition, is asynchronous, magnifying transmission delay. Delay is also inconsistent across the network. It can be a function of time (i.e., a moment in time), network design/switch design (such as buffer size), and traffic characteristics at that moment of time. The result is cell delay variation (CDV).

CDV can have several deleterious effects. The dispersion effect, or spreading out, of cell interarrival times can impact signaling functions or the reassembly of cell user data. Another effect is called *clumping*. This occurs when the interarrival times between transmitted cells shorten. One can imagine how this could affect the instantaneous network capacity and how it can impact other services using the network.

There are two performance parameters associated with CDV: 1-point CDV and 2-point CDV.

The 1-point CDV describes variability in the pattern of cell arrival events observed at a single measurement point with reference to the negotiated peak rate $1/T$ as defined in ITU-T Rec. I.371 (Ref. 23).

The 2-point CDV describes variability in the pattern of cell arrival events as observed at the output of a connection portion (MP_2) with reference to the pattern of the corresponding events observed at the input to the connection portion (MP_1).

16.10.5 Cell Loss Ratio

Cell loss may not be uncommon in an ATM network. There are two basic causes of cell loss: error in cell header or network congestion.

Cells with header errors are automatically discarded. This prevents misrouting of errored cells, as well as the possibility of privacy and security breaches.

Switch buffer overflow can also cause cell loss. It is in these buffers that cells are held in prioritized queues. If there is congestion, cells in a queue may be discarded selectively in accordance with their level of priority. Here enters the cell loss priority (CLP) bit discussed in Section 16.4. Cells with this bit set to 1 are discarded in preference to other, more critical cells. In this way buffer fill can be reduced to prevent overflow.

Cell loss ratio is defined for an ATM connection as

$$(\text{Lost cells})/(\text{Total transmitted cells})$$

Lost and transmitted cells counted in severely errored cell blocks should be excluded from the cell population in computing cell loss ratio.

16.10.6 Mean Cell Transfer Delay

Mean cell transfer delay is defined as the arithmetic average of a specified number of cell transfer delays for one or more connections.

16.10.7 Cell Error Ratio

Cell error ratio is defined as follows for an ATM connection:

$$(\text{Errored cells})/(\text{Successfully transferred cells} + \text{Errored cells})$$

Successfully transferred cells and errored cells contained in cell blocks counted as severely errored cell blocks should be excluded from the population used in calculating cell error ratio.

16.10.8 Severely Errored Cell Block Ratio

The severely errored cell block ratio for an ATM connection is defined as

$$(\text{Severely errored cell blocks})/(\text{Total transmitted cell blocks})$$

A cell block is a sequence of N cells transmitted consecutively on a given connection. A severely errored cell block outcome occurs when more than M errored cells, lost cells, or misinserted cell outcomes are observed in a received cell block.

For practical measurement purposes, a cell block will normally correspond to the number of user information cells transmitted between successive OAM cells. The size of a cell block is to be specified.

16.10.9 Cell Misinsertion Rate

The cell misinsertion rate for an ATM connection is defined as

$$(\text{Misinserted cells})/(\text{Time interval})$$

Severely errored cell blocks should be excluded from the population when calculating the cell misinsertion rate. Cell misinsertion on a particular connection is most often caused by an undetected error in the header of a cell being transmitted on a different connection. This performance parameter is defined as a rate (rather than a ratio) because the mechanism producing misinserted cells is independent of the number of transmitted cells received on the corresponding connection.

Section 16.10 is based on ITU-T Rec. I.356, Ref. 24.

16.11 TRAFFIC CONTROL AND CONGESTION CONTROL

16.11.1 Generic Functions

The following functions form a framework for managing and controlling traffic and congestion in ATM networks and are to be used in appropriate combinations:

1. *Network Resource Management* (*NRM*). Provision is used to allocate network resources in order to separate traffic flows in accordance with service characteristics.

2. *Connection Admission Control (CAC)*. This is defined as a set of actions taken by the network during the call setup phase or during the call renegotiation phase in order to establish whether a VC or VP connection request can be accepted or rejected, or whether a request for reallocation can be accommodated. Routing is part of connection admission control actions.

3. *Feedback Controls*. These are a set of actions taken by the network and by users to regulate the traffic submitted on ATM connections according to the state of network elements.

4. *Usage Network Parameter Control (UPC/NPC)*. This is a set of actions taken by the network to monitor and control traffic, in terms of traffic offered and validity of the ATM connection, at the user access and network access, respectively. Their main purpose is to protect network resources from malicious as well as unintentional misbehavior that can affect the QoS of other already established connections by detecting violations of negotiated parameters and taking appropriate actions.

5. *Priority Control*. The user may generate different priority traffic flows by using the CLP. A congested network element may selectively discard cells with low priority, if necessary, to protect as far as possible the network performance for cells with higher priority.

Figure 16.30 is a reference configuration for traffic and congestion control.

16.11.2 Events, Actions, Time Scales, and Response Times

Figure 16.31 shows the time scales over which various traffic control and congestion control functions can operate. The response time defines how quickly the controls react. For example, cell discarding can react on the order

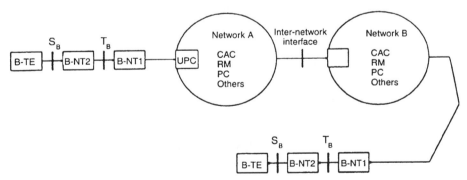

Figure 16.30. Reference configuration for traffic control and congestion control. *Note:* NPC may apply as well at some intranetwork NNIs, and the arrows indicate the direction of the cell flow. UPC, usage parameter control; CAC, connection admission control; PC, priority control; NPC, network parameter control; RM, resource management; Others, for further study. (From Figure 1/I.371, page 7, ITU-T Rec. I.371, Ref. 23.)

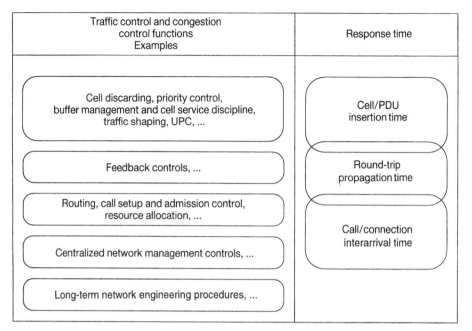

Traffic control and congestion control functions Examples	Response time
Cell discarding, priority control, buffer management and cell service discipline, traffic shaping, UPC, ...	Cell/PDU insertion time
Feedback controls, ...	Round-trip propagation time
Routing, call setup and admission control, resource allocation, ...	Call/connection interarrival time
Centralized network management controls, ...	
Long-term network engineering procedures, ...	

Figure 16.31. Control response times. (From Figure 2/I.371, page 8, ITU-T Rec. I.371, Ref. 23.)

of the insertion time of a cell. Similarly, feedback controls can react on the time scale of round-trip propagation times. Because traffic control and resource management functions are needed at different time scales, no single function is likely to be sufficient.

16.11.3 Quality of Service, Network Performance, and Cell Loss Priority

QoS at the ATM layer is defined by a set of parameters such as cell delay, cell delay variation sensitivity, cell loss ratio, and so forth.

A user requests a specific ATM layer QoS from the QoS classes which a network provides. This is part of the *traffic contract* at connection establishment. It is a commitment for the network to meet the requested QoS as long as the user complies with the traffic contract. If the user violates the traffic contract, the network need not respect the agreed-upon QoS.

A user may request at most two QoS classes for a single ATM connection, which differ with respect to the cell loss ratio objectives. The CLP bit in the ATM header allows for two cell loss ratio objectives for a given ATM connection.

Network performance objectives at the ATM SAP are intended to capture the network ability to meet the requested ATM layer QoS. It is the role of upper layers, including the AAL, to translate this ATM layer QoS to any specific application requested QoS.

16.11.4   Traffic Descriptors and Parameters

Traffic parameters describe traffic characteristics of an ATM connection. Traffic parameters are grouped into source traffic descriptors for exchanging information between the user and the network. Connection admission control procedures use source traffic descriptors to allocate resources and derive parameters for the operation of the user parameter control/network parameter control (UPC/NPC).

We now define the terms *traffic parameters* and *connection traffic descriptors*.

Traffic Parameters. A traffic parameter is a specification of a particular traffic aspect. It may be qualitative or quantitative. Traffic parameters, for example, may describe peak cell rate, average cell rate, burstiness, peak duration, and source type (such as telephone, videophone). Some of these traffic parameters are interdependent, such as burstiness with average and peak cell rate.

Traffic Descriptors. The ATM traffic descriptor is the generic list of traffic parameters which can be used to capture the intrinsic traffic characteristics of an ATM connection. A *source traffic descriptor* is the set of traffic parameters belonging to the ATM traffic used during the connection setup to capture the intrinsic traffic characteristics of the connection requested by the source.

Connection Traffic Descriptor. This specifies the traffic characteristics of the ATM connection at the public or private UNI. The connection traffic descriptor is the set of traffic parameters in the source traffic descriptor, CDV tolerance, and the conformance definition that is used to unambiguously specify the conforming cells of an ATM connection. Connection admission control (CAC) procedures will use the connection traffic descriptor to allocate resources and to derive parameter values for the operation of the UPC. The connection traffic descriptor contains the necessary information for conformance testing of cells of the ATM connection at the UNI.

Any traffic parameter and the CDV tolerance in a connection traffic descriptor should fulfill the following requirements:

1. They should be understandable by the user or terminal equipment, and conformance testing should be possible as stated in the traffic contract.
2. They should be useful in resource allocation schemes, meeting network performance requirements as described in the traffic contract.
3. They should be enforceable by the UPC.

16.11.5 User–Network Traffic Contract

16.11.5.1 Operable Conditions.
CAC and UPC/NPC procedures require the knowledge of certain parameters to operate efficiently. For example, they should take into account the source traffic descriptor, the requested QoS, and the CDV tolerance (defined below) in order to decide whether the requested connection can be accepted.

The source traffic descriptor, the requested QoS for any given ATM connection, and the maximum CDV tolerance allocated to the CEQ (customer equipment) define the traffic contract at the T_B reference point (see Figure 16.30). Source traffic descriptors and QoS are declared by the user at connection setup by means of signaling or subscription. Whether the maximum allowable CDV tolerance is also negotiated on a subscription or on a per connection basis is for further study by the ITU-T organization.

The CAC and UPC/NPC procedures are operator specific. Once the connection has been accepted, the value of the CAC and UPC/NPC parameters are set by the network on the basis of the network operator's policy.

ITU-T Rec. I.371 (Ref. 23) notes that all ATM connections handled by network connection related functions (CRF) have to be declared and enforced by the UPC/NPC. ATM layer QoS can only be assured for compliant ATM connections. As an example, individual VCCs inside user end-to-end VPC are neither declared nor enforced at the UPC and hence no ATM layer QoS can be assured for them.

16.11.5.2 Source Traffic Descriptor, Quality of Service, and Cell Loss Priority.
If a user requests two levels of priority for an ATM connection, as indicated by the CLP bit value, the intrinsic traffic characteristics of both cell flow components have to be characterized in the source traffic descriptor. This is by means of a set of traffic parameters associated with the CLP = 0 component and a set of traffic parameters associated with the CLP = 0 + 1 component.

16.11.5.3 Impact of Cell Delay Variation on UPC/NPC Resource Allocation.
ATM layer functions such as cell multiplexing may alter the traffic characteristics of ATM connections by introducing cell delay variation, as shown in Figure 16.32. When cells from two or more ATM connections are multiplexed, cells of a given ATM connection may be delayed while cells of another ATM connection are being inserted at the output of the multiplexer. Similarly, some cells may be delayed while physical layer overhead or OAM cells are inserted. Therefore, some randomness affects the time interval between reception of ATM-cell data requests at the endpoint of an ATM connection and the time that an ATM-cell data indication is received at the UPC/NPC. Besides, AAL multiplexing may cause CDV.

The UPC/NPC mechanism should not discard or tag cells in an ATM connection if the source conforms to the source traffic descriptor negotiated

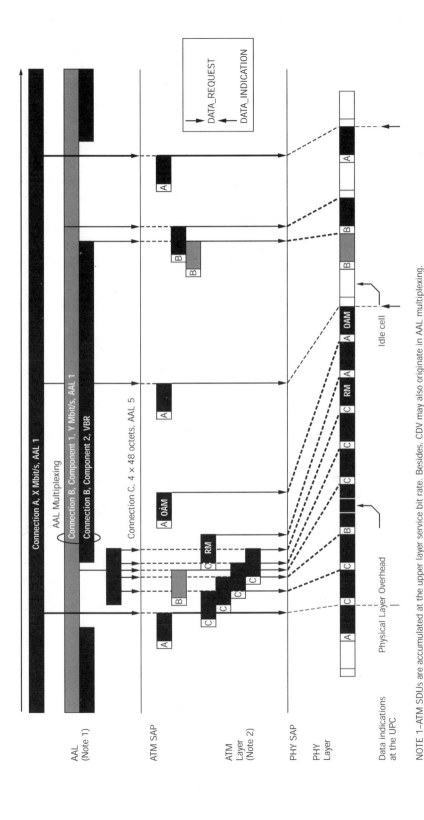

Figure 16.32. Origins of cell delay variation. (From Figure 3/I.371, page 13, ITU-T Rec. I.371, Ref. 23.)

NOTE 1–ATM SDUs are accumulated at the upper layer service bit rate. Besides, CDV may also originate in AAL multiplexing.
NOTE 2–GFC delay and delay variation is part of the delay and delay variation introduced by the ATM layer.
NOTE 3–CDV may also be introduced by the network because of random queuing delays which are experienced by each cell in concentrators, switches and cross-connects.

at connection establishment. However, if the CDV is not bounded at a point where the UPC/NPC function is performed, it is not possible to design a suitable UPC/NPC mechanism and to allocate resources properly. Therefore, it is required that a maximum allowable value of CDV be standardized edge-to-edge (e.g., between the ATM connection endpoint and T_B, between T_B and an internetwork interface, and between internetwork interfaces; see Figure 16.30).

UPC/NPC should accommodate the effect of the maximum CDV allowed on ATM connections within the limit resulting from the accumulated CDV allocated to upstream subnetworks including customer equipment (CEQ). Traffic shaping partially compensates for the effects of CDV on the peak cell rate of the ATM connection. Examples of traffic-shaping mechanisms are re-spacing cells of individual ATM connections according to their peak cell rate or suitable queue service schemes.

Values of the CDV are network performance issues. The definition of a source traffic descriptor and the standardization of a maximum allowable CDV may not be sufficient for a network to allocate resources properly. When allocating resources, the network should take into account the worst-case traffic passing through the UPC/NPC in order to avoid impairments to other ATM connections. The worst-case traffic depends on the specific implementation of the UPC/NPC. The tradeoffs between UPC/NPC complexity, worst-case traffic, and optimization of network resources are made at the discretion of network operators. The quantity of available network resources and the network performance to be provided for meeting QoS requirements can influence these tradeoffs.

16.11.5.4 Cell Conformance and Connection Compliance. Conformance applies to the cells as they pass the UNI and are in principle tested according to some combination of generic cell rate algorithms (GCRAs). The first cell of the connection initializes the algorithm, and from then on each cell is either conforming or not conforming. Because in all likelihood even with the best intentions a cell or two may be nonconforming, thus it is inappropriate for the network operator to only commit to the QoS objectives for connections whose cells are conforming. Thus the term *compliant*, which is not precisely defined, is used for connection in which some of the cells may be nonconforming.

The precise definition of a compliant connection is left to the network operator. For any definition of a compliant connection, a connection for which all cells are conforming are identified as compliant.

Based on action of the UPC function, the network may decide whether a connection is compliant or not. The commitment by the network operator is to support the QoS for all connections that are compliant.

For compliant connections at the public UNI, the agreed QoS is supported for at least the number of cells equal to the conforming cells according to the conformance definition. For noncompliant connections, the network need not respect the agreed QoS class. The conformance definition that defines conformity at the public UNI of the cells of the ATM connection uses a GCRA configuration in multiple instances to apply to particular combinations of the CLP = 0 and CLP = 1 + 0 cell streams with regard to the peak cell rate and to particular combinations of CLP = 0, CLP = 1, and CLP = 0 + 1 cell streams with regard to the sustainable cell rate and burst tolerance. For example, the conformance definition may use the GCRA twice, once for peak cell rate of the aggregate (CLP = 0 + 1) cell stream and once for the sustainable cell rate of the CLP = 0 cell stream. The network operator may offer a limited set of alternative conformance definitions (all based on GCRA) from which the user may choose for a given ATM connection.

16.11.5.5 *Generic Cell Rate Algorithm (GCRA).*
The GCRA is a virtual scheduling algorithm or a continuous-state leaky bucket algorithm as defined by the flow chart in Figure 16.33. The GCRA is used to operationally define the relationship between peak cell rate (PCR) and the CDV tolerance and the relationship between sustained cell rate (SCR) and the burst tolerance. In addition, for the cell flow of an ATM connection, the GCRA is used to specify the conformance at the public or private UNI to declared values of the above two tolerances, as well as declared values of the traffic parameters "PCR" and "SCR and burst tolerance."

For each cell arrival, the GCRA determines whether the cell is conforming with the traffic contract of the connection, and thus the GCRA is used to provide the formal definition of traffic conformance to the traffic contract. Although traffic conformance is defined in terms of the GCRA, the network provider is not obligated to use this algorithm (or this algorithm with the same parameter values) for the UPC. Rather, the network provider may use any UPC as long as the operation of the UPC does not violate the QoS objectives of a compliant connection.

The GCRA depends only on two parameters: the increment I and the limit L. These parameters have been denoted by T and τ, respectively, and in ITU-T Rec. I.371 (Ref. 23), but have been given more generic labels (by the ATM Forum) herein because the GCRA is used in multiple instances. We will now use the ATM Forum notation "GCRA (I, L)," which means "the GCRA with the value of the increment parameter set equal to I and the value of the limit parameter set equal to L."

The GCRA is formally defined in Figure 16.33, which the ATM Forum took as a generic version of Figure 1 in Annex A of I.371. The two algorithms in Figure 16.33 are equivalent in the sense that for any sequence of cell arrival times $[t_a(k), k \geq 1]$, the two algorithms determine the same cells to be conforming and thus the same cells to be nonconforming. The two algorithms are easily compared if one notices that at each arrival epoch, $t_a(k)$, and after the algorithms have been executed, TAT = X + LCT (see Figure 16.33).

The virtual scheduling algorithm updates a theoretical arrival time (TAT), which is the "nominal" arrival time of the cell assuming equally spaced cells

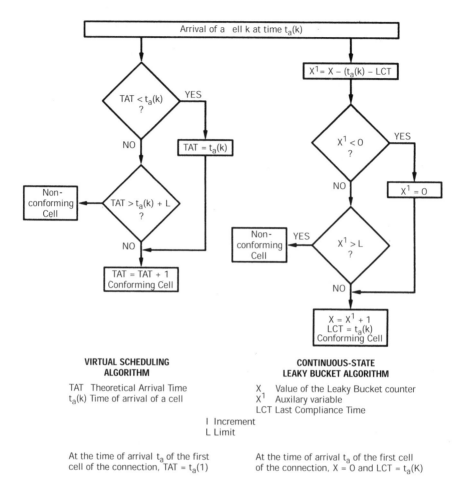

Figure 16.33. Equivalent versions of the generic cell rate algorithm (GCRA). (From Figure A-1/I.371, page 62, ITU-T Rec. I.371, Ref. 23.)

when the source is active. If the actual arrival time of a cell is not "too" early relative to the TAT, in particular if the actual arrival time is after TAT-L, then the cell is conforming; otherwise the cell is nonconforming.

The continuous-state leaky bucket algorithm can be viewed as a finite-capacity bucket whose real-valued content drains out at a continuous rate of 1 unit of content per time unit and whose content is increased by the increment I for each conforming cell. Equivalently, it can be viewed as the work load in a finite-capacity queue or as a real-valued counter. If at a cell arrival the content of the bucket is less than or equal to the limit value, L, then the cell is conforming; otherwise the cell is nonconforming. The capacity of the bucket (i.e., the upper bound on the counter) is $L + I$.

16.11.5.6   Traffic Contract Parameter Specification. Peak cell rate for CLP = 0 + 1 is a mandatory traffic parameter to be explicitly or implicitly declared in any source traffic descriptor. In addition to the peak cell rate of an ATM connection, it is mandatory for the user to declare either explicitly or implicitly the cell delay variation tolerance τ within the relevant traffic contract.

Peak Cell Rate (PCR). The following definition applies to ATM connections supporting both CBR and VBR services (Ref. 23):

> The peak cell rate in the source traffic descriptor specifies an upper bound on the traffic that can be submitted on an ATM connection. Enforcement of this bound by the UPC/NPC allows the network operator to allocate sufficient resources to ensure that the performance objectives (e.g., for cell loss ratio) can be achieved.

For switched ATM connections, the PCR for CLP = 0 + 1 and the QoS class must be explicitly specified for each direction in the connection-establishment SETUP message.

The CDV tolerance must be either explicitly specified at subscription time or implicitly specified.

The SCR and burst tolerance comprise an optional traffic parameter set in the source traffic descriptor. If either SCR or burst tolerance is specified, then the other must be specified within the relevant traffic contract.

16.12   TRANSPORTING ATM CELLS

16.12.1   In the DS3 Frame

One of the most popular higher-speed digital transmission systems in North America is DS3 operating at a nominal transmission rate of 45 Mbps. It is also being widely implemented for transport of SMDS. The system used to map ATM cells into the DS3 format is the same as used for SMDS.

DS3 uses the physical layer convergence protocol (PLCP) to map ATM cells into its bit stream. A DS3 PLCP frame is shown in Figure 16.34. There are 12 cells in a frame. Each cell is preceded by a 2-octet framing pattern (A1, A2), to enable the receiver to synchronize to cells. After the framing pattern there is an indicator consisting of one of 12 fixed bit patterns used to identify the cell location within the frame (POI). This is followed by an octet of overhead information used for path management. The entire frame is then padded with either 13 or 14 nibbles (a nibble = 4 bits) of trailer to bring the transmission rate up to the exact DS3 bit rate. The DS3 frame has 125-μs duration.

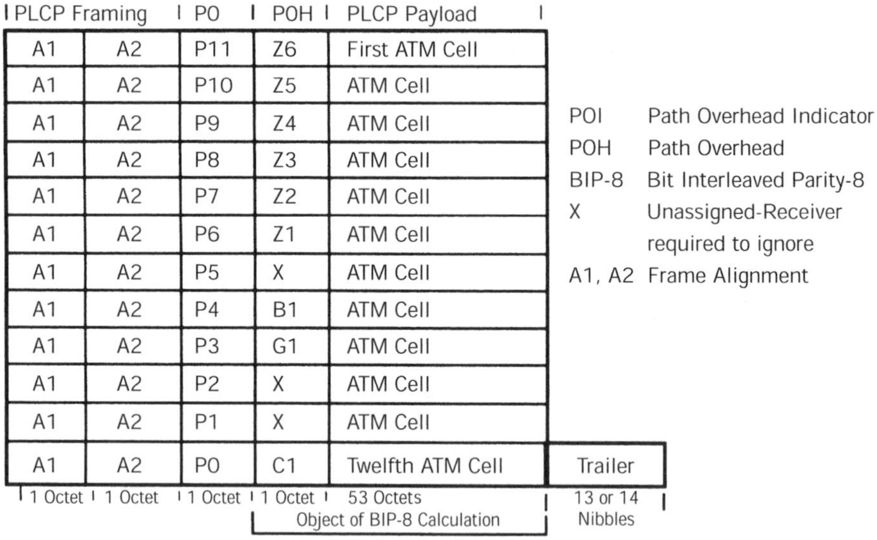

	PLCP Framing		PO		POH		PLCP Payload		

A1	A2	P11	Z6	First ATM Cell
A1	A2	P10	Z5	ATM Cell
A1	A2	P9	Z4	ATM Cell
A1	A2	P8	Z3	ATM Cell
A1	A2	P7	Z2	ATM Cell
A1	A2	P6	Z1	ATM Cell
A1	A2	P5	X	ATM Cell
A1	A2	P4	B1	ATM Cell
A1	A2	P3	G1	ATM Cell
A1	A2	P2	X	ATM Cell
A1	A2	P1	X	ATM Cell
A1	A2	P0	C1	Twelfth ATM Cell	Trailer

| 1 Octet | 1 Octet | 1 Octet | 1 Octet | 53 Octets | 13 or 14 |
Object of BIP-8 Calculation — Nibbles

POI   Path Overhead Indicator
POH   Path Overhead
BIP-8 Bit Interleaved Parity-8
X   Unassigned-Receiver
       required to ignore
A1, A2 Frame Alignment

Figure 16.34. Format of DS3 PLCP frame. (From Ref. 25. Courtesy of Hewlett-Packard Company.

DS3 has to contend with network slips (added/dropped frames to accommodate synchronization alignment). Thus PLCP is padded with a variable number of stuff (justification) bits to accommodate possible timing slips. The C1 overhead octet indicates the length of padding. The bit interleaved parity (BIP) checks the payload and overhead functions for errors and performance degradation. This performance information is transmitted in the overhead.

Figure 16.35 illustrates a 680-bit DS3 frame transporting ATM cells. The figure also shows 84 bits being grouped into 21 nibbles. Remember a nibble is half an octet or byte; it is a group of 4 bits.

ATM cells can also be mapped into a DS3 frame by what is called HEC-based mapping. In this case the ATM cells are mapped into the payload with the octet structure of the cells aligned with the nibble structure of the multiframe. The multiframe is organized such that 84 bits of payload follow every overhead bit. The 84 bits can be assumed to be organized into 21 consecutive nibbles. The ATM cell is placed such that the start of cell always coincides with the start of nibble. As in most ATM cell mapping situations, the cells may cross multiframe boundaries. See Figure 16.35.

16.12.2 DS1 Mapping

One approach to mapping ATM cells into a DS1 frame is to use a similar procedure as used on DS3 with PLCP. In this case only 10 cells are bundled into a frame, and two of the Z overheads are removed. The padding in the frame is set at 6 octets. The entire frame takes 3 ms to transmit and spans

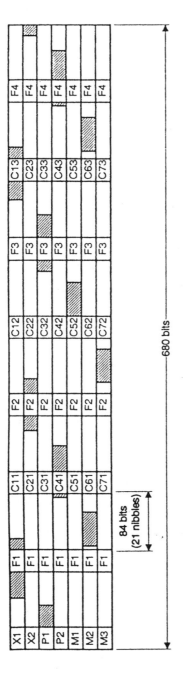

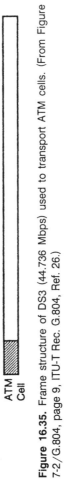

Figure 16.35. Frame structure of DS3 (44.736 Mbps) used to transport ATM cells. (From Figure 7-2/G.804, page 9, ITU-T Rec. G.804, Ref. 26.)

1	1	1	1	◄——53 Octets——►
A1	A2	P9	Z4	ATM Cell
A1	A2	P8	Z3	ATM Cell
A1	A2	P7	Z2	ATM Cell
A1	A2	P6	Z1	ATM Cell
A1	A2	P5	F1	ATM Cell
A1	A2	P4	B1	ATM Cell
A1	A2	P3	G1	ATM Cell
A1	A2	P2	M2	ATM Cell
A1	A2	P1	M1	ATM Cell
A1	A2	P0	C1	ATM Cell

OH Byte	Function
A1, A2	Framing Bytes
P9-P0	Path Overhead Identifier Bytes

PLCP Path Overhead Bytes

Z4-Z1	Growth Bytes
F1	PLCP Path User Channel
B1	BIP-8
G1	PLCP Status
M2-M1	SMDS Conyrol Onformation
C1	Cycle/Stuff Counter Byte

Trailer = 6 Octets

3 msec

Figure 16.36. DS1 mapping with PLCP. (From Ref. 25. Courtesy of Hewlett-Packard Company.)

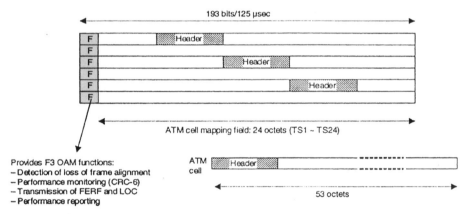

Figure 16.37. Frame structure for 1.544 Mbps (DS1) used to transport ATM cells. (From Figure 2-1/G.804, page 8, ITU-T Rec. G.804, Ref. 26.)

many DS1 ESF (extended superframe) frames. The mapping is shown in Figure 16.36. One must consider the arithmetic. There are 24 time slots (octets) in a DS1 frame, which breaks down to 192 bits. There are 424 (8 × 53) bits in an ATM cell. This, of course, leads to the second method of carrying ATM cells in DS1 by directly mapping ATM cells octet for octet (time slots). This is done in groups of 53 octets (1 cell) and would, by necessity, cross DS1 frame boundaries to accommodate an ATM cell. See Figure 16.37.

16.12.3 E1 Mapping

E1 PCM has a 2.048-Mbps transmission rate. An E1 frame has 256 bits representing 32 channels or time slots, 30 of which carry traffic, and which are available to carry ATM cells. Thus, a frame has 32 time slots (TS) where

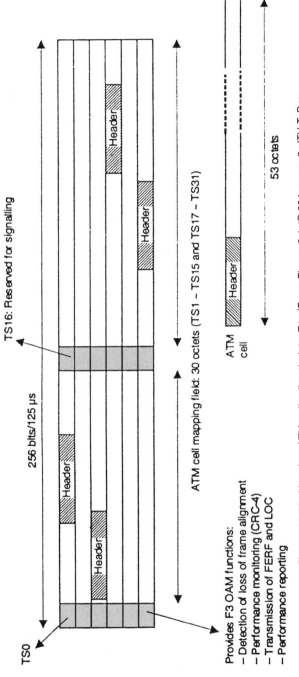

Figure 16.38. Mapping ATM cells directly into E1. (From Figure 3-1/G.804, page 3, ITU-T Rec. G.804, Ref. 26.)

663

TS0 and TS16 are reserved. TS0 is used for synchronization of the frame and TS16 is used for signaling. The E1 frame is shown in Figure 16.38. Bit positions 9–128 and 137–256 may be used for ATM cell mapping. ATM cells can also be directly mapped into special E3 and E4 frames. The first has 530 octets available for cells (thus it fits 10 cells) and the second has 2160 octets (not evenly divisible).

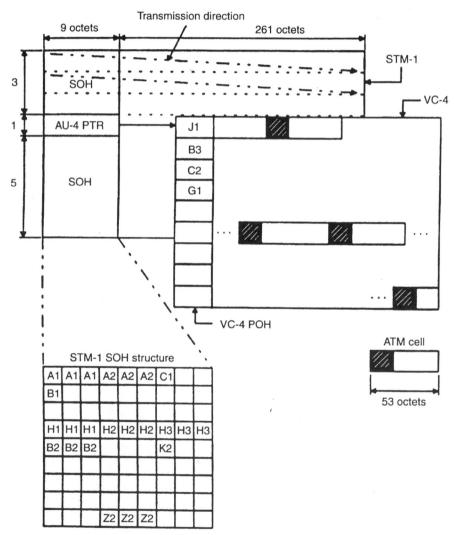

Figure 16.39. Mapping ATM cells into the 155.520 Mbps frame structure of SDH-based UNI. (From Figure 8/I.432.1, page 13, ITU-T Rec. I.432.1, Ref. 5.)

16.12.4   Mapping ATM Cells into SDH

16.12.4.1   At STM-1 (155.520 Mbps). SDH is described in Chapter 15. Figure 16.39 shows the mapping procedure. The ATM cell stream is first mapped into the C-4 and then mapped into the VC-4 container along with the VC-4 path overhead. The ATM cell boundaries are aligned with the STM-1 octet boundaries. Because the C-4 capacity (2340 octets) is not an integer multiple of the cell length (53 octets), a cell may cross a C-4 boundary. The AU-4 pointer (octets H1 and H2 in the SOH) is used for finding the first octet of the VC-4.

16.12.4.2   At STM-4 (622.080 Mbps). As shown in Figure 16.40, the ATM cell stream is first mapped into C-4-4c and then packed into the VC-4-4c container along with the VC-4-4c path overhead. The cell boundaries are

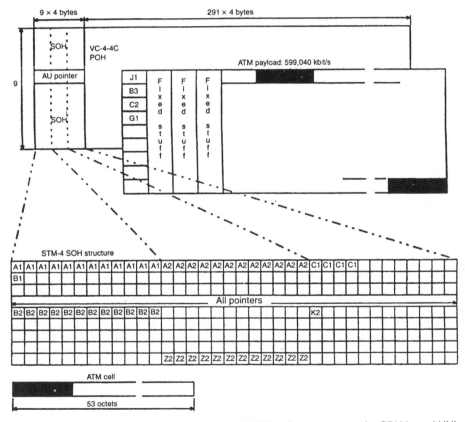

Figure 16.40. Mapping ATM cells into the 622.080-Mbps frame structure for SDH-based UNI. (From Figure 8/I.432.1, page 13, ITU-T Rec. I.432.1, Ref. 5.)

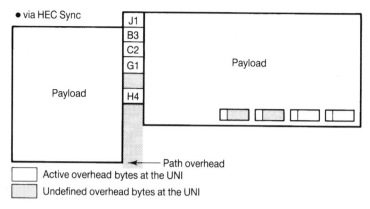

Figure 16.41. Mapping ATM cells into a SONET STM-1 frame at the UNI.

aligned with STM-4 octet boundaries. Because the C-4-4c capacity (9360 octets) is not an integer multiple of the cell length (53 octets), a cell may cross a C-4-4c boundary. The AU pointers are used for finding the first octet of the VC-4-4c.

16.12.5   Mapping ATM Cells into SONET

ATM cells are mapped directly into the SONET payload (49.54 Mbps). As with SDH, the payload in octets is not an integer multiple of the cell length, and thus a cell may cross an STS frame boundary. This mapping concept is shown in Figure 16.41. The H4 pointer indicates where the cells begin inside an STS frame. Another approach is to identify cell headers and, thus, identify the first cell in the frame.

16.12.6   Cell Rates for Various Typical Connections

Table 16.6 shows connection cell rates for various familiar user information bit rate capacities. By choosing appropriate cell rate values for the traffic parameters associated with a connection, a user defines the bit rate capacity associated with an ATM connection. The cell rates listed in the table include an allowance for the AAL header and operations, administration, and maintenance (OAM) cell overhead.

These rates are used to assign the peak cell rate (PCR) and sustainable cell rate (SCR) traffic parameters defined in the ATM Forum UNI Specification Version 3.1.

TABLE 16.6   Connection Cell Rates for Various Familiar Bit Rate Capacities

User Information Bandwidth	Cell Rate (Cells / Second)	Comments
$N \times 64$ Kbps (DS0)	$N \times 173$	$N = 1$ to 23 (i.e., sub-DS1 rates)
$N \times 384$ Kbps	$N \times 1035$	$N = 1$ to 31 (i.e., up to 12 Mbps)
$N \times 1.544$ Mbps (DS1)	$N \times 4140$	$N = 1$ to 85 (i.e., up to 131 Mbps)
$N \times 44.736$ Mbps (DS3)	$N \times 119{,}910$	$N = 1, 2$ (i.e., 1 or $2 \times$ DS3 rate)
$N \times 1$ Mbps	$N \times 2689$	$N = 1$ to 131 (i.e., up to 131 Mbps)
DS3 PLCP	96,000	DS3 line rate, PLCP option
DS3 direct mapping	104,268	DS3 line rate, direct mapping (HEC)
Full STS-3c	353,207	STS-3c line rate
Full STS-12c	1,412,828	STS-12c line rate

Source: Table 9-1, page 9-1, Bellcore Special Report, SR-3330, Ref. 27.

16.13   ATM OPERATION, ADMINISTRATION, AND MAINTENANCE (OAM)

16.13.1   OAM Levels in the B-ISDN

OAM functions in the network are performed on five OAM hierarchical levels associated with the ATM and physical layers of the protocol reference model. (See Figures 16.2 and 16.3.) The functions result in corresponding bidirectional information flows F1, F2, F3, F4, and F5 referred to as OAM flows. Not all of these need to be present. The OAM functions of a missing level are performed at the next higher level. The levels are as given in Table 16.7

TABLE 16.7   OAM Levels in the B-ISDN (in Decreasing Order)

Level	Description
F5	*Virtual channel level:* Extends between network elements performing virtual channel connection termination functions and can extend through one or more path connections.
F4	*Virtual path level:* Extends between network elements performing virtual path connection termination functions and may extend through one or more transmission paths.
F3	*Transmission path level:* Extends between network elements assembling/disassembling the payload of a transmission system and associating it with its OAM functions. Cell delineation and header error control (HEC) functions are required at the endpoints of each transmission path. The transmission path is connected through one or more digital sections.
F2	*Digital section level:* Extends between section endpoints and comprises a maintenance entity according to the definition in ITU-T Rec. M.20, Section 3.
F1	*Regenerator section level:* A regenerator section is a portion of a digital section and as such is a maintenance subentity.

OAM functions related to OAM levels are independent from the OAM functions of other layers and have to be provided at each layer.

Each layer, where OAM functions are required, is able to carry out its own processing to obtain quality and status information. OAM functions are performed by the layer management. These results may be provided to the plane management or to the adjacent higher layer. Higher-layer functions are not necessary to support the OAM of the lower layer.

16.13.2 Mechanisms to Provide OAM Flows

16.13.2.1 Physical Layer Mechanisms. The physical layer contains the three lowest OAM levels as outlined in Table 16.7. The allocation of the OAM flows is as follows:

F1: regenerator section level
F2: digital section level
F3: transmission path level

The mechanisms to provide OAM functions and to generate OAM flows F1, F2, and F3 depend on the format of the transmission system as well as on the supervision functions contained in B-NT1 and B-NT2 for the section crossing the T_B reference point. There are three types of transmission for customer access: SDH-based, cell-based, and PDH-based.

SDH-Based Transmission Systems. Flows F1 and F2 are carried on bytes in the section overhead (SOH); flow F3 is carried in the path overhead (POH) of the transmission frame.

Cell-Based Transmission Systems. Such transmission systems may use an interface structure as specified in Section 4.2 of ITU-T Rec. I.432.1. OAM flows F1 and F3 are carried through maintenance cells for the physical layer using a specific pattern in the header for F1 and F3. F2 flows are not provided, but the associated functions are supported by F3 flows. These cells are not passed to the ATM layer. The occurrence of a PLOAM* cell is determined by the requirements of the supported OAM functions. For each type (F1 and F3) of PL-OAM cell, maximum spacing is applied. If maximum spacing is exceeded, loss of OAM flow (LOM) will occur.

PDH†-Based Transmission Systems. These systems may only be used on the network side of the B-NT1. Specific means to monitor the section performance (e.g., violation code counting CRC, etc.) are specified for these systems. The capability to carry OAM information other than bit-oriented messages is very limited.

---

*PLOAM or PL-OAM stands for physical layer OAM.
†PDH stands for plesiochronous digital hierarchy (e.g., E1, DS1, DS3, etc.).

16.13.3   ATM Layer Mechanism: F4 and F5 Flows

The ATM layer contains the two highest OAM levels shown in Table 16.7. These are:

F4:   virtual path level

F5:   virtual channel level

These OAM flows are provided by cells dedicated to ATM layer OAM functions for both virtual channel connections (VCC) and virtual path connections (VPC). In addition, such cells are usable for communication within the same layers of the management plane.

16.13.3.1   F4 Flow Mechanism. The F4 flow is bidirectional. OAM cells for the F4 flow have the same VPI value as the user cells of the VPC and are identified by one or more preassigned VCI values. The same preassigned VCI value shall be used for both directions of the F4 flow. The OAM cells for both directions of the F4 flow must follow the same physical route so that any connecting points supporting that connection can correlate the fault and performance information from both directions.

There are two kinds of F4 flows, which can simultaneously exist in a VPC. These are as follows:

- *End-to-End F4 Flow.* This flow, identified by a standardized VCI (see Rec. I.361), is used for end-to-end VPC operations communications.
- *Segment F4 Flow.* This flow, identified by a standardized VCI (see Rec. I.361), is used for communicating operations information within the bounds of one VPC link or multiple interconnected VPC links, where all of the links are under the control of one administration or organization. Such a concatenation of VPC links is called a VPC segment. A VPC segment can be extended beyond the control of one administration by mutual agreement.

F4 flows must be terminated only at the endpoints of a VPC or at the connecting points terminating a VPC segment. Intermediate points (i.e., connecting points) along the VPC or along the VPC segment may monitor OAM cells passing through them and insert new OAM cells, but they cannot terminate the OAM flow. The F4 flow will be initiated at or after connection setup.

The administration/organization that controls the insertion of OAM cells for operations and maintenance of a VPC segment must ensure that such OAM cells are extracted before they leave the span of control of that administration/organization (Ref. 28).

16.13.3.2 F5 Flow Mechanism. The F5 flow is bidirectional. OAM cells for the F5 flow have the same VCI/VPI values as the user cells of the VCC and are identified by the payload type identifier (PTI). The same PTI value shall be used for both directions of the F5 flow. The OAM cells for both directions of the F5 flow must follow the same physical route so that any connecting points supporting that connection can correlate the fault and performance information from both directions.

There are two kinds of F5 flows, which can simultaneously exist in a VCC. These are as follows:

- *End-to-End F5 Flow.* This flow, identified by a standardized PTI (see Rec. I.361), is used for end-to-end VCC operations communications.
- *Segment F5 Flow.* This flow, identified by a standardized PTI (see Rec. I.361), is used for communicating operations information within the bounds of one VCC link or multiple interconnected VCC links, where all of the links are under the control of one administration or organization. Such a concatenation of VCC links is called a VCC segment. A VCC segment can be extended beyond the control of one administration by mutual agreement.

F5 flows must be terminated only at the endpoints of a VCC or at the connecting points terminating a VCC segment. Intermediate points (i.e., connecting points) along the VCC or along the VCC segment may monitor OAM cells passing through them and insert new OAM cells, but they cannot terminate the OAM flow. The F5 flow will be initiated at or after connection setup.

The administration/organization that controls the insertion of OAM cells for operations and maintenance of a VCC segment must ensure that such OAM cells are extracted before they leave the span of control of that administration/organization.

16.13.4 OAM Functions of the Physical Layer

16.13.4.1 OAM Functions. Two types of OAM functions need to be distinguished:

1. OAM functions supported solely by the flows F1, F2, and F3.
   - Dedicated to detection and indication of unavailability state
   - Requiring "real time" failure information transport toward the affected endpoints for system protection
2. OAM functions with regard to the system management
   - Dedicated to performance monitoring and reporting, or for localization of failed equipment
   - May be supported by the flows F1 to F3 or by other means

OAM Functions Supported Solely by the Flows F1 to F3. Table 16.8 gives an overview of the OAM functions and the related OAM flows. It also lists the different failures to be detected together with the failure indications for the SDH-based physical layer. Table 16.9 illustrates the same information for the cell-based physical layer (Ref. 28).

16.13.5   OAM Functions of the ATM Layer

The F4 flow relates to the *virtual path* and the F5 flow relates to the *virtual channel*. In both cases the fault management functions consist of monitoring of the path/channel for availability and other, overall performance monitoring. A path or channel is either not available or has degraded performance.

16.13.5.1   VP-AIS and VP-FERF Alarms. The VP-AIS (virtual path–alarm indication signal) and the VP-FERF (virtual path–far-end reporting failure) alarms are used for identifying and reporting VPC (virtual path connection) failures.

VP-AIS. VP-AIS cells are generated and sent downstream to all affected VPCs from the VPC connecting point (e.g., ATM cross-connect) which detects the VPC failure. VP-AIS results from failure indications from the physical layer as shown in Tables 16.8 and 16.9.

VP-AIS Cell Generation Condition. VP-AIS cells are generated and transmitted as soon as possible after failure indication, and transmitted periodically during the failure condition in order to indicate VPC unavailability. The generation frequency of VP-AIS cells is nominally one cell per second and is the same for each VPC concerned. VP-AIS cell generation is stopped as soon as the failure indications are removed.

VP-AIS Cell Detection Condition. VP-AIS cells are detected at the VPC endpoint and VP-AIS status is declared after the reception of one VP-AIS cell. VP connecting points may monitor the VP-AIS cells.

VP-AIS Release Condition. The VP-AIS state is removed under either of the following conditions:

- Absence of VP-AIS cell for nominally 3 s
- Receipt of one valid cell (user cell or continuity check cell)

VP-FERF. VP-FERF is sent to the far-end from a VPC endpoint as soon as it has declared a VP-AIS state or detected VPC failure.

TABLE 16.8 OAM Functions of the SDH-Based Physical Layer (Failures Occurring on the B-NT2 ↔ B-NT1 Section)

Level	Function	Failure Detection	System Protection and Failure Information Transmitted in the Flow		
			F2 on the B-NT2 ↔ B-NT1 Section	B-NT1 ↔ LT Section (Note 2)	F3 on the B-NT2 ↔ Transmission Path Termination
Regenerator section	Signal detection frame alignment	Loss of signal or loss of frame into B-NT1 (from B-NT2)	MS-FERF toward the B-NT2 (Note 3)	Note 1	Path-AIS toward the transmission path termination (generated by the B-NT1)
		Loss of signal or loss of frame into B-NT2 (from B-NT1)	MS-FERF toward the B-NT1 (Note 3)		Path-FERF toward the transmission path termination (generated by the B-NT2)
Digital section	Section error monitoring (B2)	Unacceptable error performance into B-NT1	MS-FERF toward the B-NT2 (Notes 3, 4)	Note 1	Path-AIS toward the transmission path termination (generated by the B-NT1) (Note 4)
		Unacceptable error performance into B-NT2	MS-FERF toward the B-NT1 (Notes 3, 4)		—

Transmission path				
Cell rate decoupling	Failure of insertion/ suppression of idle cells in B-NT2	—	Note 1	For further study
Cell delineation	Loss of cell sync into B-NT2	—		Path-FERF
CN status monitoring	CN not available	—		Path-AIS
AU pointer operation	Loss of AU pointer or path-AIS into B-NT2	—		Path-FERF toward the transmission path termination

Note 1: Capabilities for reporting faults from the T_B reference point to the relevent Q-interface must be accommodated by the transmission equipment specification.

Note 2 In accordance with the OAM Recommendation of the transmission system.

Note 3: In accordance with the SDH Recommendations, the term MS (multiplex section) is used.

Note 4: Can be disabled (see Rec. G.783).

Source: Table 1 / I.610, page 9, Ref. 28.

TABLE 16.9 OAM Functions of the Cell-Based Physical Layer (Failures Occurring on the B-NT2 ↔ B-NT1 Section)

Level	Function	Failure Detection	System Protection and Failure Information Transmitted in the Flow		
			F1 on the B-NT2 ↔ B-NT1 Section	B-NT1 ↔ LT Section (Note 2)	F3 on the B-NT2 ↔ Transmission Path Termination
Regenerator section	Signal detection PL-OAM cell recognition	Loss of signal or loss of F1 PL-OAM cell recognition into B-NT1 (from B-NT2)	Section-FERF toward the B-NT2	Note 1	Path-AIS toward the transmission path termination (generated by the B-NT1) (Note 3)
		Loss of signal or loss of F1 PL-OAM cell recognition into B-NT1 (from B-NT2)	Section-FERF toward the B-NT1		Path-FERF toward the transmission path termination (generated by the B-NT2)
	Section error monitoring	Unacceptable error performance into B-NT1	Section-FERF toward the B-NT2		Path-AIS toward the transmission path termination (generated by the B-NT1)
		Unacceptable error performance into B-NT2	Section-FERF toward the B-NT1		—

Transmission path					For further study
	Cell rate decoupling	Failure of insertion / suppression of idle cells in B-NT2	—	Note 1	
	PL-OAM cell recognition	Loss of F3 PL-OAM cell recognition into B-NT2	—		Path-FERF
	Cell delineation	Loss of cell sync into B-NT2	—		Path-FERF
	CN status monitoring	CN not available	—		Path-AIS

Note 1: Capabilities for reporting faults from the T_B reference point to the relevant Q-interface must be accommodated by the transmission equipment specification.

Note 2: In accordance with the OAM Recommendation of the transmission system.

Note 3: The B-NT1 as a connecting point can insert a path-AIS at the F3 level.

Source: Table 2/I.610, page 10, Ref. 28.

VP-FERF Cell Generation Condition. VP-FERF cells are generated and transmitted periodically during the failure condition in order to indicate VPC unavailability. Generation frequency of VP-FERF cells is nominally one cell per second and shall be the same for all VPCs concerned.

VP-FERF cell generation shall be stopped as soon as the failure indications are removed.

VP-FERF Cell Detection Condition. VP-FERF cells are detected at the VPC endpoint and VP-FERF state is declared after the reception of one VP-FERF cell. VPC connecting points may monitor the VP-FERF cells.

VP-FERF Release Condition. The VP-FERF state is removed when no VP-FERF cell is received during a nominally 3-s period.

VPC Continuity Check. The continuity check cell is sent downstream by a VPC endpoint when no user cell has been sent for a period of t, where Ts $< t <$ 2Ts and no VPC failure is indicated. If the VPC endpoint does not receive any cell within a time interval Tr (Tr $>$ 2Ts), it will send VP-FERF to the far end.

This mechanism can also be applied to test continuity across a VPC segment. The need for supporting this mechanism for all VPCs simultaneously is for further study.

VP Performance Management Functions. Performance monitoring of a VPC or VPC segment is performed by inserting monitoring cells at the ends of the VPC or VPC segment, respectively. In the procedure supporting this function, forward error detection information (e.g., the error detection code) is communicated by the endpoints using the forward (outgoing) F4 flow. The performance monitoring results, on the other hand, are received on the reverse (incoming) F4 flow. Note that when monitoring VPCs that are entirely within one span of control or when monitoring VPC segments, the monitoring result may be reported using the reverse F4 flow or via some other means [e.g., TMS (telecommunications management network)] (Ref. 28).

Performance monitoring is done by monitoring blocks of user cells.

A performance monitoring cell insertion request is initiated after every N user cells. The monitoring cell is inserted at the first free cell location after the request.

The block size N may have the values 128, 256, 512, and 1024. These are nominal block size values, and the actual size of the monitored cell block may vary. The cell block size may vary up to a maximum margin of 50% of the value of N for end-to-end performance monitoring. However, for end-to-end performance monitoring, the monitoring cell must be inserted into the user cell stream no more than $N/2$ user cells after an insertion request has been initiated. The actual monitoring block size averages out to approximately N cells.

TABLE 16.10   OAM Type Identifiers

OAM Type	4-bit	Function Type	4-bit
Fault management	0001	AIS	0000
	0001	FERF	0001
	0001	Continuity check	0100
Performance management	0010	Forward monitoring	0000
	0010	Backward reporting	0001
	0010	Monitoring/reporting	0010
Activation/deactivation	1000	Performance monitoring	0000
		Continuity check	0001

Source: Table 4/I.610, page 17, Ref. 28.

To eliminate forced insertions when monitoring VPC segment performance, the actual monitoring block size may be extended until a free cell is available after the insertion request. However, in this case, the actual monitoring block size may not average out to N cells. Forced insertion at the segment level remains as an option.

16.13.5.2   *OAM Functions for the VCC (F5 Flows).*   VCC F5 functions are similar to the VPC F4 functions.

16.13.6   ATM Layer OAM Cell Format

The ATM layer OAM cells contain fields common to all types of OAM cells, as well as specific fields for each specific type of OAM cell. See Table 16.10. The coding for unused common and specific fields is as follows:

- Unused OAM cell information field octets are coded 0110 1010 (6AH)
- Unused cell information field bits (incomplete octets) are coded all zeros

16.13.6.1   *Common OAM Cell Fields.*   All ATM layer OAM cells have the following five common fields: (see Figure 16.42).

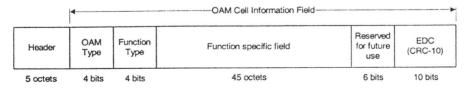

Figure 16.42. Common OAM cell format. (From Figure 8/I.610, page 18, Ref. 28.)

1. *Header.* Details of this field are in Rec. I.361. For F4 flow identification, two preassigned VCIs are used to distinguish OAM cells for VPCs and VPC segments. These two values are defined in Rec. I.361. For F5 flow identification, two PTI values are used to distinguish OAM cells for VCCs and VCC segments. These two values are defined in Rec. I.361.

2. *OAM Cell Type (4 bits).* This field indicates the type of management function performed by this cell, e.g., fault management, performance management, and activation/deactivation.

3. *OAM Function Type (4 bits).* This field indicates the actual function performed by this cell within the management type indicated by the OAM Cell Type field.

4. *Reserved Field for Future Use (6 bits).* Default value coded all zero.

5. *Error Detection Code (EDC) (10 bits).* This field carries a CRC-10 error detection code computed over the information field of the OAM cell. The CRC-10 generating polynomial is

$$G(x) = 1 + x + x^4 + x^5 + x^9 + x^{10}$$

REFERENCES

1. John Atkins and Mark Norris, *Total Area Networking*, John Wiley & Sons, Chichester, UK, 1995.

2. *Broadband Aspects of ISDN*, CCITT Rec. I.121, CCITT Geneva, 1991.

3. *ATM User–Network Interface Specification, Version 3.1*, The ATM Forum, Prentice-Hall, Englewood Cliffs, NJ, 1996.

4. *B-ISDN ATM Layer Specification*, ITU-T Rec. I.361, ITU Geneva, November 1995.

5. *B-ISDN User–Network Interface—Physical Layer Specification: General Characteristics*, ITU-T Rec. I.432.1, ITU Geneva, August 1996.

6. *B-ISDN Protocol Reference Model and Its Application*, CCITT Rec. I.321, ITU Geneva, April 1991.

7. *B-ISDN ATM Adaptation Layer Specification: Type 1 AAL*, ITU-T Rec. I.363.1, ITU Geneva, August 1996.

8. *B-ISDN ATM Adaptation Layer (AAL) Specification*, CCITT Rec. I.363, ITU Geneva, 1991.

9. *B-ISDN ATM Adaptation Layer Specification: Type 2 AAL*, ITU I.363 (para 3) Geneva, March 1993.

10. Radia Perlman, *Interconnections*, 2nd edition, Addison-Wesley, Reading, MA, 2000.

11. *B-ISDN ATM Adaptation Layer Specification: Type $\frac{3}{4}$ AAL*, ITU-T Rec. I.363.3, ITU Geneva, August 1996.

12. David E. McDysan and Darren L. Spohn, *ATM: Theory and Application*, McGraw-Hill, New York, 1995.

13. *B-ISDN ATM Adaptation Layer Specification: Type 5 AAL*, ITU-T Rec. I.363.5, ITU Geneva, August 1996.

14. *Support of the Broadband Connectionless Data Bearer Service by the B-ISDN*, ITU-T Rec. I.364, ITU Geneva, November 1995.

15. *The International Public Telecommunication Numbering Plan*, ITU-T Rec. E.164, ITU Geneva, May 1997.

16. *B-ISDN General Network Aspects*, ITU-T Rec. I.311, ITU Geneva, August 1996.

17. *Digital Subscriber Signaling System No. 2 (DSS2)—User–Network Interface (UNI) Layer 3 Specification for Basic Call / Connection Control*, ITU-T Rec. Q.2931, ITU Geneva, February 1995.

18. *Interim Inter-Switch Signaling Protocol (IISP) Specification, Version 1.0*, ATM Forum, Mt. View, CA, December 1994.

19. *Functional Description of the B-ISDN User Part (B-ISUP) of Signaling System No. 7*, ITU-T Rec. Q.2761, ITU Geneva, February 1995.

20. *Signaling System No. 7 B-ISDN User Part (B-ISUP)—Basic Call Procedures*, ITU-T Rec. Q.2764, ITU Geneva, February 1995.

21. *B-ISDN Inter-Carrier Interface (B-ICI) Specification, Version 2.0*, ATM Forum, Mt. View, CA, December 1995.

22. *ATM User–Network Interface (UNI) Signaling Specification, Version 4.0*, ATM Forum, Mt. View, CA, July 1996.

23. *Traffic Control and Congestion Control in B-ISDN*, ITU-T Rec. I.371, ITU Geneva, August 1996.

24. *B-ISDN ATM Layer Cell Transfer Performance*, ITU-T Rec. I.356, October 1996.

25. *Broadband Testing Technologies*, an H-P seminar, Hewlett-Packard Company, Burlington, MA, October 1993.

26. *ATM Cell Mapping into Plesiochronous Digital Hierarchy (PDH)*, ITU-T Rec. G.804, ITU Geneva, February 1998.

27. *Cell Relay Service Core Features*, Bellcore Special Report SR-3330, Issue 2, Bellcore, Red Bank, NJ, 1996.

28. *B-ISDN Operation and Maintenance Principles and Functions*, ITU-T Rec. I.610, ITU Geneva, March 1993.

17

LAST-MILE DATA DISTRIBUTION SYSTEMS

17.1 NEW APPROACHES TO OUTSIDE PLANT DISTRIBUTION

Outside plant is a term used to describe all the equipment involved in taking a telephone signal from the local serving exchange (local switch) and delivering it to the subscriber. In a traditional sense, this "equipment" mostly encompassed wire-pair distribution systems. Outside plant represented as much as one-third of the total investment in plant and equipment of telephone companies. Historically and through the present, it is wire-pair systems that served the "last mile" to the subscriber.

Twisted wire pair is an excellent transmission medium when employed in some kind of controlled environment. In this last-mile application it transported a simple, band-limited analog signal that interfaced with a telephone handset on the customer premises. On most of its lay from the local switch to its termination in a telephone subset, it was encased in a multipair cable often carrying many hundreds of such pairs. There was access to the cable along its lay route in distribution boxes and at a pedestal outside the customer location or at a small distribution frame in an office building. In nearly all cases, the telephone company could not afford to "pamper" or "doctor" the cable: "Here it is, take it as you find it." We find that some pairs did a better job than others; and as time went by, the cable tended to deteriorate due to simple aging, splicing, and the effects of the environment (e.g., freezing, flooding, and drying out) (Ref. 1).

We take these same pairs from a cable and put them in a laboratory, and we can demonstrate the ability of carrying megabits of digital data. Still and yet, it can easily be shown that bit rate on a wire pair varies with

- Length of the pair
- Diameter of the wire
- Capacitance between the pair members

681

Place that pair back in its cable sheath, bury it and leave it some years and it just will not work as well as it did in the laboratory.

Now in this day and age of the Internet, we would like to deliver some megabits/second to the customer premise on that wire pair (downstream). However, because of their condition, some of the pairs force us to reduce the bit rate to the 128-kbps range or even less. Of course, the longer the subscriber line, the lower the bit rate that is achievable.

The category of system we describe here is the *digital subscriber line* (DSL). Early on we had ISDN BRI (Chapter 13) which provided 64-kbps full-duplex access to the serving exchange from a subscriber on one logical line.* Placing the two available logical lines together in the BRI, 128 kbps was achievable with only some constraints. Then HDSL (high-speed digital subscriber line) appeared, providing 784 kbps or a full DS1 rate (1.544 Mbps). To achieve this transmission rate requires two pair, each carrying 784 kbps in each direction. The specified reach of HDSL was 12,000 ft (3700 m).

Up to this point we have described *symmetrical operation*. All this means is that there is the same bit rate operating in each direction. Then *asymmetrical operation* came along. This is where the bit rate in one direction is different than in the other direction. Consider the Internet. Where we'd like the high bit rate is from the ISP (internet service provider) toward the customer. The difficult part of this connectivity is from the local serving switch to the customer. The connection provided from the ISP to the local serving switch often is at the DS1 rate (1.544 Mbps) or greater. The connectivity from the customer to the ISP need not have a bit rate any greater than 56 kbps or even 28 kbps. Thus we have asymmetrical service: megabits downstream (from the switch to the customer) and kilobits of bit rate in the other direction (i.e., toward the ISP or upstream).

17.2   INTRODUCING LAST-MILE DISTRIBUTION SYSTEMS

The last-mile distribution, of course, involves the connectivity from some node or local serving switch to the customer premise. Whether the distance is exactly one mile is immaterial. In some cases it is much less than a mile, while in other cases it is more than a mile. Certainly in the rural situation it may be many miles. In this case the connectivity should be handled in some other manner.

Currently, there are four ways to deliver a megabit signal to customer premises:

1. By DSL (digital subscriber line)
2. By modified cable television (CATV) plant

---

*Remember that in North America, ISDN BRI is a 2-wire, full-duplex system that carried two B channels and a D channel. Its bit rate in either direction was 160 kbps, of which 16 kbps was overhead.

3. By LMDS (local multipoint distribution system)
4. By other wideband radio (wireless) systems

We will briefly discuss DSL. In many cases we believe its weaknesses outweigh its benefits. The three basic weaknesses were described in Section 17.1. Its single benefit is economy, and the wire pairs are now in place to provide POTS (plain old telephone service). The proprietary or semiproprietary schemes of comparatively high bit rates on wire pair usually take on an acronym ending in SL. We described the generic term DSL and the HDSL. There is ADSL (asymmetrical digital subscriber line) supported by ANSI on one hand and a proprietary scheme on the other. They do not interface and are quite different, one from the other. The ANSI ADSL supports some 6 Mbps downstream (distance limited) and 640 kbps upstream. As the subscriber line gets longer, the supported bit rates drop. Of course, if the distance is limited to about 1 mile, some 6000 ft (1829 m), the full data rate should be achieved if the subscriber loop is in good condition and without resorting to specialized line coding techniques.

XDSL is a generalized acronym for any one of the several semiproprietary or proprietary schemes available.

As we have discussed, the shorter the wire-pair loop between the DSL terminal and the user installation, the greater the bit rate that can be supported. One way to achieve this goal is to bundle a group of wire pairs and replace the bundle with a fiber-optic pair of strands. This fiber-optic pair has one strand for downstream and one for upstream. It carries a multiplexed configuration to a remote terminal near the subscribers to be served. The remote terminal unit (RTU) demultiplexes the downstream and distributes the signals on to the appropriate pairs. In the other direction, upstream, the RTU takes the electrical signals from each pair and multiplexes them for transport to the local serving exchange. Here we can say that the last mile segment is from the RTU to user sites. With good planning of the system, these derived copper pair loops are shortened to 6000 ft (1829 m) or less, permitting the highest bit rate the DSL design allows.

As we have covered in several instances, the principal constraint of data rate is loop attenuation. When planning a DSL installation, one must consider that loop attenuation increases with frequency. Considering the bandwidth of a DSL system and allowing 1 bit per hertz, a DS1 configuration would have a bandwidth of 1.544 MHz (equivalent to 1.544 Mbps). If, somehow, we can reduce the signal bandwidth of a DSL system, we can effectively stretch the *reach* of the system. In North America, a 2B1Q waveform has been adopted for ISDN BRI. For every electrical pulse transmitted with 2B1Q, 2 bits of data are sent. Thus, effectively, the bandwidth is cut in half. This will more than double the reach of a typical DSL system such as DS1. The 2B1Q line code is discussed in Section 13.6.2.6.

There is also a proprietary line code developed by Paradyne (Ref. 2) called *CAP*. CAP stands for carrierless amplitude and phase modulation. This is a

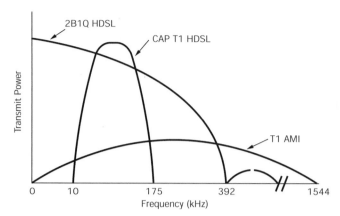

Figure 17.1. A sketch of an idealized spectra of three DSL waveforms. (From Paradyne, Figure 6, page 14, Ref. 2.)

line coding technique that allows multiple bits per hertz of bandwidth, up to 9 bits per hertz. CAP further reduces bandwidth requirements on a wire pair providing considerably greater range (or bit rate) than a 2B1Q system provides.

Figure 17.1 is an idealized sketch of the occupied spectrum on a wire-pair subscriber line of three transmission techniques: 2B1Q HDSL, CAP T1 HDSL, and conventional T1 AMI.* This latter waveform is described in Section 7.2.4.

Crosstalk is a dominant factor in the performance of many DSL systems. *Crosstalk* is the coupling of signals from adjacent wire pairs in a cable bundle into the signal bit stream of the individual desired pair under consideration. Crosstalk can seriously degrade bit error rate.

In a fully configured, traditional DSL installation, data traffic bypasses the local serving switch. The data traffic derived from the DSL network is concentrated and handed off to an interswitch digital cross-connect (DACS).

A better way of handling this data aggregate is to use a digital subscriber line access multiplexer (DSLAM). This device groups data channels before handoff. It uses packet and cell multiplexing technology in addition to conventional T1(DS1)/E1 introduced into the DSLAM, which results in a more efficient use of bit stream flow resources.

In the local serving switch which houses the DSL/DSLAM equipment we will expect to find transport system resources such as T1/E1, T3/E3, OC-1, OC-3, STS-1, and STS-3 facilities being used to support the switch. These

---

*AMI stands for alternate mark inversion. This is the specialized waveform used in the transmission on wire pair of T1 (DS1) PCM.

resources interface the switch to produce interswitch trunks. They will also interface the DSLAM facilities.

The local access network utilizes the local carrier interswitch network as a foundation for services. Additional equipment may be required in order to provide connectivity between multiple service providers and multiple service users. This equipment may involve frame relay switches, ATM switches and/or routers and digital data cross-connects.

We consider the DSLAM the cornerstone of the DSL solution. Functionally, the DSLAM concentrates data traffic from multiple DSL loops onto the backbone network for connection to the rest of the network. The DSLAM provides backhaul services for packet, cell, and/or circuit-based applications through concentration of DSL lines onto T1/E1, T3/E3, or ATM outputs. The DSLAM facility may be called upon to carry out routing functions such as supporting dynamic IP address assignment using dynamic host control protocol (see Section 11.4.3).

Figure 17.2 shows a DSL system reference model for full service deployment. ADSL terminology is used throughout. On the left side of the drawing are the network service providers (NSPs). These could represent frame relay

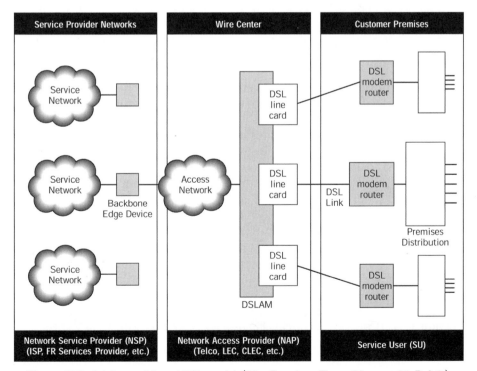

Figure 17.2. A fully provisioned DSL model. (After Paradyne Figure 32, page 35, Ref. 2.)

service provider, ISP (internet service provider), ATM service provider, and so forth. The edge device may be a switch.

ATU stands for ADSL transceiver unit. One is called an ATU-C and is usually housed in the DSLAM facility which is located in the local service switch complex with access to its MDF (main distribution frame). Of course "C" stands for central office, a distinctly North American term for a telephone switching center. The ATU-C has a companion remote unit called an ATU-R, which is located on the user premises. The DSL connects the two.

In the center of the drawing is the network access provider (NAP). The NAP may belong to the local telephone company or LEC or to the CLEC or, possibly, to an independent operator that provides a DSL service. Often the DSLAM houses the ATU-C devices. It will have at least one port for N64 services to its associated local switch. Otherwise, it bypasses the local switch accessing the appropriate networks for service such as (a) a SONET network or (b) a PSTN PDH digital network.

There is another aspect to "last-mile" connectivity. This is a North American phenomenon. Historically, the outside plant distribution system, better defined as last-mile delivery, was a monopoly, held by the local telephone company. That has now changed in the United States. The local telephone company is officially known as a *local exchange carrier* or LEC. The monopoly has been broken and now appearing on the scene is the *CLEC* or competitive local exchange carrier. A CLEC may share the outside plant, probably originally installed and maintained by the LEC; install their own wire-pair distribution system, or use some other means to provide an electrical signal to the subscriber.

Cable Television (CATV)* is a last-mile delivery system of entertainment, principally television. It is a broadband system that is already in place. We reiterate those important points: CATV is a broadband system, and it is already in place in the United States and other nations. With some modifications, it has the potential to be an excellent broadband delivery system for voice (POTS), data/internet, and conventional television entertainment. It can provide 10-Mbps or greater data rates for downstream internet for many users simultaneously as well as be a transporter of ATM, frame relay at E1/DS1 rates. It can create VPNs and set up IP connectivity. It is much more versatile than DSL and suffers few of its shortcomings. CATV is discussed in Section 17.3.

A second very viable "other" means is to transport and distribute wideband signals via LMDS, which stands for Local Multipoint Distribution System. This is a wireless or radio system that operates in the 28 to 31-GHz band and provides better than 1 GHz of bandwidth. LMDS is described in Section 17.4. It provides straight, head-on competition with CATV.

---

*CATV stands for community antenna television, the more formal name for cable television.

17.3   INTRODUCTION TO CATV

The original concept of CATV was to deliver quality TV program service to the home and office. TV in itself is a broadband signal (6 MHz). It is broadband when compared to the familiar voice channel which has a bandwidth of some 4 kHz. Whereas a wire pair in the conventional outside plant only delivers just one nominal 4-kHz signal to a customer, CATV delivers a choice of many TV channels to its customers. In essence, CATV is much more complex than its wire-pair counterpart. It also has much greater service potential.

The concept of CATV is fairly simple. In its earlier days (ca. 1960–1970), a high location was selected near or outside a city which had excellent line-of-sight conditions with the city's various TV emitting antennas. Their signals were received at this location, which we choose to call a *head-end*, then amplified and coupled into a low-loss coaxial cable system. The coaxial cable distributed this very broadband signal to customers along a preplanned route. At various locations along the route the composite TV signal aggregate was amplified. Each broadband amplifier along the route added noise to the signal, placing a limit on the number of amplifiers in tandem that may be employed. It was a neat concept because a regular TV set could couple into the coaxial cable with a tap and tune to any one of some fair number of channels. The set needed no modification. In fact the set could be simply switched from a rabbit-ear (or other) antenna to the CATV tap and back again if a user so desired.

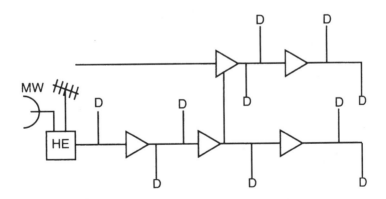

HE = head-end
D = drop wire to residence
MW = microwave connectivity

Figure 17.3. An early CATV distribution system.

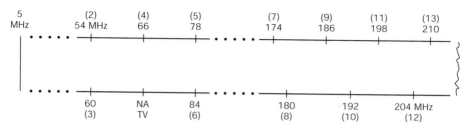

Figure 17.4. The radio-frequency spectrum as seen by a TV set, whether off-the-air or off the cable. *Note:* The familiar channel numbers are in parentheses. Frequencies are in megahertz.

Figure 17.3 shows a simplified block diagram of the original CATV concept (ca. 1965). That concept had changed little up to several years ago, when optical fiber entered the picture. Note that the head-end (HE) is where off-the-air signals are received (captured); added to these may be other TV signals brought to the head-end by LOS microwave. Later, of course, many of the channels derived from satellite transport, probably the bulk of the TV channels.

Figure 17.4 shows the radio-frequency (RF) spectrum. Only TV channels are shown. This is the same spectrum that appears on the coaxial cable, and only the TV channels are operable.

CATV system designers soon learned not to waste bandwidth. For example, the off-the-air spectrum, which is assigned for usage by the FCC, is not a continuous spectrum from 54 MHz upwards. Other services such as aeronautical mobile, public safety, marine, and others are assigned frequencies (i.e., from 88 to 174 MHz). There is the familiar FM broadcast band from 88 to 106 MHz as well. The coaxial cable is well-shielded from unwanted off-the-air emissions, so then why not use that vacant spectrum (from a CATV operator perspective)?

What the operator did was simply take TV channels from other sources such as those brought in by microwave or by satellite, and it translated them in frequency to the desired vacant 6-MHz-wide frequency slots. Typically, we might find channel 25 off-the-air appearing on our cable in the channel 14 slot.

17.3.1   The Essentials of TV Signal Distribution on a CATV System

The CATV system designer deals with a broadband system based on coaxial cable for signal transport. The frequency response of coaxial cable is nonlinear, more approaching the quadratic. On the other hand, from our perspective, fiber-optic cable presents a flat response to the user. Figure 17.5 shows a typical frequency response curve of coaxial cable. For viable TV signal distribution we would like as flat a response as possible. A hypothetical frequency response curve for a section of fiber-optic cable is shown in Figure 17.6. Note that the curve is flat. We can flatten the frequency response

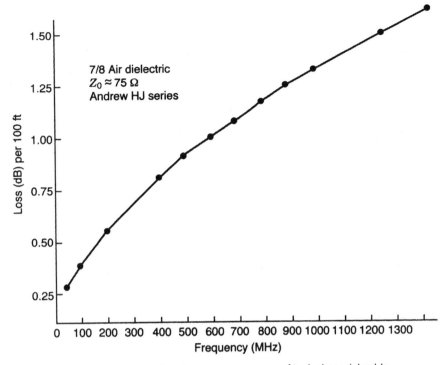

Figure 17.5. Attenuation–frequency response of typical coaxial cable.

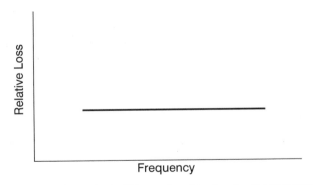

Figure 17.6. A length of hypothetical fiber-optic cable showing flat frequency response.

characteristics of coaxial cable by the use of equalizers. An equalizer, in this case, adds loss in a mirror image of the response curve in Figure 17.5, making the response appear flat. Of course, the insertion loss of an equalizer must be taken into account. The more we extend the useful upper frequency of an equalizer, the greater the insertion loss. Thus there is a practical limit to how far we can extend the useful upper frequency. Right now the frequency lies between 800 and 1000 MHz. This gives us a transmission medium nearly 1 GHz wide.

To distribute CATV TV signals to customer premises, we need to flatten the medium with equalizers and present a signal of sufficient level to the TV receiver input to achieve acceptable picture quality. The TV picture must also have a signal-to-noise ratio (S/N) of at least 40 dB, preferably > 46 dB. Now remember that we end up with a system with some number X of wideband amplifiers in tandem. Each amplifier adds noise to the system. The number X therefore must be limited or we will not achieve our S/N objective. This puts a definite limit on the CATV coverage area of customer premises.

17.3.2   Extending CATV Coverage Area

CATV system designers began to consider the application of fiber-optic cable transmission to extend coverage area. One practical approach is to transmit the entire RF spectrum from 54 MHz to some maximum frequency, say 750 MHz for argument's sake, along with the artificially inserted channels along the fiber. Remember that optical fiber is a very broadband medium. This frequency configuration would originate at the head-end and would be transmitted out on sets of fiber-optic supertrunks to nodes somewhere near groups of residences or office buildings. Each node converts the signal from a light signal to an electrical RF signal on a conventional CATV coaxial cable system. As we have configured our system, there is a maximum capacity of 116 NTSC* TV channels [$(750 - 54)/6$].

Such systems described here are called HFC (hybrid fiber-coax) systems. They offer many advantages to the CATV operator (among them are improved carrier-to-noise ratio C/N), and they result in a better S/N value at the premises TV set, less self-interference, improved reliability because there are fewer net active components, and the potential to greatly increase a serving area. Another extremely important advantage is to provide a means of implementing a very viable two-way digital service.

One important service in high demand is access to the Internet. The configuration is asymmetrical in that the downstream bit rate will be considerably greater than the companion upstream data rate. We probably can allow the upstream data rate to be on the order of 56 or 64 kbps per active user, along with the companion downstream rate on the order of 1–6 Mbps.

Let's now take the RF spectrum shown in Figure 17.4, and let's reconfigure it for two-way digital operation. The spectrum layout to meet the requirements outlined above is shown in Figure 17.7. Note the asymmetry where the bandwidth assigned for downstream voice and data services is

---

*NTSC stands for National Television System Committee. This is an analog, color TV configuration where each channel is assigned a 6-MHz bandwidth. NTSC is the North American analog color TV transmission standard.

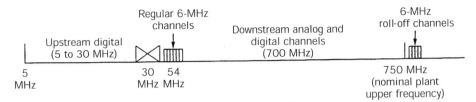

Figure 17.7. CATV spectrum assigned for upstream and downstream operation. The frequency assignment is somewhat arbitrary but probably will be encountered in practice. The 5- to 42-MHz segment for upstream operation is specified by DOCSIS. (See Section 17.3.3.) Table 17.1 shows DOCSIS downstream operation is in 6-MHz segments from 88 MHz upwards.

contained in one or two 6-MHz segments way up at the top of the band, in what are called roll-off channels. This is where filter roll-off occurs. The channel response will not support conventional TV, but will support downstream data. If more bandwidth is required, the CATV operator can assign one or several of the top 6-MHz TV segments to provide additional downstream service. Here we have 40 MHz available upstream and 6, 12, and 18 MHz (or more) for downstream data/voice.

Curing the Ills of Bandwidth Scarcity. If we were to implement the CATV plant as we describe, where the entire plant is served by one coaxial cable configuration, we would never be able to serve the demand for Internet service. This is so, leaving aside other desirable data services. Probably the first shortage would show up on the downstream side. If we were to allow 5 Mbps per user for internet downstream, including guardband, each simultaneous user would take up about 1 TV channel (i.e., 6 MHz) of bandwidth. Either we cut back the bandwidth per user in periods of high demand, thus cutting bit rate, or we may look to another method of signal distribution.

One method being adopted widely is to bring the serving node closer to the customer. CATV users are grouped in units of 50 to 200. The group is served by a node. The node takes the downstream frequency segment from an FDM configuration on the fiber of some 1000 MHz in bandwidth—say in the band 7 to 8 GHz—and translates the signals to the 54- to 750-MHz band. Note that some additional bandwidth is used for guard band. A nearby node's downstream configuration is in the band 6 to 7 GHz, and so forth, for each serving node. Note that the same fiber strand is used for each node. A passive optical signal splitter delivers the downstream signal for grooming and downconversion at each node.

The nodal processing for the upstream direction is shown in Figure 17.8. Here the frequency band of 5 to 30 MHz is used for upstream traffic, a 25-MHz bandwidth. Three active groupings of 50 to 200 residences are shown in the figure. In this direction we call the device residing in the node a coax-to-fiber translator. Because of the richness of bandwidth availability on

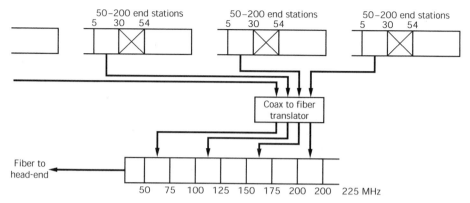

Figure 17.8. Upstream concept breaking up a user community into groups of from 50 to 200 end-user stations. This figure relates to Figure 17.7, where frequency band assignments are somewhat arbitrary. The upstream band will carry POTS, data, and internet carriers. The last mile of operation will be on coaxial cable. Downstream internet, POTS, and data service are provided on separate frequency segments as well. The downstream frequency assignment is in 6-MHz segments from about 50 MHz upward to 750 MHz, and in some plants to 860 MHz. Expect to find the data, internet, and POTS channels at the top of the band for downstream.

the fiber, only every other 25-MHz segment carries upstream traffic where the other segments are guardbands. Of course these fiber strands derive from the head-end via supertrunks.

For internet application, upstream operation will be the direction from end-user to the ISP. In the case of POTS, operation will be symmetrical; assume 64 kbps per voice channel. The CATV operator will probably also offer data network connectivity, possibly IP, frame relay, or ATM at differing bit rates. One common approach is to turn to DOCSIS (Data-over-Cable Service Interface Specifications) developed by Cable Labs and SCTE. Modems designed to meet DOCSIS requirements are available from the principal CATV equipment vendors such as GI Jerrold and Motorola.

17.3.3 The DOCSIS Specification

Figure 17.9 shows the reference architecture for the data-over-cable specification. The following acronyms and abbreviations are defined to assist the reader in understanding DOCSIS operation.

MCNS. Multimedia Cable Network System (consortium).

HFC. Hybrid fiber-coax (described above).

IP. Internet Protocol—see Section 11.4.

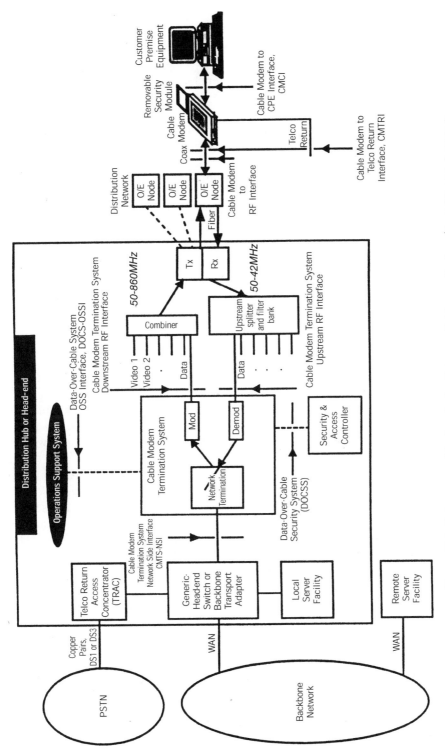

Figure 17.9. Reference architecture for the Data-over-Cable specification. (From Figure 1-2, page 3, DOCSIS Specification, Ref. 3.)

CMTS. Cable Modem Terminator System. DOCSIS modem located at CATV head-end. See Figure 17.9.

CM. Cable modem. The DOCSIS modem located at the customer premises installation. See Figure 17.9.

It should be noted from Figure 17.9 that the upstream spectrum occupies the band from 5 to 42 MHz, and the downstream spectrum occupies the band from 50 to 860 MHz.

The key functional characteristics assumed in the DOCSIS specification are as follows:

- Two-way transmission.
- A maximum optical/electrical spacing between the CMTS and the most distant customer terminal of 100 miles, although typical maximum separation may be 10–15 miles (16–24 km).
- A maximum differential optical/electrical spacing between CMTS and the closest and most distant modems of 100 miles (161 km), although this would typically be limited to 15 miles (24 km).

17.3.3.1 *RF Channel Assumptions.* The Data-over-Cable system, configured with at least one set of defined physical-layer parameters (e.g., modulation, forward error correction, symbol rate, etc.) from the range of configuration settings described in that specification, must be capable of operating with a 1500-byte packet where the packet loss rate will be less than 1% while forwarding at least 100 packets per second on cable networks having characteristics defined in this section.

Transmission Downstream. The RF channel transmission characteristics of the cable network in the downstream direction assumed for the purposes of minimal operating capability are described in Table 17.1. This assumes nominal analog video carrier level (peak envelope power) in a 6-MHz channel bandwidth. All conditions are present concurrently.

Transmission Upstream. The RF channel transmission characteristics of the cable network in the upstream direction assumed for the purposes of minimal operating capability are described in Table 17.2. All conditions are present concurrently.

Transmission Levels. The nominal power level of the downstream CMTS 64-QAM* signal(s) within a 6-MHz channel is targeted to be in the range of −10 dBc to −6 dBc relative to analog video carrier level. The nominal power level of the upstream CM signal(s) will be as low as possible to achieve

---

*Note that 64-QAM will give us theoretically 6 bits/Hz of bit packing ($2^6 = 64$).

TABLE 17.1   Downstream RF Channel Transmission Characteristics

Parameter	Value
Frequency range	Cable system normal downstream operating range is from 50 MHz to as high as 860 MHz. However, the values in this table apply only at frequencies > 88 MHz.
RF channel spacing (design bandwidth)	6 MHz
Transit delay from head-end to most distant customer	$\leq$ 0.800 ms (typically much less)
Carrier-to-noise ratio in a 6-MHz band (analog video level)	Not less than 35 dB (Note 4)
Carrier-to-interference ratio for total power (discrete and broadband ingress signals)	Not less than 35 dB within the design bandwidth
Composite triple beat distortion for analog modulated carriers	Not greater than -50 dBc within the design bandwidth
Composite second-order distortion for analog modulated carriers	Not greater than -50 dBc within the design bandwidth
Cross-modulation level	Not greater than -40 dBc within the design bandwidth
Amplitude ripple	0.5 dB within the design bandwidth
Group delay ripple in the spectrum occupied by the CMTS	75 ns within the design bandwidth
Micro-reflections bound for dominant echo	-10 dBc @ $\leq$ 0.5 μs, -15 dBc @ $\leq$ 1.0 μs -20 dBc @ $\leq$ 1.5 μs, -30 dBc @ > 1.5 μs
Carrier hum modulation	Not greater than -26 dBc (5%)
Burst noise	Not longer than 25 μs at a 10-Hz average rate
Seasonal and diurnal signal level variation	8 dB
Signal level slope, 50 \750 MHz	16 dB
Maximum analog video carrier level at the CM input, inclusive of above signal level variation	17 dB mV
Lowest analog video carrier level at the CM input, inclusive of above signal level variation	-5 dB mV

Note 1: Transmission is from the head-end combiner to the CM input at the customer location.

Note 2: For measurements above the normal downstream operating frequency band (except hum), impairments are referenced to the highest-frequency NTSC carrier level.

Note 3: For hum measurements above the normal downstream operating frequency band, a continuous-wave carrier is sent at the test frequency at the same level as the highest-frequency NTSC carrier.

Note 4: This presumes that the digital carrier is operated at analog peak carrier level. When the digital carrier is operated below the analog peak carrier level, this C/N may be less.

Note 5: Measurement methods defined in [NCTA] or [CableLabs2] documents.

Source: Table 2-1, page 8, DOCSIS Specification, Ref. 3.

TABLE 17.2   Upstream RF Channel Transmission Characteristics

Parameter	Value
Frequency range	5–42 MHz edge to edge
Transit delay from the most distant CM to the nearest CM or CMTS	≤ 0.800 ms (typically much less)
Carrier-to-noise ratio	Not less than 25 dB
Carrier-to-ingress power (the sum of discrete and broadband ingress signals) ratio	Not less than 25 dB (Note 2)
Carrier-to-interference (the sum of noise, distortion, common-path distortion, and cross-modulation) ratio	Not less than 25 dB
Carrier hum modulation	Not greater than −23 dBc (7.0%)
Burst noise	Not longer than 10 μs at a 1-kHz average rate for most cases (Notes 3, 4, and 5)
Amplitude ripple	5–42 MHz: 0.5 dB/MHz
Group delay ripple	5–42 MHz: 200 ns/MHz
Micro-reflections—single echo	−10 dBc @ ≤ 0.5 μs −20 dBc @ ≤ 1.0 μs −30 dBc @ > 1.0 μs
Seasonal and diurnal signal level variation	Not greater than 8 dB min to max

Note 1: Transmission is from the CM output at the customer location to the head-end.
Note 2: Ingress avoidance or tolerance techniques MAY be used to ensure operation in the presence of time-varying discrete ingress signals that could be as high as 0 dBc [CableLabs1].
Note 3: Amplitude and frequency characteristics sufficiently strong to partially or wholly mask the data carrier.
Note 4: CableLabs report containing distribution of return-path burst noise measurements and measurement method is forthcoming.
Note 5: Impulse noise levels more prevalent at lower frequencies (< 15 MHz).
Source: Table 2-2, page 9, DOCSIS Specification, Ref. 3.

the required margin above noise and interference. Uniform power loading per unit bandwidth is commonly followed in setting upstream signal levels, with specific levels established by the cable network operator to achieve the required carrier-to-noise and carrier-to-interference ratios.

17.3.3.2 *Communication Protocols.* The CM and CMTS operate as forwarding agents and also as end-systems (hosts). The protocol stacks used in these modes differ as described below.

The principal function of the cable modem system is to transmit Internet Protocol (IP) packets transparently between the head-end and the subscriber location. Several management functions also ride on IP, so that the protocol stack on the cable network is as shown in Figure 17.10. These management functions include, for example, supporting spectrum management functions and the downloading of software.

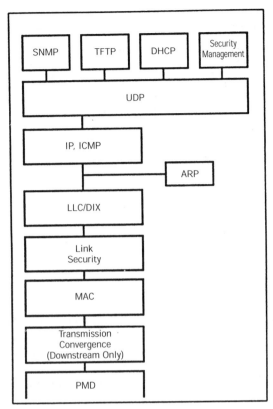

Figure 17.10. The protocol stack on the DOCSIS RF interface. (From Figure 3-1, page 11, DOCSIS Specification, Ref. 3.)

The CMs and CMTSs will operate as IP and LLC hosts in terms of IEEE Standard 802 for communication over the cable network. The protocol stack at the CM and CMTS RF interfaces is shown in Figure 17.10.

The CM and CMTS function as IP hosts. As such, the CM and CMTS support IP and ARP over DIX link-layer framing. The CM and CMTS may also support IP and ARP over SNAP framing.

CM and CMTS as Hosts. The CM and CMTS also function as LLC hosts. As such, the CM and CMTS respond appropriately to TEST and XID requests per ISO 8802-2 (LLC).

Data Forwarding Through CM and CMTS

General. Data forwarding through the CMTS may be transparent bridging, or may employ network-layer forwarding (routing, IP switching) as shown in Figure 17.11.

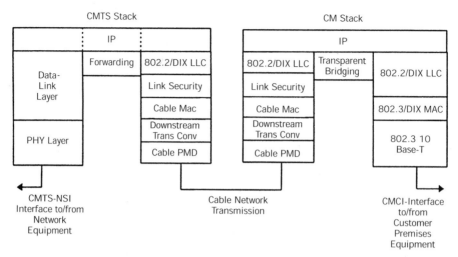

Figure 17.11. Data forwarding through CM and CMTS. (From Figure 3-2, page 12, DOCSIS Specification, Ref. 3.)

Data forwarding through the CM is link-layer transparent bridging, as shown in Figure 17.11. Forwarding rules are similar to ISO/IEC 10038 with modifications described in CMTS and CM *Forwarding Rules* below. This allows support of multiple network layers.

Forwarding of IP traffic is supported. However, support of other network-layer protocols is optional. The ability to restrict the network layer to a single protocol is required.

Support for the IEEE 802.1d spanning tree protocol of ISO/IEC-10038 with the modifications described in *CM Forwarding Rules* is optional for CMs intended for residential use. CMs intended for commercial use and bridging CMTSs support this version of the spanning tree. CMs and CMTSs include the ability to filter (and disregard) 802.1d BPDUs. The DOCSIS specification assumes that CMs intended for residential use will not be connected in a configuration which would create network loops such as that shown in Figure 17.12.

The MAC Forwarder. The MAC forwarder is a MAC sublayer that resides on the CMTS just below the MAC service access point (MSAP) interface, as shown in Figure 17.13. It is responsible for delivering upstream frames to:

- One or more downstream channels
- The MSAP interface

In Figure 17.13, the LLC sublayer and link security sublayers of the upstream and downstream channels on the cable network terminate at the MAC Forwarder.

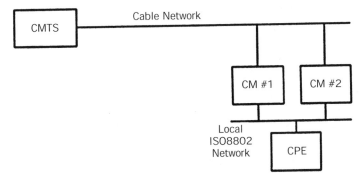

Figure 17.12. Example condition for network loops.

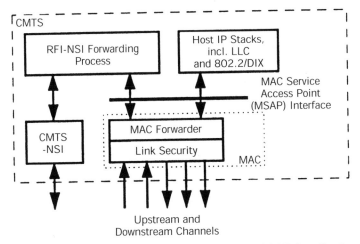

Figure 17.13. MAC Forwarder. (From Figure 3-4, page 15, DOCSIS Specification, Ref. 3.)

The MSAP interface user may be the NSI-RFI Forwarding process or the CMTS's host protocol stack.

Delivery of frames may be based on data-link-layer (bridging) semantics, network-layer (routing) semantics, or some combination thereof. Higher-layer semantics may also be employed (e.g., filters on UDP port numbers). The CMTS should provide IP connectivity between hosts attached to cable modems, and it must do so in a way that meets the expectations of Ethernet-attached customer equipment. For example, the CMTS must either forward ARP packets or must facilitate a proxy ARP service. The CMTS MAC Forwarder may provide service for non-IP protocols.

Note that there is no requirement that all upstream and downstream channels be aggregated under one MSAP as shown above. The vendor could just as well choose to implement multiple MSAPs, each with a single upstream and downstream channel.

Network Layer. As stated at the outset, the purpose of the DOCSIS system is to transport IP traffic transparently through the system. The network layer protocol is the Internet Protocol (IP) version 4, as defined in RFC-791, and migrating to IP version 6. The DOCSIS specification imposes no requirements for reassembly of IP packets.

Above the Network Layer. Users of DOCSIS will be able to use the transparent IP capability as a bearer of higher-layer services. Use of these services will be transparent to the CM.

In addition to the transport of user data, there are several network management capabilities that depend upon the network layer. These include:

- SNMP (Simple Network Management Protocol) based on RFC-1157. See Chapter 18 for a general discussion of network management and of SNMP in particular.
- TFTP (Trivial File Transfer Protocol), RFC-1350, a file transfer protocol, for downloading software and configuration information.
- DHCP (Dynamic Host Configuration Protocol, RFC-1541), a framework for passing configuration information to hosts on a TCP/IP network.

Data-Link Layer. The data-link layer is divided into sublayers in accordance with IEEE 802 family of standards (see Chapters 9 and 10). The sublayers from the top are:

- Logical link control (LLC) sublayer (Class 1 only)
- Link-layer security sublayer
- Media Access Control (MAC) sublayer

17.3.3.3 *Logical Link Control.* The LLC sublayer is provided in accordance with Chapter 9 (IEEE Std. 802.2). Address resolution is used as defined in Chapter 11 and RFC-826. The MAC-to-LLC service definition is specified in ISO/IEC-10039.

17.3.3.4 *Link-Layer Security Sublayer.* Link-layer security is provided in accordance with specifications MCNS2 (DOCSIS) and MCN8.

17.3.3.5 *MAC Sublayer.* The MAC sublayer defines a single transmitter for each downstream channel—the CMTS (cable modem termination system). The CMs (cable modems) listen to all frames transmitted on the downstream channel upon which they are registered and accept those where the destinations match the CM itself or CPEs reached via the CMCI (cable modem-to-customer interface) port. CMs can communicate with other CMs only through the CMTS. (See Figure 17.9.)

The upstream channel is characterized by many transmitters (CMs) and one receiver (the CMTS). Time in the upstream channel is slotted, providing for TDMA at regulated time ticks. The CMTS provides the time reference and controls the allowed usage for each interval. Intervals may be granted for transmissions by particular CMs, or for contention by all CMs. CMs may contend to request transmission time. To a limited extent, CMs may also control to transmit actual data. In both cases, collisions can occur and retries are used.

MAC Overview. Some of the MAC protocol highlights include:

* Bandwidth allocation controlled by CMTS
* A stream of mini-slots in the upstream
* Dynamic mix of contention- and reservation-based upstream transmit opportunities
* Bandwidth efficiency through support of variable-length packets
* Extensions provided for future support of ATM or other Data PDU
* Class-of-service support
* Extensions provided for security at the data–link layer
* Support for a wide range of data rates

Physical Layer. The physical (PHY) layer comprises two sublayers:

* Transmission Convergence sublayer (present in the downstream direction only)
* Physical media-dependent (PMD) sublayer

17.3.3.6 *Downstream Transmission Convergence Sublayer.* The
Downstream Transmission Convergence sublayer exists in the downstream direction only. It provides an opportunity for additional services over the physical-layer bitstream. These additional services, for example, might include digital video.

The sublayer is defined as a continuous series of 188-byte MPEG (see ITU-T Rec. H.222.0) packets, each consisting of a 4-byte header followed by 184 bytes of payload. The header identifies the payload as belonging to the data-over-cable MAC. Other values of the header may indicate other payloads. The mixture of payloads is arbitrary and controlled by the CMTS.

17.3.3.7 *PMD Sublayer Overview.* The PMD sublayer involves digitally
modulated RF carriers on the analog cable network. In the downstream direction, the PMD sublayer is based on ITU-T Rec. Q.83-B, with the exception of the "Scalable Interleaving to Support Low Latency Capability."

The PMD includes the following features:

- 64- and 256-QAM modulation formats*
- 6-MHz occupied spectrum coexists with all other signals on the planet
- Concatenation of Reed–Solomon block code and Trellis code supports operation in a higher percentage of North American cable networks
- Variable-depth interleaver supports both latency sensitive and insensitive data.

The features in the upstream direction are as follows:

- Flexible and programmable CM under control of the CMTS
- Frequency agility
- Time division multiple access
- QPSK and 16 QAM modulation formats
- Support of both fixed-frame and variable-length PDU formats
- Multiple symbol rates
- Programmable Reed–Solomon block coding
- Programmable preambles

17.3.3.8 Interface Points. Three RF interface points are defined at the PMD sublayer:

(a) Downstream output on the CMTS
(b) Upstream input on the CMTS
(c) Cable in/out at the cable modem

Separate downstream output and upstream input interfaces on the CMTS are required for compatibility with typical downstream and upstream combining and splitting arrangements in head-ends.

17.3.3.9 Physical-Media-Dependent Sublayer Specification. The DOCSIS standard defines the electrical characteristics and protocol for a cable modem (CM) and cable modem termination system (CMTS). Some of the high points of the PMD are covered in the following text.

Upstream

Overview. The upstream PMD sublayer uses FDMA/TDMA burst modulation format, which provides five symbol rates and two modulation formats (QPSK and 16-QAM). The modulation format includes pulse shaping for

---

*QAM modulation, where QAM stands for quadrature amplitude modulation, an efficient method of getting more bits per hertz. 16-QAM achieves 4 bits per Hz bit packing; 64-QAM, 6 bits per Hz; and 256-QAM achieves 8-bits/Hz bit packing. For example, with 256-QAM, 6-MHz bandwidth can accommodate 6×8 or 48-Mbps bit rate in theory; in practice the actual bit rate is considerably less.

spectral efficiency, is carrier-frequency agile, and has selectable output power level. The PMD format also includes a variable-length modulated burst with precise timing beginning at boundaries spaced at integer multiples of 6.25 μs apart (which is 16 symbols at the highest data rate). Each burst supports a flexible modulation, symbol rate, preamble, randomization of the payload, and programmable FEC encoding.

At the upstream transmission, parameters associated with burst transmission outputs from the CM are configurable by the CMTS via MAC messaging. Many of the parameters are programmable on a burst-by-burst basis.

The PDM sublayer can support a near-continuous mode of transmission, wherein ramp-down of one burst *may* overlap the ramp-up of the following burst, so that the transmitted envelope is never zero. The system timing of the TDMA transmission from the various CMs will provide that the center of the last symbol of one burst and the center of the first symbol of the preamble of an immediately following burst are separated by at least the duration of five symbols. The guard time will be greater than or equal to the duration of five symbols plus the maximum timing error. Timing error is contributed by both the CM and CMTS. Maximum timing error and guard time vary with CMTSs from different vendors.

The upstream modulator is part of the cable modem that interfaces with the cable network. The modulator contains the actual electrical-level modulation function and the digital signal processing function. The latter provides the FEC, preamble prepend, symbol mapping, and processing steps. The specification is written with the idea of buffering the bursts in the signal processing portion, along with the signal processing portion (1) accepting the information stream a burst at a time, (2) processing this stream into a complete burst of symbols for the modulator, and (3) feeding the properly timed bursted symbol stream to a memoryless modulator at the exact burst time. The memoryless portion of the modulator only performs pulse shaping and quadrature upconversion.

At the demodulator, similar to the modulator, there are two basic functional components: the demodulation function and the signal processing function. Unlike the modulator, the demodulator resides in the CMTS and the specification is written with the concept that there will be one demodulation function (not necessarily an actual physical demodulator) for each carrier frequency in use. The demodulation function would receive all bursts on a given frequency.

Note: The unit design approach should be cognizant of the multiple-channel nature of the demodulation and signal processing to be carried out at the head-end, and partition/share functionality appropriately to optimally leverage the multichannel application. A demodulator design supporting multiple channels in a demodulator unit may be appropriate.

The demodulation function of the demodulator accepts a varying-level signal centered around a commanded power level and performs symbol timing and carrier recovery and tracking, burst acquisition, and demodulation.

Additionally, the demodulation function provides an estimate of burst timing relative to a reference edge, an estimate of received signal power, and an estimate of signal-to-noise ratio and may engage adaptive equalization to mitigate the effects of (a) echoes in the cable plant, (b) narrowband ingress, and (c) group delay. The signal-processing function of the demodulator performs the inverse processing of the signal-processing function of the modulator. This includes accepting the demodulated burst data stream and decoding, and so on, and possibly multiplexing the data from multiple channels into a single output stream. The signal-processing function also provides the edge-timing reference and gating-enable signal to the demodulators to activate the burst acquisition for each assigned burst slot. The signal-processing function may also provide an indication of successful decoding, decoding error, or fail-to-decode for each codeword and the number of corrected Reed–Solomon symbols in each codeword. For every upstream burst, the CMTS has a priori knowledge of the exact burst length in symbols.

17.3.3.10 Modulation Formats and Rates. The upstream modulator/ demodulator provides/supports both QPSK and 16-QAM modulation formats. The upstream modulator using QPSK provides 160, 320, 640, 1280, and 2560 kilo-symbols per second, and at 16-QAM it provides 160, 320, 1280, and 2560 kilo-symbols per second. The variety of modulation rates and flexibility in setting upstream carrier frequencies permits operators to position carriers in gaps in the pattern of narrowband ingress. This is discussed in Appendix F of the reference publication (DOCSIS, Ref. 3).

17.3.3.11 Symbol Mapping. The modulation mode, whether QPSK or 16-QAM, is programmable. The symbols transmitted in each mode and the mapping of the input bits to the I and Q constellation is defined in Table 17.3. In the table, I_1 is the MSB of the symbol map, Q_1 is the LSB for QPSK, and Q_0 is the LSB for 16-QAM. Q_1 and I_0 have intermediate bit positions in 16-QAM. The MSB is the first bit in the serial data into the symbol mapper. The upstream QPSK symbol mapping is shown in Figure 17.14. The 16-QAM noninverted (Gray-coded) symbol mapping is shown in Figure 17.15. The 16-QAM differential symbol mapping is shown in Figure 17.16.

If differential quadrant encoding is enabled, the currently transmitted symbol quadrant is derived from the previously transmitted symbol quadrant and the current input bits are derived in accordance with Table 17.4.

TABLE 17.3 I/Q Mapping

QAM Mode	Input Bit Definitions
QPSK	I_1 Q_1
16-QAM	I_1 Q_1 I_0 Q_0

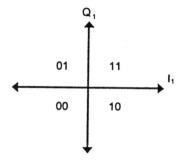

Figure 17.14. QPSK symbol mapping.

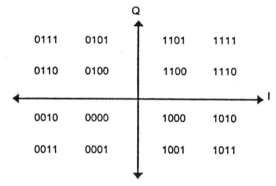

Figure 17.15. 16-QAM Gray coded symbol mapping. (From Figure 4-3, page 23, DOCSIS Specification, Ref. 3.)

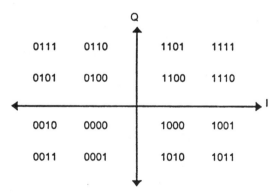

Figure 17.16. 16-QAM differential-coded symbol mapping. (From Figure 4-3, page 23, DOCSIS Specification, Ref. 3.)

TABLE 17.4 Derivation of Currently Transmitted Symbol Quadrant

Current Input Bits $I_1 Q_1$	Quadrant Phase Change	MSBs of Previously Transmitted Symbol	MSBs for Currently Transmitted Symbol
00	0°	11	11
00	0°	01	01
00	0°	00	00
00	0°	10	10
01	90°	11	01
01	90°	01	00
01	90°	00	10
01	90°	10	11
11	180°	11	00
11	180°	01	10
11	180°	00	11
11	180°	10	01
10	270°	11	10
10	270°	01	11
10	270°	00	01
10	270°	10	00

Source: Table 4-2, page 24, Ref. 3.

Spectral Shaping. The upstream PMD layer shall support 25% Nyquist square root cosine shaping. The occupied spectrum shall not exceed the channel widths shown in Table 17.5. (Note that the DOCSIS specification defines the channel width as the -30 dB bandwidth.)

Upstream Frequency Agility and Range. The upstream PMD sublayer supports operation over the frequency range of 5–42 MHz edge to edge. Offset frequency resolution should be supported with a range of ± 32 kHz (increment = 1 Hz; implement within ± 10 Hz).

TABLE 17.5 Maximum Channel Width

Symbol Rate (ksym / s)	Channel Width (kHz)
160	200
320	400
640	800
1280	1600
2560	3200

Source: Table 4-3, page 24, Ref. 3.

Spectrum Format. The upstream modulator *must* provide operation with the format $s(t) = I(9t)^* \cos(\omega t) - Q(t)^* \sin(\omega t)$, where t denotes time and ω denotes angular frequency.

FEC Encode

FEC Encode Modes. The upstream modulator *must* be able to provide the following selections: Read–Solomon codes over GF(256) with $T = 1$ to 10 or no FEC coding.

The following Reed–Solomon generator polynomial *must* be supported:

$$g(x) = (x + \alpha^0)(x + \alpha^1) \cdots (x + \alpha^{2T-1})$$

where the primitive element alpha is 0×02 hex.

The following Reed–Solomon primitive polynomial *must* be supported:

$$p(x) = x^8 + x^4 + x^3 + x^2 + 1$$

The upstream modulator *must* provide codewords from a minimum size of 18 bytes (16 information bytes [k] plus two parity bytes for $T = 1$ error correction) to a maximum size of 255 bytes (k-bytes plus parity-bytes). The uncoded word size can have a minimum of one byte.

In Shortened Last Codeword mode, the CM *must* provide the last codeword of a burst shortened from the assigned length of k data bytes per codeword as described in Section 4.2.10.1.2 of the reference document.

The value of T *must* be configured in response to the Upstream Channel Descriptor from the CMTS.

FEC Bit-to-Symbol Ordering. The input to the Reed–Solomon Encoder is logically a serial bit stream from the MAC layer of the CM, and the first bit of the stream is mapped into the MSB of the first Reed–Solomon symbol into the encoder. The MSB of the first symbol out of the encoder is mapped into the first bit of the serial bit stream fed to the Scrambler.

Note: The MAC byte-to-serial upstream convention calls for the byte LSB to be mapped into the first bit of the serial bit stream per Section 6.2.1.3 of the reference document.

Scrambler. The upstream modulator will implement a scrambler (shown in Figure 17.17) where the 15-bit seed value should be arbitrarily programmable.

At the beginning of each burst, the register is cleared and the seed value is loaded. The seed value should be used to calculate the scrambler bit that is combined in an XOR with the first bit to data of each burst (which is the MSB of the first symbol following the last symbol of the preamble).

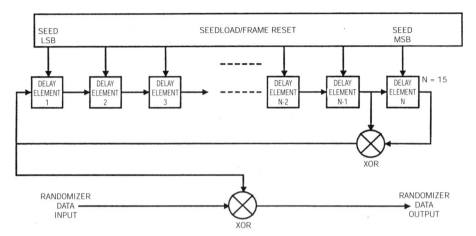

Figure 17.17. Scrambler block diagram. (From Figure 4-4, page 26, DOCSIS Specification, Ref. 3.)

The scrambler seed value is configured in response to the Upstream Channel Descriptor from the CMTS.

The polynomial is $x^{15} + x^{14} + 1$.

Preamble Prepend. The upstream PMD sublayer supports a variable-length preamble field that is prepended to the data after they have been randomized and Reed–Solomon encoded. The first bit of the Preamble Pattern is the first bit into the symbol mapper (see Figure 17.18) and is I1 in the first symbol of the burst. The first bit of the Preamble Pattern is designated by the Preamble Value Offset as described in Table 6-15 and Section 6.3.2.2 of the reference document.

The value of the preamble that is prepended is programmable and the length is 0, 2, 4..., or 1024 bits for QPSK and 0, 4, 8,..., or 1024 for 16-QAM. Thus, the maximum length of the preamble is 512 QPSK symbols or 256 QAM symbols.

Figure 17.18 is a block diagram of the signal processing sequence.

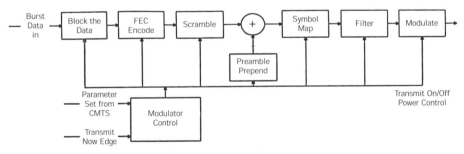

Figure 17.18. Functional block diagram showing signal processing sequence. (From Figure 4-8, page 35, DOCSIS Specification, Ref. 3.)

*17.3.3.12 **Burst Profiles.*** The transmission characteristics are separated into three portions:

(a) Channel parameters

(b) Burst profile attributes

(c) User unique parameters

The channel parameters include: (1) the symbol rate (there are rates from 160 ksym/s to 2.56 Msym/s in octave steps), (2) center frequency (Hz), and (3) the 1024-bit preamble superstring. The channel parameter characteristics are shared by all users on a given channel. The burst profile attributes are listed in Table 17.6. These parameters are shared attributes corresponding to a burst type. The user-unique parameters may vary for each user even when using the same burst type on the same channel as another user (for example, power level) and are listed in Table 17.7.

The CM generates each burst at the appropriate time as conveyed in the mini-slot grants provided by the CMTS MAPs (MAP stands for multiaccess protocol). The CM supports all burst profiles that are commanded by the CMTS via burst descriptors in the UCD (upstream channel descriptor) and that are subsequently assigned for transmission in a MAP.

The CM implements the offset frequency to within ± 10 Hz.

The CM is capable of switching burst profiles with no reconfiguration time required between bursts except for changes in the following parameters: (1) output power, (2) modulation, (3) symbol rate, (4) offset frequency, (5) channel frequency, and (6) ranging offset.

TABLE 17.6 Burst Profile Attributes

Burst Profile Attributes	Configuration Settings
Modulation	QPSK, 16 QAM
Diff Enc	On/Off
Preamble length	0 to 1024 bits (Note Section 4.2.5 of Ref. 3)
Preamble value offset	0 to 1022
FEC error correction (T bytes)	0 to 10 (0 implies FEC = off)
FEC codeword information bytes (k)	Fixed: 16 to 253 (assuming FEC on) Shortened: 16 to 253 (assuming FEC on)
Scrambler seed	15 bits
Maximum burst length (minislots)[a]	0 to 255
Guard time	5 to 255 symbols
Last codeword length	Fixed, shortened
Scrambler on/off	On / Off

[a]A burst length of 0 mini-slots in the channel profile means that the burst length is variable on that channel for that burst type. The burst length, while not fixed, is granted explicitly by the CMTS to the CM in the MAP.

Source: Table 4-4, page 27, Ref. 3.

TABLE 17.7 User Unique Burst Parameters

User Unique Parameter	Configuration Settings
Power level[a]	+8 to +55 dBmV (16QAM)
	+8 to +58 dBmV (QPSK)
	1-dB steps
Offset frequency	Range = ±32 kHz; increment = 1 Hz; implement within ±10 Hz
Ranging offset[b]	0 to (216 − 1), increments of 6.25 s/64
Burst length (mini-slots) if variable on this channel (changes burst-to-burst)	1 to 255 mini-slots
Transmit equalizer coefficients 1 (advanced modems only)	Up to 64 coefficients; 4 bytes per coefficient: 2 real and 2 complex ·

[a]Values in table apply for this given channel and symbol rate.
[b]Ranging offset is the delay correction applied by the CM to the CMTS upstream frame time derived at the CM, in order to synchronize the upstream transmissions in the TDMA scheme. The ranging offset is an advancement equal to roughly the round-trip delay of the CM from the CMTS. The CMTS *must* provide feedback correction for this offset to the CM, based on reception of one or more successfully received bursts (i.e., satisfactory result from each technique employed: error correction and/or CRC), with accuracy within 1/2 symbol and resolution of 1/64 of the frame tick increment (6.25 μs/64 = 0.09765625 μs = 1/4 the symbol duration of the highest symbol rate = 10.24 MHz^{-1}). The CMTS sends adjustments to the CM, where a negative value implies that the ranging offset is to be decreased, resulting in later times of transmission at the CM. CM *must* implement the correction with resolution of at most 1 symbol duration (of the symbol rate in use for a given burst), and (other than a fixed bias) with accuracy within ±0.25 μs plus ±1/2 symbol owing to resolution. The accuracy of CM burst timing of ±0.25 μs plus ±1/2 symbol is relative to the mini-slot boundaries derivable at the CM based on an ideal processing of the timestamp signals received from the CMTS.
Source: Table 4-5, page 27, Ref. 3.

For symbol rate, offset frequency, and ranging offset, the CM transmits consecutive bursts as long as the CMTS allocates at least 96 symbols in between the last symbol center of one burst and the first symbol center of the following burst. The maximum reconfiguration time of 96 symbols should compensate for the rampdown time of one burst and the rampup time of the next burst as well as the overall transmitter delay time including the pipeline delay and optional pre-equalizer delay. For modulation-type changes, the CM transmits consecutive bursts as long as the CMTS allocates at least 96 symbols in between the last symbol center of one burst and the first symbol center of the following burst. Output power, symbol rate, offset frequency, channel frequency, and ranging offset should not be changed until the CM is provided sufficient time between bursts by the CMTS. Transmitted output power, symbol rate, offset frequency, channel frequency, and ranging offset are not changed while more than −30 dB of any symbol's energy of the previous burst remains to be transmitted, or more than −30 dB of any symbol's energy of the next burst has been transmitted, excluding the effect of the transmit equalizer (if present in the CM). Negative ranging offset adjustments will cause the 96 symbol guard to be violated. The CMTS assures that this does not happen by allowing extra guard time between bursts that is at least equal to the amount of negative ranging.

If the channel frequency is to be changed, then the CM changes between bursts as long as the CMTS allocates at least 96 symbols plus 100 ms between the last symbol center of one burst and the first symbol of the following burst. The output transmit power is maintained constant within a TDMA burst to within less than 0.1 dB (excluding the amount theoretically present due to pulse shaping, and amplitude modulation in the case of 16-QAM).

Burst Timing Convention. Figure 17.19 illustrates the nominal burst timing.

Figure 17.20 shows worst-case burst timing. In this example, burst N arrives 1.5 symbols late, and burst $N + 1$ arrives 1.5 symbols early, but separation of 5 symbols is maintained; 8-symbol guardband is shown.

At a symbol rate of Rs, symbols occur at a rate of one each $Ts = 1/Rs$ seconds. Rampup and rampdown are the spread of a symbol in the time domain beyond Ts duration owing to the symbol-shaping filter. If only one symbol were transmitted, its duration would be longer than Ts due to the shaping filter impulse response being longer than Ts. The spread of the first and last symbols of a burst transmission effectively extends the duration of the burst to longer than $N \times Ts$, where N is the number of symbols in the burst.

Frame Structure. Figure 17.21 shows two examples of frame structure: one where the packet length equals the number of information bytes in a codeword, and another where the packet length is longer than the number of information bytes in one codeword, but less than two codewords. Example 1 illustrates the fixed codeword-length mode, and Example 2 illustrates the shortened last codeword mode. These modes are defined below.

Codeword Length. The CM operates in a fixed-length codeword mode or with the shortened codeword capability enabled. The minimum number of information bytes in a codeword, for fixed or shortened mode, is 16 bytes. Shortened codeword capability is available with $k \geq 16$ bytes, where k is the number of information bytes in a codeword. With $k < 16$, shortened codeword capability is not available.

Upstream Demodulator Input Power Requirements. The maximum total input power to the upstream demodulator should not exceed 35 dBmV in the 5- to 42-MHz frequency range of operation. The intended receive power in each carrier should be within the values shown in Table 17.8. The demodulator operates within its defined performance specifications with received bursts within ± 6 dB of the nominal commanded received power.

Upstream Electrical Output of the CM. The CM operates with an RF output modulated signal with the characteristics given in Table 17.9.

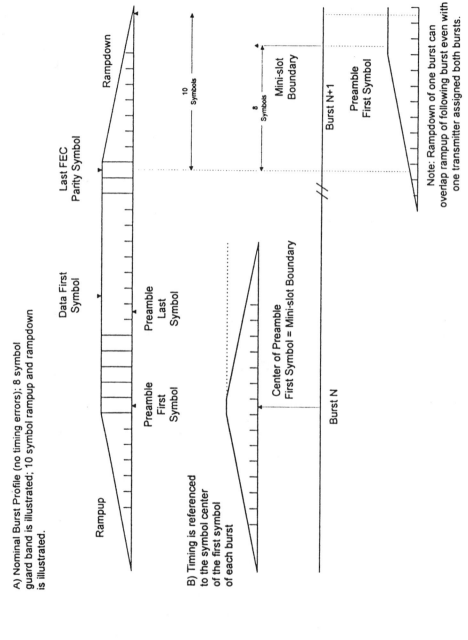

A) Nominal Burst Profile (no timing errors); 8 symbol guard band is illustrated; 10 symbol rampup and rampdown is illustrated.

Rampup

Preamble First Symbol

Data First Symbol

Preamble Last Symbol

Last FEC Parity Symbol

Rampdown

10 Symbols

B) Timing is referenced to the symbol center of the first burst

Center of Preamble First Symbol = Mini-slot Boundary

Burst N

8 Symbols

Mini-slot Boundary

Burst N+1

Preamble First Symbol

Note: Rampdown of one burst can overlap rampup of following burst even with one transmitter assigned both bursts.

Figure 17.19. Nominal burst timing. (From Figure 4-5, page 29, DOCSIS Specification, Ref. 3.)

712

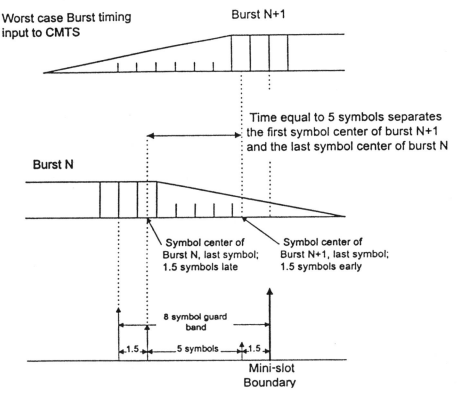

Figure 17.20. Worst-case burst timing. (From Figure 4-6, page 30, DOCSIS Specification, Ref. 3.)

Example 1. Packet length = number of information bytes in codeword = k

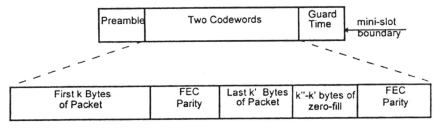

Example 2. Packet length = k + remaining information bytes in 2nd codeword = k + k' ≤ k + k" ≤ 2k

Figure 17.21. Example frames structure with flexible burst length mode. (From Figure 4-7, page 34, DOCSIS Specification, Ref. 3.)

TABLE 17.8   Maximum Range of Commanded Nominal Receive Power in Each Carrier

Symbol Rate (ksym/s)	Maximum Range (dBmV)
160	−16 to +14
320	−13 to +17
640	−10 to +20
1280	−7 to +23
2560	−4 to +26

Source: Table 4-9, page 37, Ref. 3.

TABLE 17.9   Electrical Output From CM

Parameter	Value
Frequency	5 to 42 MHz edge to edge
Level range (one channel)	+8 to +55 dBmV (16-QAM)
	+8 to +58 dBmV (QPSK)
Modulation type	QPSK and 16-QAM
Symbol rate (nominal)	160, 320, 640, 1280, and 2560 ksym/s
Bandwidth	200, 400, 800, 1600, and 3200 kHz
Output impedance	75 ohms
Output return loss	> 6 dB (5–42 MHz)
Connector	F connector per [IPS-SP-401] (common with the input)

Source: Table 4-10, page 37, Ref. 3.

Downstream. The downstream PMD sublayer conforms to ITU-T Rec. J.83, Annex B, for low-delay video applications with the exception of the scalable interleaving discussed below.

Scalable Interleaving to Support Low Latency. The downstream PMD sublayer supports a variable-depth interleaver with the characteristics defined in Table 4-11 of the DOCSIS reference publication. It should be noted that the table contains a subset of the interleaver modes found in ITU-T Rec. J.83-B.

Downstream Frequency Plan. The downstream frequency plan should comply with harmonic related carrier (HRC), incremental related carrier (IRC), or standard (STD) North American plans in accordance with EIA-542. However, operation below a center frequency of 91 MHz is not required.

CMTS Output—Electrical. The CMTS has an output that is an RF modulated signal with the characteristics given in Table 17.10.

TABLE 17.10 CMTS Output

Parameter	Value
Center frequency (fc)	91 to 857 MHz ± 30 kHz[a]
Level	Adjustable over the range 50 to 61 dBmV
Modulation type	64 QAM and 256 QAM
Symbol rate (nominal)	
64 QAM	5.056941 Msym/s
256 QAM	5.360537 Msym/s
Nominal channel spacing	6 MHz
Frequency response	
64 QAM	~ 18% Square root raised cosine shaping
256 QAM	~ 12% Square root raised cosine shaping
Total discrete spurious inband (fc ± 3 MHz)	< −57 dBc
Inband spurious and noise (fc ± 3 MHz)	< −48 dBc; where channel spurious and noise includes all discrete spurious, noise, carrier leakage, clock lines, synthesizer products, and other undesired transmitter products. Noise within ±50 kHz of the carrier is excluded.
Adjacent channel (fc ± 3.0 MHz) to (fc ± 3.75 MHz)	< −58 dBc in 750 kHz
Adjacent channel (fc ± 3.75 MHz) to (fc ± 9 MHz)	< −62 dBc, in 5.25 MHz, excluding up to 3 spurs, each of which must be < −60 dBc when measured in a 10-kHz band
Next adjacent channel (fc ± 9 MHz) to (fc ± 15 MHz)	< −65 dBc in 6 MHz, excluding up to three discrete spurs. The total power in the spurs must be < −60 dBc when each is measured with 10-kHz bandwidth
Other channels (47 MHz to 1000 MHz)	< −12 dBmV in each 6-MHz channel, excluding up to three discrete spurs. The total power in the spurs must be < −60 dBc when each is measured with 10-kHz bandwidth.
Phase noise	1 to 10 kHz: −33 dBc double-sided noise power
	10 to 50 kHz: −51 dBc double-sided noise power
	50 to 3 MHz: −51 dBc double-sided noise power
Output impedance	75 ohms
Output return loss	> 14 dB within an output channel up to 750 MHz; > 13 dB in an output channel above 750 MHz
Connector	F connector per [IPS-SP-401]

[a] ± 30 kHz includes an allowance of 25 kHz for the largest FCC frequency offset normally built into upconverters.

Source: Table 4-12, page 39, Ref. 3.

TABLE 17.11 CM BER Performance

Item	BER	Image Reject	Adjacent Channel
64-QAM	Implementation loss such that CM achieves a post-FEC BER better than or equal to 10^{-8} when $E_b/N_0 = 23.5$ dB or better.	Meeting BER performance requirements, with analog or digital signal at $+10$ dBc in any portion of the band other than adjacent channels.	Meeting BER performance requirements with digital signal at 0 dBc in adjacent channels. With analog signal in adjacent channels at $+10$ dBc. With additional 0.2 dB allowance with ±10 dBc in adjacent channels.
256-QAM	Implementation loss of CM is such that CM achieves a post-FEC BER of equal or better than 10^{-8} with an E_s/N_0 of 30 dB or more.	Meeting BER performance requirements, with analog or digital signal at $+10$ dBc in any portion of RF band other than adjacent channels.	Meeting BER performance requirements with analog or digital signal at 0 dBc in adjacent channels. Meeting BER performance requirements with an additional 0.5 dB allowance, shall be met with analog signal at $+10$ dBc in adjacent channels. Meeting BER performance requirements with an additional 1.0 dB allowance, shall be met with digital signal at $+10$ dBc in adjacent channels

Source: Derived from material in Section 4.3.6, Ref. 3.

Downstream Electrical Input to the CM. The CM accepts an RF-modulated signal with the characteristics shown in Table 17.10.

CM BER Performance. The bit error rate performance of a CM is described in Table 17.11. The requirements apply to the $I = 128$, $J = 1$ mode of interleaving.

17.4 LOCAL MULTIPOINT DISTRIBUTION SYSTEM (LMDS)

LMDS as we know it today evolved out of a one-way radio (wireless) broadband cellular system that delivered conventional 6-MHz analog TV channels to residential and business customers. This system was developed by

Bernard (Bernie) Bossard whose company became Cellular Vision USA. Its beta tests were carried out in a test bed in Brooklyn, New York in the period 1989 to 1995. The system carried 49 TV channels in an FDM-type configuration where the RF modulation was FM. By digitizing and then applying MPEG compression techniques, it was found that the number of operational TV channels was tripled over the original system for a given RF bandwidth (Ref. 4).

The operational frequency band was the 28- to 30-GHz band. In this band, LMDS competes for usage with conventional LOS microwave and communication satellite feeder uplinks. The U.S. FCC broke up the band so all parties would be satisfied. Keep in mind that LMDS was really a wireless CATV technique. With implementations being fielded today it serves as a two-way configuration. Here, of course, downstream means from an LMDS hub to end-user, and upstream operation is from the end-user to the LMDS hub.

The specific LMDS frequency bands in the United States are 27.5 to 28.35 GHz (850-MHz bandwidth), 29.1 to 29.25 GHz (150-MHz bandwidth), and 31.075 to 31.225 GHz (150-MHz bandwidth). The range from an LMDS hub to the furthermost end-user depends on rainfall analysis. It can vary from 1 mile (1.6 km) to 5 miles (8.0 km) (Refs. 9 and 10.)

The LMDS concept is shown in Figure 17.22. In the figure a hub is a facility that houses a radio (wireless) transmitter and receiver(s) which provide connectivity with end-users. The hub connects through a fiber-optic cable or LOS microwave radiolink (wireless) with a central processing and routing unit. This unit may be colocated with a local switching center providing interface with the PSTN.

The antenna at each hub will normally be omnidirectional with +10- to +15-dBi gain, whereas the antenna at an end-user location will be directional with a +35-dBi gain. Cost and performance dictate that the RF front end be colocated with the antenna. The antenna unit feeds and accepts IF from an inside unit which may be likened to a CATV set-top box. The inside unit accepts the IF signal from the antenna and provides, in the transmit direction, an IF signal to the antenna unit.

LMDS provides different levels of services. A residence accepts a more modest service capability, such as 1 Mbps of downstream internet, symmetric upstream and downstream data in increments of 64 kbps, and from one to eight POTS ("plain old telephone service") lines and equivalent CATV downstream TV service using MPEG-2 frame video. Depending on requirements, a business may require one or more DS1 or E1 symmetric line equivalents, several 1-Mbps downstream internet connections, or one 5-Mbps connectivity to an on-premises server as well as several frame relay connectivities and/or packet service based on TCP/IP. They may not need or want the entertainment TV. Many of the configurations in North America have turned to using a version of DOCSIS (Section 17.3) providing POTS, TV, data, and, of course, internet.

As we see from this discussion, the LMDS frequency band is split into three segments. We could make the assignment highly asymmetrical with

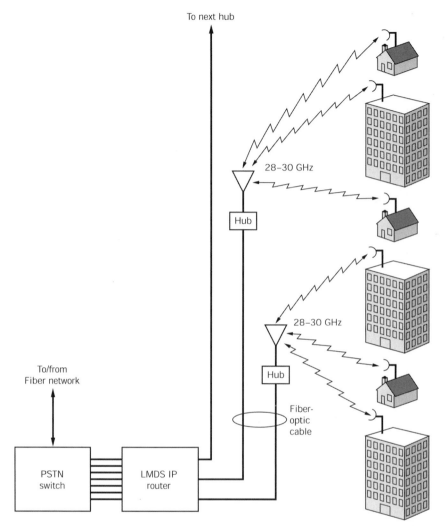

Figure 17.22. Sketch illustrating the LMDS concept. Both residential and business services are shown.

1000 MHz assigned for downstream and 150 MHz for upstream. For efficient operation, frequency reuse is mandatory.

17.4.1 Frequency Reuse

The original LMDS attempted frequency reuse, with some success. It depended on free space loss, polarization isolation, and some frequency offset of television radio carriers. Although the value of polarization isolation was

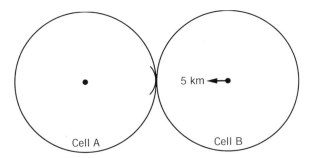

Figure 17.23. Cell B hub transmitter antenna faces the back of closest cell A end-user antenna. Cell radius is 5 km. Gain at the back of receiving antenna is −10 dBi. (Note: −10 dBi based on ITU-R model.)

claimed to be some 36 dB, it was more on the order of 30 dB. Polarization isolation works on the premise that adjacent cells will be on opposite polarizations. Cell A, for example, would transmit its signal with horizontal polarization, and adjacent cell B would be on vertical polarization. If we measured the level of cell A's signal at cell B's receiver, it would be down (attenuated) at least 30 dB due to the polarization alone. Actually it would be down much more because of the free space loss between A's location and B's. Remember, in this model cell A and cell B transmit on the same frequency.

In this first case the isolation at the receiver at cell A from the transmit signal of cell B, which is on the same frequency but different polarization, depends on (see model in Figure 17.23)*:

- Free space loss at 30 GHz assuming that cell radius of 5 km is −135.9 dB
- Polarization isolation   −30   dB
- Receive antenna discrimination (back of antenna)   −10   dB
- Total "path" losses   −175.9   dB
- Transmit power is 1 watt for a 10-Mbps bitstream; transmit antenna gain is +14 dBi, thus the EIRP is +14 dBW.
- The isotropic receive level (IRL) is the difference between two values (i.e., the EIRP and the FSL): −161.9 dBW.
- Add the receive antenna gain (i.e., −10 dBi). This is the RSL (receive signal level): −171.9 dBW.
- The receiver has a 6-dB noise figure. Its thermal noise threshold is −198 dBW/Hz (N_0).

---

*For a tutorial on the design of wireless (radio) systems, consult Roger L. Freeman, *Radio System Design for Telecommunications*, 2nd edition, John Wiley, New York, 1997 (Ref. 5).

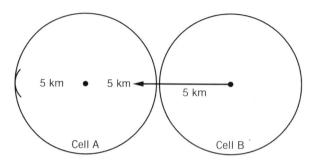

Figure 17.24. Interference level into receiver in cell A where cell B's transmit antenna main beam is oriented such that it directly enters cell A receive antenna main beam. Distance is 15 km.

- The energy per interference bit $= -171.9$ dBW $- 10 \log 10 \times 10^6 = -241.9$ dBW/Hz $= I_0$.
- The $I_0/N_0 = -241.9 - (-198) = -43.9$ dB. This value shows that co-channel interference based on the model presents a very acceptable value.

(For a description of calculation of interference levels, consult Ref. 5, Chapter 13.)

This tells us that same frequency operation is feasible with this geometry where the offending transmit antenna faces the back of the victim receive antenna, distance 5 km.

We must examine another model that may not be so forgiving. The model is illustrated in Figure 17.24. This is where the offending transmitter of cell B is on the other side of the cell circle and it is looking right down the beam of the receive antenna in an adjacent cell A circle.* Here the victim receive antenna is at a 15-km distance from the offending transmitter. It is on the far side of the 5-km-range circle; and the offending transmitter, of course, is in the center of its range circle. Thus, the total range is 15 km (5 + 5 + 5). Now the offending transmitter's potentially interfering signal has a free space loss at the victim receiver of 145.5 dB. Consult Table 17.12 which is used to determine the interference potential in this particular case.

The interference signal would essentially reduce the system margin of the wanted signal by 1.5 dB. It would only seriously affect the system where, by happenchance, a hub transmit antenna was facing directly into the aperture of the customer receiver antenna. The chances are very slim that this would occur. Thus we still hold that same frequency operation is quite feasible.

Note: A tutorial on decibels and their applications may be found in Appendix C of Ref. 6.

---

*Commonly, hexagons are used to represent cells because they can be arranged without overlap. We use circles because they are more realistic for propagation studies.

TABLE 17.12 Calculation Table

Parameter	Wanted Signal	Interference Signal
EIRP (effective isotropic radiated power)[a]	+14 dBW	−14 dBW
FSL (free-space loss)	−135.97 dB	−145.51
IRL (isotropic rec level)	−123.97 dBW	−131.5 dBW
Polarization loss	0.0 dB	−30 dB
Sum	−123.97 dBW	−161.5 dBW
Antenna gain	+35 dBi	+35 dBi
RSL	−88.97 dBW	−126.5 dBW
(1) E_b, (2) I_0	−158.97 dBW/bit	−196.5 dBW/Hz
N_0	−198 dBW	−198 dBW
(1) E_b/N_0, (2) $E_b/(N_0 + I_0)$	39.03 dB	35.21 dB
Required $E_b/(N_0 + I_0)$, 16 QAM, BER = 1×10^{-8}	16 dB	
Margin	23.03 dB	

[a]Note that the EIRP is the sum in dB units of the transmit power and the antenna gain. Transmission line losses have been neglected.

Other designers take a different position and divide the 800-MHz downstream band into four 200-MHz bands. One we'll call frequency family A, the next one we'll call frequency family B, the next C, and the last D. They will occupy a single group of adjacent cells with the same nomenclature (i.e., A, B, C, and D). This concept is illustrated in Figure 17.25.

For improved isolation, 90-degree sectorized antennas are used which would appear more or less as in Figure 17.26. Here we would have a 90-degree pattern group on the line between cells D and C.

Propagation is a most challenging problem in LMDS design. In the models we used above, it is assumed that the receive antenna is oriented correctly toward the transmit antenna, and they are in *line-of-sight* of each other. This means that the two antennas are visible one-to-the-other, and there is sufficient clearance of the ray beam to obstacles in its path. This feature alone is difficult to achieve for all receive antennas being served by this nodal transmit antenna. Verdure (e.g., tree leaves) can block the signal; there will be diffraction paths and signal reflections. These latter conditions can cause multipath conditions and will disperse the signal.

Rainfall is yet another matter. In areas of heavy downpour rain such as in Florida and Louisiana, path lengths can be severely limited, to about some 2-km maximum. In other areas of the country which do not consistently have such conditions, maximum path lengths can be lengthened to up to 5 or 6 km (3.1 or 3.7 mi.), typically Los Angeles and Phoenix. When designing an LMDS system, it is mandatory that a rainfall study be carried out. Methodology for these studies can be found in *Radio System Design for Telecommunications*, 2nd edition (Ref. 5).

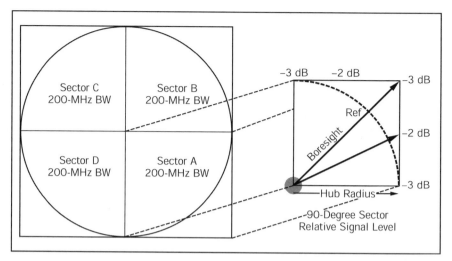

Figure 17.25. Splitting the downstream 800-MHz bandwidth into four 200-MHz bandwidths for improved frequency reuse and to minimize co-channel interference. A 90-degree sector-ized pattern is shown. Four transmitters, each with its own 90-degree sector antenna, sit at the center of the pattern. (From Figure 5, page 4, H-P Application Note 1296, Ref. 7.)

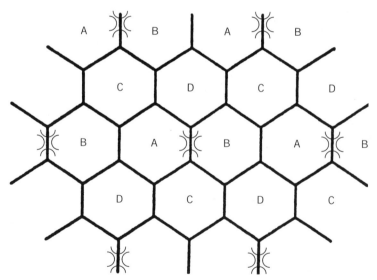

Figure 17.26. A large grouping of frequency reuse cells using a family of four center frequencies for downstream: A, B, C, and D. This is identical to a cellular radio pattern where $N = 4$. For our model, the cells have a 5-km radius.

The only way to improve system time availability is to optimize link margin. *Time availability* may be defined as the percentage of time that a system is operational meeting its BER performance requirements. *Link margin* represents those extra, surplus decibels left over when doing the link budget. One way to increase link margin is to shorten link length; another way is to use larger power amplifiers, reduce bit rate, use forward error correction (FEC), and increase antenna aperture at the user site.

Suppose we were to use some of the link parameters from the interference link budget in Table 17.12 and call it the signal link budget; however, the maximum distance required for signal transmission is 10 km (2 radii or 2×5). In this case, the transmit and receive antennas are on the same polarization, so the polarization isolation loss drops out. We have

EIRP	$+14$ dBW
FSL	-142 dB (30 GHz, 10 km)
Receive antenna gain	$+35$ dBi
RSL	-93 dBW

Convert to energy spectral density for 10 Mbps:

$$\mathrm{RSL}_0 = -93 - 10 \log 10 \times 10^6$$

$$= -163 \ \mathrm{dBW/Hz} = E_b$$

The receiver noise figure is 6 dB. Then

$$N_0 = -204 \ \mathrm{dBW} + 6 \ \mathrm{dB}$$

$$= -198 \ \mathrm{dBW/Hz}$$

$$E_b/N_0 = -163 \ \mathrm{dBW/bit} - (-198 \ \mathrm{dBW/Hz})$$

$$= 35 \ \mathrm{dB}$$

The required E_b/N_0 is 11 dB for QPSK modulation and BER $= 1 \times 10^{-10}$.

$$\text{The calculated } E_b/N_0 = 35 \ \mathrm{dB}$$

$$\text{The required } E_b/N_0 = 11 \ \mathrm{dB}$$

$$\text{Remainder (subtract)} = 24 \ \mathrm{dB} \text{ (link margin)}$$

The link margin in this case is 24 dB. These are the "dBs" left over, which we discussed previously. They are put to very good use. First, we include them in the calculations for rain margin and they become the rain margin. Simultaneously they help to burn the signal through verdure, to penetrate buildings, to help in refracting a signal around corners, and for the overall improvement of coverage. The question arises: Why not increase the margin? We are using a

1-watt HPA (high power amplifier) now; increase this to 2 watts, and our margin goes up 3 dB. A 2-watt amplifier costs something in the order of three times the cost of a 1-watt amplifier. We can see that the price tends to go up exponentially. There is a point where increasing amplifier power is not cost-effective any more. If we were to cut the radius of the cell in half, we would increase the margin about 6 dB.

As the cost of the system increases, the cost per CPE (customer premise equipment) goes up. One goal may be to limit the cost of CPE to $300 or thereabouts. So every system design change has implications on CPE cost. Suppose we increase antenna height and HPA power. This we find adds two thousand more possible customers to the system. How does this system change affect the cost per CPE?

Business customers most assuredly will be handled differently than residential customers. They will demand more telephone service, more data service, and more internet ports. The $300 goal must be discarded. Other cost goals should be established depending on the level of business. For larger businesses—say on the order of > 200 FX telephone lines, ten internet ports, a 45-Mbps frame relay port, and two 10-Mbps IP ports—thought must be given to the use of another service type such as LOS microwave at 23 GHz—for example, with a link to the hub. This separate radiolink would serve all the data/telephony requirements of the business in question. It may also be shared with other larger businesses in the same building, and other radio carriers can be added as required.

17.5 OTHER METHODS TO BREACH THE LAST MILE

17.5.1 Remote Wireless (Radio) Bridges

These systems originally operated in single, comparatively long-range point-to-point connections. Such simple architectures still predominate; however, point-to-multipoint products are now appearing in the marketplace. ISPs providing wireless last-mile connections have formed the largest market for such systems. Some systems provide as much as 100-Mbps connectivities, and many systems provide up to 30 Mbps. There is essentially no standardization. Compliance with IEEE 802.11 (wireless LAN, Section 10.1.5) is not even contemplated by equipment manufacturers of these systems.

The 5.8-GHz ISM band is widely used for wireless bridge applications. FCC Part 18.301 stipulates that operation shall be in the band 5.725 to 5.875 GHz, providing a 150-MHz bandwidth. Even with this wide bandwidth, frequency reuse sectorized antennas is commonplace.

Point-to-multipoint systems are beginning to appear on scene using some of the concepts discussed in Section 17.4. There are at least two manufacturers of multipoint-to-multipoint systems. The principal market addressed is that of ISP connectivities. No base stations are used, and each node is a router. Network intelligence is distributed throughout, and the network adapts instantaneously to changing interference and traffic levels. These nodes are connected in a mesh network where every and all stations are connected one to the other (Ref. 8).

17.5.2   In-Building Wireless Telephone Systems

This is a most pragmatic application of wireless systems where offices have many moves and rearrangements. The wireless PABX, along with the wireless LAN (WLAN—see Section 10.1.5), makes moves and rearrangements a breeze. Of course we assume telephone number portability. This means that when you are moved to another office in this or a nearby building, you keep the same telephone number.*

The *IP telephone* takes on wider applications allowing voice to be used over IP along with numerous data applications. It consolidates voice and data into unified network. Wireless IP telephony operates in the 2.4-MHz ISM band rather than the conventional 1.9-MHz PCS band.

REFERENCES

1. Roger L. Freeman, *Telecommunications Transmission Handbook*, John Wiley & Sons, New York, 1998.
2. *The DSL Sourcebook*, Paradyne Corporation, Lago, FL, 1997.
3. *Data-Over-Cable Service Interface Specification*, Revision 1, Radio Frequency Interface Specification, SP-RFI-980202, SCTE, Exton, PA, February 1998.
4. Various private communications between Bernard Bossard and Roger Freeman dealing with early design issues of LMDS as well as formulation of FCC Position Papers for Cellular Vision New York and USA, February 1988 through December 1995.
5. Roger L. Freeman, *Radio System Design for Telecommunications*, 2nd edition, John Wiley, New York, 1997.
6. Roger L. Freeman, *Fundamentals of Telecommunications*, John Wiley & Sons, New York, 1999.
7. *LMDS—The Wireless Interactive Broadband Access*, HP Application Note 1296, Hewlett-Packard Co., Englewood, CO, 1997.
8. *Wireless Integration*, Buyer's Guide, Pennwell, Nashua, NH, 2000.
9. Peter B. Papazian et al., *Study of the Local Multipoint Distribution Service Radio Channel*, *IEEE Transactions on Broadcasting*, Vol. 43, No. 2, IEEE, New York, June 1997.
10. Gene Heftman, LMDS Set to Challenge for Last-Mile Supremacy, *Microwaves and RF*, April 1999.

---

*Telephone number portability can be extended to an entire area code, statewide, countrywide, and even worldwide. It can take on this connotation.

18

NETWORK MANAGEMENT
FOR ENTERPRISE NETWORKS

18.1   WHAT IS NETWORK MANAGEMENT?

Effective network management optimizes a telecommunication network's operational capabilities. The key word here is *optimizes*. How does this function optimize?

- It keeps the network operating at peak performance.
- It informs the operator of deteriorating performance.
- It provides the tools for pinpointing cause(s) of performance deterioration or failure.
- It provides easy alternative routing and work-arounds when deterioration and/or failure take place.
- It serves as the front-line command post for network survivability.

There are numerous secondary functions of network management. They are important but, in our opinion, still secondary. Among these items are:

- It informs in quasi-real time regarding network performance.
- It maintains and enforces network security, such as link encryption (code changes, state variables), issuance, and use of passwords.
- It gathers and files data on network usage.
- It performs a configuration management function.
- It also performs an administrative management function.

18.2   THE BIGGER PICTURE

Many seem to view network management as a manager of data circuits only. There is a much bigger world out there. Numerous enterprise and government networks serve for the switching and transport of *multimedia communi-*

cations. The underlying network will direct (switch) and transport voice, data, and image traffic. Each will have a traffic profile notably differing from the other. Nevertheless, they should be managed as an entity. It is more cost effective to treat the whole than to treat the parts.

There is a tendency in the enterprise scene to separate voice telephony (calling it *telecommunications*) and data communications. This is unfortunate and a major error on the part of management. Perhaps that is why network management seems to often operate on two separate planes. One is data and very sophisticated, and the other is voice, which may have no management facilities at all. This section will treat network management as a whole consisting of its multimedia parts: voice, image, and data, which includes facsimile, telemetry, and CAD/CAM.

18.3 TRADITIONAL BREAKOUT BY TASKS

There are five tasks traditionally involved with network management:

- Fault management
- Configuration management
- Performance management
- Security management
- Accounting management

18.3.1 Fault Management

This is a facility that provides information on the status of the network and subnetworks. The "information on the status" should not only display faults (i.e., failures) and their location, but should also provide information on deteriorated performance. One cause of deteriorated performance is congestion. Thus, ideally, we would like to isolate the cause of the *problem.*

Fault management also includes the means to bypass troubled sections of a network—that is, the means to patch in new equipment for deteriorated or failed equipment.

The complexity of modern telecommunication networks is such that as many network management tasks as possible should be automated. All displays, readouts, and hard copy records should be referenced to a network time base down to 0.1 s. This helps in correlating events, an important troubleshooting tool.

18.3.2 Configuration Management

Configuration management establishes an inventory of the resources to be managed. It includes resource provisioning (timely deployment of resources to satisfy an expected service demand) and service provisioning (assigning services and features to end users). It identifies, exercises control over, collects data from, and provides data to the network for the purpose of preparing for, initializing, starting, and providing for the operation and termination of services. Configuration management deals with equipment and services, subnetworks, networks, and interfaces. Its functions are closely tied to fault management, as we have defined it previously.

18.3.3 Performance Management

Performance management is responsible for monitoring network performance to ensure that it is meeting specified performance. Some literature references (Ref. 1) add growth management. They then state that the objective of performance and growth management is to ensure that sufficient capacity exists to support end-user communication requirements.

Of course, there is a fine line defining *network capacity*. If too much capacity exists, there will probably be few user complaints, but there is excess capacity. Excess capacity implies wasted resources, thus wasted money. Excess capacity, of course, can accommodate short-term growth. Therefore performance/growth management provides vital information on network utilization. Such data provide the groundwork for future planning.

18.3.4 Security Management

Security management controls access to and protects both the network and the network management subsystem against intentional or accidental abuse, unauthorized access, and communication loss. It involves link encryption, changes in encryption keys, user authentication, passwords, and unauthorized usage of telecommunication resources.

18.3.5 Accounting Management

Accounting management processes and records service and utilization records. It generates customer billing reports for services rendered. It can identify costs and establish charges for the use of services and resources in the network. In other words, accounting management measures the usage of an enterprise's one, several, or many networks. It is a repository of data dealing with enterprise network plant-in-place information, along with usage of that plant, both internal (LANs) and external (WANs). The network usage data is vital for planning future upgrades and expansion.

18.4 SURVIVABILITY—WHERE NETWORK MANAGEMENT REALLY PAYS OFF

The network management center is the front-line command post for the battle for network survivability. We can model numerous catastrophic events affecting a telecommunications network, whether the public switched telecommunication network (PSTN) or a private/enterprise network. Among such events are fires, earthquakes, floods, hurricanes, terrorism, and public disorder. Telecommunications has brought revolutionary efficiencies to the way we do business and is thus very necessary, even for life itself. Loss of these facilities could destroy a business, even possibly destroy a nation. A properly designed network management system could mitigate losses, even save a network almost in its entirety.

Cite the World Trade Center explosion in New York City. The entire PABX telephone system survived, except for extensions in the immediate area of the bomb, simply because the PABX was installed on the sixth floor. All offices in the building had communications within the building and with the outside world. Thus vital communications were never lost.

This brings in the first rule toward survivability. A network management center is a place of point failure. Here we mean that if the center is lost, probably nearly all means of reconfiguring the network are lost. To avoid such a situation, a second network management center should be installed. This second center should be geographically separate from the principal center. It is advisable that the second center share the network management load and be planned and sized to be able to take the entire load with the loss of the principal center. There should be a communications orderwire between the two centers. Both centers should be provided with no-break power and backup diesel generators.

One simple expedient for survivability is to back up circuits with an arrangement with the local telephone company. This means that there must be compatibility between the enterprise network and the PSTN as well as one or more points of interface. A major concern is the network clock. One easy solution is to have the enterprise network derive its clock from the digital PSTN.

In the following sections we describe means to enhance survivability still further.

18.5   AVAILABILITY ENHANCEMENT—RAPID TROUBLESHOOTING

Availability is defined by the IEEE (Ref. 1): "The ability of an item—under combined aspects of its reliability, maintainability and maintenance support —to perform its required function at a stated instant or over a stated period of time." In the availability formula, only two terms may be varied: reliability, measured as mean time between failures (MTBF), and time to repair, measured as "mean time to repair" (MTTR). Both are measured in hours.

$$\text{Availability } (A) = \text{MTBF}/(\text{MTBF} + \text{MTTR})$$

Availability values for a network should well exceed 99.9%. The 99.9%

figure means that the network will be down (i.e., out of operation) 0.1% of the time or 0.001 × 8760 hours per year or 8.76 hours per year. This value may be acceptable to a network operator, but for many it is unacceptable. Suppose we require an availability of 99.999%. In this case the network will be down 0.001% of the time or 0.0001. Multiply this value by 8760 hours, the number of hours in a year, and it will be out of service 0.0876 hours/year or 5.256 minutes/year. To achieve such an excellent availability will be expensive. No way can a fault be recognized, isolated, and repaired in 5 minutes if we work on yearly increments. To achieve such an availability value, all circuits must be redundant, all signal paths must also be redundant, and redundant no-break power would be a requirement.

A well-designed network management system can notably reduce MTTR, thus improving the availability value. The system can do this in two ways:

1. Warn of deterioration
2. Identify exact location of a fault down to the card level

Telephone companies use two values for MTTR; 2 hours when a fault occurs in a facility with a craftsperson on duty (e.g., at a switch) and 4 hours when a fault occurs at a remote device such as at a fiber-optic repeater or at a microwave relay site (Ref. 2). The U.S. Navy often uses 0.3 hours because of the proximity of servicing personnel and the availability of spare parts.

It should be noted that much of the modern equipment is either equipped with BITE (built-in test equipment) or carries a software function to isolate and report faults. The BITE information, which is commonly "go" "no-go," can usually be remoted to a network management facility.

Another approach, which is described later, is the use of software, such as SNMP (simple network management protocol), which carries out a query–answer regime automatically regarding equipment status and performance.

An ideal network management system will advise the operator of one or more events,* and will reveal where in the network they occurred, as well as provide handy troubleshooting data. This should include the ability to pinpoint a fault down to the circuit board level. Ideally, the network management system should, in most cases, warn the operator in advance of impending fault/faults. This is a boon to improving network availability and its survivability during disaster. In the case of the latter, the network data reporting system can quickly identify circuits that are still up and functional and those that are down.

---

*I define an *event* as something out of the ordinary that occurred. In the telecommunication management arena an event might be a line dropout, switchover to backup power, loss of frame alignment, and so on.

18.5.1 Troubleshooting

Many network faults will be correctly isolated by the network management system, and in some cases no immediate action may be required by a network technician. However, other cases may prove to be more difficult. In such cases the network management system becomes an adjunct to the human craftsperson assisting in the troubleshooting effort.

We will assume that the troubleshooter will have available certain units of test equipment to aid in pinpointing the cause of an event.

We consider four steps in the "finding fault" process:

1. Observing symptoms
2. Developing a hypothesis
3. Testing the hypothesis
4. Forming conclusions

Observing Symptoms. It should be kept in mind that often many symptoms will appear at once, usually in a chain reaction. We must be able to spot the real causal culprit or we may spend hours or even days chasing effects, not the root cause.

We thus offer four reminders (Ref. 3):

1. Find the range and scope of the symptoms. Does the problem affect all stations (all users?). Does it affect random users or users in a given area? We are looking for a pattern here.
2. Are there some temporal conditions to the problem? How often does it occur per day, per hour, and so on. Is the problem continuous or intermittent? What is its regularity. Can we set our watches by its occurrence, or is it random in the temporal context?
3. Have there been recent rearrangements, additions to the network, or reconfigurations?
4. Software and hardware have different vintages that are called *release dates* in the case of software. My computer is using Windows 2000, but what is its release date? The question one should ask then would be: Are all items of a certain genre affected the same way? Different release dates of a workstation operating system may be affected differently, some release dates not at all. The problem may be peculiar to a certain release date.

Before we can move forward in the troubleshooting analysis, we must have firmly at hand the troublesome network's *baseline* performance. What is meant here is that we must have a clear idea of "normal" operation of the network so that we can really qualify and quantify its anomalous operation. For example, what is the expected BER, and is the value related to a time distribution? Can we express error performance in error-free seconds (EFS), errored seconds (ES), and severely errored seconds (SES)? These are defined in ITU-T Rec. G.821 (Ref. 4).

Five network-specific characteristics that the troubleshooter should have familiarity with or data on are as follows:

1. *Network Utilization.* What is the average network utilization? How does it vary through the workday? Characteristics of congestion, if any, should be known, as well as where and under what circumstances congestion might be expected.

2. *Network Applications.* What are the dominant applications on the network? What version numbers is it running?

3. *Network Protocol Software.* What protocols are running on the network? What are the performance characteristics of the software, and are these characteristics being achieved?

4. *Network Hardware.* Who manufactured the network interface controllers, media attachment units, servers, hubs, and other connection hardware? What versions are they? What are their performance characteristics? Expected? Met?

5. *Internetworking Equipment.* Who manufactured the repeaters, bridges, routers, and gateways on the network? What versions of software and firmware are they running? What are the performance characteristics? What are the characteristics of the interfaced network that are of interest?

Developing a Hypothesis. In this second step, we make a statement as to the cause of the problem. We might say that T1 or E1 frame alignment is lost because of deep fades being experienced on the underlying microwave transport network. Or we might say that excessive frames being dropped on a frame relay network is due to congestion being experienced at node B.

Such statements cannot be made without some strong bases to support the opinion. Here is where the knowledge and experience of the troubleshooter really pay. Certainly there could be other causes of E1 or T1 loss of frame alignment; but if underlying microwave is involved, that would be a most obvious place to look. There could be other reasons for dropping frames in a frame relay system. Errored frames could be one strong reason.

Testing the Hypothesis. We made a statement, and now we must back it up with tests. One test I like is correlation. Are the fades on the microwave correlated with the fade occurrences? That test can be done quite easily. If they are correlated, we have some very strong indication that the problem is with the microwave. The frame relay problem may be another matter. First, we could check the FECN and BECN bits to see if there were a change of state passing node B. If there is no change, assuming that flow control is implemented, then congestion may not be the problem. Removing the frame relay from the system and carrying out a bit error rate test (BERT) over some period of time would prove or disprove the noise problem.

A network analyzer is certainly an excellent tool in assisting in the localization of faults. Some analyzers have preprogrammed tests which can save the troubleshooter time and effort. Many networks today have some sort of network monitoring equipment incorporated. This equipment may be used in lieu of, or in conjunction with, network analyzers. Again we stress the importance of separating cause and effects. Many times network analyzers or network management monitors and testers will only show effects. The root cause may not show at all and must be inferred, or separate tests must be carried out to pinpoint the cause.

We might digress here to talk about what is often called in Spanish *tonterias*. This refers to "silly things." Such *tonterias* are often brought about by careless installation or careless follow-up repair. Coaxial cable connectors are some of my favorites. Look for intermittents and cold solder joints. A good (but not necessarily foolproof) tool is a time-domain reflectometer (TDR). It can spot where a break in a conductor is down to a few feet or less. It can do the same for an intermittent, when in the fault state. In fact, intermittents can prove to be a nightmare to locate. An electrically noisy environment can also be very troublesome.

Forming Conclusions. A conclusion or conclusions are drawn. As we say, "the proof of the pudding is in the eating." The best proof that we were right in our conclusion is to fix the purported fault. Does it disappear? If so, our job is done, and the network is returned to its "normal" (baseline) operation.

What conclusions can we draw from this exercise? There are two basic ingredients to network troubleshooting: (1) expertise built on experience of the troubleshooter and (2) the availability of basic test equipment. Trouble-shooting time can be reduced (degraded operation or out-of-service time reduction) by having on-line network management equipment. With ideal network management systems, this time can be cut to nearly zero.

18.6   SYSTEM DEPTH—A NETWORK MANAGEMENT PROBLEM

An isolated LAN is a fairly simple management problem. There is only a singular transmission medium, and, under normal operating conditions, only one user is transmitting information to one or several recipients. It is limited to only two OSI layers. For troubleshooting, often a protocol analyzer will suffice, although much more elaborate network management schemes and equipment are available.

Now connect that LAN to the outside world by means of a bridge or router and network management becomes an interesting challenge. One example from experience was a VAX running DECNET which was a station on a CSMA/CD LAN. The LAN was bridged to a frame relay box which fed

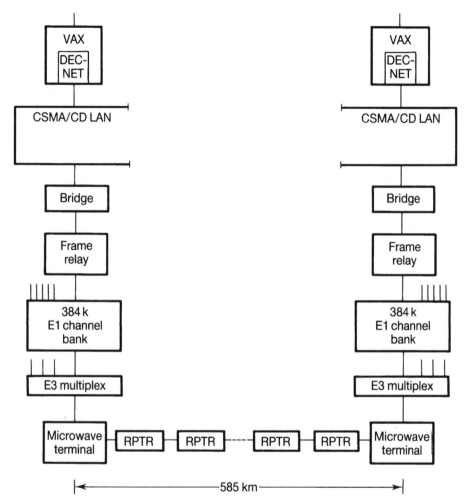

Figure 18.1. A typical multilevel network that employs a network management system. Note the multiple convergences.

a 384-kbps channel in an E1 hierarchy (i.e., 6 E0 channels) via tandemed microwave links to a large facility at the distant end (550 km) with similar characteristics. This connectivity is shown diagrammatically in Figure 18.1. Such is typical of a fairly complex network where we would apply a network management system. To make it even more difficult, portions of the network were leased from the local common carrier but were soon cut over to its own ownership.

18.6.1   Aids in Network Management Provisioning

Modern E1 and DS1 digital systems are provided with a means of operational monitoring of performance. The monitoring is done in quasi-real time and while operational (i.e., in-traffic).

In Chapter 7 we discussed PCM systems and the digital network including T1(DS1) and E1 hierarchies. Let us quickly review their in-service performance monitoring capabilities.

DS1 or T1 has a frame rate of 8000 frames per second. Each frame is delineated with a framing bit, or F bit. With modern alignment and synchronization algorithms, it is excessive and unnecessary to maintain framing repetition of the F bit 8000 times per second. Advantage was taken of F-bit redundancy by the development of the extended superframe (ESF). The ESF consists of 24 consecutive DS1 frames. With 24 frames we expect to have 24 framing bits. Of these, only 6 bits need to be used for framing, 6 are used for a cyclic redundancy check (CRC-6) on the frame, and the remaining 12 bits form a 4-kbps data link for network control and maintenance. It is this channel that can serve as transport for network management information. It can also serve as an ad hoc test link.

Our present concern is in-service monitoring. For instance, we can get real-time error performance with the CRC-6, giving us a measure of errored seconds and severely errored seconds in accordance with ITU-T Rec. G.821 (Ref. 4). Such monitoring can be done with test equipment such as the HP 37702A or HP 37741A.*

Permanent monitoring can be carried out using Newbridge network monitoring equipment, which can monitor an entire T1 or E1 network for frame alignment loss, errored seconds, and severely errored seconds. One can "look" at each link and examine its performance in 15-minute windows for a 24-hour period. The Newbridge equipment can also monitor selected other data services such as frame relay, which, in our preceeding example, rides on E1 aggregates. Such equipment can be a most important element in the network management suite.

The E1 digital network hierarchy also provides capability of in-service monitoring and test. We remember from Chapter 7 that E1 has 32 channels or time slots: 30 are used for the payload, and 2 channels or time slots serve as support channels. The first of these is channel (or time slot) 0, and the second is channel (time slot) 16. This latter is used for signaling. Time slot (channel) 0 is used for synchronization and framing. Figure 18.2 shows E1 multiframe structure. It also shows the sequence of bits in the frame alignment (TS0) signal of successive frames. In frames not containing the frame alignment signal, the first bit is used to transmit the CRC multiframe signal (001011) which defines the start of the submultiframe (SMF). Alternate frames contain the frame alignment word (0011011) preceded by one of the CRC-4 bits. The CRC-4 remainder is calculated on all the 2048 bits of the previous SMF, and the 4-bit word sent as $C1, C2, C3, C4$ of the current SMF. Note that the CRC-4 bits of the previous SMF are set to zero before the calculation is made.

---

*HP stands for Hewlett-Packard Company, a well-known manufacturer of electronics test equipment.

Multiframe	Submultiframe (SMF)	Frame number	Bits 1 to 8 of the frame in time slot 0							
			1	2	3	4	5	6	7	8
Multiframe	I	0	C_1	0	0	1	1	0	1	1
		1	0	1	A	S_{a4}	S_{a5}	S_{a6}	S_{a7}	S_{a8}
		2	C_2	0	0	1	1	0	1	1
		3	0	1	A	S_{a4}	S_{a5}	S_{a6}	S_{a7}	S_{a8}
		4	C_3	0	0	1	1	0	1	1
		5	1	1	A	S_{a4}	S_{a5}	S_{a6}	S_{a7}	S_{a8}
		6	C_4	0	0	1	1	0	1	1
		7	0	1	A	S_{a4}	S_{a5}	S_{a6}	S_{a7}	S_{a8}
	II	8	C_1	0	0	1	1	0	1	1
		9	1	1	A	S_{a4}	S_{a5}	S_{a6}	S_{a7}	S_{a8}
		10	C_2	0	0	1	1	0	1	1
		11	1	1	A	S_{a4}	S_{a5}	S_{a6}	S_{a7}	S_{a8}
		12	C_3	0	0	1	1	0	1	1
		13	E	1	A	S_{a4}	S_{a5}	S_{a6}	S_{a7}	S_{a8}
		14	C_4	0	0	1	1	0	1	1
		15	E	1	A	S_{a4}	S_{a5}	S_{a6}	S_{a7}	S_{a8}

Figure 18.2. E1 multiframe, CRC-4 structure, *Notes:* E = CRC-4 error indication bits. S_{a4} to S_{a8} = space bits. May be used for maintenance (network management) data link. C_1 to C_4 = cyclic redundancy check-4 (CRC-4) bits. A = remote alarm indication. (From Table 4b/G.704, page 81, CCITT Rec. G.704, Ref. 5.)

At the receive end, the CRC remainder is recalculated for each SMF, and the result is compared with the CRC-4 bits received in the next SMF. If they differ, then the checked SMF is in error. What this is telling us is that a block of 2048 bits had one or more errors. One thousand CRC-4 block error checks are made every second. It should be noted that this in-service error detection scheme does not indicate BER unless one assumes a certain error distribution (random or burst errors) to predict the average errors per block. Rather it provides a block error measurement.

This is very useful for estimating percentage of errored seconds (%ES), which is usually considered the best indication of quality for data transmission—itself a block or frame transmission process. CRC-4 error checking is fairly reliable, with the ability of detecting 94% of errored blocks even under poor BER conditions (see Ref. 6).

Another powerful feature of E1 channel 0 (when equipped) is the provision of local indication of alarms and errors detected at the far end. When an errored SMF is detected at the far end, one of the E bits (see Figure 18.2) is changed from a 1 to a 0 in the return path multiframe (TS0). The local end, therefore, has exactly the same block error information as the far-end CRC-4 checker. Counting E-bit changes is equivalent to counting CRC-4 block errors. Thus the local end can monitor the performance of both the go and return paths. This can be carried out by the network equipment itself, or by a test set such as the HP 37722A monitoring the E1 2.048-Mbps data bit

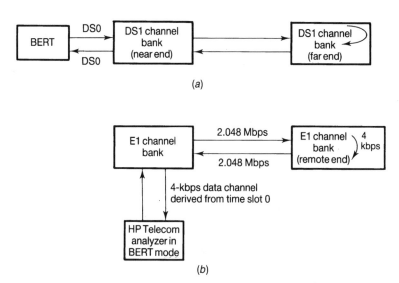

Figure 18.3. (a) Loopback of a DS0 channel with BERT test in place (intrusive or on a spare DS0). (b) Loopback of 8-kbps data channel derived from ESF on DS1 or from channel 0 of E1 (nonintrusive testing).

stream. In the same way, the A bits return alarm signals for loss of frame or loss of signal from the remote end.

Loopback testing is a fine old workhorse in our toolbox of digital data troubleshooting aids. There are two approaches for DS1 (T1) and E1 systems: intrusive and nonintrusive. Intrusive, of course, means that we interrupt traffic by taking one DS0 or E0 channel out of service, or the entire aggregate. We replace the channel with a pseudorandom binary sequence (PRBS) or other sequence specifically designed to stress the system. Commonly we use conventional bit error rate test (BERT) techniques when looping back. The concept of loopback is illustrated in Figure 18.3. ESF and Channel 0 data channel testing is nonintrusive. It does not interfere with customer traffic. Trouble can also be isolated, whether in the "go" or "return" channel of the loopback. Both intrusive and nonintrusive testing is commonly automated in the network management suite.

Many frame relay equipment also have forms of in-service monitoring as well as a system of fault alarms. In fact, most complex telecommunication equipment has built-in monitoring and test features. The problem often is that these features are proprietary, whereas our discussion of DS1/E1 systems reveals that they have been standardized by ITU-T and Telcordia recommendations and publications. This is probably the stickiest problem facing network management systems—that is, handling, centralizing, and controlling network management features in a multivendor environment.

18.6.2   Communications Channels for the Network Management System

A network management facility is usually centrally located. It must monitor and control distant communications equipment. It must have some means of communicating with this equipment, which may be widely dispersed geographically. We have seen where DS1 and E1 systems provide a data channel for OAM operations and maintenance. Higher levels of the DS1 and E1 hierarchies have ready communication channel(s) for OAM. So do SONET and SDH.

The solution for a LAN is comparatively straightforward. The network management facility/LAN protocol analyzer becomes just another active station on the LAN. Network management traffic remains as any other revenue-bearing traffic on the LAN. Of course, the network management traffic should not overpower the LAN with message unit quantity which we might call network management overhead.

WANs vary in their capacity to provide some form of communicating network management information. X.25 provides certain types of frames or messages dedicated for network control and management. However, these frames/messages are specific to X.25 and do not give data on, say, error rate at a particular point in the network. Frame relay provides none with the exception of flow control and the CLLM, which are specific to frame relay. For a true network management system, a separate network management communication channel may have to be provided. It would have to be sandwiched into the physical layer. However, SNMP (described below) was developed to typically use the transport services of TCP/IP. (See Section 11.4 for a discussion of TCP/IP and related protocols.) It is additional overhead, and care must be taken of the percentage of such overhead traffic compared to the percentage of "revenue-bearing" traffic. Some use the term "in-band" when network management traffic is carried on separate frames on the same medium, and the term "out-of-band" is used when a separate channel or time slot is used such as with E1/T1.

18.7   AN INTRODUCTION TO NETWORK MANAGEMENT PROTOCOLS

18.7.1   Two Network Management Protocols

Two separate communities have been developing network management protocols:

- The TCP/IP community: SNMP
- The ISO/OSI community: CMIP

The most mature and most implemented is the Simple Network Management Protocol (SNMP). Certain weak points arose in the protocol, and a version called SNMPv.2 has been developed and fielded.

The Common Management Information Protocol (CMIP) has been developed for the OSI environment. It is more versatile but requires about five times the memory of SNMP.

18.7.2 An Overview of SNMP

SNMP is the dominant method for devices on a network to relay network management information to centralized management consoles that are designed to provide a comprehensive operational view of the network. Having come on-line in about 1990, literally thousands of SNMP systems have been deployed.

There are three components of the SNMP protocol:

- The management protocol itself
- The management information base (MIB)
- The structure of management information (SMI)

Figure 18.4 shows the classic client–server model. The client runs at the *managing* system. It makes requests and is typically called the *network management system* (NMS) or *network operations center* (NOC). The server is in the *managed* system. It executes requests and is called the *agent*.

Structure of Management Information (SMI). This defines the general framework within which an MIB can be defined. In other words, SMI is the set of rules which define MIB objects, including generic types used to describe management information. The SNMP SMI uses a subset of Abstract Syntax Notation One (ASN.1) (Ref. 7) specification language that the ISO (International Standards Organization) developed for communications above the OSI presentation layer. Layer 7, for example, may use ASN.1 standards and ITU-T Recs. X.400 and X.500. It was designed this way so that SNMP could

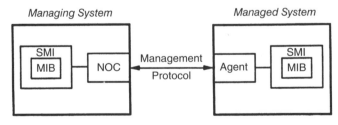

Figure 18.4. SNMP basic management architecture. SMI, structure of management information; NOC, network operations center; MIB, management information base.

be aligned with the OSI environment. The SMI organizes MIB objects into an upside-down tree for naming purposes.

Management Information Base (MIB). This is the set of managed objects or variables that can be managed. Each data element, such as a node table, is modeled as an object and given a unique name and identifier for management purposes. The complete definition of a managed object includes its naming, syntax, definitions, and access method (such as read-only or read-write) which can be used to protect sensitive data and status. By allowing a status of "required" or "optional," the SNMP formulating committee allows for the possibility that some vendors may not wish to support optional variables. Products are obligated to support required objects if they wish to be compliant with the SNMP standard (Ref. 8).

Figure 18.4 also shows an *agent* in every equipment to be managed, and Figure 18.5 illustrates how a management console manages "agents."

SNMP utilizes an architecture that depends heavily upon communication between one or a small number of managers and a large number of remote agents scattered throughout the network. Agents use the MIB to provide a view of the local data that are available for manipulation by the network management station. In order for a variable, such as the CPU utilization of a remote Sun workstation, to be monitored by the network management station, it must be represented as an MIB object.

Yemini (Ref. 9) introduces the MIB in an interesting manner; and as many other authors in the field, he portrays it as a *tree*, as shown in Figure 18.6. It is located in the expanded agent's box. The MIB is organized as a directory tree of managed data located at the leaves, and it is shown as shaded cells. The internal nodes of the tree are used to group managed variables by related categories. Dr. Yemini shows that all information associated with a given protocol entity [e.g., Internet Protocol (IP), TCP, User Datagram

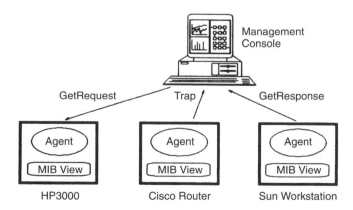

Figure 18.5. The network management console manages agents.

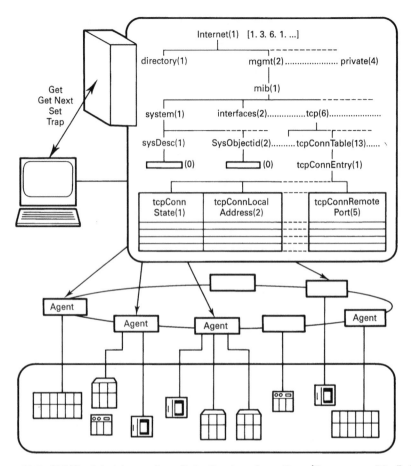

Figure 18.6. SNMP pictorial overview. Note the tree formation. (From page 29, Ref. 9, reprinted with permission of the IEEE Press.)

Protocol (UDP)] is recorded in a given subtree. The tree structure provides a unique path identifier (index) for each leaf. This identifier is formed by concatenating the numerical labels of its path from the root. The managed information stored at the leaves can be accessed by passing the respective identifier to the agent.

The management station sends *get* and *set* requests to remote agents. They provide a query mechanism for platform programs (managers) to retrieve or modify MIB data. *GET* retrieves data, *SET* changes the data, and *GET-NEXT* provides a means to traverse the MIB tree and retrieve data in order.

When an unexpected event takes place, these agents initiate *traps* to the management station. In such a configuration, most of the burden for retrieving and analyzing data rests on the management application. Unless data are

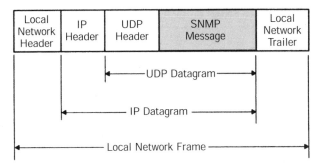

Figure 18.7. An SNMP "message" embedded in a local network frame.

requested in a proactive way, little information will be displayed at the management station. This poll-based approach increases network traffic, especially on the backbone where some users report a 5–10% increase in traffic due to SNMP packets (or messages). SNMP is a connectionless protocol that initially was designed to run over a UDP/IP (see Sections 11.5 and 11.4, respectively) stack. Because of its design, it is traditionally high in overhead with the ratio of overhead to usable data running about 10 octets to one. A typical SNMP "message" (PDU) embedded in a local network frame is illustrated in Figure 18.7.

Inside the frame in Figure 18.7 we find an internet protocol (IP—see Section 11.4) datagram that has a header. The header has an IP destination address that directs the datagram to the intended recipient. Following the IP header, there is the UDP datagram that identifies the higher-layer protocol process. This is the SNMP message, shown in Figure 18.7 as embedded in the UDP. The application here is typically for LANs. If the IP fragment is too long for one frame, it is fragmented (segmented) into one or more additional frames.

Figure 18.8 shows a typical SNMP PDU structure that is valid for all messages but the *trap* format. This structure is embedded as the "message" field in Figure 18.7. The SNMP message itself is divided into two sections: (1) a version identifier plus community name and (2) a PDU. The version identifier and community name are sometimes referred to as the SNMP *authentication header*.

The *version* field assures that all parties in the management transaction are using the same version of the SNMP protocol. We must remember that the origins of SNMP evolved from TCP/IP, and they were originally designed for the exclusive use on TCP/IP networks. In Chapter 11, we became familiar with the version field.

Each SNMP message contains a *community name*, which is one of the only security mechanisms in SNMP. The agent examines the community name to ensure that it matches one of the authorized community strings loaded in its configuration fields or novolatile memory. Each SNMP PDU is one of five

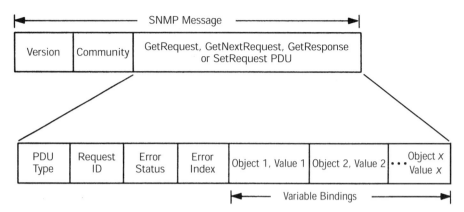

Figure 18.8. An SNMP PDU structure for GetRequest, GetNextRequest, GetResponse, and SetResponse. [From *Managing Technology Overview*, Chapt. 52, CISCO, June 1999 and from RFC 1067 (J. Case, Univ. Tenn., Aug. 1988) and RFC 1157 (J. Case, Univ. Tenn., May 1990).]

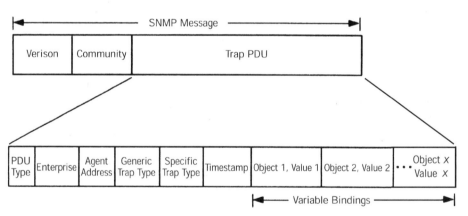

Figure 18.9. SNMP V.1 trap PDU format. [From *Internetworking Technology Overview*, Chapt. 52, CISCO, June 1999 and from RFC 1067 (J. Case, Univ. Tenn., Aug. 1988) and RFC 1157 (J. Case, Univ. Tenn., May 1990).]

types (sometimes called *verbs*): GetRequest, GetNextRequest, SetRequest, GetResponse, and Trap. The trap PDU is shown in Figure 18.9.

The PDU shown in Figure 18.8 has five initial fields. The first field is the *PDU* type. There are five types of PDU as we discussed previously. These types are shown in Table 18.1.

The *Request ID* is the second field of the PDU field. It is an INTEGER-type field that correlates the manager's request with the agent's response. INTEGER-type is a primitive type used in ASN.1.

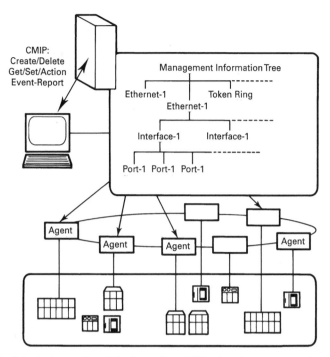

Figure 18.11. A typical overall architecture of an OSI network management system. (From Figure 2-12, page 48, Ref. 9.)

18.8 TELECOMMUNICATIONS MANAGEMENT NETWORK (TMN)

18.8.1 Objective and Scope

TMN provides a framework for telecommunications management. By introducing the concept of generic network models for management, it is then possible to perform general management of diverse equipment using generic information models and standard interfaces.

TMN was initially defined by the ITU in 1988. It is designed to facilitate supplier independence for the customer regarding telecommunication management systems. The objective of TMN's architecture is to allow interoperable interfaces between managing and managed systems and between network management systems themselves. TMN is a set of standards whose goal is to provide unambiguous definitions for the information in support of management functions and how they are exchanged across an interface between communicating parties. TMN is still in a stage of evolvement (Ref. 13).

Figure 18.12 shows the relationship between TMN and the telecommunication network that it manages.

TMN uses concepts from OSI Systems Management architecture and applies them in the context of telecommunications network management.

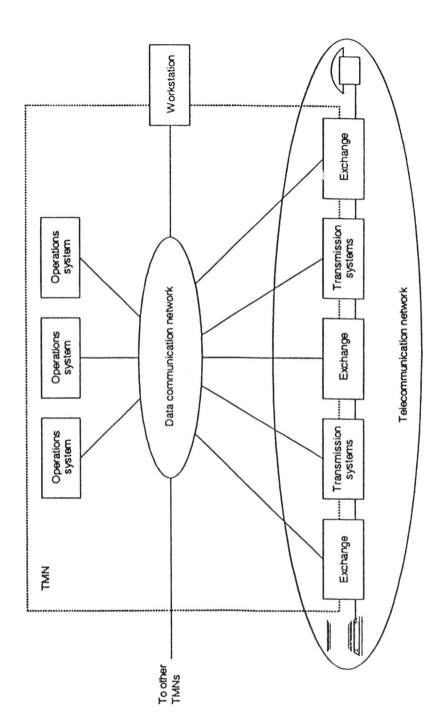

NOTE – The TMN boundary may extend to and manage customer/user services and equipment.

Figure 18.12. Relationship between TMN and the telecommunication network that it manages. (From Figure 1/M.3010, page 2, Ref. 16.)

With the OSI system management regime we find the familiar manager–agent role that is to some degree at odds with the OSI peer-to-peer basic relationship, thus we have peer-to-peer versus master-slave.

There are five independent functional blocks comprising the TMN:

1. OSF—operations system function
2. MF—mediation function
3. WSF—workstation function
4. NEF—network element function
5. QAF—Q adapter function

The processes associated with the management of the telecommunication network are represented by the operations system function (OSF). This includes gathering alarm information, processing received management information, and directing the managed entities to take appropriate action.

The mediation function (MF) facilitates the information exchange between two other functional blocks, typically between the OSF and the NEF, the network element function. This function may include the capabilities to store, filter and adapt data to meet the requirements of the two sides being connected.

The WSF functional block takes machine type data and makes it meaningful to a human user; it facilitates the man-machine interface. The QAF bridges the non-TMN system to the TMN (Refs. 15, 16, and 18).

18.8.2   Network Management Functions Carried Out in Upper OSI Layers

Crucial network management activity is accomplished in the application layer. This layer is further structured using application service elements (ASEs). There are two types of activities that concern the application layer: information processing and the communication of this information to a remote system.

Our concern here is the *application entity*, which is composed of one or more application service elements (ASEs). An ASE is a building block that meets a specific functional need of the applications. There are four distinct ASEs that make up an application entity. These are Association Control Service Element (ACSE), Remote Operations Service Element (ROSE), Common Management Information Service Element (CMISE), and Systems Management Application Service Element (SMASE). Figure 18.13 shows how these four ASEs tie together in the TMN environment.

Service definitions are used in the OSI Reference Model framework. In the TMN context, the application service elements also contain service definitions. Here the services offered by the request–reply application service element are used by the CMISE, and in turn the services of the CMISE are used by SMASE.

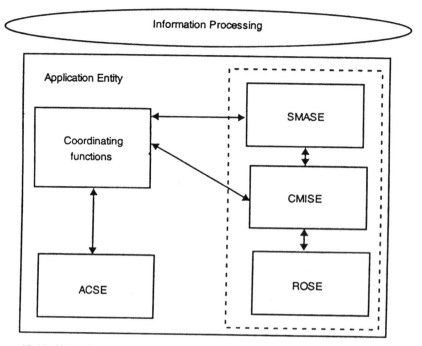

Figure 18.13. Network management application entity. (From Figure 3.4, page 67, Ref. 15.)

In Figure 18.13, the coordination functions are used in order to provide for the flexibility of introducing different ASEs and allowing for negotiation of the features on an association. It will be noted that there is a dotted line around the ASEs used for data transfer. The management message is a result of combining protocol definitions for these ASEs. This is briefly covered below.

Association Control Service Element (ACSE). This is used by any connection-oriented application to set up and release an application. This application element is used to set up and release the association. The association is set up by specifying a context for the association. There are four service primitives beside authentication, used to exchange security information. These four are:

- A-Associate Request and Confirmation
- A-Release Request and Confirmation
- A-Abort Request and Indication
- A-P-Abort Indication

The A-Abort is issued by the user of ACSE, whereas the Provider Abort (A-P-Abort) is issued by the protocol device because the event received is invalid for the current state.

The ACSE only sets up and releases an association. The actual data transfer is performed by the three ASEs described below.

Remote Operations Service Element (ROSE). ROSE is a simple protocol that defines the structure for invoking operations remotely and responding to the invocation. The services support correlation of responses to requests. When a response is received, it can be correlated with the corresponding request. ROSE is further augmented with CMISE and SMASE to achieve the management data transfer.

Common Management Information Service Element (CMISE). CMISE supports ROSE in that it refines the structure offered by the request/reply framework of ROSE. Its specialization is in the context of operations common to all management functions. Irrespective of the functional areas and the resources managed, a basic set of operations is defined by CMISE. CMISE offers services and features grouped together into functional units. The requirements for both Q3 and X interfaces specified in ITU-T Rec. Q.812 are specified in terms of profiles for network management.

Systems Management Application Service Element (SMASE). SMASE covers all management functions in a generic sense. This makes it different from other service elements in its structure, and there is no singular document providing protocol definition. SMASE represents a collection of one or more System Management Functions (SMFs) and, based on which functions are used on an association, the corresponding services to be used. These SMFs further refine the generic framework set by CMISE. Here, in this case, CMISE is not complete until further refined by functions that define specific event types. The generic event report service then is augmented by service definitions and protocol specification. Thus each specific function has service definitions and protocol specifications included.

TMN Management Message Structure Description. L. G. Raman, in Ref. 15, describes the structure of a management message to appreciate the basic differences between the TMN approach and existing message-based approaches (e.g., SNMP) in a high-level view.

In many current approaches, messages are specified in a readable format using character strings. Formulating the message using the TMN interface is different. The message itself is not completely laid out as a string of characters; instead, using level abstractions, different parts of messages are supplied by different application service elements.

Figure 18.14 shows how the flow of information occurs for this case. For understanding the application level messages, assume that the lower layer

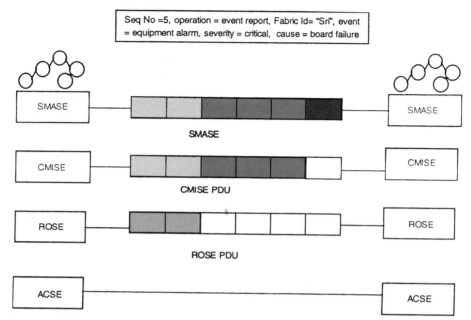

Figure 18.14. System management message components. (From Figure 3.5, page 70, Ref. 15. Reprinted with permission.)

functions are available. The first step is establishing a virtual connection between the two application entities using ACSE as shown in the figure.

Assume that a fabric in the network element has a failure and as a result an alarm is released. The message is assembled by using the structure defined in ROSE. The figure indicates that the filled fields are defined within that protocol according to the shades. The filled boxes are shown to illustrate the point that the message is formed by using the three service elements mentioned above. The Remote Operations Protocol defines the field for performing request/reply correlation and identifying the operations. The unfilled box indicates that ROSE leaves the field to be populated by its user. CMISE defines fields that are common to all management functions. A System Management function such as alarm reporting completes the other fields. In addition, the appropriate values for these fields are determined by the information model for the management information. This is shown by a tree of circles on both ends. It illustrates that there exists a shared management knowledge between the two systems on the resources and their characteristics so that the exchanged information can be interpreted appropriately (Refs. 15 and 17).

Figure 18.15 defines the completed structure for the message by combining definitions from the application service elements and the model for

Sequence Number = "789"	Operation Type = "Event Report"	Managed Object Class = "Fabric"	Managed Object Instance = ("Fabric Id = "Sri")	Event Type = "Equipment Alarm"	Time = "10:35 AM May 15,1998"	Event Info =	
						Severity = "Critical"	Probable Cause = "Board Failure"

Figure 18.15. Typical system management message. (From Figure 3.6, page 71, Ref. 15. Reprinted with permission.)

management information. The message for management application is then encoded by the presentation layer functions into the set of octets transferred over the established transport and network level connections using standard OSI structure.

REFERENCES

1. *The IEEE Standard Dictionary of Electrical and Electronics Terms*, 6th edition, IEEE Std 100-1996, IEEE, New York, 1996.
2. *Transport Systems Generic Requirements: Common Requirements*, GR-499-CORE, Issue 2, Bellcore, Piscataway, NJ, December 1998.
3. Steven M. Dauber, "Finding Fault," *BYTE Magazine*, McGraw-Hill, New York, March 1991.
4. *Error Performance of an International Digital Connection Forming Part of an Integrated Services Digital Network*, CCITT Rec. G.821, Fascicle III.5, IXth Plenary Assembly, Melbourne, 1988.
5. *Synchronous Frame Structures Used at Primary and Secondary Hierarchical Levels*, CCITT Rec. G.704, Fascicle III.4, IXth Plenary Assembly, Melbourne, 1988.
6. *Frame Alignment and Cyclic Redundancy Check (CRC) Procedures Relating to Basic Frame Structures Defined in Recommendation G.704*, CCITT Rec. G.706, CCITT Geneva, 1991.
7. *Information Processing Systems: Open Systems Interconnection—Abstract Syntax Notation One* (ASN.1), ISO 8824, Geneva, 1987.
8. *A Simple Network Management Protocol*, RFC 1157, DDN Network Information Center, STI International, Menlo Park, CA, 1990.
9. "A Critical Survey of Network Management Protocol Standards," Yechiam Yemini, Chapter 2, *Telecommunications Network Management into the 21st Century*, Edited by Salah Aidarous and Thomas Plevyak, IEEE Press, New York, 1994.
10. Mark A. Miller, *Managing Internetworks with SNMP*, M & T Books, New York, 1993.
11. William Stallings, *Network Management*, IEEE Computer Society Press, Los Alamitos, CA, 1993.

12. *Information Technology—Open Systems Interconnection—Common Management Information Protocol Specification*, ISO/IEC 9596, 1st edition, ISO Geneva, 1990.

13. *Information Technology—Open Systems Interconnection—Systems Management Overview*, ISO/IEC 10040-1998, ISO Geneva, 1998.

14. *Information Technology—Open Systems Interconnection—Common Management Information Service*, ISO/IEC 9595, ISO Geneva, 1998.

15. Lakshmi G. Raman, *Fundamentals of Telecommunications Network Management*, IEEE Press, New York, 1999.

16. *Principles for Telecommunications Management Network*, ITU-T Rec. M.3010, ITU Geneva, May 1996.

17. *TMN Interface Specification Methodology*, ITU-T Rec. M.3020, ITU Geneva, 1995.

18. *TMN Management Services and Telecommunications Managed Areas: Overview*, ITU-T Rec. M.3200, ITU Geneva, April 1997.

APPENDIX I

ADDRESSING CONVENTIONS

I.1 TRANSFORMING DECIMAL NUMBERS TO BINARY NUMBERS AND TRANSFORMING BINARY NUMBERS TO DECIMAL NUMBERS

When working with data transmission, more often than not we work with octets or bytes. A byte in this book contains 8 bits. Unless we tell you otherwise, we work from right to left. Then the rightmost digit is the "zero" digit and will be the decimal number raised to the zero power. A decimal number is often used as shorthand for a binary number..

Consider the decimal number 457,623. In these exercises we must take into account the positional location of each digit starting with the "0" location. Thus we have a zero location, a one location, a two location, and so on. Again, right-to-left, the first digit will be 10 to the zero power, the second digit will be 10 to the "one" power, and the third digit will be 10 to the "two power" (i.e., squared), as follows, but write the number conventionally. To assist in understanding what we are doing, the positional location is written above each digit grouping in the line of digit calculations:

Positional
location: 5   4   3   2   1   0

$$(4 \times 10^5) + (5 \times 10^4) + (7 \times 10^3) + (6 \times 10^2) + (2 \times 10^1) + (3 \times 10^0)$$

Let's do the addition the long way to see if we derive 457,623:

$$
\begin{array}{r}
400,000 \\
+\,50,000 \\
+\ 7,000 \\
+\ \ \ 600 \\
+\ \ \ \ 20 \\
+\ \ \ \ \ 3 \\
\hline
457,623
\end{array}
$$

Binary-to Decimal. Given a binary sequence of

$$01100111$$

convert to its binary equivalent. There are eight binary digits, which means that my first decimal number will be 2 to the 7th power, the second will be 2 to the sixth power, and so on. Here we assume that transmission is from left to right—the 0 will be the first digit out of the pipe. Of course, in some situations the first digit out of the pipe is the rightmost digit. In the translation process, we skip each positional location where a binary zero appears and only include those positional locations where a binary one appears. Let's translate:

$$01100111 \text{ would be}$$

$$(0 \times 2^7) + (1 \times 2^6) + (1 \times 2^5) + (0 \times 2^4) + (0 \times 2^3) + (1 \times 2^2) + (1 \times 2^1) + (1 \times 2^0)$$

Convert each digit to its decimal equivalent, as follows and sum:

$$0 + 64 + 32 + 0 + 0 + 4 + 2 + 1 = 103$$

In some protocols you are told that the least significant bit (LSB) is transmitted first. Let's use the same binary sequence, namely 01100111. The bit order now must be reversed, or

$$11100110$$

The binary 1 on the left end is going out the pipe first. It is the least significant bit, and has the positional location of 0. By placing the positional location above each binary digit, we get

0	1	2	3	4	5	6	7
1	1	1	0	0	1	1	0

$$(1 \times 2^0) + (1 \times 2^1) + (1 \times 2^2) + (0 \times 2^3) + (0 \times 2^4) + (1 \times 2^5) + (1 \times 2^6) + (0 \times 2^7)$$

$$1 + 2 + 4 + 0 + 0 + 32 + 64 + 0 = 103$$

In most situations the least significant bit is transmitted first. For example, according to ITU-T Rec. X.25, serial bit stream derived from a data frame based on LAPB, "addresses, commands, responses, and sequence numbers shall be transmitted with the low-order bit first," (for example, the first bit of the sequence number that is transmitted shall have the weight of 2^0). Be careful, in some texts and protocol descriptions, the rightmost bit goes out the pipe first, rather than the left.

To further add to this discussion, the Sixth edition of the IEEE Dictionary defines *most significant bit* (*MSB*) as "the bit having the greatest effect on the value of a binary numeral, usually the leftmost bit.—Logic values expressed in binary form are shown with their most significant bit on the left." In our example above, namely 11100110, with the positional location as given, the rightmost bit, the binary 0, is the most significant digit, not the leftmost. We follow the IEEE convention in the remainder of this appendix with the leftmost digit being the MSB.

Decimal Number to Binary Sequence. This is the reverse procedure as above. In this case we are given a decimal number and we are to find the binary sequence that matches that number.

For example, find the binary sequence that matches the decimal number 245. (It would be helpful to remember $2^8 = 255$, not 256, because our first digit is 0.) We find the largest value of n so that $2^n \leq 245$. It is 7. That is, $2^7 = 128$. Thus, in the 7th bit position (the 8th bit) there will be a 1. Subtract 128 from 245 and we have 117. Now again we find the largest value of n such that $2^n \leq 117$. The value of n will be 6, because $2^6 = 64$. Thus, the 6th position will be a 1. Subtracting 64 from 117 leaves 53. Still again we find the largest value of 2^n such that $2^n \leq 53$. Since $2^5 = 32$, in the 5th bit position we have a binary 1. Then, $53 - 32 = 21$. Once again find the largest value of 2^n such that $2^n \leq 21$. Since $2^4 = 16$, in the 4th bit position, we have a 1, then, $21 - 16 = 5$. Still again find the value of 2^n such that $2^n \leq 5$. Since $2^2 = 4$, then, $5 - 4 = 1$. $2^0 = 1$.

By inspection we see that this is an 8-bit binary number. We know this because $245 < 255$ but is more than 127. If it were less than 127, it would be a 7-bit binary number. Using a guide for positional notation to define our equivalent binary number:

$$
\begin{array}{cccccccc}
7 & 6 & 5 & 4 & 3 & 2 & 1 & 0 \\
1 & 1 & 1 & 1 & 0 & 1 & 0 & 1
\end{array}
$$

Wherever there is a valid power of two, insert a 1; when there is a blank in the series 76543210, insert a 0. We put in binary 0s because there was no 2^3 value nor a 2^1 value. For all the rest, we inserted binary 1s.

I.2 DECIMAL DIGIT REPRESENTATION IN IP ADDRESSES

IP4 addresses have 32 bits or 4 octets. We might see an IP address something like the following:

241.14.6.88

These decimal digits represent the first, second, third, and fourth octet of an IP address. How do we convert these digits to the more familiar binary sequences?

Each digit group is taken by itself and treated as any other decimal number to be converted to binary. In other words, what 8-bit binary sequence derives from decimal 241? We use the same methodology as in Section I.1.

What if 2^n is ≤ 241? Then $2^7 = 128$. This tells us that the first binary digit of the 8-bit sequence is a 1. Subtract $241 - 128 = 113$. If $2^n \leq 113$, then $2^6 = 64$. This tells us that the second place from the left is a binary 1.

Subtract again: $113 - 64 = 49$. If $2^n \leq 49$. Then, $2^5 = 32$. This tells us that in the third place from the left there is a binary 1.

Subtract still again. $49 - 32 = 17$. If, $2^n \leq 17$. Then, $2^4 = 16$. This tells us that the fourth place from the right contains a binary 1.

Where are we so far?: 1111xxxx.

Subtract still again. $17 - 16 = 1$. If, $2^n \leq 1$, then, $2^0 = 1$.

This tells us that the "0" position from the right or the 8th position from the left is a binary 1. Because we have no valid $2n$ values for positions three, two, and one, we then insert binary 0s in those positions. Thus we now have

11110001

Let's check it to see if we are right.

7 6 5 4 3 2 1 0 position number from the right

1 1 1 1 0 0 0 1	
1	$2^7 = 128$
1	$2^6 = 64$
1	$2^5 = 32$
1	$2^4 = 16$
0	$2^3 = $ no value present
0	$2^2 = $ no value present
0	$2^1 = $ no value present
1	$2^0 = 1$

Add   241 our original decimal value.

The largest decimal value we can have for an 8-bit binary sequence, of course, is 11111111 or 255 decimal. The smallest value is 00000000 or decimal 0.

This addressing technique, so commonly found in IP, is called *dotted decimal notation system*.

TABLE I.1   Decimal, Binary Octal, and Hexadecimal Numbers

Decimal	Binary	Octal	Hexadecimal	4-Bit String
0	0	0	0	0000
1	1	1	1	0001
2	10	2	2	0010
3	11	3	3	0011
4	100	4	4	0100
5	101	5	5	0101
6	110	6	6	0110
7	111	7	7	0111
8	1000	10	8	1000
9	1001	11	9	1001
10	1010	12	A	1010
11	1011	13	B	1011
12	1100	14	C	1100
13	1101	15	D	1101
14	1110	16	E	1110
15	1111	17	F	1111

Convert the second address octet 14 to binary: If, $2^n \leq 14$ then, $2^3 = 8$. Subtract $14 - 8 = 6$. If, $2^n \leq 6$ then, $2^2 = 4$. Subtract $6 - 4 = 2$. Then $2^1 = 2$. In the 8-digit binary number there will be 1s in the 3rd, 2nd, and 1st places (from the right). Insert 0s in the remainder of the positions as follows:

$$\text{Positional location:} \quad 7\,6\,5\,4\,3\,2\,1\,0$$
$$0\,0\,0\,0\,1\,1\,1\,0$$

I.3   BINARY, OCTAL, AND HEXADECIMAL NUMBERS

Most computer software that we deal with uses either the *octal number system*, which has a base of 8, or the *hexadecimal number system*, which uses a number base of 16. Table I.1 shows the decimal numbers 0 through 1 and their binary, octal, and hexadecimal equivalents.

To convert a binary sequence to an octal sequence, starting from the left, separate the binary digits into groups of three; to convert to a hexadecimal sequence, starting from the left, separate the binary sequence into groups of four digits. For example:

$$101011000110 = 101\quad 011\quad 000\quad 110_2 = 5306_8$$
$$= 1010\quad 1100\quad 0110_2 = AC6_{16}$$

ACRONYMS AND ABBREVIATIONS

2B1Q	A baseband waveform, 2 binary 1 quaternary
4B5B	Baseband waveform coding (4 bits to 5 bits)
8B6T	Baseband waveform coding (8 binary to 6 ternary)
8B10T	Baseband waveform coding (8 binary to 10 ternary)
A	
AAL	ATM adaptation layer
ABM	Asynchronous balanced mode
AC	Access control
ACEG	AC equipment ground
ACF	Access control field
ACK	"Acknowledge(ment)"
ACSE	Association control service element
ADM	Add–drop multiplexer; asynchronous disconnect mode
ADCCP	Advanced data communications control procedures
ADSL	Asymmetric digital subscriber line
AFI	Authority and format identifier
AIS	Alarm indication signal
ALBO	Automatic line build out
AM	Amplitude modulation
AMI	Alternate mark inversion
AM-VSB	Amplitude modulation—vestigial sideband
AMP	Auxiliary multiplexing pattern
ANSI	American National Standards Institute
AP	Access point
APD	Avalanche photodiode
AR	Access rate
ARPA	Advanced Research Projects Agency
ARPANET	ARPA network
ARM	Asynchronous response mode
ARP	Address resolution protocol

ARQ	Automatic repeat request
ASCII	American Standard Code for Information Interchange
ASE	Application service element(s)
ASK	Amplitude shift keying
ASN.1	Abstract syntax notation—1
ASTM	American Society of Testing and Materials
AT & T	American Telephone and Telegraph (Company)
AU	Access unit
AUG	Administrative unit group
AUI	Attachment unit interface
AutoVoN	Automatic voice network (DoD)
AUU	ATM user-to-ATM user
AWG	American wire gauge

B

B3ZS, B6ZS, B8ZS	Binary 3, 6, 8 zeros substitution
BASE	Baseband
BAsize	Buffer allocation size
BBE, BBER	Background block error, ratio
BCC	Block check count
BCD	Binary coded decimal
BECN	Backward explicit congestion notification
Bellcore	Bell Communications Research (formerly BTL), now Telcordia
BEOFD	Broadband end-of-frame delimiter
BER	Bit error rate
BERT	Bit error rate test
BEtag	Beginning-end tag
BICI	Broadband–intercarrier interface
BIP	Bit interleaved parity
B-ISDN	Broadband-ISDN
BISUP	Broadband ISDN User Part (SS#7)
BITE	Built-in test equipment
BIU	Baseband interface unit
BLF	Basic low layer functions
B-NT1, B-NT2	Broadband (reference interface) NT1 and NT2 (NT = network termination)
BNZS	Binary N-zeros substitution
BOM	Beginning of message
BPDU	Bridge protocol data unit
bsp	Bits per second
BPSK	Binary phase shift keying
BR	Bit rate
BRI	Basic rate interface
BROAD	Broadband, indicates "broadband"

BSA	Basic service area
BSS	Basic service set
BSSID	Basic service set identification
BT	British Telecom
C	
CAC	Connection admission control
CAD/CAM	Computer-aided design/computer-aided manufacturing
CAP	Carrierless amplitude and phase (modulation)
CAT	Category
CATV	Community antenna television
CBR	Constant bit rate
CBDS	Connectionless broadband data service
CC	Country code
CCITT	International Consultive Committee on Telephone and Telegraph (Now ITU-T)
CD	Compact disk
CDMA	Code division multiple access
CDV	Cell delay variation
CE	Connection element; collision enforcement
CEQ	Customer equipment
CF	Coordination function
CFP	Contention free period
CI	Control-in
CIB	CRC indicator bit
CIR	Committed information rate
CLEC	Competitive local exchange carrier
CLLM	Consolidated link-layer management
CLNAP	Connectionless network access protocol
CLNP	Connectionless network protocol
CLP	Cell loss priority
CLSF	Connectionless service functions
CM	Cable modem
CMCI	Cable modem customer interface
CMI	Coded mark inversion
CMIC	Concentrator medium interface connector
CMIP	Common management information protocol
CMISE	Common management information service element
CMTS	Cable modem termination system
COM	Continuation of message
CONUS	Contiguous United States
CP	Contention period
CPI	Common part indicator
CPCS	Common part convergence sublayer

CPE	Customer premise equipment
CPI	Common part indicator
CPU	Central processing unit
C/R	Command/response
CRC	Cyclic redundancy check
CRF	Connection-related functions
CRS	Carrier sense; configuration report server
CS	Convergence sublayer
CSI	Convergence sublayer indicator
CSMA	Carrier sense multiple access
CSMA/CA	Carrier sense multiple access/collision avoidance
CSMA/CD	Carrier sense multiple access with collision detection
CSU	Channel service unit
CTS	Clear to send
CW	Carrier wave (sometimes, affectionately, refers to "Morse" operation)

D

DA	Destination address
DACS	Digital automatic cross-connect system
DAMA	Demand assignment multiple access
DARPA	Defense Advanced Research Projects Agency
dB	Decibel
dB, dBm, dBW, dBi, dBc, dBmV	Decibel-related terms
DBPSK	Differential binary PSK
DC	Direct current
DCC	Data country code
DCE	Data communication equipment
DCF	Distributed coordination function
DDS	Digital data system
DE	Discard eligibility
DECnet	Digital Equipment Corporation network
DIFS	DCF interframe space
DISC	Disconnect
DLCI	Data-link connection identifier
DLL	Data-link layer
DM	Disconnect mode
DMPDU	Derived MAC protocol data unit
DNHR	Direct nonhierarchical routing
DNIC	Data network identification code
DoD	Department of Defense, U.S.
DPTE	Data processing terminal equipment (= DTE)
DQDB	Distributed queue dual bus
DQPSK	Differential quadrature phase shift keying

DS	Direct sequence (spread spectrum)
DS0, DS1	Referring to multiplex levels of U.S. digital network
DSAP	Destination service access point
DSE	Data switching equipment
DSL	Digital subscriber line
DSLAM	Digital subscriber line access multiplexer
DSM	Distribution system medium
DSP	Domain-specific part
DSS	Distribution system service(s)
DSS1, DSS2	Digital signaling system No. 1, No. 2
DSSS	Direct sequence spread spectrum
DSU	Data service unit
DTE	Data terminal equipment
DWDM	Dense wave-division multiplex

E

E0, E1	Multiplex levels of "European" digital network
EA	Address extension bit
EB	Errored block
EBCDIC	Extended Binary Coded Decimal Interchange Code
ED	End delimiter
EDAC	Error detection and correction
EDC	Error detection code
EDD	Envelope delay distortion
EDFA	Erbium-doped fiber amplifier
EFS	End-of-frame sequence
EIA	Electronic Industries Alliance
EIFS	Extended interframe space
EF	Entrance facility
EIRP	Equivalent isotropic radiated power
EMC	Electomagnetic compatibility
EMI	Electromagnetic interference
eoc	Embedded operations channel
EPO	Errored performance objective
ER	Equipment room
ES	Errored second
ESA	Extended service area
ESF	Extended superframe
ESI	End system identifier
ESR	Errored second ratio
ESS	Extended service set
ETD	End of transmission delimiter
ETSI	European Telecommunications Standardization Institute
EUTP	Enhanced unshielded twisted pair

F

FC	Frame control
FCS	Frame check sequence
FDDI	Fiber distributed data interface
FDM	Frequency division multiplex
FDMA	Frequency division multiple access
FEBE, febe	Far-end block error
FEC	Forward error correction
FECN	Forward explicit congestion notification
FER	Frame error ratio
FERF	Far-end receive failure
FH	Frequency hop
FHSS	Frequency hop spread spectrum
FIFO	First in, first out
FM	Frequency modulation
FMIF	Frame mode information field
FR	Frame relay
FRAD	Frame relay access device
FRF	Frame Relay Forum
FRMR	Frame reject response
FRS	Frame relay switch
FS	Frame status
FSK	Frequency shift keying
FSL	Free-space loss
FTP	File transfer protocol

G

Gbps	Gigabits per second
GBSVC	Generic broadcast signaling virtual channel
GCRA	Generic cell rate algorithm(s)
GFC	Generic flow control
GFSK	Gaussian frequency shift keying
GHz	Gigahertz
GMII	Gigabit medium-independent interface
GTE	General Telephone & Electronics

H

HC	Horizontal cross-connects
HCS	Header check sequence
HDB3	High density binary 3
HDLC	High-level data-link control
HDR	Header
HDSL	High-speed digital subscriber line
HE	Head-end
HEC	Header error control

HFC	Hybrid fiber coax
HIC	Highest incoming channel
HLPI	Higher-layer protocol identifier
HOC	Highest outgoing channel
HPA	High-power amplifier
HRC	Harmonic-related carrier; hybrid ring control; horizontal redundancy check
HRP	Hypothetical reference path
HRX	Hypothetical digital connection
HTC	Highest two-way channel
Hz	Hertz

I

IA5	International alphabet no. 5
IBSS	Independent basic service set
ICD	Interface control document; international code designator
ICMP	Internet control message protocol
IDI	International domain identifier
IDL, idl	Idle
IDU	Interface data unit
IEC	International Electrotechnical Commission
IEEE	Institute of Electrical and Electronics Engineers
IFG	Interframe gap
I/G	Individual/group
IHL	Internet header length
IISP	Interim interswitch signaling protocol
ILMI	Interim local management interface
IM	Intermodulation
IMPDU	Initial MAC protocol data unit
INFO	Information (field)
I/O	input/output (device)
IP	Internet protocol
IPX	Internet packet exchange
I/Q	In-phase quadrature
IR	Infrared
IRC	Incremental related carrier
IRL	Isotropic receive level
IRP	Internal reference point
ISDN	Integrated services digital network(s)
ISM	In-service measurements; industrial, scientific and medical
ISP	Internet service provider
ISO	International Standards Organization
ISUP	ISDN user part (SS#7)

ITA	International telegraph alphabet
ITU	International Telecommunication Union
IXC	Interexchange carrier (typically ATT, Worldcom, BT, etc.)

K

kbps	Kilobits per second
kHz	Kilohertz
km	Kilometer

L

LAN	Local area network
LAPB	Link access protocol—B-channel
LAPD	Link access protocol—D-channel
LAPF	Link access protocol—frame relay
LCN	Logical channel number
LCT	Last compliance time
LD	Laser diode
LEC	Local exchange carrier
LED	Light-emitting diode
LEO	Low earth-orbit (satellite)
LI	Length indicator
LLC	Logical link control
LLCDU	LLC data unit
LMDS	Local multipoint distribution system
LNA	Low-noise amplifier
LO	Local oscillator
LOC	Lowest outgoing channel; loss of cell (delineation)
LOS	Line-of-sight
LRC	Longitudinal redundancy check
LSAP	Link service access point
LSB	Least significant bit
LSDU	Link-layer service data unit
LSDV	Link segment delay value
LSI	Large-scale integration
LTE	Line terminating equipment; link terminating equipment
LX-PMD	Long-wavelength PMD

M

MAC	Medium access control
MAN	Metropolitan area network
MAP	Multiaccess protocol
M-ary	Gives modulation level of waveform (e.g., 16-QAM, where $m = 16$)

MAU	Medium attachment unit
Mbps	Megabits per second
MCNS	Multimedia cable network system
MDI	Medium-dependent interface
MDL	Management data link
MF	Mediation function
MIB	Management information base
MIC	Medium interface connector
MID	Message identification
MIT	Management information tree
MLC	Multilink control (field)
MLP	Multilink procedure
MMPDU	MAC management PDU
MO	Managed object
MODEM	Modulator-demodulator
MN(S)	Multilink N(S)
MPDU	MAC protocol data unit
MPEG	Motion picture experts group
ms	Millisecond(s)
MSAP	MAC service access point
MSB	Most significant bit
MSDU	MAC service data unit
MSVC	Metasignaling virtual channel
MTP	Message transfer part
mV	Millivolt
N	
NA	Numerical aperture
NACK	Negative acknowledgment
NANP	North American Numbering Plan
NAV	Net allocation vector
NDC	National destination code
NDM	Normal disconnect mode
NE	Network element
NEF	Network element function
NEXT	Near-end crosstalk
NF	Noise figure
NHDR	Network header
nm	Nanometer(s)
NMS	Network management system
NMT	Network management
NNI	Network–network interface or network–node interface
NOC	Network operations center
NPC	Network parameter control
NPI	Numbering plan identification

$N(R)$	Receive sequence number
NRM	Normal response mode
NRZ	Non-return to zero
NRZI	NRZ inverted ones
$N(S)$	Send sequence number
ns	Nanosecond(s)
NSAP	Network service access point
NSDU	Network service data unit
NSN	National significant number
NSP	Network service provider
NT (e.g., NT1, NT2)	Network termination
NTSC	National Television System Committee

O

OAM, OA & M	Operation, administration, and maintenance
OC	Optical carrier (used with SONET)
OSF	Operations system function
OSI	Open systems interconnection
OSIE	OSI environment
OSP	Outside plant
OSPF	Open shortest path first

P

PABX	Private automatic branch exchange
PAD	Meaning pad or padding, extra false bits added; packet assembler-disassembler
PAM	Pulse amplitude modulation
PAR	Positive acknowledgment with retransmission
PC	Personal computer; point coordinator; priority control
PCF	Point coordination function
PCI	Protocol control information
PCM	Pulse code modulation
PCR	Peak cell rate
PCS	Physical coding sublayer; personal communication services
PDH	Plesiochronous digital network
PDU	Protocol data unit
PDV	Path delay value
P/F (bit)	Poll/final
PH	Packet handling (handler)
PHY	Physical (layer)
PIFS	PCF interframe space
PIN	Positive-intrinsic-negative
PLCP	Physical layer convergence protocol
PLL	Phase lock loop

PLOAM	Physical layer OA & M
PLS	Physical layer signaling
PM	Phase modulation; physical medium
PMA	Physical medium attachment
PMC	Physical medium components
PMD	Physical medium dependent
P-NNI	Private network—NNI
POH	Path overhead
POI	Path overhead indicator
POTS	Plain old telephone service
PPM, ppm	Parts per million
$P(R)$	Receive sequence number
PRBS	Pseudo-random binary sequence
PRI	Primary rate interface
$P(S)$	Send sequence number
PSC	Physical signaling components
PSDN	Public switched data network
PSF	Packet switching facilities
PSH	Push function
PSK	Phase shift keying
PSPDN	Packet-switched public data network
PSTN	Public switched telecommunications network
PTI	Payload type identifier
PTR	Pointer
PTT	Post-Telegraph-Telephone
PVC	Permanent virtual circuit; polyvinylchloirde

Q

QAF	Q adapter function
QAM	Quadrature amplitude modulation
QoS	Quality of service
QPSK	Quadrature phase shift keying

R

RARP	Reverse address resolution protocol
REJ	Reject
REM	Ring error monitor
RF	Radio frequency
RFC	Request for comment
RFI	Radio frequency interference; request for information
RI	Routing information
RIM	Request initialization
RIP	Routing information protocol
RM	Resource management
rms	Root mean square

RNR	Receive not ready
ROSE	Remote operations service element
RPS	Ring parameter server
RR	Receive ready
RSL	Receive signal level
RSU	Remote switching unit
RTS	Request to send
RTU	Remote terminal unit
RX	Receive
RX_BITS	Receive bits
RX_CLK	Receive clock
RXD	Receive data
RZ	Return-to-zero

S

SA	Source address
SABME	Set asynchronous balanced mode (extended)
SAP	Service access point
SAPI	Service access point identifier
SAR	Segmentation and reassembly
SARM, SARME	Set asynchronous response mode, extended
SBS	Selective broadcast signaling
SCR	Sustained cell rate
SCSI	Small computer system interface
SCTE	Society of Cable Telecommunication Engineers
SD	Starting delimiter
SDLC	Synchronous data-link control
SDH	Synchronous digital hierarchy
SDP	Severely disturbed period
SDU	Service data unit
SEAL	Simple (and) efficient AAL layer
SES	Severely errored second(s)
SESR	Severely errored second ratio
SFD	Start frame delimiter
SFS	Start frame sequence
SIFS	Short interframe space
SIM	Set initialization mode
SIR	Sustained information rate
SLA	Service level agreement
SLP	Single-link procedure
SMASE	Systems management application service element
SMDS	Switched multimegabit data service
SMF	Submultiframe; system management information
SMI	Structure of management information
SMT	Station management

SMTP	Simple mail transfer protocol
SN	Subscriber number
S/N	Signal-to-noise ratio
SNAP	Subnetwork access protocol
SNMP	Simple network management protocol
SNP	Sequence number protection
SNRM	Set normal response mode
SOH	Section overhead; start-of-heading
SOM	Start of message
SONET	Synchronous optical network
SPE	Synchronous payload envelope
SREJ	Selective reject
SRL	Structural return loss
SRTS	Synchronous residual timestamp
SS7, SSN7	Signaling system No. 7
SSAP	Source service access point
SSD	Start of stream delimiter
SSM	Single segment message
ST	Segment type
STA	Station management
STE	Station terminating equipment
STL	Standard Telephone Laboratories
STM	Synchronous transport module
STP	Shielded twisted pair; signal transfer point
STS	Space time space
SVC	Switched virtual circuit; signaling virtual channel
SWP	Switching pattern
SX-PMD	Short-wavelength PMD
SYNC, sync	Synchronization

T
TA	Terminal adapter; transmission address
TBB	Telecommunication bonding backbone
TBBIC	Telecommunication bonding backbone interconnecting conductor
TBI	Ten-bit interface
TC	Transmission convergence; telecommunications closet
TCM	Trellis-coded modulation
TCP	Transmission control protocol
TCU	Trunk coupling unit
TDM	Time division multiplex
TDMA	Time division multiple access
TDR	Time-domain reflectometer
TE	Terminal equipment; terminal endpoint
Tel, te2	Terminal equipment 1, 2

TEF	Telecommunications entrance facility
TEI	Terminal endpoint identifier
Telex	A trade name (telegraph exchange)
TFTP	Trivial File Transfer Protocol
TGB	Telecommunications grounding bus/bar
THDR	Transport header
THT	Token holder timer
THz	Terahertz
TIA	Telecommunications Industry Association
TMGB	Telecommunications main grounding bar
TMN	Telecommunication management network
TOA	Type of address
TOS	Type of service
TPDU	Transport protocol data unit
TS	Time slot
TSI	Time slot interchange®
TSN	Transmit sequence number
TST	Time-space-time
TTL	Time to live
TU	Tributary unit
TUG	Tributary unit group
TWT	Traveling wave tube
TX	Transmit
TX_bits	Transmit bits
TX_clk	Transmit clock
U	
UA	Unnumbered acknowledge(ment) (command); user agent
UCD	Upstream channel descriptor
UDP	User datagram protocol
UHF	Ultrahigh frequency
UI	Unit interval; unnumbered information
ULP	Upper-layer protocol
UMD	Unscrambled mode delimiter
UNI	User−network interface
UPC	Usage parameter control
UPS	Uninterruptible power supply; United Parcel Service
μs	Microsecond
USAF	United States Air Force
UTP	Unshielded twisted pair
UU	User-to-user
V	
VBR	Variable bit rate
VC	Virtual container; virtual connection; virtual channel

VCC	Virtual channel connection
VCI	Virtual channel identifier
VHF	Very high frequency
VHSIC	Very high-speed integrated circuit
VLAN	Virtual local area network
VLSI	Very large-scale integration
VP	Virtual path
VP-AIS	Virtual path−alarm indication signal
VPC	Virtual path connection
VPI	Virtual path identifier
VPN	Virtual private network
V(R)	Receive state variable
VRC	Vertical redundancy check
V(S)	Send state variable
VSAT	Very small aperture terminal
VT	Virtual tributary
VTAM	Virtual telecommunications access method

W

WAN	Wide area network
WEP	Wired equivalent privacy
WLAN	Wireless LAN
WDM	Wavelength division multiplex
WSF	Workstation function

X

XDSL	Generic DSL
XID	Exchange identification (command)

INDEX

A **bold** page number indicates extensive coverage of that topic.
An *italic* page number indicates where a definition of that topic may be found.

WILEY SERIES IN TELECOMMUNICATIONS AND SIGNAL PROCESSING

John G. Proakis, Editor
Northeastern University

Introduction to Digital Mobile Communications
Yoshihiko Akaiwa

Digital Telephony, 3rd Edition
John Bellamy

ADSL, VDSL, and Multicarrier Modulation
John A. C. Bingham

Biomedical Signal Processing and Signal Modeling
Eugene N. Bruce

Elements of Information Theory
Thomas M. Cover and Joy A. Thomas

Practical Data Communications, 2nd Edition
Roger L. Freeman

Radio System Design for Telecommunications, 2nd Edition
Roger L. Freeman

Telecommunication System Engineering, 3rd Edition
Roger L. Freeman

Telecommunications Transmission Handbook, 4th Edition
Roger L. Freeman

Introduction to Communications Engineering, 2nd Edition
Robert M. Gagliardi

Optical Communications, 2nd Edition
Robert M. Gagliardi and Sherman Karp

Active Noise Control Systems: Algorithms and DSP Implementations
Sen M. Kuo and Dennis R. Morgan

Mobile Communications Design Fundamentals, 2nd Edition
William C. Y. Lee

Expert System Applications for Telecommunications
Jay Liebowitz

Polynomial Signal Processing
V. John Mathews and Giovanni L. Sicuranza

Digital Signal Estimation
Robert J. Mammone, Editor

Digital Communication Receivers: Synchronization, Channel Estimation, and Signal Processing
Heinrich Meyr, Marc Moeneclaey, and Stefan A. Fechtel